NYLON PLASTICS

SPE MONOGRAPHS

Injection Molding

Irvin I. Rubin

Nylon Plastics

Melvin I. Kohan

NYLON PLASTICS

Edited by

MELVIN I. KOHAN
Plastics Department
E.I. du Pont de Nemours and Co., Inc.

A Wiley-Interscience Publication

JOHN WILEY & SONS, New York · London · Sydney · Toronto

Copyright © 1973, by John Wiley & Sons, Inc.

Library of Congress Cataloging in Publication Data

Kohan, Melvin Ira, 1921-
Nylon Plastics.

(SPE monographs)
"A Wiley-Interscience Publication."
Includes bibliographical references.
1. Nylon. I. Title. II. Society of Plastics
Engineers. SPE monographs.

TP1180.P55K64   677'.473   73-9606
ISBN   0-471-49780-0

Printed in the United States of America

10 9 8 7 6 5 4 3 2 1

CONTRIBUTORS

C. M. Barañano
Injectronics, Inc.
1 Union Street
Clinton, Massachusetts   01510

R. M. Bonner
E.I. du Pont de Nemours and Co., Inc.
Plastics Department
Technical Services Laboratory
Chestnut Run
Wilmington, Delaware   19898

W. M. Bruner
E. I. du Pont de Nemours and Co., Inc.
Plastics Department
Technical Services Laboratory
Chestnut Run
Wilmington, Delaware 19898

G. Carlyon
Cadillac Plastic and Chemical Co.
15111 Second Avenue
Detroit, Michigan   48203

E. S. Clark*
E. I. du Pont de Nemours and Co., Inc.
Plastics Department
Experimental Station
Wilmington, Delaware   19898

J. H. Crate
E. I. du Pont de Nemours and Co., Inc.
Plastics Department
Technical Services Laboratory
Chestnut Run
Wilmington, Delaware   19898

E. T. Darden
E. I. du Pont de Nemours and Co., Inc.
Plastics Department
1007 Market Street
Wilmington, Delaware   19898

G. L. Graf, Jr.
E. I. du Pont de Nemours and Co., Inc.
Plastics Department
Technical Services Laboratory
Chestnut Run
Wilmington, Delaware 19898

J. R. Harrison
E. I. du Pont de Nemours and Co., Inc.
Industrial Chemicals Department
Experimental Station
Wilmington, Delaware   19898

E. M. Lacey
E. I. du Pont de Nemours and Co., Inc.
Plastics Department
Technical Services Laboratory
Chestnut Run
Wilmington, Delaware 19898

M. I. Kohan
E. I. du Pont de Nemours and Co., Inc.
Plastics Department
Experimental Station
Wilmington, Delaware   19898

J. Mengason
E. I. du Pont de Nemours and Co., Inc.
Plastics Department
Technical Services Laboratory
Chestnut Run
Wilmington, Delaware 19898

P. N. Richardson
E. I. du Pont de Nemours and Co., Inc.
Plastics Department
Experimental Station
Wilmington, Delaware 19898

D. S. Richart
Polymer Corporation
2120 Fairmont Avenue
Reading, Pennsylvania 19603

T. M. Roder
E. I. du Pont de Nemours and Co., Inc.
Plastics Department
Technical Services Laboratory
Chestnut Run
Wilmington, Delaware 19898

L. T. Sherwood, Jr.
E. I. du Pont de Nemours and Co., Inc.
Plastics Department
Technical Services Laboratory
Chestnut Run, Wilmington, Delaware 19898

J. W. Sprauer
E. I. du Pont de Nemours and Co., Inc.
Plastics Department
Experimental Station
Wilmington, Delaware 19898

H. W. Starkweather, Jr.
E. I. du Pont de Nemours and Co., Inc.
Plastics Department
Experimental Station
Wilmington, Delaware 19898

F. C. Wilson
E. I. du Pont de Nemours and Co., Inc.
Plastics Department
Experimental Station
Wilmington, Delaware 19898

*Currently, Professor, Department of Chemical and Metallurgical Engineering, University of Tennessee, Knoxville, Tennessee, 37916

The Society of Plastics Engineers is dedicated to the promotion of scientific and engineering knowledge of plastics and to the initiation and continuation of educational activities for the plastics industry.

An example of this dedication is the sponsorship of this and other technical books about plastics. These books are commissioned, directed, and reviewed by the Society's Technical Volumes Committee. Members of this committee are selected for their outstanding technical competence; among them are prominent authors, educators, and scientists in the plastics field.

In addition, the Society publishes *Plastics Engineering, Polymer Engineering and Science (PE&S)*, proceedings of its Annual, National and Regional Technical Conferences *(ANTEC, NATEC, RETEC)* and other selected publications. Additional information can be obtained by writing to the Society of Plastics Engineers, Inc., 656 West Putnam Avenue, Greenwich, Connecticut 06830.

William Frizelle,

Chairman, Technical Volumes Committee
Society of Plastics Engineers
St. Louis, Missouri

PREFACE ──────────────────────────────

The word "nylon" immediately brings to mind the world's first synthetic fiber, and many books exist that deal with nylon and related fiber technology. Nylon also was, however, the first engineering thermoplastic, that is, a material that can be readily processed as a melt via injection molding or extrusion to yield products having the strength, toughness, rigidity, and durability demanded of mechanical parts. Plastics applications for nylons have become extensive and are the result of a broadening technology marked by the appearance of new compositions, new synthetic procedures, and new fabricating methods. A need has existed for a text concerned with nylons from the plastics point of view. This book is directed toward that need.

The book follows a roughly chronological sequence from monomer synthesis to polymerization, characterization, processing, properties, and applications. Emphasis is placed on subjects of special concern to the plastics industry, such as thermal degradation and melt flow, but these topics as well as the discussions of polymer characterization, physical structure, transition phenomena, and others will also be helpful to many people outside the plastics industry.

In covering an industry from raw material to ultimate use, we have called upon the talents of many people of diverse technical skills and owe them many thanks. This book would not have been possible without the support of the Plastics Department, E. I. du Pont de Nemours and Co., Inc., and in particular many members of the Commercial Resins Division. Virtually all the manufacturers of nylon polymers for the plastics industry in the United States have generously provided information on their products and so permitted discussion of a very broad spectrum of nylons. To them all we extend our sincere thanks.

MELVIN I . KOHAN

Wilmington, Delaware
June 18, 1973

CONTENTS _____

CONVERSION
OF UNITS _____

Units are given as far as practicable in both the English and metric systems. Conversion factors are provided where appropriate. The symbols employed are those recommended by the American Chemical Society in *Handbook for Authors,* 1967, pp 97-99. Commonly encountered metric units are used rather than those proposed by the International Organization for Standardization as in ISO Recommendation R1000, Ed. 1, 1969. The table below permits interconversion of the English, common metric, and ISO units for the quantities most likely to be of concern.

	To convert from	To	Multiply by
Length	in.	cm	2.54
	cm	in.	0.394
Mass	lb	kg	0.454
	kg	lb	2.20
Force	lb (wt)	dyn ($= 1$ g cm sec^{-2})	4.45×10^{-6}
	dyn	N ($= 1$ kg m sec^{-2})*	10^{-5}
	lb (wt)	N*	0.445
	N*	lb (wt)	2.25
	kg (wt)	N*	9.81
	N*	kg (wt)	0.102
Density	lb ft^{-3}	g cm^{-3}	0.0160
	g cm^{-3}	lb ft^{-3}	62.4
	g cm^{-3}	kg m^{-3} *	10^3
Pressure, stress,	lb (wt) in.$^{-2}$	dyn cm^{-2}	6.89×10^4
or modulus	dyn cm^{-2}	lb (wt) in.$^{-2}$	1.45×10^{-5}
	lb (wt) in.$^{-2}$	kg (wt) cm^{-2}	0.0703
	kg (wt) cm^{-2}	lb (wt) in.$^{-2}$	14.2
	dyn cm^{-2}	N m^{-2} *	10^{-1}
	kg (wt) cm^{-2}	N m^{-2} *	9.81×10^4

	To convert from	To	Multiply by
Energy	ft-lb (wt)	kg (wt) - cm	13.8
	kg (wt)-cm	ft-lb (wt)	0.0723
	kg (wt)-cm	J (= 1 N-m)*	0.0981
Impact strength	ft-lb (wt) in.$^{-1}$	cm-kg (wt) cm^{-1}	5.45
	cm-kg (wt) cm^{-1}	ft-lb (wt) in.$^{-1}$	0.183
	ft-lb (wt) in.$^{-1}$	J m^{-1}*	53.3
	ft-lb (wt) in.$^{-2}$	cm-kg (wt) cm^{-2}	2.14
	cm-kg (wt) cm^{-2}	ft-lb (wt) in.$^{-2}$	0.466
Viscosity	lb (wt) sec in.$^{-2}$	poise (= 1 dyn sec cm^{-2})	6.89×10^4
	poise	lb (wt) sec in.$^{-2}$	1.45×10^{-5}
	poise	N sec m^{-2}*	10^{-1}

= ISO unit

NYLON PLASTICS

CHAPTER 1

Introduction

M. I. KOHAN

HISTORY

Synthetic materials have developed on a truly grand scale since the 1920s. They graduated from the ranks of substitutes for natural products to preferred materials with unique performance capabilities. Their annual volume grew from less than fifty million pounds to over twenty billion pounds. Many people contributed to this remarkable achievement, but among the most prominent was Wallace Hume Carothers, who began his now classic studies on condensation polymerization in 1928 at the behest of the Du Pont Company. Condensation polymerization links molecules together via a reaction that involves the loss of a small molecule. Carothers' work led to the preparation, in 1935, of a truly high-molecular-weight polyamide from hexamethylenediamine and adipic acid:

$$H_2NCH_2CH_2CH_2CH_2CH_2CH_2NH_2 + HOOCCH_2CH_2CH_2CH_2COOH$$

hexamethylenediamine   adipic acid

$$\downarrow -H_2O$$

$$\cdots HN(CH_2)_4NHCO(CH_2)_4COHN(CH_2)_6NHCO(CH_2)_4CO \cdots$$

The development of a wholly man-made fiber from this new, proteinlike material was announced to the public on October 27, 1938. The name nylon was

1

coined by Du Pont for this kind of polymer, one which included the carbonamide group —CONH— as a recurring unit of the main chain and which could be drawn into fibers. Two years later the first commercial plant for the production of nylon fiber was in operation at Seaford, Delaware.

The first nylon product to be marketed in 1938 was not a yarn but a continuous, large-diameter filament used as a bristle for toothbrushes. Like the fiber, it depends for its utility on the enhancement of properties realized by stretching the initial filament several fold, but large-diameter monofilament is normally considered a plastic rather than fiber application (see Table 1-1). This, then, was the origin of nylon, the synthetic-fiber industry, and a new concept in plastics. This very brief comment on the birth of nylon does justice neither to Carothers' research nor the fiber development; the interested reader is directed to the collected papers of Carothers (1) and to E. K. Bolton's review on the development of nylon (2).

Nylon was a new concept in plastics for several reasons; for one, it was the first crystalline plastic. Its crystallinity meant a sharp transition from solid to melt, unlike polystyrene or poly(methyl methacrylate); it also meant a much higher service temperature than previously known thermoplastics. Further, nylon provided a combination of toughness, rigidity, and lubrication-free performance which led to mechanical uses such as bearings and gears, applications heretofore denied to plastics. Nylon acquired the reputation of a quality material by showing that a thermoplastic could be tough as well as stiff and could do some jobs better than metals, which had previously defined standards of performance. This performance capability gave nylon the label "an engineering thermoplastic."

Nylon molding powder was first offered for sale by Du Pont in 1941. Beginning in 1954 with the introduction by the Allied Chemical Co. of extracted polycaprolactam, new to the American but not the European market, the number of United States manufacturers gradually grew. A tabulation of current United States suppliers and trade names is given in the Appendix.

The susceptibility of nylon to modification was clear from the outset. Different acids and amines could be reacted to provide a variety of nylons and nylon copolymers. Structural limitations on reactivity existed, but the polycondensation method of synthesis was an extension of classical organic chemistry more generally applicable and amenable to manipulation than the chain mechanisms by which other plastics such as styrene, ethylene, and methyl methacrylate were polymerized. In 1948 Du Pont's line of nylon molding and extrusion compounds included six products and twelve colors. In the next twenty years the Du Pont line alone increased by an order of magnitude in both the number of formulations and the number of standard and service colors. It is not surprising that a 1964 article cited diversity as the key to nylon and suggested that a nylon might actually be designed to meet the specific needs of an application if it were big enough (3).

Table 1-1. Nylon market analysis, 1962-1970

Market	Consumption in millions of pounds[a]								
	1962	1963	1964	1965	1966	1967	1968	1969	1970
Appliances	3.1	3.6	3.8	4.2	3.0- 4.9	2.5- 4.8	5.5- 8.0	9.0	9.0
Autos and trucks	11.5	11.0	12.5	15.1	11.0-18.5	10.3	11.8-13.4	20.0	26.0
Consumer products	5.0	5.8	6.5	7.0	7.4- 7.9	7.0	7.9	2.0	6.0
Electrical and communications	4.0	4.5	5.2	6.2	7.4- 8.0	7.5	8.7	10.0	10.0
Machinery parts	5.0	5.3	5.5	5.8	5.5- 6.3	5.0	5.6	7.0	7.0
Reinforced compound[b]	--	--	--	--	4.5	5.0- 6.0	7.5	8.5-12.0	13.0
Extrusion markets	12.0	13.5	15.0	21.0	19.1-26.2	19.9-20.4	24.0	26.0	28.0
Filaments, bristle	--	--	--	5.6	6.0- 6.1	6.1	7.0	10.0	11.0
Film	--	--	--	7.9	3.8-10.4	4.5	5.4	3.0	4.0
Sheet, rod, and tube	--	--	--	1.9	2.3- 3.3	3.3- 3.8	4.2	5.0	5.0
Wire and cable, other	--	--	--	5.6	6.0- 7.4	6.0	7.4	8.0	8.0
Miscellaneous	1.0	1.3	1.5	1.6	1.8- 3.3	1.0- 3.5	2.0	4.0	4.0
Total	41.6	45.0	50.0	60.9	61.8-73.0	60.7-62.0	73.0-77.1	86.5-90.0	103.0

[a] U.S. estimates from *Mod. Plast.*, January issues, 1963-1971.
[b] Also included in other categories.

The decade of the 1960s witnessed the commercial development of several new engineering thermoplastics such as polyformaldehyde, polycarbonate, polysulfone, and poly(phenylene oxide). It was a period also of well-known growth for olefin plastics including polypropylene and high-density polyethylene, which were landmark materials because they represented a new degree of control in chain polymerization, that is, control of the stereochemistry via coordination polymerization. It is a testimony to the viability of nylon plastics that its rate of growth has matched that of the whole plastics industry (4,5). The data on nylon are admittedly estimates which have occasionally varied widely, but the growth pattern for both nylon and all plastics is essentially exponential. This is shown for the years 1950 to 1970 by the linear relationship of sales to time on a semilogarithmic plot (Fig. 1-1). The slope of this line corresponds to an annual growth rate of 12.2%. Based on estimates of nylon fiber production (6), nylon plastics sales have averaged almost 6% of fiber production over the interval 1954 to 1969.

A breakdown of the markets in which nylon plastics is used is shown in Table 1-1. Diversity is again the keynote. Extrusion applications have amounted to about 30% of the total for the period covered, 1962 to 1970. The last twenty years have seen the price of nylon decrease by over 50% (Table 1-2). In summary, performance, versatility, a broad range of markets, and decreasing cost

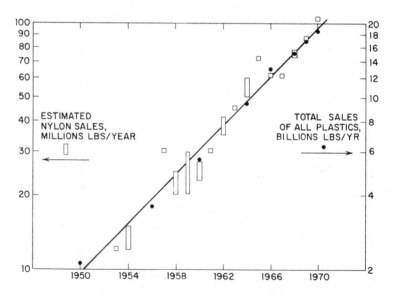

Fig. 1-1.   Growth in United States sales of nylon and all plastics for the years 1950-1970 (data from Ref. 5).

have characterized nylon plastics and have produced a growth which has kept pace with the rapid expansion of the whole plastics industry. These same factors predict a continued steady growth for the first engineering thermoplastic.

Table 1-2. Selling price of nylon molding powder[a]

Year	$/lb	¢/in.^{3}
1949	1.60	6.57
1955	1.45	5.95
1957	1.18	4.85
1960	0.98	4.03
1963	0.90	3.70
1965	0.875	3.60
1967	0.75	3.08
1971	0.73	3.00

[a] Basic volume price of first quality nylon-66 or nylon-6 (Ref. 5).

NOMENCLATURE

Nylon was the name given by Du Pont to these new fiber-forming polyamides. There are other polyamides, for example, the so-called fatty polyamides (7), which lack the rigidity, hydrocarbon resistance, or other properties associated with the fiber-forming varieties. Such polyamides are not identified as nylons and are not discussed in this book. The distinction is more clearly drawn in Chapter 16, which concerns nylons as binder polymers.

●The first nylons made were aliphatic and differed only in the ratio and arrangement of the methylene ($-CH_2-$) and amide ($-NHCO-$) groups in the polymer. These structures are readily identified by a simple, numeral system:

$$H_2N(CH_2)_nNH_2 + HOOC(CH_2)_{m-2}COOH \rightarrow$$
$$\text{diamine} \qquad\qquad \text{dibasic acid}$$

$$-NH(CH_2)_nNHCO(CH_2)_{m-2}CO- + H_2O$$
$$\text{nylon-}nm$$

$$H_2N(CH_2)_{n-1}COOH \rightarrow -NH(CH_2)_{n-1}CO- + H_2O$$
$$\text{amino acid} \qquad\qquad \text{nylon-}n$$

$$(CH_2)_{n-1}\begin{array}{c} C=0 \\ | \\ NH \end{array} \longrightarrow -NH(CH_2)_{n-1}CO-$$
$$\text{lactam} \qquad\qquad\qquad \text{nylon-}n$$

The first number, n, indicates the number of carbon atoms in the diamine; the second, m, the number of carbon atoms in the dibasic acid. A single number indicates the number of carbon atoms in the amino acid or lactam. The original nylon made from hexamethylenediamine and adipic acid is, therefore, nylon-66. Attempts have been made to separate the digits in the designation of nylons from diamines and diacids to avoid ambiguity or confusion with nylons from amino acids or lactams with more than nine carbon atoms, for example, nylon-6,6, nylon-6,12, and nylon-11. In fact, confusion rarely exists because there are no reasonable alternatives: the amino acid or lactam with 66 carbon atoms is hardly likely, nor is nylon-1,1 from methylene diamine likely, and 612 is certainly not 61,2. For reason of readability and convenience, particularly in the case of copolymers, the numerals will not be separated except in special instances where a useful purpose is served. Copolymers are indicated by use of the solidus with polymer weight fractions indicated parenthetically; for example, nylon-66/6 (50/50) describes a polymer in which the weight fraction derived from hexamethylenediamine and adipic acid equals that derived from caprolactam.

Pronunciation should make clear the identity of the polymer. Thus, nylon-66 is "six six" not "sixty-six," nylon-612 is "six twelve" not "six one two" or "six hundred and twelve," and nylon-11 is "eleven" not "one one."

Reactants involving rings and side groups have been employed in making nylons, but no rules for designating them have been promulgated. Abbreviations of the chemical names of the reactants are commonly resorted to, and some of these are listed in Table 1-3. Thus, 2,5-DiMePipT (pronounced "two five dimethyl pip tee") is the polymer from 2,5-dimethylpiperazine and terephthalic acid. In complex titles of this sort it is sometimes helpful to separate the amine and acid designations with a hyphen. It is always necessary to explain the symbols employed because of the absence of standardized designations.

The naming of nylons as subsequently described reflects common practice. As recommended by the International Union of Pure and Applied Chemistry (8), parentheses are used for names involving more than one word, but the awkward "amidamer" terminology, also recommended, is almost universally avoided. Nylons from diamines and diacids are most often named as polyamides:

poly(hexamethylene adipamide) for nylon-66
poly(decamethylene oxamide) for nylon-102
poly(m-xylylene adipamide) for MXD6
poly(piperazine isophthalamide) for PipI

Because of the absence of a generally accepted standard, the reader should be warned to expect other usage in the literature. The name of a nylon from an amino acid or lactam is usually based on the monomer:

Table 1-3. Abbreviations for nylon ingredients

Formula	Name	Abbreviation
HOOC—⬡—COOH	Terephthalic acid	T
HOOC—⬡—COOH	Isophthalic acid	I
H_2N—⬡—NH_2	*meta*-Phenylenediamine	MPD
H_2NCH_2—⬡—CH_2NH_2	*meta*-Xylylenediamine	MXD
HN(CH$_2$-CH$_2$)(CH$_2$-CH$_2$)NH	Piperazine	Pip
2,5-Dimethylpiperazine structure	2,5–Dimethylpiperazine	2,5-DiMePip
$H_2NCH_2\text{-CHCH}_2\overset{CH_3}{\underset{}{C}}\text{-CH}_2CH_2NH_2$ (CH$_3$, CH$_3$)	2,4,4-Trimethylhexa-methylenediamine	TMD (50/50 mixture)
$H_2NCH_2\text{-C-CH}_2\text{-CHCH}_2CH_2NH_2$ (CH$_3$, CH$_3$, CH$_3$)	2,2,4-Trimethylhexa-methylenediamine	

polycaprolactam for nylon-6
poly(11-aminoundecanoic acid) for nylon-11

These nylons are also named as polyamides:

polycaproamide for nylon-6
polyundecanoamide for nylon-11

Occasionally, the fact that the amine group is on the carbon atom remote from the acid group is emphasized in the name:

poly-ε-caproamide or poly-ω-caproamide
poly-11-undecanoamide or poly-ω-undecanoamide

The burdensome naming rather than shorthand designation of copolymers is almost always avoidable. Where naming of a copolymer is absolutely necessary, the combining form, -co-, is recommended, and brackets should be used if one component requires a parenthesis. Consistent terminology in any one name is obviously desirable:

poly(hexamethylene-co-decamethylene oxamide) for nylon-62/102
poly(hexamethylene adipamide-co-sebacamide) for nylon-66/610
poly(caproamide-co-heptanoamide) for nylon-6/7
poly[(7-aminoheptanoic acid)-co-(11-aminoundecanoic acid)] for nylon-7/11
poly[(hexamethylene adipamide)-co-caproamide] for nylon-66/6

Each of the ingredients used to make a nylon from a diamine and dibasic acid such as hexamethylenediamine and adipic acid is called a monomer because either one can be replaced with another reactive diamine or diacid and a polymer would still be obtained. However, the polymer resulting from one diamine and one diacid is a homopolymer, not a copolymer. This recognizes the fact that both ingredients are necessary for polymerization and they yield a structure in which the amine and acid fragments must alternate without exception. This is consistent with the internationally recommended definition of a homopolymer (8).

Nylons made from diamines and diacids are sometimes conveniently referred to as AABB polymers [as are others similarly derived from two monomers such as the polyester, poly(ethylene terephthalate)]. Nylons from single reactants such as amino acids or lactams are AB polymers.

TYPICAL NYLONS

The currently commercial nylon homopolymers and others that have received attention as candidate nylon plastics are summarized in Table 1-4. Not included are the many commercial nylon copolymers the compositions of which are for the most part proprietary. Other monomers not previously mentioned but known to be used in commercial products are PACM, bis(*para*-aminocyclo-hexyl)methane, an ingredient in a copolymer made by Badische Anilin- und Soda-Fabrik AG, and "dimer-acid," the 36-carbon atom dimer of 18-carbon atom unsaturated monobasic acids, used in a recently introduced series of specialty resins made by the General Mills Company.

Table 1-4. Commercial and conjectured nylon plastic homopolymers

Nylon-	Structure	Status[a]
AABB-aliphatic		
66	$H\left[\ HN(CH_2)_6NHCO(CH_2)_4CO\ \right]_n \cdot OH$	Comm.
69	$H\left[\ HN(CH_2)_6NHCO(CH_2)_7CO\ \right]_n \cdot OH$	Conj.
610	$H\left[\ HN(CH_2)_6NHCO(CH_2)_8CO\ \right]_n OH$	Comm.
612	$H\left[\ HN(CH_2)_6NHCO(CH_2)_{10}CO\ \right]_n \cdot OH$	Comm.
1313	$H\left[\ HN(CH_2)_{13}NHCO(CH_2)_{11}CO\ \right]_n OH$	Conj.
AB-aliphatic		
6	$H\left[\ HN(CH_2)_5CO\ \right]_n OH$	Comm.
7	$H\left[\ HN(CH_2)_6CO\ \right]_n \cdot OH$	Conj.
8	$H\left[\ HN(CH_2)_7CO\ \right]_n \cdot OH$	Conj.
9	$H\left[\ HN(CH_2)_8CO\ \right]_n OH$	Conj.
11	$H\left[\ HN(CH_2)_{10}CO\ \right]_n \cdot OH$	Comm.
12	$H\left[\ HN(CH_2)_{11}CO\ \right]_n OH$	Comm.
13	$H\left[\ HN(CH_2)_{12}CO\ \right]_n \cdot OH$	Conj.

(continued)

9

Table 1-4. Commercial and conjectured nylon plastic homopolymers (continued)

AABB-ring containing

MXD6	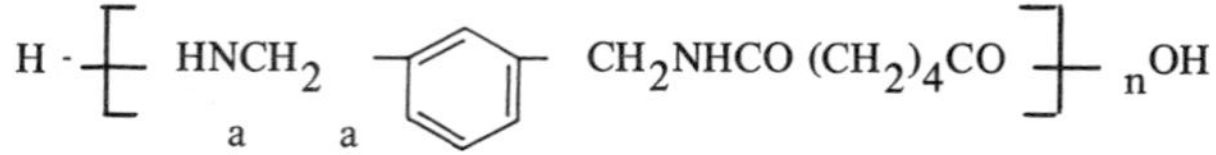 $H \cdot \left[HNCH_2 - \bigcirc - CH_2NHCO(CH_2)_4CO \right]_n OH$	Conj.
TMDT[b]	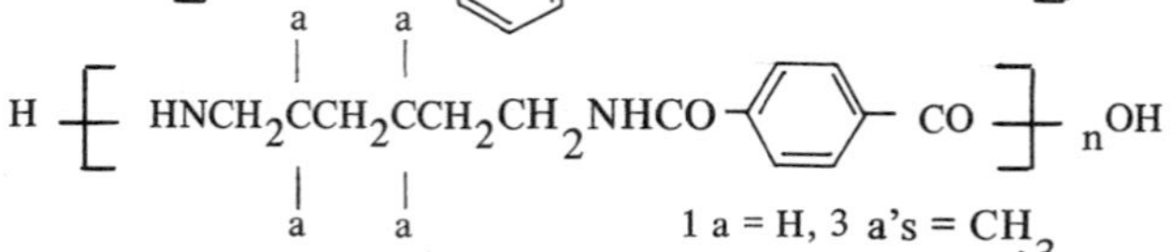 $H \left[HNCH_2CCH_2CCH_2CH_2NHCO - \bigcirc - CO \right]_n OH$ 1 a = H, 3 a's = CH_3	Comm.
HPXD8[c]	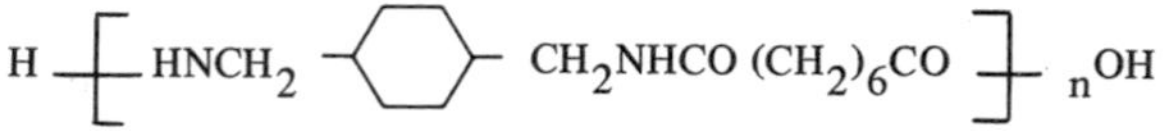 $H \left[HNCH_2 - \bigcirc - CH_2NHCO(CH_2)_6CO \right]_n OH$	Conj.

---

[a] As of June, 1970.

[b] Actually a copolymer because TMD is a mixture of 2,2,4- and 2,4,4-trimethylhexamethylenediamines. A head-to-head, head-to-tail isomerism along the chain would exist even if TMD were a single isomer.

[c] HPXD = hexahydro-*para*-xylylenediamine. This exists in cis and trans isomeric forms.

REFERENCES

1. Mark, H., and G. S. Whitby, Eds., *Collected Papers of W. H. Carothers,* Interscience Publishers, Inc., New York, 1940.
2. Bolton, E. K., *Ind. Eng. Chem.* **34**, 53 (1942).
3. Anon., *Mod. Plast.* **42** (3), 92 (Nov., 1964).
4. Simonds, H. R., A. J. Weith, and M. H. Bigelow, *Handbook of Plastics,* 2nd ed., D. Van Nostrand Co., Inc., New York, 1949, p. 16.
5. Anon., *Mod. Plast.* January issues, 1949-1970.
6. Anon., *Text. Organon* **41** (2), 28 (1970).
7. Peerman, D. E., in *Encyclopedia of Polymer Science and Technology,* Vol. 10, Interscience Publishers, a division of John Wiley and Sons, Inc., New York, 1969, p. 597.
8. Int. Union Pure Appl. Chem., *J. Polym. Sci.* **8**, 257 (1952).

CHAPTER 2

Preparation and Chemistry of Nylon Plastics

M. I. KOHAN

PREPARATION OF NYLON PLASTICS

Discussion of the synthetic routes to commercial and candidate nylon plastics logically begins with nylon-66 and nylon-6. As the commercially most important nylon plastics as well as fibers, they have received the most study. The methods employed to synthesize the monomers (often referred to as intermediates because of their mid-position between raw materials and polymers) and their conversion into useful nylons are representative of synthetic methods and polymerization techniques applicable to other nylons. The preparation of other nylon plastics is subsequently reviewed, but excluded are some nylons that have been mentioned as commercial possibilities such as nylons-3, -4, and -5 and wholly aromatic nylons. Such nylons are fiber (1) but not plastic candidates because they are melt unstable or have excessively high melt viscosities, difficulties which can be overcome in fiber technology by solution spinning.

Intermediates for Nylon-66 and Nylon-6

The multiplicity of processes that lead to intermediates for nylon-66 and nylon-6 is illustrated in Figure 2-1. Adipic acid is manufactured principally by the nitric acid oxidation of a cyclohexanol/cyclohexanone mixture obtained by the air oxidation of cyclohexane, although the oxidation of cyclohexanol obtained by the hydrogenation of phenol is also used (2,3,4). Hexamethylenediamine is made by the hydrogenation of adiponitrile which can be derived from adipic acid, butadiene, furfural, propylene, or acetylene. Receiving particular attention in recent years is the propylene route because it affords low-cost acrylonitrile which has been hydrodimerized electrolytically to adiponitrile (5,6,7). The reaction of butadiene and hydrogen cyanide in the presence of selected catalysts has been described as a new route to adiponitrile (220).

   Most processes for the manufacture of caprolactam involve as a final step the rearrangement of cyclohexanoneoxime to caprolactam (8):

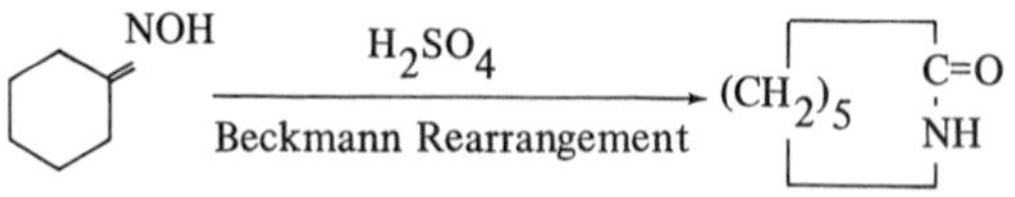

The oxime can be obtained by hydrogenation of nitrocyclohexane and by treatment of cyclohexane with nitrosyl chloride in the presence of light (9,10),

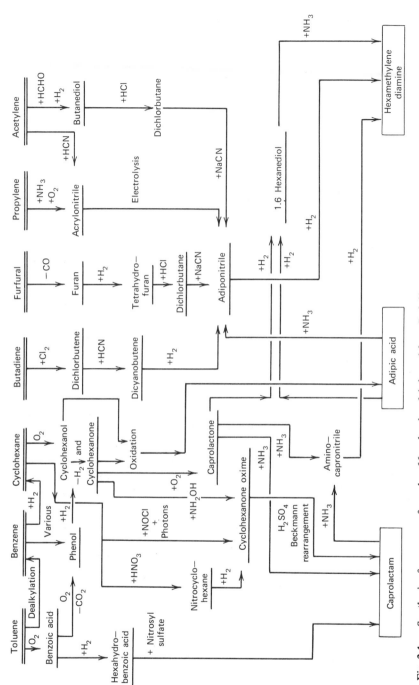

Fig. 2-1. Synthesis of monomers for nylon-66 and nylon-6 (adapted from Ref. 2).

but most of the oxime has been and is still made by reaction of cyclohexanone with hydroxylamine:

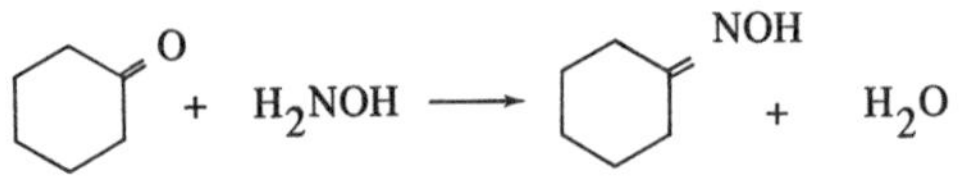

The cyclohexanone is made by dehydrogenation of cyclohexanol obtained by hydrogenation of phenol or, together with cyclohexanone, by air oxidation of cyclohexane. The air oxidation is analogous to the first step in the synthesis of adipic acid, but conditions are altered to favor ketone formation (8). In the Snia Viscosa process (11) which involves oxidation of toluene to benzoic acid followed by hydrogenation to hexahydrobenzoic acid, the treatment of hexahydrobenzoic acid with nitrosyl sulfate in sulfuric acid leads directly to the lactam without isolation of the oxime:

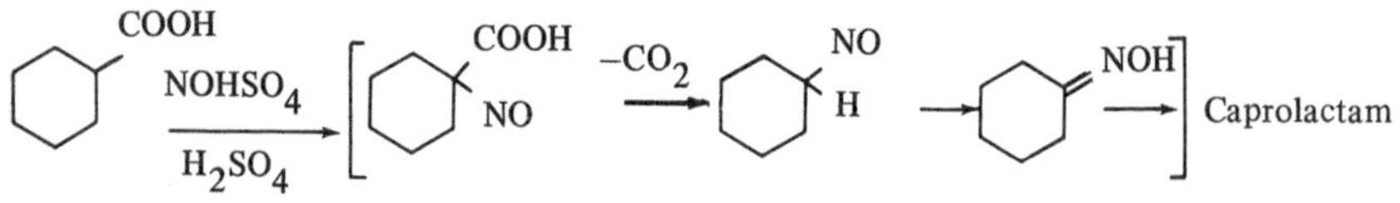

A Union Carbide process(12) avoids the oxime altogether and involves amination of caprolactone which is prepared by oxidation of cyclohexanone with peracetic acid (13). The ether oxygen atom is replaced with a nitrogen atom to yield caprolactam directly. The cost of hydroxylamine and a market for by-product ammonium sulfate are important factors in the synthesis of caprolactam (8). A recent report (219) cites the use of N-acetylcaprolactam in the Beckmann rearrangement to avoid by-product formation.

A recently proposed process claims conversion of cyclohexanone to the 2-nitro compound by reaction with acetic anhydride in concentrated nitric acid and subsequent hydrolysis/hydrogenation of the nitro compound to the linear aminocaproic acid (14):

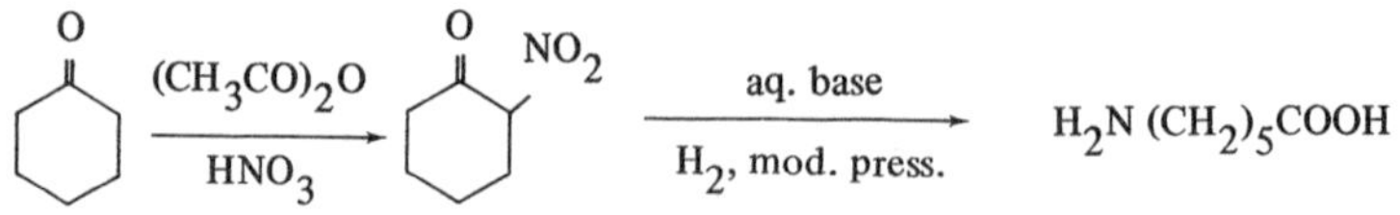

The amino acid can be cyclized to caprolactam. The status of the alternative processes to caprolactam was reviewed in 1967 and again in 1970 (12).

The conversion of cyclohexane to adipic acid and caprolactam typifies processes applicable to hydrocarbon rings, the cycloalkanes, for the synthesis of higher acids and lactams. The only ring sizes currently available commercially, however, are the eight and twelve carbon rings made from butadiene or acetylene (see comments on nylon-8 and -12, pages 26 and 28).

Preparation of Nylon-66

Hexamethylenediamine and adipic acid have been indicated to be the monomers for nylon-66. In general, amide-forming derivatives of the diacid such as a diester, dihalide, or diamide could also be used (15), but the acid itself is normally preferred. Use of the free acid allows formation of the salt, hexamethylenediammonium adipate or $^+H_3N(CH_2)_6NH_3^+$ $^-OOC(CH_2)_4COO^-$, as a first step, and checking the pH of a salt solution is an excellent way of assuring the necessary stoichiometry. The equivalence pH is that pH which corresponds to an exact 1:1 mole ratio. It is determined by observing the inflection point in a titration curve, that is, the pH at which the pH changes most rapidly when a strong acid or base is added to the salt solution. The discrepancy between the pH of a salt solution and the equivalence value measures the imbalance of reactants which will limit the achievable molecular weight. For example, as shown on p. 39, an excess of one monomer over the other by 1.25 mole percent limits the theoretical maximum molecular weight to 18,000 and imposes severe restrictions on moisture sensitivity during processing. The equivalence pH is approximately equal to $(\log K_B - \log K_w - \log K_A)/2$, where K_w is the dissociation constant for water and K_B and K_A are the second dissociation constants of the acid and base (93).

Another factor in polymerization is the loss of diamine because of its relative volatility (16). The manufacturer has to assess this factor for his particular equipment and adjust the pH of the salt solution to take this into account. The equivalence pH for the 66 salt and for the 69, 610, and 612 salts as well is about 7.6 in 1 to 10% aqueous solutions (16).

The salt may be made by dissolving each of the reactants in a suitable solvent such as methanol, mixing and collecting the precipitated salt which is then dissolved in water. It may also be made by adding an aqueous solution of the diamine to a stirred aqueous dispersion of the acid. In the latter case, which does not involve the purification inherent in a crystallization step, it is desirable to treat the solution with activated charcoal to assure freedom from impurities.

In a typical batch polymerization, the salt solution is charged to evaporators where it is concentrated to a liquor containing about 75% solids before transfer to a pressure vessel purged of oxygen. The concentrated solution is then heated to about 210°C (410°F) or until a pressure of about 17 atm is reached. Pressure is maintained while bleeding steam, and the temperature is gradually increased. Finally, the pressure is reduced to atmospheric and the temperature increased to about 280°C (535°F). The melting point of dry nylon-66 is 269°C (516°F); this is depressed by the presence of water, and the polymerization technique is designed to remove the water at a rate commensurate with the increase in temperature so that the melt does not solidify. The extent of polymerization determines the molecular weight and is determined by the residual moisture content. A vacuum may be applied in order to achieve a high molecular weight. Excessive molecular weight can be avoided by use of an excess of adipic acid or hexamethylenediamine or by adding a material such as acetic acid which, by having only one reactive site, prevents further growth(17). The polymer is forced out of the bottom of the autoclave as a wide ribbon by application of pressure with an inert gas. The ribbon is quenched with water which is subsequently removed by jet blowers, cut into chips, blended, and packaged(18). Hermetically sealed containers are required and most often are 25-lb cans or 50-lb bags, but polyolefin-lined drums and cartons of 250 to 1200-lb capacity are also used.

The literature contains other relatively brief descriptions of the preparation described above which include flow diagrams(19,20,21) and the properties of the monomers(19), but the precise operational details are submerged in proprietary know-how and the voluminous patent literature.

Catalysts are not needed for the polymerization that yields nylon-66. Catalysts cited in other instances of AABB polyamide preparations are oxides, carbonates, and halogen salts of polyvalent metals(15), strong acids(22), litharge(23), and organotin compounds(24). Also unnecessary for nylon-66 but occasionally useful is polymerization in the presence of high-boiling fluids which may or may not be solvents for the nylon(16,25,26).

Nylon-66 is also made by continuous processes. Provision must be made to remove water and to avoid dead spots where delayed flow may lead to excessive variations in molecular weight. Suitable equipment has been described, largely in the patent literature(27–31).

The reaction of an amine and an acid to produce an amide requires a high temperature but may occur in the solid phase at temperatures below the melting point(32). Processes have been described which take advantage of this fact(33,34). Differences between solid-phase and melt-phase polymerized nylons (35) tend to disappear once the solid-phase product is melted.

The molecular weight of a nylon can also be increased by melt extrusion through a device equipped with an extraction zone purged with an inert gas(94).

Preparation of Nylon-6

The batch manufacture of nylon-6 from caprolactam is similar to the process outlined above for nylon-66 except that caprolactam and up to 10% water are directly charged to the pressure vessel. The nature of the monomer precludes the need for concern about stoichiometry, but the presence of an exactly equal amount of residual amine and acid groups in the polymer is not necessarily assured(36). The polymerization reactions of nylons-66 and -6 differ. Dry caprolactam will not polymerize(37), but in the presence of water reaction occurs at a reasonable rate above 200°C (392°F). Actually, several reactions occur(37–40):

Hydrolysis
$$(CH_2)_5 \underset{\substack{| \\ NH}}{\overset{C=0}{}} + H_2O \rightarrow H_2N(CH_2)_5COOH$$

Condensation
$$H\!-\!\!\left[NH(CH_2)_5CO\right]_m\!\!OH + H\!-\!\!\left[NH(CH_2)_5CO\right]_n\!\!OH \rightarrow$$

$$H\!-\!\!\left[NH(CH_2)_5CO\right]_{m+n}\!\!OH + H_2O$$

Addition
$$(CH_2)_5 \underset{\substack{| \\ NH}}{\overset{C=0}{}} + H\!-\!\!\left[NH(CH_2)_5CO\right]_m\!\!OH \rightarrow$$

$$H\!-\!\!\left[NH(CH_2)_5CO\right]_{m+1}\!\!OH$$

All of these reactions were found to be acid catalyzed although recent work(63) contests the catalysis of the condensation reaction. The addition reaction, not the condensation reaction by which nylon-66 is made, is the major growth mechanism. Because of the importance of the addition reaction, it is not surprising that the polymerization is expedited by adding aminocaproic acid at the start. Hexamethylenediammonium adipate (or similar amine-acid combination) is equally effective, but the product is then a 6/66 copolymer of composition depending on the amount of 66 salt used. Adjusting the acidity of the initial charge such that the pH of a 20% solution is 5.7 is reported to result in a fast, smooth polymerization(41).

The melting point of polycaprolactam is 40°C (72°F) lower than that of nylon-66, but polymerization temperatures are normally not very different because of the demands of adequate melt flow and reasonable rates of reaction. As for 66, molecular weight can be controlled by adding amine or acid to cause an imbalance in the amine/acid ratio or by adding a monofunctional reactant such as acetic acid. The same dependence of molecular weight on residual moisture obtains, and the same need exists to hermetically package the product.

Nylon-6 contains about 10% of extractable material which acts as a plasticizer and is preferably removed for most applications. This is accomplished by application of a vacuum or by water washing of the cut polymer which must then be carefully dried. The extractables include monomer and cyclic oligomers:

$$\left[\begin{array}{c} \text{— NH(CH}_2\text{)}_5\text{CO —} \\ \text{—}\left[\text{CO(CH}_2\text{)}_5\text{NH}\right]_n \end{array}\right]$$

Monomer ($n = 0$) is removed with a vacuum of 5 to 10 mm Hg, but removal of oligomers ($n \geqslant 1$) requires prolonged heating at less than 1 mm Hg(42).

The polymerization of caprolactam was rapidly adapted to a continuous process because the relative volatility of two different reactants is not a factor. Continuous processes have been described in the literature(19,20,43,44,45). A recent patent cites the advantage of intermittent injection of inert gas for more efficient removal of water from the melt(46). Another suggests limiting initial conversion to 45% in order to obtain a product containing less than 2% of the cyclic oligomers with a vacuum finish that does not require prolonged heating at less than 5 mm Hg(42).

Solid-phase polymerization is applicable to low-molecular-weight polycapro-lactam but is preferably applied to polymer free of the extractable fraction in order to achieve optimum rates of reaction(47). Successful polymerization of ϵ-aminocaproic acid (melting point 200°C) at 150 to 180°C has been reported(48). As previously noted, melt-phase and solid-phase polymerized nylon-6 differs(35).

The anhydrous polymerization of caprolactam using strong base (anionic) or strong acid (cationic) as catalyst is possible. The cationic procedure has yet to yield a polymer of useful molecular weight, but the anionic polymerization leads to very high molecular weight in a matter of minutes or even seconds, particularly if a suitable cocatalyst is added. The cocatalyst can be any of a wide variety of N-substituted lactams (or precursors thereof) where the substituent on the nitrogen atom is of the electron-withdrawing type(49). Anionic polymeriza-tion is used in monomer casting which is discussed in Chapter 13. The polymer made anionically is not as thermally stable as that made in the presence of water and is not directly applicable to injection molding or extrusion. Procedures to adapt anionic polymerization to the manufacture of melt processable nylon-6 have been described(50,51) but are not yet in use. Two recent reviews discuss the chemistry of polymerization of lactams(52,53).

Other AABB-Nylons

In addition to hexamethylenediamine and adipic acid, the intermediates for the currently commercial and conjectured AABB-nylon plastics (Table 1-4) include the 8, 9, 10, 12, 13, and 36-carbon aliphatic dibasic acids, terephthalic and isophthalic acids, the xylylenediamines, trimethylhexamethylenediamine, bis(4-aminocyclohexyl)methane, and the 13-carbon diamine.

Suberic acid, $HOOC(CH_2)_6COOH$, is not an article of commerce like azelaic or sebacic acid. The most often cited raw material is the 8-membered ring, cyclooctane, obtained from acetylene or butadiene (see discussion on nylon-8 on p. 26). Conversion to suberic acid would follow the same route described above for the conversion of cyclohexane to adipic acid. Dodecanedioic acid, $HOOC(CH_2)_{10}COOH$, is similarly made from cyclododecane (see discussion of nylon-12 on p. 28).

Azelaic acid is made by ozonolysis of oleic acid:

$$CH_3(CH_2)_7CH{=}CH(CH_2)_7COOH \xrightarrow[\text{solvent}]{O_2, O_3, H_2O}$$

$$\underset{\text{pelargonic acid}}{CH_3(CH_2)_7COOH} + \underset{\text{azelaic acid}}{HOOC(CH_2)_7COOH}$$

Sebacic acid is made from castor oil by treatment with hot caustic. Castor oil is 80 to 90% the glyceride of ricinoleic acid:

$$\left[CH_3(CH_2)_5CHOHCH_2CH{=}CH(CH_2)_7COO\right]_3C_3H_6 + 6NaOH + 2H_2O \xrightarrow{\text{heat}}$$

$$3\ CH_3(CH_2)_5\underset{\underset{\text{2-octanol}}{\overset{|}{OH}}}{\overset{|}{C}}HCH_3 + 3NaOOC(CH_2)_8COONa + C_3H_6(OH)_3 + 3H_2$$

$$\Big|\ HCl$$

$$\underset{\text{sebacic acid}}{HOOC(CH_2)_8COOH} \qquad \text{glycerol}$$

The 13-carbon diacid, brassylic acid, is potentially available from crambe oil which yields on hydrolysis fatty acids containing 55 to 60% of the precursory erucic acid(54,55):

$$\underset{\text{erucic acid}}{CH_3(CH_2)_7CH{=}CH(CH_2)_{11}COOH} \xrightarrow[\text{acetic acid}]{O_2, O_3, H_2O}$$

$$\underset{\text{pelargonic acid}}{CH_3(CH_2)_7COOH} + \underset{\text{brassylic acid}}{HOOC(CH_2)_{11}COOH}$$

Dimer acids are formed by the combination of two molecules of unsaturated fatty acids, especially oleic and linoleic acids. The structure of the 36-carbon dimer depends on the monomeric acid and the method of dimerization(56) and is unspecified in a commercially available product reported to be high in purity and low in color(57). Because the unsaturation occurs near the middle of the original fatty acids and is the site of interaction, the dimer acid can be pictured as a dibasic acid about 18 carbon atoms long with a pair of side chains near the middle, each about 9 carbon atoms long. Possible structures for the dimers of oleic acid and linoleic acid are, respectively,

$$\begin{array}{c} (CH_2)_7CH_3 \\ | \\ HOOC(CH_2)_7CH-CH \\ |\quad\quad | \\ \quad CH-CH(CH_2)_7COOH \\ \quad | \\ \quad CH_3(CH_2)_7 \end{array}$$

and

$$\begin{array}{c} HOOC(CH_2)_7CH-CHCH_2CH=CH(CH_2)_7COOH \\ \diagup \quad\quad \diagdown \\ CH \quad\quad CH(CH_2)_4CH_3 \\ \diagdown\diagdown \quad \diagup \\ CH-CH \\ | \\ CH_3(CH_2)_4 \end{array}$$

These structures involve isomeric alternatives, and dimers involving the addition of one molecule of oleic acid to one molecule of linoleic acid are also possible. Either the dimer acid is hydrogenated or the amount of linoleic acid used is very small because the iodine number reported for the commercial dimer acid(57) corresponds to less than 0.02 carbon-carbon double bonds per mole of acid.

Terephthalic acid is manufactured on a large scale for use in making the polyester, poly(ethylene terephthalate). Isophthalic acid is available at low cost via oxidation of *meta*-xylene. The xylenes can also be converted to xylylenediamines through the dinitriles(58):

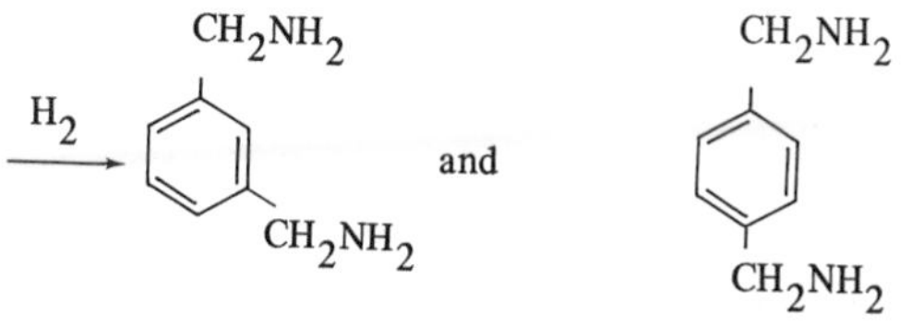

$$\xrightarrow{\text{H}_2}$$

CH$_2$NH$_2$ ... CH$_2$NH$_2$ and   CH$_2$NH$_2$ ... CH$_2$NH$_2$

meta-xylylenediamine   *para*-xylylenediamine

Trimethylhexamethylenediamine is made from isophorone, which is a trimer of acetone(59):

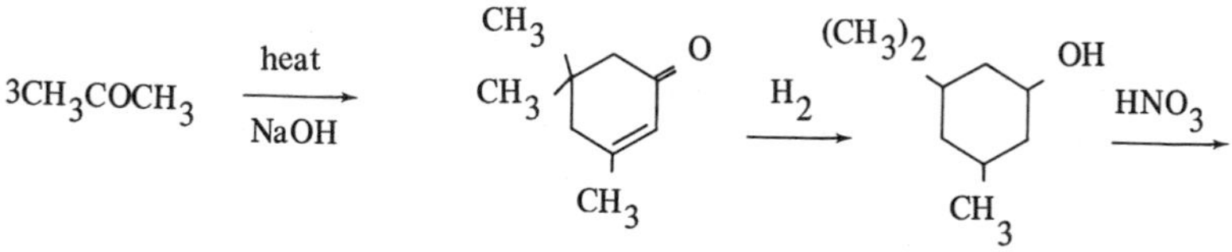

$$3CH_3COCH_3 \xrightarrow[\text{NaOH}]{\text{heat}}$$

CH$_3$ CH$_3$ ... O

$$\xrightarrow{\text{H}_2}$$

(CH$_3$)$_2$ ... OH ... CH$_3$

$$\xrightarrow{\text{HNO}_3}$$

isophorone

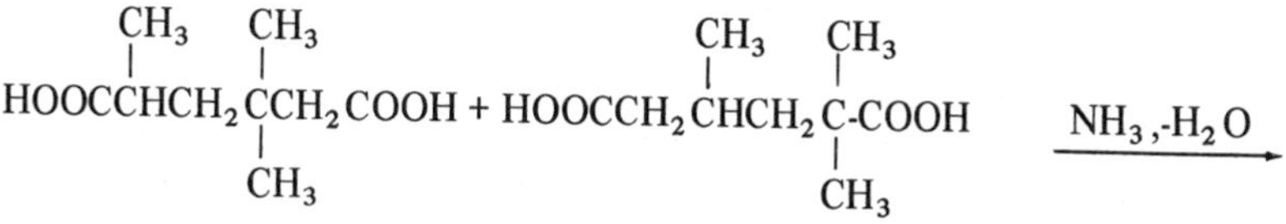

$$\begin{array}{c} CH_3 \quad CH_3 \\ | \qquad | \\ HOOCCHCH_2CCH_2COOH + HOOCCH_2CHCH_2C\text{-}COOH \\ | \qquad\qquad | \\ CH_3 \qquad\qquad CH_3 \end{array} \xrightarrow{\text{NH}_3\text{,-H}_2\text{O}}$$

trimethyladipic acids

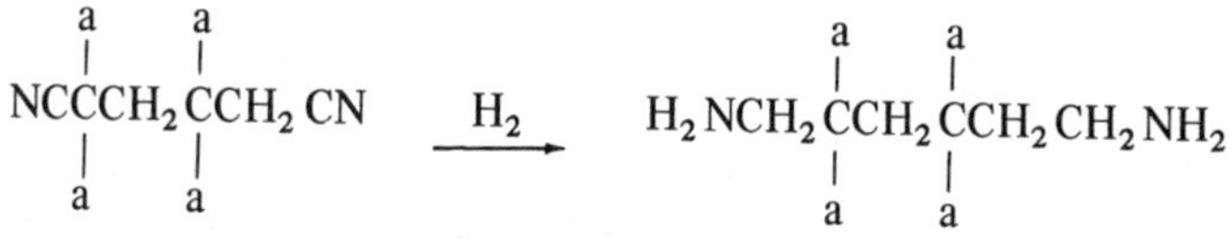

$$\begin{array}{c} a \quad a \\ | \quad | \\ NCCCH_2CCH_2CN \\ | \quad | \\ a \quad a \end{array} \xrightarrow{\text{H}_2} \begin{array}{c} a \quad a \\ | \quad | \\ H_2NCH_2CCH_2CCH_2CH_2NH_2 \\ | \quad | \\ a \quad a \end{array}$$

la = H
3a = CH$_3$

50/50 mixture of 2,2,4- and 2,4,4-
trimethylhexamethylenediamine, "TMD"

Bis(4-aminocyclohexyl)methane is made by reaction of aniline with formaldehyde:

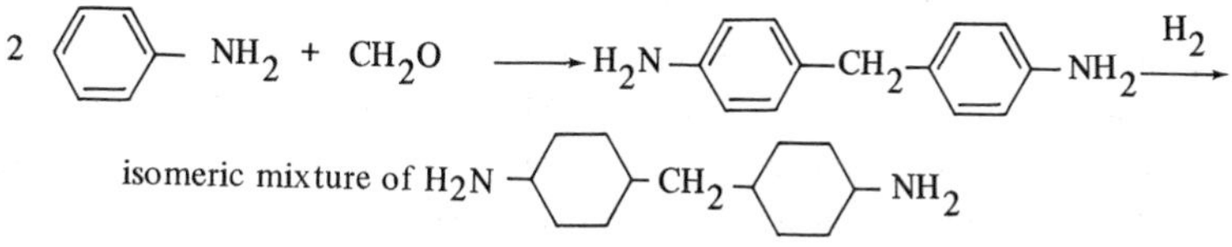

$$2 \; \langle\rangle\text{-}NH_2 + CH_2O \longrightarrow H_2N\text{-}\langle\rangle\text{-}CH_2\text{-}\langle\rangle\text{-}NH_2 \xrightarrow{\text{H}_2}$$

isomeric mixture of H$_2$N-⟨⟩-CH$_2$-⟨⟩-NH$_2$

The 13-carbon diamine is made from brassylic acid which, as noted on p. 21, is derived from crambe oil(55):

$$HOOC(CH_2)_{11}COOH \xrightarrow[\text{polyphosphoric acid}]{NH_3} NC(CH_2)_{11}CN$$

$$\xrightarrow{H_2, \text{ Co cat.}} \underset{\text{1,13-diaminotridecane}}{H_2N(CH_2)_{13}NH_2}$$

Conversion of the AABB-salt into a polymer such as nylon-610, -612, or -1313 is usually accomplished in a batch process as described above for nylon-66. The solubility of the salt in water decreases as the hydrocarbon chain increases so that preparation using a solvent which dissolves both the amine and acid but not the salt may be preferred over synthesis in the presence of water. For example, 1313-salt is made using absolute ethanol. Increasing the hydrocarbon chain also means reduced volatility of the diamine, which permits conversion of the 1313-salt to high-molecular-weight polymer simply by heating at 225°C under nitrogen at atmospheric pressure. The greater hydrocarbon character also means a lower melting point for the nylon (nylon-610 = 225°C or 437°F, nylon-1313 = 175°C or 347°F), so by the time a temperature for a reasonable rate of reaction is reached there is no need for the presence of water to depress the melting point. The temperature in a commercial process is then dictated by viscosity considerations because the viscosity depends on molecular weight and chain stiffness, not proximity to the melting point (see Chapter 4).

The preparation of nylons from *meta*-xylylenediamine and the 6 to 10 C-atom dibasic acids requires a specific level of steam pressure (4 to 7 atm) in the early stages of polymerization to assure the presence of enough water to depress the melting point but not enough to promote a decomposition of the diamine(60). The nylon from TMD and terephthalic acid is partially polymerized before discharging to an extruder in which the polymerization is completed. Melt viscosity is obviously a factor here because a viscosity of 100,000 poise at 250°C (482°F) has been cited for the commercial product(61).

The AABB-nylons we have been considering do contain an extractable fraction, but this is normally less than 2%(62,121), so that an extraction step analogous to that for nylon-6 is not required.

It is clear that if all the monomers were to become commercial realities, other combinations such as nylon-613, MXD10, TMDI, and others, including copolymers, would be more extensively studied and might be shown to provide unique combinations of properties of special utility.

Other AB-Nylons

Nylon-7 is not commercial at present except in the USSR. The monomer, 7-aminoheptanoic acid or *zeta*-aminoenanthic acid, is made together with the monomer for nylon-9 via telomerization of ethylene:

$$nCH_2{=}CH_2 + CCl_4 \xrightarrow{\text{catalyst}} Cl(CH_2CH_2)_nCCl_3$$

$$\xrightarrow[\text{acid}]{H_2O} Cl(CH_2CH_2)_nCOOH \xrightarrow{NH_3} H_2N(CH_2CH_2)_nCOOH$$

n = mostly 3 = $H_2N(CH_2)_6COOH$, 7-aminoheptanoic acid or
  ζ-aminoenanthic acid
 + some 4 = $H_2N(CH_2)_8COOH$, 9-aminononanoic acid or
  ω-aminopelargonic acid
 + lesser amounts of others

Limiting the number of products of telomerization and separation of these products are major problems in this synthesis(64,65). Polymerization resembles that for nylon-66 in that the reaction of amine and acid groups is involved, but it also shares the built-in stoichiometry of nylon-6. The special advantage of much less concern with the volatility of monomer obtains, and polymerization has been carried out at atmospheric as well as higher pressure(65,66,67).

More recently a process for making nylon-7 from cyclohexanone has been developed by Union Carbide(66). This route involves as an intermediate an ester which is first partially hydrolyzed to expedite the polymerization:

cyclohexanone   caprolactone

$$Cl(CH_2)_5COOH \xrightarrow[\text{acid}]{C_2H_5OH} Cl(CH_2)_5COOC_2H_5 \xrightarrow[\text{solvent}]{NaCN}$$

$$NC(CH_2)_5COOR \xrightarrow[\text{Ni cat.}]{H_2} H_2N(CH_2)_6COOC_2H_5$$

$$\xrightarrow[\text{3 hr}]{\text{H}_2\text{O, 90-100}^\circ\text{C}} \quad \text{"precursor"} = \text{mixture of acid and ester, some dimers}$$

$$\xrightarrow[\text{4 hr}]{270^\circ\text{C, N}_2} \quad \text{H}\!\left[\text{HN(CH}_2)_6\text{CO}\right]_n\!\text{OH} \quad \text{nylon-7}$$

Nylon-8 has been cited as a potential new nylon, but the combination of high cost and modest difference in properties has to date prevented commercialization(68,69,70). Capryllactam is the monomer, and it is made from cyclooctane by standard procedures. The cyclooctane is accessible from either acetylene(71) or butadiene(72):

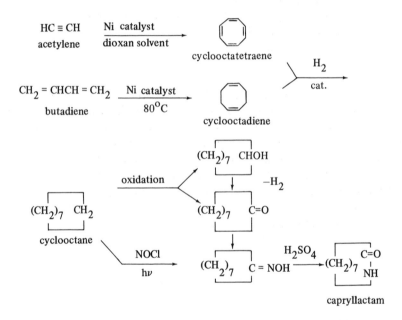

Polymerization is analogous to that of caprolactam and occurs in the presence of water at 240 to 270°C at a rate comparable to that of caprolactam or enantholactam(73).

As noted above, nylon-9 has been made in the USSR by the same process that produces nylon-7. A process for making the monomer, 9-aminononanoic acid, from soybean oil is under development in the United States and is said to promise a less expensive route to nylon-9(74,75). High yields in the ozonolysis and reductive amination are claimed(76,77,78):

Soybean oil, 85% glycerides of:

$CH_3(CH_2)_7CH=CH(CH_2)_7COOH$ oleic acid

$CH_3(CH_2)_4CH=CHCH_2CH=CH(CH_2)_7COOH$ linoleic acid

$CH_3(CH_2CH=CH)_3(CH_2)_7COOH$ linolenic acid

$$\xrightarrow[\text{alkali}]{\text{methanol}} \text{methyl soyate, } C_8H_nCH=CH(CH_2)_7COOCH_3, n = 13,15,17$$

$$\xrightarrow{O_3,O_2} \xrightarrow[\text{Pd cat.}]{H_2} \begin{array}{l} OCH(CH_2)_7COOCH_3 + \text{by-products} \\ \text{methyl azelaaldehyde} \\ \text{``maz''} \end{array}$$

$$\xrightarrow[\text{acid}]{\text{methanol}} \begin{array}{l} (CH_3O)_2CH(CH_2)_7COOCH_3 + \text{by-products} \\ \text{methyl azelaaldehyde-} \\ \text{dimethylacetal ``mazda''} \end{array}$$

$$\xrightarrow{\text{fractionate}} \text{purified mazda} \xrightarrow{\text{acid}} \text{purified maz}$$

$$\xrightarrow[\text{Ni cat.}]{H_2,\,NH_3,\,\text{solvent}} H_2N(CH_2)_8COOCH_3 \xrightarrow{H_2O} \begin{array}{l} H_2N(CH_2)_8COOH \\ \text{9-aminononanoic acid} \end{array}$$

Polymerization falls into the same category as that for nylon-7, and high molecular weight has been realized after 4 to 8 hr at 225 to 260°C under nitrogen at atmospheric pressure(78).

Nylon-11 was first manufactured commercially by Societe-Organico in France in 1955 but was available as a pilot plant product for several years before then(79). As for sebacic acid, castor oil is the starting material. Methyl ricinoleate obtained from the oil by treatment with methanol is pyrolyzed to yield an 11-carbon fragment. A key step is the addition of hydrogen bromide which must occur as indicated to yield the 11-bromo and not the 10-bromo compound:

$$CH_3(CH_2)_5CH \cdots CH(CH_2)_7COOCH_3 \xrightarrow[\text{water}]{550°C} CH_3(CH_2)_5CHO +$$

methyl ricinoleate

$$CH_2 = CH(CH_2)_8 COOCH_3 \xrightarrow{\text{hydrolysis}} CH_2 = CH(CH_2)_8 COOH$$

undecylenic acid

$$\xrightarrow{\text{HBr, air}} Br(CH_2)_{10} COOH \xrightarrow[\text{water}]{NH_3} H_2 N(CH_2)_{10} COOH$$

11-aminoundecanoic acid

The 11-aminoundecanoic acid is obtained after recrystallization as an aqueous paste which can be charged directly to a continuous polymerization unit(80). The use of phosphoric acid and hypophosphoric acid as catalysts has been described(81). Branching is involved in the case of hypophosphoric acid because it induces insolubility after 1 to 5 hr at 218°C depending on the amount added (2.27 to 9.1 mole %).

Nylon-12 was first produced in a commercial plant in Germany in 1966 although pilot plant quantities have been available since 1963. Chemische Werke Huels AG convert butadiene to dodecanolactam or laurolactam, the monomer for nylon-12(70,72,82,83,84).

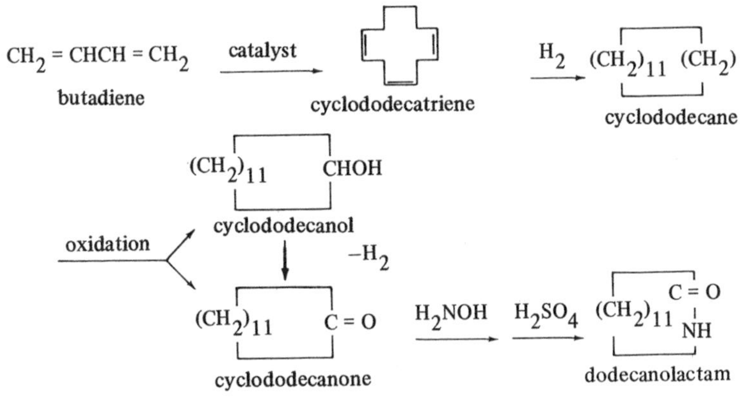

Other routes to the lactam from the butadiene trimer, cyclododecatriene, have been attributed to Badische Anilin- und Soda-Fabrik(85) and Aquitaine-Organico(86). The BASF procedure has so far been applied only to the 12-membered ring, but it would appear to be technically feasible whenever an unsaturated ring is available as the starting material:

BASF

$$\text{cyclododecatriene} \xrightarrow[\text{peracetate}]{\text{acetaldehyde}} \text{epoxide} + O \xrightarrow{H_2} (CH_2)_{10} \begin{array}{c} CH \\ | \\ CH \end{array} \hspace{-2mm} \diagdown O$$

$$\xrightarrow[100^{\circ}C]{MgI_2} (CH_2)_{11} \begin{array}{c} \\ C=O \end{array} \xrightarrow{H_2NOH \ H_2SO_4} (CH_2)_{11} \begin{array}{c} C=O \\ | \\ NH \end{array}$$

Aquitaine-Organico (86) or Mitsubishi-Torary (86a)

$$\text{cyclododecatriene} \xrightarrow{H_2} (CH_2)_{11} \ CH_2 \xrightarrow[h\nu]{HNOSO_4, \ H_2SO_4}$$

$$(CH_2)_{11} \ \ C=NOH \longrightarrow (CH_2)_{11} \begin{array}{c} C=O \\ | \\ NH \end{array}$$

Polymerization is slow because of the stability of the ring(36). Acidic catalysts are used, but temperatures of 300 to 350°C are necessary to complete the polymerization(87,88).

Nylon-13 has been mentioned as a possible derivative of crambe oil(54,89). The synthetic route has not been detailed but could presumably be analogous to that suggested for nylon-9 from soybean oil (see above), the difference residing in the use of erucic acid which would yield a 13-carbon fragment after ozonolysis.

All of the AB-monomers that correspond to ring structures containing more than the seven atoms of caprolactam yield polymers that contain only small amounts of extractable material and do not require an extraction step:

Nylon-	Approx. wt. % extractables	Ref.
7	1.5	66,90
8	2.0	69,91
11	0.5	79
12	0.5	92

Modified Nylons

Modified nylons include copolymers, compositions containing additives to impart specific properties, filled and reinforced materials, and chemically modified nylons (see Chapter 11). The preparation of a copolymer requires only that the desired ingredients be added to the polymerization vessel. Any of the diamines, diacids, aminoacids, and lactams mentioned above can be used in any kind of combination as long as the necessary balance of amine and acid groups is maintained.

Direct addition to the polymerizer of modifiers such as antioxidants, nucleating agents, plasticizers, and carbon black is feasible. The additives must, of course, be stable and inert to the nylon at the high temperature of polymerization (225 to 280°C or 437 to 536°F). The preferred mode of addition may vary with the additive and nylon. It is convenient to charge modifier and monomer at the same time, and this is done, for example, in the case of the copper salt-halide system for better resistance to thermal oxidation(95). Best results are sometimes achieved by introduction after the polymerization has proceeded part way as in the case of adding carbon black to nylon-66(96). Carbon black may be added at the start of the polymerization of caprolactam if the specific conductance of the lactam solution is kept below 2.3 $\times$ 10^{-3} mho cm^{-1} (97).

Incorporation of modifiers by mixing with the nylon melt is widespread and occurs in the plant of the processor as well as the manufacturer. Melt blending permits preparation of batches of any desired size, allows for mixing of materials that cannot be added during polymerization, and provides operability over a wide range of melt viscosities. The high melting point of most nylons, which may mean relatively low melt viscosity and certainly leads to relatively rapid oxidation on open rolls, normally precludes use of mixing rolls, but screw extruders serve the purpose very well. It is possible to pump an additive stream into the mixing zone of a screw, but satisfactory results are commonly obtained by feeding a dry blend of the nylon granules and modifiers to the hopper of the extruder. It is sometimes helpful in attempting to achieve uniform coating of the granules with a solid additive to use a binder such as a silicone oil or "Santicizer" 8 (mixture of N-ethyl *ortho-* and *para*-toluenesulfonamides)(98). Careful mixing of the dry blend may be necessary, particularly when using colorants, and one set of details for accomplishing this has been published(99) although any technique that imparts a uniform surface coating of the colorant on the nylon granules is adequate. Moisture must not be allowed to contact the polymer during blending lest melt behavior be affected (see Chapter 4).

The choice of colorants is discussed in Chapter 11. Selecting the right colorant, methods of incorporation, and a formula for concentrates have been reviewed in the literature(100).

Many materials, for example, fillers such as titanium dioxide or a phenol-formaldehyde resin, may be added either to the polymerization or to the nylon melt depending on the convenience and purpose of the user(101).

In recent years reprocessed nylons have become available. Textile fiber or other nylon waste is extruded with or without colorants or other additives into ribbons which are cut to provide the commercially available granules. Titanium dioxide is widely used in fibers to obtain a dull rather than glossy appearance.

Sizes are also necessary to facilitate handling of the yarn which varies in molecular weight but is often low compared to injection molding compounds. The high surface-to-volume ratio of yarn waste increases the possibility of chance contamination. These are all factors that pose problems for the reprocessor.

Glass-reinforced nylon is an important modification of growing interest. Long fibers, short fibers, and spheres have all been incorporated in nylon, and techniques for accomplishing this include impregnation of roving, extrusion compounding of short fibers, in-plant compounding, and blending with concentrates (102-104).

If the nylon is a soluble copolymer to be cast from solution, it is possible to add modifiers such as plasticizers directly to the nylon solution(105).

"Type 8" nylon is a name sometimes applied to Belding Heminway Company's "BCI" 800 series of nylon resins which are, however, not related to nylon-8 but are alkoxyalkylated nylon-66(106). These are made by acid catalyzed addition of formaldehyde to nylon-66 in the presence of an alcohol(107):

$$-NHCO- + CH_2O + ROH \rightarrow -NCO- + H_2O$$
$$\mid$$
$$CH_2OR$$

alkoxyalkyl nylon

The various resins in the 800 series correspond to different degrees of substitution on the nitrogen atom. Thermal stabilization adequate for molding at 165 to 210°C (329 to 410°F) can be achieved by adding 2 to 15% dicyandiamide, $H_2NC(=NH)NHCN(108)$.

A newly introduced product is "microcrystalline" nylon which involves particle sizes of 50 to 100 Å and exhibits properties of a colloid. Preparation involves a specific history of acid hydrolysis, treatment with a low molecular weight, aliphatic acid, and mechanical agitation(109).

Laboratory Methods

A number of techniques, not yet applied commercially, have been found to yield nylons. These include reaction of formaldehyde with dinitriles, the reaction of dicarboxylic acids and diisocyanates, the low-temperature condensation of oxalic esters with diamines, and others(110). Examples of the cited reactions follow:

$$NC(CH_2)_4CN + CH_2O \xrightarrow[H_2O]{acid} -NHCO(CH_2)_4CONHCH_2-$$

nylon-16

$$OCN(CH_2)_6NCO + HOOC(CH_2)_4COOH \longrightarrow$$

$$-NH(CH_2)_6NHCO(CH_2)_4CO- + CO_2$$
nylon-66

$$H_2N(CH_2)_6NH_2 + ROOCCOOR \longrightarrow -NH(CH_2)_6NHCOCO- + ROH$$
nylon-62

Details of a variety of laboratory syntheses are now available in convenient recipe form(111,112).

The reaction of acid chlorides and amines is widely useful(113) and is the only way to make many stiff polyamides that are unstable in the melt, for example,

2,5-DiMePipT

CHEMISTRY OF NYLONS

Polycondensation and Chemical Equilibria in Nylons

A condensation reaction is one which occurs with the loss of a small molecule, for example,

$$RCOOH + R'OH \rightleftharpoons RCOOR' + H_2O$$
acid   alcohol   ester

$$RX + R'ONa \longrightarrow ROR' + NaX$$
halide   alkoxide   ether

$$RCOOH + R'NH_2 \rightleftharpoons RCONHR' + H_2O$$
acid   amine   amide

The formation of large molecules by polycondensation requires that the reacting molecules contain at least two reactive or functional groups which may be the same in each of two different monomers (AABB-type) or may be different in a single molecule (AB-type):

$$n \text{ aA}-\text{Aa} + n \text{ bB}-\text{Bb} \longrightarrow \text{a} + (\text{A}-\text{AB}-\text{B}) {}_n\text{b} + (2n-1)\text{ab}$$

$$n \text{ aA}-\text{Bb} \longrightarrow \text{a} + (\text{A}-\text{B}) {}_n\text{b} + (n-1)\text{ab}$$

Examples of commercial polymers other than nylons made by AABB polycondensation are poly(ethylene terephthalate), polysulfones, and polycarbonates. In nonsolvent, bulk polymerizations, which are typical of nylons as noted in the preceding section, the small molecule (ab) lost in the reaction must be small enough to diffuse readily through and volatilize from the reaction mass. It also makes good economic sense not to lose too large a weight fraction of the monomers.

The reaction of an amine and acid to form an amide can be represented in a general way as follows:

$$\overline{\qquad} \text{NH}_2 + \text{HOOC}- \underset{\text{amidation}}{\overset{\text{hydrolysis}}{\rightleftharpoons}} -\text{NHCO}\overline{\qquad} + \text{H}_2\text{O}$$

Arrows point in both directions because the reverse reaction, hydrolysis, can also occur. Whether condensation or hydrolysis takes place depends on how the system is displaced from equilibrium. The equilibrium constant according to the mass action law is defined as the product of the concentration of the groups produced by the reaction divided by the concentration of the original species:

$$K_c = \frac{\left[\text{NHCO}\right]\left[\text{H}_2\text{O}\right]}{\left[\text{NH}_2\right]\left[\text{COOH}\right]} \tag{2-1}$$

An exact expression would involve the activities rather than concentrations, indicated by the brackets, but concentrations are readily accessible and activities are difficult to determine. If the ratio of concentrations on the right-hand side of Eq. (2-1) is different from K_c, then that reaction will occur which causes the ratio to approach K_c - condensation if the ratio is less than K_c and hydrolysis if the ratio is greater than K_c.

A nylon with useful properties as a plastic will normally have a molecular weight in excess of 10,000. For reason of simplicity, consider a nylon-6 with a molecular weight of exactly 11,334, which corresponds to 100 moles of caprolactam (molecular weight = 113.16) plus one mole of combined water to provide the end groups. The formula for this polymer is $\text{H}+\text{NH(CH}_2)_5\text{CO}+{}_{100}\text{OH}$. Free, uncombined water in a melt-processable nylon will be below 0.36 wt % or 2.3 moles of water per 100 moles of amide group. In a useful nylon, therefore, the ratio of amide groups to free and combined water is always high so that the concentration of amide groups is essentially constant for all practical purposes. Thus, it is whether the amount of water in the nylon is more or less than the equilibrium value that determines whether condensation to a higher molecular weight or hydrolysis to a lower molecular weight takes place.

Concentrations are conveniently expressed as the number of equivalents of each group present in one million grams of polymer. For the polymer described in the above paragraph, containing 0.36 wt % water, the amide-group concentration is $(10^6 \times 0.9964/11,334)$ 99 or 8705. $[COOH] = [NH_2] = 0.01$ $[CONH] = 87$, and $[H_2O] = 0.36 \times 10^4/18 = 200$ equiv./10^6 g. If the concentration of free water were zero and the combined water were ignored, the theoretically maximum amide-group concentration of $[CONH]_{max}$ would then be $10^6/113.16$ or 8840. In the simple aliphatic nylons, which constitute most of the commercial resins, the ratio of CH_2-groups to amide-group determines $[CONH]$:

Nylon	$CH_2/CONH$	Molecular weight per CONH	$[CONH]_{max}$ equiv./10^6g
6	5	113.16	8840
66	5	113.16	8840
610	7	141.21	7080
612	8	155.24	6440
11	10	183.30	5460
12	11	197.33	5070

The ratio $CH_2/CONH$ is n-1 in an AB-polymer, nylon-n, and $(n + m$-2$)/2$ in an AABB-polymer, nylon-nm.

At any given molecular weight the concentration of end groups is a constant, but $[CONH]$ decreases as $CH_2/CONH$ increases. Thus, if K_c is constant, $[H_2O]$ at equilibrium will decrease as $CH_2/CONH$ increases. At first glance this may appear to pose a problem for the nylons of high $CH_2/CONH_2$, but their lower affinity for water and their lower processing temperature tend to offset the decrease in the value of $[CONH]/[NH_2][COOH]$.

The amidation reaction is exothermic, which means that K_c decreases as temperature increases. The heat of amidation, ΔH_c, is (-2.303) $\times$ gas constant $\times$ slope of the log K_c versus $1/T^\circ$ (Kelvin) line (a negative sign indicates the evolution of heat). Values for ΔH_c cited in the literature vary(52), but Fukumoto's value(117) of -6.8 kcal mole^{-1} for the interval 235 to 272°C (445 to 522°F) was obtained for water concentrations below 0.5 wt % in nylon-6 and presumably applies to similar aliphatic nylons in real processing situations. This implies a change in K_c by a factor of 1.134 for a 10°C (18°F) change in temperature. The precise value of K_c is open to question because of discrepant results in the literature. A complication is the observation that K_c varies with the water concentration presumably because of changes in activities with the change in the polar nature of the medium(63,114-116). Fukumoto's work with nylon-6(117) indicates K_c at $272°C$ and 0.09 to 0.14 wt % water to be 270. Wiloth(114) indicated K_c for nylon-7 is greater than that for nylon-66, but Van Velden and co-workers(116) saw little difference in K_c for nylons-7 and -11.

The amidation equilibrium is of concern with any nylon whether made from a salt, a lactam, or otherwise, as long as amide functions and water are present. Other equilibria such as that of polycaprolactam with its monomer, discussed in detail in a later section on thermal degradation (p. 63), may also apply. In an earlier section on the preparation of nylon-6 from caprolactam (p. 19), three reactions were cited for consideration in lactam polymerizations — hydrolysis of the monomeric ring (M) to the linear aminoacid (L_1), condensation of one linear molecule (L_m) with another (L_n) to form a larger molecule (L_{m+n}) plus water, and addition of monomer to a linear molecule (L_m) forming a new linear molecule (L_{m+1}) larger by one monomer unit. The following equilibria apply:

$$K \text{ (hydrolysis)} = K_1 = \left[L_1\right] / \left[M\right]\left[H_2O\right] \tag{2-2}$$

$$K \text{ (condensation)} = K_2 = \left[L_{m+n}\right]\left[H_2O\right] / \left[L_m\right]\left[L_n\right] =$$

$$\left[L_{m+1}\right]\left[H_2O\right] / \left[L_m\right]\left[L_1\right] \tag{2-3}$$

$$K \text{ (addition)} = K_3 = \left[L_{m+1}\right] / \left[M\right]\left[L_m\right] \tag{2-4}$$

$$K_2 = K_3/K_1 \tag{2-5}$$

The use of $\left[L_1\right]$ in one expression for K_2 is based on the assumption that reactivity is independent of the size of the linear molecule and leads to the conclusion that K_2 equals the ratio of the lactam addition and hydrolysis equilibrium constants, K_3/K_1. K_2 is identical to the amidation equilibrium constant, K_c. All of the equilibrium constants depend on the water concentration to a greater or lesser degree, and Reimschuessel(52) has calculated the heats of reaction for caprolactam based on the maximum values for each constant at each temperature:

Reaction	K	ΔH
Ring hydrolysis	K_1	2.11 kcal mole^{-1}
Condensation	K_2 or K_c	-6.14
Addition	K_3	-4.03

The tendency of the linear nylon plastics to equilibrate with cyclic compounds is small except for polycaprolactam. This is associated with the difficulty with which rings containing more than seven atoms are formed and not with the stability of such rings. For example, dodecanolactam polymerizes to nylon-12 slowly because of ring stability, but there is little tendency of nylon-12 to regenerate the lactam monomer. Cyclic compounds are found, however, in small concentrations even in AABB nylons(62).

Kinetics

The rate of a reaction is proportional to the concentrations of the reacting species, and the constant of proportionality is the rate constant. The rate of disappearance of end groups in amidation, taking into account regeneration via hydrolysis, would appear to be second order:

$$-\frac{d\left[NH_2\right]}{dt} = -\frac{d\left[COOH\right]}{dt}$$

$$= k_a\left[NH_2\right]\left[COOH\right] - k_h\left[CONH\right]\left[H_2O\right] \qquad (2\text{-}6)$$

One actually needs to know the details of the steps involved in the reaction so that the slowest or rate-controlling step can be identified. If, for example, the amidation is acid catalyzed, then the acid functions in a dual role and the rate expression becomes third order:

$$-\frac{d\left[NH_2\right]}{dt} = k_a\left[NH_2\right]\left[COOH\right]^2 - k_h\left[CONH\right]\left[H_2O\right]\left[COOH\right]$$

$$(2\text{-}7)$$

In either case K_c equals the ratio of the rate constants for amidation and hydrolysis:

$$\text{at equilibrium,} \quad -\frac{d\left[NH_2\right]}{dt} = 0$$

$$\frac{k_a}{k_h} = \frac{\left[CONH\right]\left[H_2O\right]}{\left[NH_2\right]\left[COOH\right]} = K_c \qquad (2\text{-}8)$$

The rate constants, like K_c, vary with the water contents, and the literature contains conflicting claims. Evidence in support of catalyzed and uncatalyzed reactions and combinations of both can be found(63,128,129). Acid catalysis appears to be favored particularly at the high levels of conversion and low water contents characteristic of nylon plastics. The rate constant for addition of caprolactam to a linear molecule is about one hundred times faster than the rate constant for hydrolysis of the lactam and comparable to the rate constant for condensation at $220°C$ ($428°F$)(37,52,130). Polymerization of caprolactam proceeds therefore largely via the addition reaction.

Arrhenius plots of the logarithm of the rate constants versus reciprocal temperature yield activation energies. Values for amidation rate constants ranging from 11 kcal mole^{-1} for the second-order condensation of nylon-11 to 30 kcal mole^{-1} for an acid catalyzed reaction yielding nylon-66 in the presence of meta-cresol have been reported in a review provided by Korshak and Frunze(128). Still higher values are cited for amidations involving reactants containing benzene rings. The activation energy for hydrolysis of caprolactam and a polyamide resin in sulfuric acid solution at 70 to $118°C$ (158 to $244°F$) has been found to be 20 kcal mole^{-1} by Myagkov and Pakshver(131).

Extent of Reaction, Molecular Weight, and Imbalance

A frequently used parameter in condensation polymerization is the extent of reaction which is the fraction of A or B groups that has reacted. Consider our example of H [NH(CH$_2$)$_5$CO]$_{100}$OH again. Here, 99 out of 100 amine (or acid) groups have reacted to form amide. The extent of reaction, p, is 0.99. In general,

If: N_0 = initial number of molecules (AB or AA + BB)
        N = number of polymer molecules
Then: N_0 = number of A or B ends at start
        N = number of A or B ends in polymer

$$p = \frac{N_0 - N}{N_0} = 1 - \frac{N}{N_0} \qquad (2\text{-}9)$$

N_0/N is the average number of monomer molecules per polymer molecule or the number-average degree of polymerization, $\overline{P}_n$;

$$p = 1 - \frac{1}{\overline{P}_n} \quad \text{or} \quad \overline{P}_n = \frac{1}{1-p} \qquad (2\text{-}10)$$

$\overline{P}_n$ is one greater than the number of condensation reactions necessary to form the polymer molecule and is 100 in our example above. Ignoring the effect of the end groups on molecular weight and letting M_s represent the average molecular weight per CONH permit the simple relation $\overline{M}_n = M_s\overline{P}_n$, where $\overline{M}_n$ is the number-average molecular weight.

For the linear nylons, $\overline{M}_n$ can be determined by end-group analysis (Chapter 3) since the number of polymer molecules is half the sum of ends: $\overline{M}_n = 2 \times 10^6/([COOH] + [NH_2])$ and, if $[COOH] = [NH_2] = E$, $\overline{M}_n = 1 \times 10^6/E$. An imbalance of end groups results when, for example, RA or AA is intentionally added to AB or balanced AA-BB or in the event of end group decomposition. RA stands for a monofunctional reactant such as acetic acid which introduces one reactive and one unreactive (U) end. In general, as long as the polymer is linear,

$$\overline{M}_n = \frac{2 \times 10^6}{\left[COOH\right] + \left[NH_2\right] + \left[U\right]} \tag{2-11}$$

Imbalance limits the molecular weight that can be obtained. The maximum attainable $\overline{P}_n$ is limited by the fractional excess of ends, e, as shown below.

$$e = \frac{\text{excess A-ends} + \text{U-ends}}{\text{original no. of B-ends}} = \frac{\text{"total excess" ends}}{N_{B_0}}$$

Total ends at start = B-ends + equal no. of A-ends + "total excess"

$$= 2N_{B_0} + eN_{B_0} = N_{B_0}(2 + e)$$

The extent of reaction is defined for B-ends because only B-ends can be completely consumed:

$$p_B = p = \frac{N_{B_0} - N_B}{N_{B_0}} = 1 - \frac{N_B}{N_{B_0}},$$

where N_B = B-ends in polymer.

$$N_B = N_{B_0}(1 - p)$$

$$\text{Total ends in polymer} = 2N_B + eN_{B_0} = N_{B_0}\left[2(1-p) + e\right]$$

$$\overline{P}_n = \frac{\text{ends at start}/2}{\text{ends in polymer}/2} = \frac{2+e}{2(1-p)+e}$$

$$\text{If } p = 1, \qquad \overline{P}_n = \text{max.}\,\overline{P}_n = \frac{2+e}{e} \qquad\qquad (2\text{-}12)$$

For the case of RA or excess AA in AABB, $e = N_{RA}/N_{BB}$ or $(N_{AA} - N_{BB})/N_{BB}$, where N is the number of moles of the indicated species. If RA or AA is added to AB, then $e = 2N_{RA}/N_{AB}$ or $2N_{AA}/N_{AB}$. Suppose 1 mole % of acetic acid is added to 66-salt. Here, $e = 0.01$, and max. $\overline{P}_n = (2 + 0.01)/.01 = 201$. One mole percent of acetic acid added to caprolactam yields max. $\overline{P}_n = (2 + 0.02)/.02 = 101$. The factor-of-two difference is not a fundamental one but is due to the way in which the added acetic acid is normally calculated, that is, mole percent based on the ingredients which for AABB salt involves two monomer units. In discussing the preparation of nylon-66 (p. 17), we noted the effect of an excess of one monomer over the other by 1.25 mole %. In this case, $e = 0.0125$; max. $\overline{P}_n = (2 + 0.0125)/0.0125$ or 161; and $\overline{M}_n = 161 \times 113$ or 18,200.

The effect of imbalance on $\left[H_2O\right]_{eq.}$ is readily estimated as follows:

$$\text{Let:} \quad \Delta \;=\; \left[COOH\right] - \left[NH_2\right] = fT$$

$$T \;=\; \left[COOH\right] + \left[NH_2\right]$$

$$P \;=\; \left[COOH\right]\left[NH_2\right]$$

$$\text{Then:} \quad \Delta \;=\; T - 2\left[NH_2\right] = fT \text{ and } \left[NH_2\right] = \frac{T}{2}(1-f)$$

$$\left[COOH\right] = \left[NH_2\right] + fT = \frac{T}{2}(1+f)$$

$$P \;=\; \frac{T^2}{4}(1-f^2) \text{ and } P_0 \text{ (no imbalance) } = \frac{T^2}{4}$$

$$\frac{P}{P_0} = 1 - f^2 = \frac{\left[H_2O\right]_{eq.} \quad \text{for } \Delta = fT}{\left[H_2O\right]_{eq.} \quad \text{for } \Delta = 0} \tag{2-13}$$

In words, for a given sum of ends, that is, constant molecular weight, the product of ends for an imbalanced nylon divided by the product for a balanced polymer is unity less the square of the ratio (f) of the difference in ends (Δ) to the sum of ends (T). Because $\left[H_2O\right]_{eq.}$ changes in direct proportion to the product of ends, P/P_0, assuming that K_c is constant, $\left[H_2O\right]_{eq.}$ is significantly affected by imbalance only if f is large.

Molecular Weight Distribution

In polycondensation every reactive end group has the same probability of reacting independent of the size of the molecule to which it is attached. This means that nylons do not contain molecules of a single size but contain molecules of many sizes, and it is necessary to define average values. As noted above the molecular weight derived from end group analysis is a number-average, $\overline{M_n}$, which is directly proportional to the average number of monomer units in the polymer chain, $\overline{P_n}$.

Let: n_1 = no. of moles of monomer of mol. wt. = M_1

n_2 = no. of moles of dimer of mol. wt. = M_2

n_i = no. of moles of i-mer of mol. wt. = M_i

N = total no. of moles of polymer

W = weight of polymer

Then: $N = n_1 + n_2 + \cdots + n_i + \cdots = \sum_{i=1}^{i=\infty} n_i$

$W = n_1 M_1 + n_2 M_2 + \cdots + n_i M_i + \cdots = \sum_{i=1}^{i=\infty} n_i M_i$

$$\overline{M}_n = \frac{W}{N} = \frac{n_1 M_1}{N} + \frac{n_2 M_2}{N} + \cdots = \frac{\sum\limits_{i=1}^{i=\infty} n_i M_i}{\sum\limits_{i=1}^{i=\infty} n_i} \tag{2-14}$$

$$= f_{n_1} M_1 + f_{n_2} M_2 + \cdots = \sum\limits_{i=1}^{i=\infty} f_{n_i} M_i \tag{2-15}$$

where f_n = mole fraction. That is, $\overline{M}_n$ is the sum of the products of the *mole fraction* for each molecular weight multiplied by its molecular weight. Similarly,

$$\overline{P}_n = \sum\limits_i f_{n_i} i \tag{2-16}$$

Another averaging procedure involves the sum of the products of the *weight fraction* for each molecular weight multiplied by its molecular weight. If f_w is weight fraction, the weight-average molecular weight, $\overline{M}_w$, is

$$\overline{M}_w = f_{w_1} M_1 + f_{w_2} M_2 + \cdots = \frac{n_1 M_1}{W} M_1 + \frac{n_2 M_2}{W} M_2 + \cdots$$

$$= \frac{\sum\limits_{i=1}^{i=\infty} n_i M_i^2}{\sum\limits_{i=1}^{i=\infty} n_i M_i} \tag{2-17}$$

$\overline{M}_w$ can be determined by light scattering or ultracentrifugation which also yields the Z-average, $\sum n_i M_i^3 / \sum n_i M_i^2$.

A measure of the molecular weight distribution is the ratio of these average values, particularly $\overline{M}_w/\overline{M}_n$. Flory (118) has shown that the "most probable distribution" for high extents of reaction ($p \cong 1$) has a $\overline{M}_w/\overline{M}_n$ of 2.0. The probability that an end reacts to form an amide group is equal to the extent of reaction for balanced polymer, and the probability that an end has not reacted is $1 - p$. Every molecule has an end so we can start at one end and assess the probability that $(i - 1)$ reactions have occurred to yield an i-mer. This is the mole fraction. Equation (2-18) follows from the fact that the probability of several events occurring together equals the product of the probabilities for each event.

$$f_{n_i} = \text{mole fraction of } i\text{-mer} = p^{i-1}(1 - p) \qquad (2\text{-}18)$$

$$\sum_{i=1}^{i=\infty} f_{n_i} = (1 + p + p^2 + p^3 + \cdots)(1 - p) = \frac{1}{1-p}(1 - p) = 1$$

as it must. Since $\overline{P}_n = \sum_{1}^{\infty} f_{n_i} i$ [Eq. (2-16)],

$$\overline{P}_n = \sum_{1}^{\infty} i p^{i-1} (1 - p) = (1 + 2p + 3p^2 + \cdots)(1 - p)$$

$$= \frac{1}{(1-p)^2} (1 - p) = \frac{1}{1-p}$$

This is the same value for $\overline{P}_n$ more simply arrived at in Eq. (2-4). The probability that a monomer unit selected at random belongs to an i-mer is similarly assessed, remembering that it may occur at the end, in the middle, or in any of i-positions along the chain. This is the weight fraction of i-mer, f_{w_i}:

$$f_{w_i} = i p^{i-1} (1 - p)^2 \qquad (2\text{-}19)$$

$$\sum_{i=1}^{i=\infty} f_{w_i} = (1 + 2p + 3p^2 + \cdots)(1 - p)^2 = \frac{1}{(1-p)^2} (1 - p)^2 = 1$$

as it must.

$$\overline{P}_w = \text{wt. av. degree of polymerization} = \sum_{i=1}^{i=\infty} f_{w_i} i \qquad (2\text{-}20)$$

$$\overline{P}_w = \sum_{i=1}^{i=\infty} i^2 p^{i-1} (1-p)^2 = (1 + 4p^2 + 9p^3 + \cdots)(1-p)^2$$

$$= \frac{(1+p)}{(1-p)^3} (1-p)^2 = \frac{1+p}{1-p}$$

Thus:

$$\frac{\overline{M}_w}{\overline{M}_n} = \frac{\overline{P}_w}{\overline{P}_n} = \frac{1+p}{1-p} \bigg/ \frac{1}{1-p} = 1 + p \cong 2 \text{ for } p \cong 1 \qquad (2\text{-}21)$$

The distributions corresponding to Eqs. (2-18) and (2-19) are shown in Figures 2-2 and 2-3. Even at $p = 0.99$ the numerically most common i-mer is monomer, but the weight fraction is very low because, from Eq. (2-19) for $i = 1$, $f_{w_1} = (1-p)^2 = 0.0001$. The peak in the weight distribution curve occurs at $\overline{P}_n$ as shown below:

$$\frac{d(f_{w_i})}{di} = p^{i-1} (1-p)^2 + i(1-p)^2 p^{i-1} \ln p$$

At peak,

$$\frac{d(f_{w_i})}{di} = 0, \qquad i_{max} = -\frac{1}{\ln p}$$

For $p \cong 1$, $p = 1 - x$ where x is small, and $\ln p = \ln(1-x) = -x = p - 1$,

$$i_{max} = \frac{1}{1-p} = \overline{P}_n \qquad (2\text{-}22)$$

Solution viscosity is often used to estimate molecular weight via Eq. (2-23) where $[n]$ is the intrinsic viscosity (Chapter 3) and K and a are constants characteristic of the polymer and solvent.

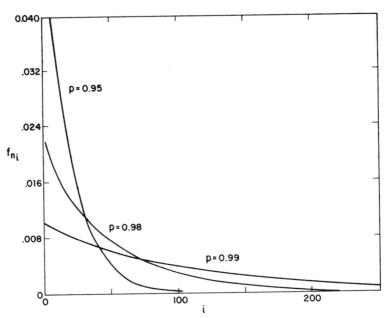

Fig. 2-2. Number fraction distributions of chain molecules in linear condensation polymers for several extents of reaction (p) (214).

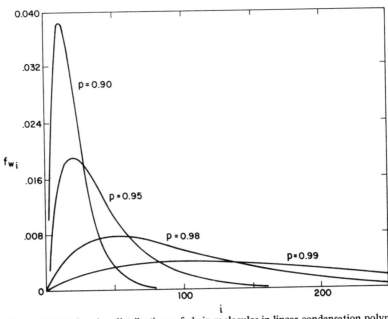

Fig. 2-3. Weight fraction distributions of chain molecules in linear condensation polymers for several extents of reaction (p) (214).

$$\left[\eta\right] = KM^a \tag{2-23}$$

Whether M is $\overline{M}_n$ or $\overline{M}_w$ depends on the technique, for example, end-group analysis or osmometry for $\overline{M}_n$ and light scattering for $\overline{M}_w$, used to establish the correlation and define the constants. The equation is limited also to the range of molecular weights for which it was originally established. If K and a have been determined from a series of polymer fractions of narrow molecular weight distribution, a viscosity-average molecular weight, $\overline{M}_v$, is defined which assumes that the viscosities of the polymer fractions are additive:

$$\overline{M}_v = \left(\sum_{i=1}^{i=\infty} f_{w_i} M_i^{a} \right)^{1/a} \tag{2-24}$$

$$\left[\eta\right] = K(\overline{M}_v)^a \tag{2-25}$$

In general, $\overline{M}_v \cong 0.90\,\overline{M}_w$(119-121), and $\left[\eta\right]$ on a whole polymer depends more nearly on $\overline{M}_w$ than $\overline{M}_n$. Nylons usually approximate the most probable distribution so that $\overline{M}_n \cong 0.5\,\overline{M}_w$(119-123).

Sweeny and Zimmerman (110) have provided a convenient summary of other considerations relating to molecular weight distribution in condensation polymers, much of which was developed by Flory (118). The distribution functions are also arrived at by a kinetic argument to show that the statistical approach given above is not limited to equilibrium situations. They also provide equations for $\overline{P}_n$ and $\overline{P}_w$ for condensation of monomers having more than two reactive groups per molecule and for condensations of an AB-monomer with a polyfunctional additive such as RB_f. In the latter case a narrower than "most probable" distribution is achieved at complete reaction and is equal to $1 + 1/f$. The distribution broadens quickly, however, as the extent of reactions departs from unity for reasonable amounts of additive, that is, amounts of RB_f which do not limit $\overline{P}_n$ excessively because of the imbalance.

Blends and Interchange

The $\overline{M}_w$ and $\overline{M}_n$ of blends are readily estimated (124).

$$\overline{M}_{n_B} = \frac{1}{(F_1/\overline{M}_{n_1}) + (F_2/\overline{M}_{n_2})} \tag{2-26}$$

$$\overline{M}_{w_B} = F_1 \overline{M}_{w_1} + F_2 \overline{M}_{w_2} \tag{2-27}$$

F_1, F_2 = wt fractions of polymers 1 and 2 in blend B. Dividing Eq. (2-27) by Eq. (2-26), assuming most probable distributions for each polymer, and rearranging lead to the following result which clearly defines the broadening of the distribution due to blending:

$$\frac{\overline{M}_{w_B}}{M_{n_B}} = 2 + \frac{F_1 F_2}{R} (R - 1)^2 \quad \text{where } R = \frac{\overline{M}_{n_2}}{\overline{M}_{n_1}} \tag{2-28}$$

The component polymers interact via amminolysis (-CONH- + H_2NR $\longrightarrow$ -CONHR + H_2N-) or acidolysis (RCOOH + -CONH- $\longrightarrow$ -COOH + RCONH-), both of which involve reaction of the amide function of one polymer with an end group of the other polymer. Even if the polymers are at equilibrium with the water present, some amide groups are hydrolyzing while others are being formed, so condensation of ends from different polymers contributes to interaction. Reaction of one amide group with another, "amide interchange," is another possibility (RCONHR + R'CONHR' $\longrightarrow$ RCONHR' + R'CONHR) which has been studied (125,126). Interaction leads ultimately to the same kind of random copolymer with a most probable distribution that results from polymerization of the mixed monomers (127).

Thermal Degradation

Thermal degradation implies different things to different people. It has meant for many investigators the rate of weight loss at temperatures far in excess of those encountered in normal processing. For some it has meant extensive analysis of volatiles and residual product after prolonged heating at temperatures more nearly comparable to processing temperatures where the goal has been elucidation of the chemistry of the degradation. For the processor it is the implications of time and temperature during fabrication for the quality of his product and for the utility of rework. A comparison of the common, commercial aliphatic nylons in this last and most practical sense is not available probably because thermal degradation is not a factor in processing these nylons except in unusual instances of excessively high (over 310°C or 590°F) temperatures or excessively long (hours or days) hold-up. The upper limit in processing temperature for the commercial nylons is not as well defined as it is for such polymers as poly(vinyl chloride), polyacetal, or poly(methyl methacrylate) where significant decomposition occurs at about 200, 240, and 270°C

(392, 464, and 518°F), respectively. An important factor with nylons is the amidation-hydrolysis equilibrium. As discussed earlier (p. 34), the equilibrium constant and the water in equilibrium with the original polymer decrease with increasing temperature. This means that hydrolysis becomes more likely with increasing molecular weight and temperature. It is frequently a lack of understanding of the interaction of moisture content, molecular weight, and processing conditions that leads to problems in the molding and extrusion of nylons.

Lack of analytical data on moisture content also mars the work on nylons subjected to degradative treatment. A summary of studies related to the thermal, non-oxidative degradation of nylons is provided in Table 2-1. Examination of these studies reveals, not surprisingly, that the two commercially most important nylons, 6 and 66, have received the most attention. Nylon-610, a few oxamides, nylon-7, and ring-containing polyamides have also been examined together with some model compounds, but systematic investigation of the effect of structural parameters on thermal stability is limited. One reason is the complexity of the problem. For example, ammonia was almost always found as a decomposition product of nylon-66 or -6 (132, 135, 136, 145, 153, 154, 157, 168) but was not observed in one case with nylon-6 (144) nor in the pyrolysis of 66/6 copolymers (134, 144), or 610 (168). Based on studies with model compounds it was asserted (142) that decomposition involves fission of each bond in and adjacent to the amide group. Others (144, 153) reported, however, that primary cleavage is at the C-N bond alpha to the amide carbonyl:

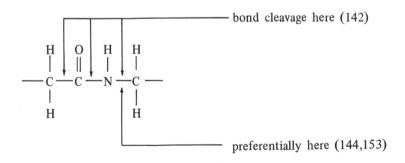

In support of preferential cleavage at the C-N bond, the bond strengths of a C-C bond and a C-N bond were cited as 80 and 66 kcal mole^{-1}, respectively. These values are, however, open to question if comparable structures are considered. For example, the bond dissociation energies for C-C in propane and C-N in the analogous dimethyl amine are reported as 82 and 84 kcal mole^{-1}, respectively (171). The CH_3-CH_2 bond strength of ethyl benzene ($C_6H_5CH_2-CH_3$) is

Table 2-1. Review of studies related to thermal degradation of nylons

Investigator (year)	Ref.	Description of experiment	Comment
Taylor (1947)	132	Analysis of off-gas from synthesis of nylon-66	Nitrogen bases and CO_2 in off-gas.
Korshak et al. (1951)	133	Determine rate of evolution of CO_2 from dibasic acids	Decomposition temp. increases with chain length. Acids with even number of carbon atoms more stable than those with odd number. See Fig. 2-4 below.
Achhammer et al. (1951)	134	Pyrolyze 30 min. at 10^{-6} mm Hg, 400°C: 66/6 (50/50 and 60/40) and 66/610/6 (60/20/20)	Mass spectrometer analysis of volatiles showed CO_2, CO, H_2O, cyclopentanone (C5K), and hydrocarbons. Ratio C5K/CO = 4/1, not 1/1. No NH_3.
Hasselstrom et al. (1952)	135	Pyrolyze nylon-66 at 280-400°C	Time to 10% wt loss of unscoured Oxford cloth: 18 hr at 280°C, 8 hr at 305°C, 1.5 hr at 330°C, 1.0 hr at 350°C. NH_3 and CO_2 in volatiles.
Hopff (1952)	136	Pyrolyze nylon-66	Volatiles included NH_3, CO_2, cyclopentanone, hexamethylenediamine.
Liquori et al. (1953)	137	Heat nylon-66 4-8 hr at 275°C	Note absorption at 290 $m\mu$ in acid solution; not due to ketodiacid or cyclopentanone.
Korshak et al. (1954)	138	Decomposition of dibasic acids in the presence of glycols	Glycols lower the decomposition temperature of dibasic acids.

Meacock (1954)	139	Heat nylons-66, -610, and extracted 6 about 20 hr in steam at 293°C	Follow change in rel. visc. and end groups. 66 and 610 relatively stable in visc. but 6 decreases due to formation of monomer. 610 and 6 relatively stable in ends but 66 shows increase in NH_2 ends and loss in COOH ends.
Hill (1954)	140	Review paper	Cites possibility of 2 NH_2 ends $\longrightarrow$ secondary amine + NH_3
Goodman (1954)	141	Pyrolyze model compounds 5 hr at 330-350°C	$RCONHR'$ and $RCONH(CH_2)_6NHCOR$ decompose very slightly to give a trace of nitriles. $BuNHCO(CH_2)_n CONHBu$ gives $BuNH_2$, hydrocarbons and CO. CO_2 formed only if $n = 4$.
Goodman (1955)	142	Analysis of pyrolysate of $BuNHCO(CH_2)_4 CONHBu$	Decomposition of N,N'dialkyladipamides yields same 290 mμ absorption as 66 (see Liquori et al. above). Pyridine, pyrrole, and cyclopentanone derivatives identified; also $BuNHCO(CH_2)_3 CH_3$, lesser amounts of others. Suggests decomposition involves fission of each bond in and adjacent to the amide group.
Edgar et al. (1958)	143	Reaction of cyclopentanone with amines at 300-350°C	Products similar in structure to those isolated by Goodman (above).

49

Table 2-1. (continued)

Investigator (year)	Ref.	Description of experiment	Comment
Straus et al. (1958)	144	Same as Achhammer (above). Also pyrolyze nylon-6 and nylon-4. Also study rate of volatilization at 310-380°C	Mass spectrometer analysis similar to that of Achhammer, no amines. For 66 containing polymers at 310-320°C max. rate occurs at 30-40% in accord with theory for random decomp; E_{act} = 14-42 kcal mole^{-1}. For 6 at 345-365°C max. rate at 20-30%; E_{act} = 27-34; max. rate lower than for 66 polymers; decomp. accelerated by acids. Suggests cleavage of C-N bond alpha to carbonyl; cites C-N and C-C bond strengths of 66 and 80 kcal, respectively.
Straus (et al.) (1959)	144	Pyrolysis and rates of volatilization (345-365°C) of nylon-6 of $\overline{M}_v$ = 60,000	Polymer was wet; reduced rate of decomp. by drying. Drying also raised E_{act} to 43 kcal mole^{-1}. Decomp. is nearly random.
Korshak et al. (1958)	145	Heat nylon-66 in N_2 at 300°C	20% wt. loss in 10 hr, about 2 wt. % NH_3, 1.5% CO_2. Mol. wt. dropped from 18,000, 35,000, or 55,000 to 7-11,000 in 4 hr.
Bailey et al. (1958)	146	Pyrolysis of $CH_3CON(Et)CH_2CH_2OCOCH_3$ at 465 and 490°C (vapor phase over glass helices)	Produce 57% $CH_3CON(Et)CH{=}CH_2$ + CH_3COOH at 465°C; 25% $CH_3CONHEt$ + $CH_2{=}CHOCOCH_3$ at 490°C. Latter decomp. involves H-atom gamma to amide carbonyl.

Bailey (1965)	147	Review article	Discusses thermal decomp. of unsaturated materials and cyclic mechanism involving H-atom gamma to unsaturation.
Smith (1958)	148	Depolymerization of nylon-6	Rate of monomer formation increases with temp., % H_2O, and lower initial mol. wt. Rate of formation in moles of monomer/ 10^6 g/hr at 250°C = 20-50 for % $H_2O \leqslant 0.2$. Acetylation reduces initial rate of formation.
Katorzhnov et al. (1959)	149	Depolymerization of nylon-6	Same conclusions as Smith above.
Cawthon et al. (1960)	150	Equilibration of nylon-6 and monomer above and below melting point (180-280°C)	Equilibrium amount of monomer varies with % crystallinity as well as temp.
Ogata (1961)	151	Depolymerization of end-capped nylon-6	Monomer not formed at 230°C and formed only slowly at 257°C if polymer end groups are COONa or NHCOCH₃.
Fukumoto (1961)	152	Equilibration of nylon-6	At equilibrium, vapor pressure of monomer (P) defined by equation: log P (in mm Hg) = $(-4.10 \times 10^3/T) + 9.6$.

51

Table 2-1. (continued)

Investigator (year)	Ref.	Description of experiment	Comment
Kamerbeek et al. (1961)	153	Heat nylon-66 and -6 under N_2 at 305°C and below	Nylon-66 gels in 5-6 hr at 305°C. NH_3 and CO_2 evolved in moles/mole of $-CONH-$ = 0.1 after 13 and 20 hr, 0.2 after 25 and 43 hr, and after 100 hr 0.57 and 0.33; thus, significantly more NH_3. No evidence of ketodiacids via reaction of two acid ends, but di-(ω-aminohexyl)amine from reaction of 2 amine ends with loss of NH_3 found. Residue shows loss of crystallinity, decrease in CONH, some $-C\equiv N$. Nylon-6 evolves after 120 hr 0.16 moles NH_3, 0.12 CO_2, and 0.14 H_2O per mole CONH at 305°C; 0.009, 0.019, and 0.040 at 270°C; 0.002, 0.005, and 0.031 at 257°C. Thus temp. dependence of formation of NH_3, CO_2, and H_2O differs. Viscosity decreases for 10 days at 281°C but get gel after 12 days. Di(ω-carboxypenty)amine does not cause gel, but 1,11-diamino-6-oxo-undecane does. Nitrile in gelled polymer about 5% of original CONH. Extensive analysis suggests nitrile formation via dehydration of amide formed by reaction involving gamma H-atom (see Bailey's work above) is principle degradation reaction.

Reference	No.	Description	Findings
Rafikov et al. (1962)	154	Pyrolysis of nylon-66 at 350-400°C in N_2	NH_3, CO, CO_2, hydrocarbons, cyclopentanone derivatives formed. Completely degraded after 5-6 hr at 400°C.
Knappe et al. (1963)	155	Determine degradation in processing and effect on mechanical strength. For nylon-610 use 68 sec cycle and 260°C nozzle.	Solution viscosity decreases with number of passes through machine. Melt index increases proportionately, but get much smaller change if hold 610 in melt indexer for length of time similar to hold-up in machine. Difference attributed to effect of shear, but water absorption and hydrolysis not clearly eliminated as factors in molding of rework.
Rafikov et al. (1964)	156	Degradation of nylon-7 at 300-350°C	"More resistant to high temp." than nylon-6 or -66.
Chelnokova et al. (1964)	157	Degradation of nylon-6 and -7 at 370-420°C	NH_3, CO_2, amines, acids, H_2O, nitriles formed. Reactions cited include depolymerization, hydrolysis, salt formation, nitrile formation, and recombination of degradation products.
Dine-Hart et al. (1964)	158	Weight loss of aromatic polyamides heated at 3°C/min	Decomposition temp. varies with orientation. For combination of diamine (o, m, or p) with diacid in order of increasing temp. of onset of rapid wt. loss: m-$o < o$-$o < p$-$o < o$-$m = o$-$p < m$-$m < p$-$m < m$-$p < p$-p.
Indest (1964)	159	Effect of metal surface on degradation	Cr plating reduces degradation of nylon-66.

53

Table 2-1. (continued)

Investigator (year)	Ref.	Description of experiment	Comment
Krasnov et al. (1964)	160	Pyrolysis at 275-400°C of nylons-102, 62, 66, 6T, MPD2, PPD2, and BB2. MPD = m-phenylene diamine PPD = p-phenylene diamine BB = 4,4'-diaminobiphenyl	Rate of decomposition depends on mode of polymerization: (a) gas phase, (b) melt, or (c) interfacial. E_{act} increases with temp: 102 (a) 13 kcal mole^{-1} below 315°C, 35 above 320°C 102 (b) 34.0 62 (a) 13.5 below 315, 66.9 above 315 66 (b) 25.8 below 295, 41 above 300 66 (c) 15.8 below 290, 37.2 above 300 6T (c) 15 below 330, 58 above 340 MPD2 (a) 34.4 PPD2 (a) 9.8 below 330, 33 above 340 BB2 (a) 11.9 below 360, 34.5 above 360
Krasnov et al. (1966)	161	Pyrolysis of isomeric aromatic polyamides	See Dine-Hart above. E_{act}, kcal mole^{-1}, as follows: m-m, 330-390°C, 33.8; p-m, 390-470°, 31.4; m-p, 410-480°, 42.4; p-p, 390-460°, 11.5 and 470-500°, 53.2. Temp. after which polymer is insoluble in H_2SO_4: 305°C for m-m, 360 for p-m or m-p, and 400 for p-p. NMR shows much greater mobility at 270-300°C in m-m. Similar decomp. products for all polymers: H_2, CO, CO_2, H_2O, HCN, benzene, toluene, benzonitrile, NH_3, others. Formation of isocyanate as intermediate in decomp. postulated.

Bruck (1965)	162	Thermal degradation of poly(*trans*-2,5-dimethyl-piperazine terephthalamide)(2,5-DiMePipT) and poly(*trans*-2,5-dimethylpiperazine oxamide)(2,5-DiMePip2)	Both polymers stable to 420°C. Above 420°C get much CO, also H_2O, CO_2, hydrocarbons, pyrazines, pyrroles. E_{act} 64 kcal mole⁻¹. Random, free radical degradation indicated. More resistant than aliphatic polyamides to acid hydrolysis at high (about 400°C) temp.
Bruck (1966)	163	Thermal degradation of poly(2-methylpiperazine terephthalamide)(2MePipT), poly(*trans*-2,5-dimethylpiperazine isophthalamide) (2,5-DiMePipI) and poly(piperazine terephthalamide)(PipT)	Order of stability: PipT > 2MePipT > 2,5-DiMePipI. As noted by Dine-Hart above T gives more stable polymer than I. Methyl substitution on Pip ring decreases stability.
Bruck (1966)	164	Pyrolysis of UV-irradiated 2,5-DiMePipI (see above) at 380-440°C	Irradiation increased thermal stability. E_{act} raised from 53 to 78 kcal mole⁻¹.
Bruck et al. (1967)	165	Pyrolysis of random and block piperazine copolyamides at 390-440°C	2,5-DiMePipT/2,5-DiMePipI random and block copolyamides show E_{act} of about 60 kcal mole⁻¹, but random polymer degrades 2-4 times faster.
Dobrokhotova et al. (1966)	166	Pyrolysis of nylon-6 up to 450°C	Visible evolution noted at 325°C. Polymer becomes branched. Based on volatiles, E_{act} = 18-21 kcal mole⁻¹. Polymer made with H_3PO_4 catalyst behaved similarly. Anionic polymer decomposed completely to monomer at 350°C.

Table 2-1. (continued)

Investigator (year)	Ref.	Description of experiment	Comment
Chevychelov (1967)	167	Degradation of nylon-6 under load	Not thermal degradation but of interest because it suggests that application of stress increases degradation.
Goldfarb et al. (1969)	168	Thermal degradation of nylons-66 and 610	Solution viscosity, end-group analysis, vapor pressure osmometry, and gel permeation chromatography used to characterize polymer. Wt. loss determined and volatiles analyzed by mass spectrometer. Inconsistencies among the analytical results prevent clear-cut conclusions. The solution viscosity data are probably the most dependable, but variations here suggest problems in drying. GPC shows significant changes: 610 at 311-327°C first shows an increase in high mol. wt. fraction but becomes bimodal with a dramatic decrease in the high mol. wt. fraction. 66 at 250-275°C shows a large increase in the high mol. wt. fraction. For both 66 and 610 rate of wt. loss depends on rate of heating. E_{act} varies with % wt. loss from 30 to 65 kcal mole^{-1}.

Author (year)	Ref.	Title	Description
Krasnov et al. (1969)	169	Thermal degradation of aromatic polyamides	Mass spec. shows NH_3, H_2O, CO, CO_2, cyclopentanone, hydrocarbons in 66 volatiles; H_2O, CO, CO_2, and hydrocarbons but no NH_3 in 610 volatiles.
Bruck (1969)	170	Pyrolysis of block piperazine copolyamides at 336-440°C	Introduction of "hinge" groups ($-SO_2-$, $-CH_2-$) and bulky side groups (phenyl, cyclohexyl) considerably decreases thermal stability. 2,5-DiMePipT/2,5-DiMePipI, 2,5-DiMePipI/2,5-DiMePip10, and 2,5-DiMePipI/2,5-DiMePip10 studied (see Refs. 162 and 163 above). Unlike homopolymers, stability influenced by mol. wt. of block copolymers.
Reimschuessel and Dege (1970)	215	Decarboxylation and desamination of nylon-6 at 250-290°C	Mechanisms for loss of carbon dioxide and ammonia from nylon-6 are discussed.
Peebles and Huffman (1971)	216	Thermal degradation of nylon-66 at 283-305°C	Rate of gel and color formation found to increase as rate of removal of volatile products was increased.
Wiloth (1971)	217	Thermal degradation of nylon-66 and nylon-610	Temperature dependence of degradation of nylons-66 and -610 compared.
Reimschuessel and Nagasubramanian (1972)	218	Rate of monomer formation in nylon-6	Rate data provided for monomer formation as function of time, temperature, and water concentration.

reported as 63 kcal mole^{-1}; the H_2N-CH_2 bond of the similar benzyl amine ($C_6H_5CH_2-NH_2$), 59 kcal mole^{-1} (172). Thus the C-C and C-N bond strengths appear to be much closer together than the 14-kcal difference cited above. Furthermore, the availability of a mechanism may be more important than the bond strength. In studying the decomposition of a compound that is both an ester and an amide, Bailey and Bird (146) observed that the C-O bond broke preferentially even though the dissociation energy of the C-O bond is about 10 kcal mole^{-1} higher than that of the C-N bond. (The dissociation energy for CH_3-OH is about 90 kcal mole^{-1}; for CH_3-NH_2, 80 kcal mole^{-1} (172).) The decomposition mechanism involves the gamma H-atom, and it can be argued that the hydrogen on the carbon alpha to the amide nitrogen is more labile than that attached to the carbon alpha to the oxygen, and it is this mechanistic feature that controls the decomposition:

$$\xrightarrow[57\%]{465°C} \quad CH_3CH_2-\overset{\overset{\displaystyle COCH_3}{|}}{N}-CH{=}CH_2 + HO\overset{\overset{\displaystyle O}{\|}}{C}CH_3$$

$$\xrightarrow[25\%]{490°C} \left[\begin{array}{c} CH_3CH_2N{=}\overset{\overset{\displaystyle}{|}}{C}-CH_3 \\ | \\ OH \end{array} \right] + CH_2{=}CHOCOCH_3$$

$$CH_3CH_2NHCOCH_3$$

The message here is that thermal decomposition is complex, and generalizations based on bond strengths or imperfectly understood mechanisms can be misleading. Many useful, empirical observations have been made, however, and these are discussed below.

Nylons that involve ingredients that cyclize readily are melt unstable. A glutaramide can imidize; for example,

$$
\begin{array}{ccc}
& \overset{\displaystyle O}{\underset{\displaystyle \parallel}{C}}\!-\!CH_2 & \\
& \diagup \qquad \diagdown & \\
-N & & CH_2 \\
\mid & & \\
\boxed{HOH} & \diagup & \\
\diagdown & & \\
& C\!-\!CH_2 & \\
& \parallel & \\
& O &
\end{array}
\qquad\longrightarrow\qquad
\begin{array}{ccc}
& CO\!-\!CH_2 & \\
& \diagup \qquad \diagdown & \\
-N & & CH_2 + H_2O \\
& \diagdown \qquad \diagup & \\
& CO\!-\!CH_2 &
\end{array}
$$

$$
\begin{array}{ccc}
& CO\!-\!CH_2 & \\
& \diagup \qquad \diagdown & \\
-N & & CH_2 \\
\mid & & \\
\boxed{H} & \diagup & \\
& & \\
\boxed{-\!NH\!-}\!CO\!-\!CH_2 & &
\end{array}
\qquad\longrightarrow\qquad
\begin{array}{ccc}
& CO\!-\!CH_2 & \\
& \diagup \qquad \diagdown & \\
-N & & CH_2 + H_2N- \\
& \diagdown \qquad \diagup & \\
& CO\!-\!CH_2 &
\end{array}
$$

This results in the generation of relatively unreactive ends and an essentially irreversible loss in molecular weight. A succinamide will behave similarly and yield a relatively stable five-membered imide ring. Polypyrrolidone, nylon-4, is unstable even below its melting point of 260°C because of its tendency to revert back to monomer, a five-membered ring (178). The decomposition temperature of dibasic acids increases with increasing size except for odd-even alternation (133, see Figure 2-4). It is interesting that this odd-even alternation familiar in melting points of dibasic acids (Figure 2-4) is also found in the melting points of nylons (170) (see Chapter 9, Figure 9-1). The presence of glycols lowers the temperature of decomposition of dibasic acids (138) and points again to the need to examine specific systems for reliable quantitative comparisons. It is not surprising in light of the acid stabilities that polyoxamides tend to decompose at temperatures of 280 to 300°C (536 to 572°F), which are permissible with nylons -66 and -610 (175).

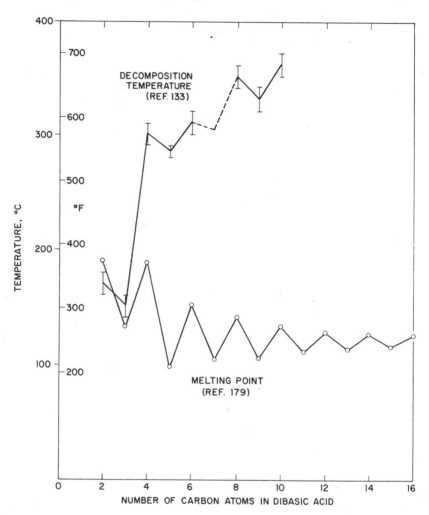

Fig. 2-4. Dependence of decomposition temperature and melting point on chain length for dibasic acids, $HOOC(CH_2)_n COOH$.

In examining simple compounds as models of nylon polymers, Goodman (141) found that N-alkylcarboxamides, $RCONHR'$, and N,N'-diacylhexamethylenediamines, $RCONH(CH_2)_6 NHCOR$, decompose very slightly at 330 to 350°C, but bisamides of dibasic acids, $RNHCO(CH_2)_n CONHR$, decompose to give RNH_2, CO, and hydrocarbon. CO_2 was formed only if $n = 4$ (adipic acid). Additional work with the adipamide revealed the presence of nitrogen-containing rings and cyclopentanone derivatives (142). Edgar (143) subsequently showed that the nitrogeneous compounds could arise from the reaction of cyclopentanone and amines. A cyclopentanone is generated by reaction of a carbonyl with a hydrogen activated by the other carbonyl of the adipic acid moiety:

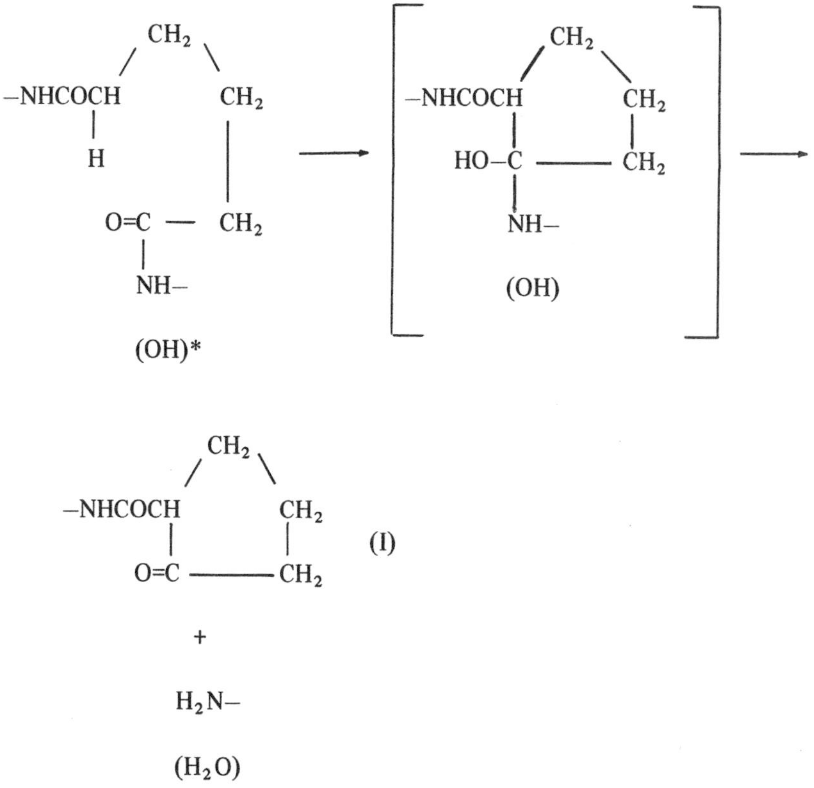

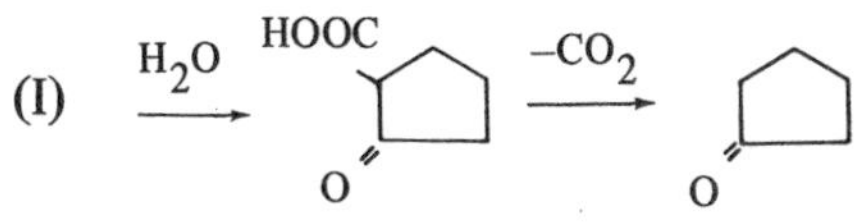

The presence of cyclopentanone or its derivatives in the degradation products of nylon-66 has been confirmed by others (144, 146, 154, 164, 178). The appearance of an absorption band at 290 mμ in acid solution is associated with a

*OH instead of NH− if at end of molecule.

reaction product of cyclopentanone (137, 142). Nylon-66 gels after about 6 hr in nitrogen at 305°C (22). There was no evidence (153) of reaction between acid ends (2 $-COOH \longrightarrow$ $-\overset{\overset{\displaystyle O}{\displaystyle \parallel}}{C}-$ $+ CO_2 + H_2O$), but there was evidence of reaction between amine ends (2 $-NH_2 \longrightarrow -NH- + NH_3$) in accord with an earlier conjecture (140). Gelation was ascribed to reaction of the thus formed secondary amine leading to branched structures:

$$-NH- + HOOC- \longrightarrow -\underset{\underset{\displaystyle \mid}{\underset{\displaystyle C=O}{\mid}}}{N}- + H_2O$$

The presence of a nitrile group ($C \equiv N$) in the degraded polymer was attributed to a C-N bond cleavage leading to an amide which then lost water (153). The C-N bond cleavage is consistent with Bailey's gamma hydrogen elimination mechanism:

$$-C \equiv N + H_2O$$

The importance of this mechanism in the decomposition of nylons is attested to also by the relative stability of polypivalolactam, which can be melt processed, and poly (β-aminoisovaleric acid), which is unstable in the melt (176):

polypivalolactam

$$
\begin{array}{ccccc}
CH_3 & & O & CH_3 & O \\
| & & \| & | & \| \\
-NHC-CH_2-C-NHC-CH_2-C- \\
| & & & | \\
CH_3 & & & CH_3
\end{array}
$$

H-atoms on C-atom gamma to C=O

poly(β-aminoisovaleric acid)

Kamerbeek et al (153) also carried out an extensive study of nylon-6 heated at 257 to 305°C (495 to 581°F) under nitrogen. The viscosity decreased for 10 days at 281°C (538°F), but gel was observed after 12 days. The reaction described above which leads to the formation of a nitrile group was thought to be the principal degradation reaction. No simple explanation of the gelation mechanism could be described although the addition of a ketodiamine such as might arise by the reaction of two carboxyl groups did induce gelation:

$$-NH(CH_2)_5COOH + HOOC(CH_2)_5NH- \longrightarrow -NH(CH_2)_5CO(CH_2)_5NH-$$

$$+ CO_2 + H_2O$$

The branching of thermally degraded nylon-6 was also noted in a subsequent study (166) which reported that the evolution of volatiles became visually observable at 325°C (617°F) and that anionically made polymer decomposed completely to monomer at 350°C (662°F). An interesting observation by Kamerbeek et al. (153) concerns the principal gaseous components after 120 hr of heating at different temperatures:

°C	NH$_3$	CO$_2$	H$_2$O
305	0.16*	0.12	0.14
270	0.009	0.019	0.04
257	0.002	0.005	0.03

*Moles per mole of CONH in the original polymer.

The change with temperature in the amount of gas evolved varies for the different gases. This indicates a multiplicity of reactions differing in activation energies and is additional evidence of the complexity of the degradation process. Consistent with this observation is the reported dependence on temperature of the activation energy for decomposition of nylon-66 and other nylons (160).

The equilibrium between nylon-6 and its cyclic monomer, caprolactam, is particularly important because nylon-6, at possible processing temperatures (for example, 250°C or 482°F), contains at equilibrium about 10% of a water extractable fraction most of which (about 8%) is monomer (148). This assumes, as is necessary in melt processing, a low moisture content because the equilibrium amount of monomer increases with increasing moisture (174). Nylon-6 is frequently extracted to remove the monomer and cyclic oligomers which amount to 20-25 wt % of the monomer depending on moisture content but not on temperature over the range, 220 to 265°C (428 to 509°F) (174). The rate of formation of monomer is, therefore, important and has been shown to increase with increasing temperature, decreasing molecular weight (because of higher concentration of end groups), and increasing water content (148, 149). The initial rate of formation is 20 to 50 moles of monomer per hour per 10^6 g of nylon-6 ($\overline{M}_n$ = 20,000 to 35,000) at 250°C or 482°F) (148). Monomer formation is drastically reduced by end capping, that is, by removing the amine and/or acid ends (148, 151). The vapor pressure of monomer in equilibrium with polymer has been defined (152; see Table 2-1). The equilibrium monomer content depends also on the degree of crystallinity (150). The variation of caprolactam monomer, not including the extractable oligomers, with temperature is shown in Figure 2-5.

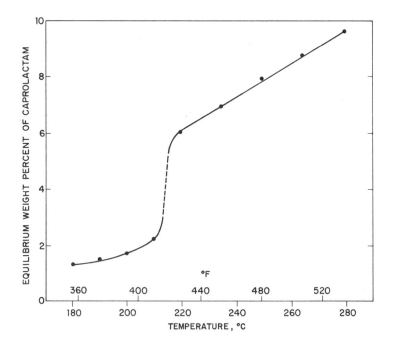

Fig. 2-5. Dependence of monomer content on temperature in polycaprolactam (data from Refs. 150 and 174).

Nylons-66, -610, and -6 were compared after prolonged heating in steam at 290°C or 559°F (139). Nylon-66 did not change much in solution viscosity, but the acid ends decreased and the amine ends increased. The loss of acid can be ascribed to cyclopentanone formation described above, and the increase in amine ends can be ascribed to hydrolysis. The retention of viscosity is perhaps attributable to the development of branching. Nylon-610 showed relatively little change. Nylon-6, which was originally low in monomer, showed no significant change in end groups but decreased in solution viscosity probably because of the formation of monomer.

A study of the effect of rework on nylon-610 has been reported (155), and an effect of shear on the rate of degradation was described. In light of the absence of data on moisture content and because of inadequate description of the handling of the rework, this study is open to some question. However, it is of interest that the deterioration of polycaprolactam is claimed to be accelerated by application of a load (167); free radical mechanisms are invoked so that oxidation may also be involved here.

Solution viscosity, end-group analysis, vapor pressure osmometry, and gel permeation chromatography (GPC) were all used in a recent study of the behavior of nylon-610 at 289 to 327°C (552 to 621°F) (168). Inconsistencies in the analytical results prevent simple conclusions; the GPC data are interesting, however, and suggest the development of a bimodal distribution after about 8 hr at 317°C or 603°F.

Nylon-7 was degraded at 370 to 420°C (698 to 788°F) and resembled nylon-6 in terms of the degradation products (157). Studies at 300 to 350°C (572 to 662°F) suggested that nylon-7 is more stable than either nylon-6 or -66 (156). This can be attributed to the absence of the cyclopentanone decomposition mechanism and to the low equilibrium monomer content.

The effect of end groups on the rate of generation of caprolactam and the total decomposition of anionically polymerized nylon-6 into monomer at 350°C have already been mentioned. The generalization that the method of synthesis can significantly affect thermal stability is supported by the work of Krasnov and co-workers (160). They showed that the rate of decomposition of poly (decamethylene oxamide), nylon-102, depended on whether it was made in the gas phase or in the melt; similarly, the decomposition of nylon-66 depended on whether it was made in the melt or by interfacial polymerization.

An additional complication worth mentioning is the question of catalysis by the material of construction of the processing equipment. It has been reported, for example, that chromium plating of extruder parts decreases the degradation of nylon-66 (159). It is possible that oxidation is a factor in this situation.

Thermogravimetric analysis (158) of interfacially prepared aromatic polyamides involving ortho, meta, and para diamines and diacids indicates the order of stability based on the temperature corresponding to the onset of catastrophic weight loss when heated in vacuum at a rate of about $3°C$ per minute as shown in Table 2-2.

Table 2-2. Onset of catastrophic weight loss in aromatic polyamides (Ref. 158)

Polyamide[a]	Temp. $(°C)^b$
m-o	200
o-o	250
p-o	260
o-m, o-p	300
m-m	400
p-m	425-550
m-p	450-550
p-p	600

[a] Letters represent orientation (ortho, meta, or para) in diamine and diacid, e.g.:

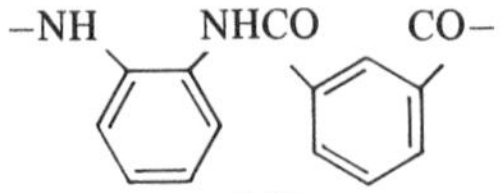

o-m

[b] Heating rate about $3°C/min.$

p-p

The temperatures are rough approximations but are adequate to show that stability increases from ortho to meta to para substitution. The data also suggest that substituent orientation in the acid part of the polymer is somewhat more important than in the diamine part. Similar experiments (168) on melt polymerized polymers indicate temperatures of about $390°C$ for nylon-66 and $410°C$ for nylon-610. The improved stability of para over meta isomers was again demonstrated in a subsequent study (161) which also ranked the polymers in order of the temperature required to make them insoluble in conc. sulfuric acid: m-m ($305°C$) < p-m or m-p ($350°C$) < p-p ($400°C$). Stability was conjectured to be associated with chain rigidity because only the m-m polymer showed considerable molecular mobility at $270-300°C$ based on nuclear magnetic resonance measurements (161, 177). The introduction of "hinge" groups such as methylene ($-CH_2-$) or sulfone ($-SO_2-$) and bulky side groups such as phenyl or cyclohexyl is said to reduce significantly the thermal stability (169).

Polyamides from piperazines and terephthalic acid are similarly more stable than those from isophthalic acid (163). 2,5DiMePip-2 and -T are stable to 420°C (162). UV irradiation increases the thermal stability of 2,5DiMePip-I (164). Random copolyamides from piperazines are somewhat less stable than block copolyamides (165), whose stability depends on molecular weight unlike the homopolymers (170).

Chemical Attack

In normal usage most nylons have excellent resistance to chemicals, but nylons can be made to participate in a variety of reactions. The site of attack may be anywhere in the molecule – the end groups, the amide nitrogen, the amide carbonyl, or the hydrocarbon portion, depending on the nature of the reaction. Examples of useful reactions and the purposes they serve are provided in Table 2-3. Other reactions such as oxidation, photolysis, and inadvertent hydrolysis are undesirable and have led to the development of modifications designed to minimize their effects.

The reactions of polyamides have been broadly covered in a recent review (180). The reaction products important to plastics applications are considered in subsequent discussions on properties (Chapter 10) and modifications (Chapter 11). Our remarks here are confined to general and mechanistic considerations on selected reactions. A discussion of solvent attack is included because it draws attention to another important aspect of the chemical structure of nylons—hydrogen bonding.

Table 2-3. Examples of reactions involving nylons

Site of attack	Reaction	Purpose
$-COOH$, $-NH_2$	Salt formation or reaction with isocyanate	Eliminate end groups to decrease rate of lactam monomer formation in nylon-6
O H ‖ ∣ ↗ $-C-N-$	Alkoxyalkylation	Lower crystallinity, increase solubility, and permit cross linking
O H ‖ ∣ $-C-N-$ ↑	Hydrolysis	Regeneration of intermediates
H ∣ ↗ $-CONHCH-$	Irradiation in the presence of a vinyl monomer	Grafting to alter properties

Solvents

The amide functions of nylons participate in hydrogen bonding, that is, the hydrogen on a nitrogen atom is associated with the oxygen atom of an adjacent molecule:

$$
\begin{array}{c}
\text{H} \\
| \\
\text{CH}_2 \diagdown \quad \diagup \text{N} \diagdown \\
\diagup \quad \diagdown \text{C} \quad \quad \text{CH}_2 \diagup \\
\quad \quad \| \\
\quad \quad \text{O} \\
\quad \quad \vdots \\
\quad \quad \text{H} \\
\quad \quad | \\
\diagdown \quad \diagup \text{N} \diagdown \quad \diagup \text{CH}_2 \diagdown \\
\text{CH}_2 \quad \quad \text{C} \\
\quad \quad \| \\
\quad \quad \text{O}
\end{array}
$$

Such bonds are relatively strong (dissociation energy $\cong 8$ kcal mole^{-1}), serve to tie the nylon molecules together, and play a very important part in the properties of nylons. They exist in the noncrystalline as well as crystalline regions of nylons and in copolymers as well as homopolymers (181). Increasing temperature facilitates the exchange of hydrogen bonding partners, particularly in the noncrystalline region, but the hydrogen bonding persists on the time scale of ordinary measurements, for example, infrared spectroscopy (181,182). Molecular configuration between hydrogen bonds is less ordered in the noncrystalline region, which is thereby more susceptible to attack. For example, equilibrium water absorption decreases with increasing crystallinity at a rate dependent upon the kind of nylon and the relative humidity (183). The sorption of alcohols by a nylon can be varied by as much as an order of magnitude depending on the crystallinity, the nylon, and the alcohol (186).

Disruption of the hydrogen bonds in the noncrystalline region is necessary for a solvent to attack nylons and is a major factor in the plasticizing effect of absorbed solvents. Solvents most likely to swell or dissolve nylons are those containing hydrogens attached to electron attracting groups which increase the positive character of the hydrogen and enable it to compete effectively with the amide hydrogen for association with the electronegative oxygen atom of the amide carbonyl group. Water, alcohols, and partially halogenated hydrocarbons such as chloroform are examples of solvents absorbed by nylons. Decreasing crystallinity via copolymerization and increasing temperature may combine to effect solution in solvents which normally only swell the homopolymers. The homopolymer, nylon-66, will dissolve in methanol if a pressure vessel is used to allow solution temperatures of 180°C (356°F), and 66 also dissolves in hot benzyl alcohol (boiling point 205°C).

Absorption of a solvent by a nylon diminishes as the hydrogen bonding capability of the solvent decreases. This accounts for the outstanding resistance of nylons to hydrocarbons, esters, ethers, ketones, and many other common solvents. On the other hand, the strongly acidic mineral acids, phenolic compounds, and fluorinated alcohols dissolve even homopolymers at room temperature unless diluted with nonsolvents such as water or acetone. Some solutions of salts such as calcium bromide, zinc chloride, or potassium thiocyanate, alcohol-salt combinations such as methanol-lithium chloride, and mixed solvents such as chloroform-methanol will attack nylons, particularly at higher temperatures. Extensive studies on the effect of metal salts on the stress cracking of nylon-6 (190) suggest that zinc, cobalt, copper, and manganese chlorides destroy the intermolecular hydrogen bonding of the nylon by forming metal complexes with the amide carbonyl. Water or alcohol is necessary to transport the metal ion to the amide, and increasing temperature hastens cracking by facilitating this transport. The structure of the complex is indicated to be

Cracking is attributed to increased stress caused by the localized swelling and plasticization arising from these complexes. The metal ion and not hydrogen ion concentration was shown to be controlling because an acid solution containing the same hydrogen ion concentration but not containing the metal ion was relatively ineffective. Different salts behave in different fashion. Lithium, calcium, and magnesium chlorides do not induce cracking in aqueous solution but are effective in alcohol solution. It is proposed by Dunn and Sansom (190) that in this latter case the metal salt is surrounded by fewer alcohol than water molecules, the charge density on each proton is therefore greater, and the alcohol-salt solution becomes a sufficiently good proton donor to attack the nylon in the same manner as a mineral acid or phenol.

Details on the solubility of individual nylons in specific solvents are available in trade literature and reviews (110, 184, 185). As expected, as the concentration of amide groups decreases, sensitivity to solvents changes. For example, nylon-6 and -66 dissolve in 90% formic acid at room temperature, and nylon-610 dissolves in hot formic acid, but nylon-11 is insoluble. On the other hand, nylon-11 absorbs at room temperature about twice as much benzene or toluene as nylon-610 which, in turn, absorbs more than nylon-6 or -66 (186).

Most solvents are reversibly absorbed and desorbed without harmful effect, but as noted above, fracture or delamination can sometimes occur under stress (186, 190).

In summary, percent crystallinity, amide group concentration, the nature of the solvent, especially its hydrogen bonding capability, temperature, and stress must all be considered in determining the precise effect of solvents on nylons.

Hydrolysis

The hydrolysis of nylon melts has been considered above in relation to the amidation-hydrolysis equilibrium. Its role in thermal degradation has also been noted, and it will reappear in discussing melt behavior in Chapter 4. Our concern here is with the hydrolysis of nylon parts at temperatures more likely to be encountered in end-use service.

One approach has been to define the reaction kinetics without the complication of a two-phase system by studying hydrolysis in acid solution. A polyamide resin hydrolyzed in 21.2M (40% by wt) sulfuric acid at 90 to 118°C (208 to 244°F) according to a first—order equation, that is, the concentrations of water and acid changed so little that only the concentration of amide groups in solution was a factor (131). The rate constants varied from 0.0025 to 0.011 and corresponded to an activation energy for hydrolysis of 20 kcal mole[-1]. The hydrolysis in 68.4M (92% by wt) sulfuric acid was too slow to be measured, presumably because of the low water concentration. Several studies, as reviewed by Ravens and Sisley (187), lead to the conclusion that the hydrolysis of nylons in acid solution is random, that is, all amide groups in the chain are equally susceptible to hydrolysis.

The same factors controlling solvent attack would be expected to influence hydrolysis. Stress corrosion studies (186) on injection molded discs in dilute acids at room temperature confirm that the $CH_2/CONH$ ratio, acid concentration, and degree of crystallization affect the loss in molecular weight observed after one year. The loss is greater with nylon-6 or -66 ($CH_2/CONH$ = 5/1) than nylon-610 or -8 ($CH_2/CONH$ = 7/1). The effect of more subtle factors such as percent crystallinity or some other parameter indicative of the degree of perfection of the molecular configuration is illustrated in the solubility of nylon-6 but not nylon-66 in 4.2M (14% by wt) hydrochloric acid (188). Similarly, nylon-6 was completely hydrolyzed, nylon-66 was less readily hydrolyzed, and nylon-610 proved the most difficult to hydrolyze in 8M (26% by wt) hydrochloric acid (189).

Injection molded bars of nylon-66 exposed to a pH 4 or pH 10 but not pH 7 solution suffer a loss in properties after a year at $70°C$ ($158°F$) (192), suggesting both acid and base catalysis. It has been reported that acid catalysis is observed only if the hydrogen ions exceed the number necessary for salt formation with the amine ends of the polymer (186). Stabilization toward hydrolysis by use of excess amine ends is claimed (193).

In general, nylons are more resistant to bases than to acids. An activation energy toward hydrolysis in aqueous sodium hydroxide of 17 kcal mole^{-1} has been reported (131). A sodium hydroxide solution (10% by wt or 2.8M) caused a loss in elongation of a nylon-66 bar after three months at $70°C$, but an ammoniacal solution (10% by wt or 5.6M) required a year at the same temperature (192).

Thermal Oxidation

The susceptibility of nylons to air oxidation at elevated temperatures in the dark as well as in the light is well documented (200). A variety of stabilizers permit extended life at higher temperatures (see Chapter 11).

A heat of combustion of -7.55 kcal g^{-1} can be calculated for nylon-66 using Goldstein's figure for the heat of formation of -784 cal g^{-1} (197). This agrees with values of -7.53 and -7.54 kcal g^{-1} determined on nylons-66 and -6 in this laboratory (202). The heats of combustions for nylon-610 and for nylon-11 are higher because of the higher $CH_2/CONH$ ratio: -8.28 and -8.79 kcal g^{-1} (202).

Several comparisons between exposure in air and inert environments have been made. For example, thermogravimetric analysis of a nylon from trimethyl-hexamethylenediamine and terephthalic acid shows a sharp break in the weight loss versus temperature curve at $300°C$ ($572°F$) in air and at $380°C$ ($716°F$) in nitrogen (194). Differential thermal analysis of nylon-66 fiber heated at $10°C/min$ in air shows the onset at $185°C$ ($365°F$) of an exotherm that is missing from the thermogram when the fiber is heated in nitrogen (201).

Injection molded bars of nylon-66 (191) became brittle after two years at $70°C$ ($158°F$) and in less than 2 hr at $250°C$ ($482°F$). An activation energy for oxidation of 22.2 kcal mole^{-1} was calculated. A value of 27 kcal mole^{-1} has also been reported (110). Embrittlement occurs in eight weeks at $70°C$ if the bar is wet. The acceleration of the effects of oxidation in the presence of moisture was also shown in fiber studies (195). The oxidized bars (191) decreased in molecular weight on the surface although an insoluble fraction was also observed at 200 and 250°C (392 and 482°F). The core of the bars often showed an increase in molecular weight. That surface attack was responsible for the embrittlement was established by demonstrating toughness in bars from which the surface was milled away. It was postulated that moisture facilitates diffusion of oxygen into the bar. The behavior of nylons-6, -610, and -11 was said to be similar to that of 66, but these nylons were described as less stable than 66. An independent study (196) indicated oxygen uptake to increase in the sequence: nylon-6 $<$ nylon-610 $<$ 66/610/6 copolymer.

Air oxidation of a nylon-66 fiber at 136 to 215°C (277 to 419°F) resulted in a loss in molecular weight although the acid end groups were unchanged and the amine ends actually decreased (199). Heating in nitrogen or in the presence of antioxidants increased molecular weight. Both chain-linking and chain-scission reactions are therefore to be expected in thermal oxidation, and the scission reactions cannot be explained on the basis of simple hydrolysis.

Valeric acid and adipic acid were the main products obtained from nylon-6 hydrolyzed after air oxidation at 200°C (392°F) (198). Both can be attributed to the decomposition of the hydroperoxide formed by attack of oxygen on the carbon atom adjacent to the nitrogen atom:

$$
\begin{array}{c}
\underset{|}{\overset{OOH}{}} \quad \underset{\parallel}{\overset{O}{}} \\
-NHCHCH_2(CH_2)_3C- \longrightarrow \underset{|}{\overset{O\cdot}{}} \quad \underset{\parallel}{\overset{O}{}} \\
-NHCHCH_2(CH_2)_3C- \quad +\cdot OH
\end{array}
$$

$-H_2O$

$$
\underset{\parallel}{\overset{O}{}} \quad \underset{\parallel}{\overset{O}{}} \\
-NHC(CH_2)_4C-
$$

$-NHCHO + \cdot CH_2(CH_2)_3C-$

hydrolysis

adipic acid

RH

$CH_3(CH_2)_3C- + R\cdot$

hydrolysis

valeric acid

The presence of amines in the hydrolyzed polymer indicated that another mode of oxidative attack also took place.

Photooxidation

The discoloration and embrittlement of unstabilized polyamides after room temperature exposure to air and sunlight is well known (211). Photodegradation occurs in the absence of oxygen if the incident radiation is sufficiently energetic (wavelength less than 3000Å) (203), but the combined effects of oxygen and light of higher wavelength are of greater interest because the solar radiation transmitted by the earth's atmosphere is nil below 3000 Å (205).

Analysis of the degradation products of N-pentylhexanamide, $CH_3(CH_2)_4CONH(CH_2)_4CH_3$, exposed to oxygen and weakly absorbed light of wavelength greater than 3000Å suggested the following chain reaction mechanism (204).

Initiation:

$$RCONHCH_2R' \xrightarrow{h\nu} RCO\cdot \ + \ \cdot NHCH_2R'$$

$$RCONHCH_2R' \begin{cases} \xrightarrow{RCO\cdot} RCON\dot{H}CHR' + RCHO \\ \xrightarrow{R'CH_2NH\cdot} RCON\dot{H}CHR' + R'CH_2NH_2 \end{cases}$$

Propagation:

Peroxide formation and regeneration of radical

$$RCON\dot{H}CHR' + O_2 \longrightarrow RCONHCHR'\overset{OO\cdot}{|}$$

$$RCONHCHR'\overset{OO\cdot}{|} + RCONHCH_2R' \longrightarrow RCONHCHR'\overset{OOH}{|} + RCON\dot{H}CHR'$$

Peroxide decomposition and regeneration of radical

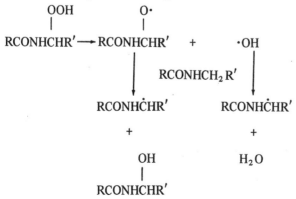

Formation of unsubstituted amide, aldehyde, and acid:

$$\underset{\overset{|}{\text{RCONHCHR}'}}{\overset{\text{OH}}{}} \longrightarrow \text{RCONH}_2 + \text{R}'\text{CHO}$$

$$\text{R}'\text{CHO} + \text{O}_2 \longrightarrow \text{R}'\text{COOH}$$

Support for the cleavage reaction postulated in the initiation step has come from low-temperature studies of the free radicals found by irradiation of nylon-6 with light of wavelength above 2500 Å (212). The existence of hydroperoxides in irradiated nylon-66 has been demonstrated quantitatively (210). Analysis of the degradation products from photooxidation of nylon-66, nylon-610, nylon-106, and nylon-6 indicate that the chemistry involved in the photooxidation of model amides applies also to the nylon polymers (203).

The above mechanism suggests that substitution on the carbon atom adjacent to the nitrogen atom should improve resistance to photooxidation, and this was shown to be true (204).

The development of a sensitive test for peroxides led to the following conclusions on photooxidized nylon-66 yarn (210): Tenacity decreases and peroxide concentration increases more rapidly as the incident light decreases in wavelength; the development of color accompanies peroxide decomposition; peroxide decomposition occurs in the absence of light above about $100°C$; and prior thermal history can affect the rate of loss in tenacity and gain in peroxide content.

Traces of iron (10 to 50 ppm) sensitize photodegradation (210), and titanium dioxide accelerates photooxidation by a chemical rather than energy transfer mechanism (209).

A series of studies (206, 207, 208) on irradiation of nylon-66 fiber with 2537-Å ultraviolet light in nitrogen and in vacuo are of interest because they suggest that the rate of deterioration and the relative amount of cross-linking and chain-scission reactions can be affected by the rate at which gaseous products of irradiation diffuse out of the polymer. The effect of high-energy radiation on polyamides has been reviewed by Zimmerman (213).

REFERENCES

1. Cook, J.C., *Handbook of Textile Fibres*, Textile Book Service, Metuchen, N.J., 1968, 4th ed., Vol. 2, pp. 324-340.
2. Hardy, J.V.E., in *The Encyclopedia of Basic Materials for Plastics*, H.R. Simonds and J.M. Church, Eds., Reinhold Publishing Corp., New York, 1967, p. 293.
3. Anon., *Chem. Eng. News* 44 (23), 17 (June 6, 1966).
4. Sherwood, P.W., *Melliand Textilber* 43, 477 (1962).

5. Anon., Chem. Eng. News 45 (42), 58 (Oct 2, 1967); Ibid. 41 (41), 69 (Oct. 14, 1963).

6. Baizer, M.M., J. Electrochem. Soc. 111, 215 (1964).

7. Hurd, R.M., Hydroc. Process. 43 (11), 154 (1964).

8. Sherwood, P.W., Mod. Textiles Mag. 43 (9), 26 (1962).

9. Ito, Y., Ann. N.Y. Acad. Sci. 147, 618 (1969).

10. Aikawa, K., Hydroc. Process. 43 (11), 154 (1964).

11. Muench, W., L. Notarbartolo, and G. Silvestri (to Snia Viscosca), U.S. Patent 3,022,291 (Feb. 20, 1962).

12. Anon., Chem. Eng. News 45 (36), 82 (Aug 28, 1967); Taverna, M., and M. Chiti, Hydroc. Process. 49 (11), 137 (1970).

13. Starcher, P.S., and B. Phillips, J. Am. Chem. Soc. 80, 4079 (1958).

14. Anon., Chem. Eng. News 45 (26), 52 (June 19, 1967).

15. Carothers, W.H. (to E.I. du Pont de Nemours and Co.), U.S. Patent 2,130,523 (Sept. 20, 1938).

16. Coffman, D.D., G.J. Berchet, W.R. Peterson, and E.W. Spanagel, J. Polym. Sci. 2, 306 (1947).

17. Peterson, W.R. (to E.I. du Pont de Nemours and Co.), U.S. Patent 2,174,527 (Oct. 3, 1939).

18. Graves, G.D. (to E.I. du Pont de Nemours and Co.), U.S. Patent 2,289,774 (July 14, 1942); Johnson, W.G. (to E.I. du Pont de Nemours and Co.), U.S. Patent 3,491,177 (Jan. 20, 1970).

19. Bannerman, D.G., and E.E. Magat in Polymer Processes, C.E. Schildknecht, Ed., Interscience Publishers, Inc., New York, 1956, pp. 247-254.

20. Kleinert, C.J., and F.F. Hoy in Manufacture of Plastics, W.M. Smith, Ed., Reinhold Publishing Corp., New York, 1964, Vol. 1, pp. 512-534.

21. Hopff, H., in Man-Made Fibers. H.F. Mark, S.M. Atlas, and E. Cernia, Eds., Interscience Publishers, a division of John Wiley and Sons, New York, 1968, Vol. 2, pp. 182-190.

22. Flory, P.J. (to E.I. du Pont de Nemours and Co.), U.S. Patent 2,244,192 (June 3, 1941).

23. Sperati, C.A. (to E.I. du Pont de Nemours and Co.), U.S. Patent 2,669,556 (Feb. 16, 1954).

24. Caldwell, J.R., and R. Gilkey (to Eastman Kodak Co.), Brit. Patent 1,137,151 (Dec. 18, 1968).

25. Carothers, W.H. (to E.I. du Pont de Nemours and Co.), U.S. Patent 2,130,948 (Sept. 20, 1938).

26. Muench, W., L. Notarbartolo, and G. Messina (to Snia Viscosa), U.S. Patent 3,017,394 (Jan. 16, 1962).

27. Lum, F.G. (to California Research Corp.), U.S. Patent 2,987,506 (June 6, 1961).

28. Taul, H., and F. Wiloth (to Vereinigte Glanzstoff-Fabriken A.G.), U.S. Patent 3,027,355 (Mar. 27, 1962).

29. Clemo, P.F., J.A. Briggs, W. Wilson, and J.A. Carter (to British Nylon Spinners Ltd.), U.S. Patent 3,185,672 (May 25, 1965); Carter, J.A. (to Brit. Nylon Spinners), U.S. Patent 3,193,535 (July 6, 1965); Griffiths, G.D. (to Brit. Nylon Spinners), U.S. Patent 3,258,313 (June 28, 1966); Carter, J.A., and G.J. Tyler (to Brit. Nylon Spinners), U.S.

Patent 3,475,387 (Oct. 28, 1969); Stanistreet, H.P. (to I.C.I. Fibres, Ltd.), Brit. Patent 1,140,526 (Jan. 22, 1969).

30. Anon., *Chem. Eng. News* 43 (26), 49 (June 28, 1965).

31. Sovereign, G.W. (to Monsanto Co.), U.S. Patent 3,218,297 (Nov. 16, 1965).

32. Flory, P.J. (to E.I. du Pont de Nemours and Co.), U.S. Patent 2,172,374 (Sept. 12, 1939).

33. Monroe, G.C., Jr. (to E.I. du Pont de Nemours and Co.), U.S. Patent 3,031,433 (April. 24, 1962)).

34. Werner, A.C. (to Celanese Corp.), U.S. Patent 3,232,909 (Feb. 1, 1966).

35. Zimmerman, J., *J. Polym. Sci.* 2B, 955 (1964).

36. Ogata, N., *J. Polym. Sci.* 1A, 3151 (1963).

37. Hermans, P.H., D. Heikens, and P.F. van Velden, *J. Polym. Sci.* 30, 81 (1958).

38. Kruissink, Ch. A., G.M. van der Want, and A.J. Staverman, *J. Polym. Sci.* 30, 67 (1958).

39. Heikens, D., and P.H. Hermans, *J. Polym. Sci.* 44, 429 (1960).

40. Heikens, D., P.H. Hermans, and G.M. van der Want, *J. Polym. Sci.* 44, 437 (1960).

41. Papero, P.V., Jr., O.E. Snider, and R.J. Duggan (to Allied Chemical Corp.), U.S. Patent 3,090,773 (May 21, 1963).

42. Twilley, I.C., D.W.H. Roth, Jr., and R.A. Lofquist (to Allied Chemical Corp.), Can. Patent 823,290 (Sep. 16, 1969).

43. Reynolds, R.J.W., in *Fibres from Synthetic Polymers*, R. Hill, Ed., Elsevier Publishing Co., New York, 1953, p. 132.

44. Anon., *Chem. Eng.* 73 (23), 178 (Nov. 7, 1966).

45. Sbrolli, W., in *Man-Made Fibers*, H.F. Mark, S.M. Atlas, and E. Cernia, Eds., Interscience Publishers, a division of John Wiley and Sons, New York, 1968, Vol. 2, pp. 239-245.

46. Bergeijk, J. V. (to American Enka Corp.), U.S. Patent 3,450,679 (June 17, 1969).

47. Kjellmark, E. W., Jr. (to E. I. du Pont de Nemours and Co.), U.S. Patent 3,015,651 (Jan. 2, 1962).

48. Oya, S., M. Tomioka, and T. Araki, *Chem. High Polymers (Japan)* 23, 415 (1966).

49. Stehlicek, J., J. Sebenda, and O. Wichterle, *Coll. Czech. Chem. Comm.* 29, 1236 (1964).

50. Saunders, J., *J. Polym. Sci.* 30, 479 (1958).

51. Illing, G., *Mod. Plast.* 46 (8), 70 (1969).

52. Reimschuessel, H. K., in *Ring-Opening Polymerization*, K. C. Frisch and S. L. Reegen, Eds., Marcel-Dekker, New York, 1969, pp. 303-326.

53. Hall, H. K., Jr., publication in preparation.

54. Wolff, I. A., in *The Encyclopedia of Basic Materials for Plastics*, H. R. Simonds and J. M. Church, Eds., Reinhold Publishing Corp., New York, 1967, p. 129.

55. Greene, J. L., Jr., E. L. Huffman, R. E. Burks, Jr., W. C. Sheehan, and I. A. Wolff, *J. Polym. Sci. Part A-1*, 5, 391 (1967).

56. Cowan, J. C., *J. Am. Oil Chem. Soc.* 39, 534 (1962).

57. Anon., Tech. Bull. No. 454A, Emery Industries, Inc., Organic Chemical Division, Dec., 1969.

58. Anon., *Hydroc. Process.* 46 (11), 244 (1967).

59. Gabler, R., H. Muller, G. E. Ashby, E. R. Agouri, H.-R. Meyer, and G. Kobas, *Chimia* **21** (2), 65 (1967).

60. Lum, F. G. (to California Research Corp.), U.S. Patent 2,997,463 (Aug. 22, 1961).

61. Bier, G., in *Addition and Condensation Polymerization Processes*, Advances in Chemistry Series, No. 91, Am. Chem. Soc., Washington, D.C., 1969, p. 612.

62. Zahn, H., and G. B. Gleitsman, *Angew. Chem. Int. Ed. Engl.* **2**, 410 (1963); Anon., U.S. Dept. Agricul. Publication CA-71-28, June, 1968.

63. Giori, C., and B. T. Hayes, *J. Polym. Sci., Part A-1*, **8**, 335 (1970).

64. Freidlina, R. K., and S. A. Karapetyan, *Telomerization and New Synthetic Materials*, Pergamon Press, London, 1961.

65. Mark, H. F., and S. M. Atlas, *Chem. Eng.* **68** (25), 143 (1961).

66. Horn, C. F., B. T. Freure, H. Vineyard, and H. J. Decker, *J. Appl. Polym. Sci.* **7**, 887 (1963).

67. Coffman, D. D., N. L. Cox, E. L. Martin, W. E. Mochel, and F. J. Van Natta, *J. Polym. Sci.* **3**, 85 (1948).

68. Anon., *Chem. Eng. News.* **39** (15), 91 (April 15, 1961).

69. Dachs, K., and E. Schwartz, *Angew. Chem. Int. Ed. Engl.* **1**, 430 (1962).

70. Krapf, H., *Angew. Chem. Int. Ed. Engl.* **5**, 652 (1966).

71. Ushakov, S. N., and O. F. Solomon, *Bull. Acad. Sci. USSR Div. Chem. Sci.* **1954**, 593; *C.A.* 49, 10868d (1955).

72. Heimbach, P., P. W. Jolly, and G. Wilke, in *Advances in Organometallic Chemistry*, F.G.A. Stone and R. West, Eds., Academic Press, New York, 1970, Vol. 8, pp. 50-59.

73. Cubbon, R. C. P., *Polymer* **4**, 545 (1963).

74. Anon., *Chem. Eng. News* **44** (53), 38 (Dec. 26, 1966).

75. Anon, *Chem. Week* **101** (13), 77 (Sep. 23, 1967).

76. Pryde, E. H., and J. C. Cowan, *J. Am. Oil Chem. Soc.* **39**, 496 (1962).

77. Anders, D. E., E. H. Pryde, and J. C. Cowan, *J. Am. Oil Chem. Soc.* **42**, 824 (1965).

78. Miller, W. R., E. H. Pryde, D. J. Moore and R. A. Awl, *Am. Chem. Soc. Div. Org. Coatings Plast. Preprints* **27** (2), 160 (1967).

79. Genas, M., *Angew, Chem.* **74**, 535 (1962).

80. Notarbartolo, L., *Ind. Plastiq. Mod.* **10**, 44 (1958).

81. Genas, M. (to Societe-Organico), Fr. Patent 951,924 (Apr. 25, 1949).

82. Wilke, G., *J. Polym. Sci.* **38**, 45 (1959).

83. Strauss, G., *Chem. Eng.* **76** (17), 106 (July 28, 1969).

84. Griehl, W., and D. Ruestein, *Ind. Eng. Chem.* **62** (3), 16 (1970).

85. Anon., *Chem. Eng. News* **40** (17), 57 (April 23, 1962).

86. Anon., *Chem. Eng. News* **47** (47), 14 (Nov. 10, 1969).

86a. Anon., *Chem. Eng. News* **48** (36), 12 (Aug. 31, 1970).

87. Anon. (to Badische Anilin- und Soda-Fabrik), Ger. OLS 1,495,147 (Jan. 30, 1969); Kunde, J. H. Wilhelm, H. Metzger, and H. Dorfel (to Badische Anilin- und Soda-Fabrik), Ger. AS 1,495,149 (Apr. 9, 1970); Anon. (to Badische Anilin- und Soda-Fabrik), Ger. OLS 1,495,199 (Feb. 13, 1969).

88. Anon. (to Imperial Chemical Industries, Ltd.), Fr. Patent 1,565,239 (Mar. 17, 1969); McGrath, H. (to Imperial Chemical Industries, Ltd.), Ger. OLS 1,907,034 (Sep. 11, 1969).

89. Kestler, J., *Mod. Plast.* **45** (2), 86 (1968).

90. Schaaf, S., *Faserforsch. Textiltech.* **10**, 328 (1959).

91. Anon. (to Badische Anilin-und Soda-Fabrik), Fr. Patent 1,104,719 (Nov. 23, 1955).

92. Aelion, R., *Ind. Eng. Chem.* **53**, 826 (1962).

93. Kolthoff, I.M., and V. A. Stenger, *Volumetric Analysis,* Interscience Publishers, Inc., New York, 1942, Vol. 1, 2nd rev. ed., p. 23.

94. Bernhardt, E.C., O. M. Hahn, and J. E. Hansen (to E. I. du Pont de Nemours and Co.), U.S. Patent 3,040,005 (June 19, 1962).

95. Stamatoff, G. S. (to E. I. du Pont de Nemours and Co.), U.S. Patent 2,705,227 (Mar. 29, 1955).

96. Foster, S. P., and R. W. Peterson (to E. I. du Pont de Nemours and Co.), U.S. Patent 2,875,171 (Feb. 24, 1959).

97. Symons, N. K. J. (to E. I. du Pont de Nemours and Co.), U.S. Patent 2,868,757 (Jan. 13, 1959).

98. Brossman, P. D., and E. H. Price (to E. I. du Pont de Nemours and Co.), Can. Patent 613,255 (Jan. 24, 1961).

99. Anon., *Plastics (London)* **27**, 135 (May, 1962).

100. Simpson, J. E., *Mod. Plast.* **40** (3), 94 and 142 (Nov., 1962); *ibid.* **40** (4), 90 (Dec., 1962) and **40** (7), 88 (Mar., 1963).

101. Vaala, G. T. (to E. I. du Pont de Nemours and Co.), U.S. Patent 2,378,667 (June 19, 1945).

102. Hunt, R. E., *Plast. Technol.* **15** (12), 38 (1969).

103. Riley, M. W., *Plast. Technol.* **13** (11), 27 and 33 (1967); Best, J. R., and R. P. Wood, *ibid.,* 30 (1967); Striebel, J. D., *ibid.,* 35 (1967).

104. Rosato, D. V., in *Handbook of Fiberglass and Advanced Plastics Composites,* G. Lubin, Ed., Van Nostrand Reinhold Co., New York, 1969, Chap. 6.

105. Vaala, G. T. (to E. I. du Pont de Nemours and Co.), U.S. Patent 2,456,344 (Dec. 14, 1948).

106. Anon., Belding Chemical Industries Technical Bulletin VIII-A.

107. Cairns, T. L., H. D. Foster, A. W. Larchar, A. K. Schneider, and R. S. Schreiber, *J. Am. Chem. Soc.* **71**, 651 (1949).

108. Strain, D. E. (to E. I. du Pont de Nemours and Co.), U.S. Patent 2,758,985 (Aug. 14, 1956).

109. Battista, O. A. (to FMC Corp.), U.S. Patent 3,299,011 (Jan. 17, 1967).

110. Sweeny, W., and J. Zimmerman in *Encyclopedia of Polymer Science and Technology,* Interscience Publishers, a division of John Wiley and Sons, New York, 1969, Vol. 10, pp. 483–597.

111. Sorenson, W. R., and T. W. Campbell, *Preparative Methods of Polymer Chemistry,* 2nd ed., Interscience Publishers, a division of John Wiley and Sons, New York, 1968.

112. Elliott, J.R., Ed., *Macromolecular Syntheses,* John Wiley and Sons., Inc., New York, 1966, Vol. 2; Overberger, C.G., Ed., *ibid.,* Vol. 1 (1963).

113. Morgan, P. W. *Condensation Polymers: By Interfacial and Solution Methods,* Interscience Publishers, a division of John Wiley and Sons, New York, 1965.

114. Wiloth, F., *Makromol. Chem.* **15**, 98 (1955).

115. Wiloth, F., *Z. Physik. Chem.* **5**, 66 (1955).

116. Van Velden, P. F., G. M. Van Der Want, D. Heikens, Ch. A. Kruissink, P. H. Hermans, and A. J. Staverman, *Rec. Trav. Chim.* **74**, 1376 (1955).

117. Fukumoto, O., *J. Polym. Sci.* **22**, 263 (1956).

118. Flory, P. J., *Principles of Polymer Chemistry*, Cornell University Press, Ithaca, 1953, Chaps. 2,3,8, and 9.

119. Taylor, G. B., *J. Am. Chem. Soc.* **69**, 638 (1947).

120. Howard, G. J., *J. Polym. Sci.* **39**, 548 (1959).

121. Burke, J. J., and T. A. Orofino, *J. Polym. Sci., Part A-2*, **7**, 1 (1969).

122. Hermans, P. H., D. Heikens, and P. F. Van Velden, *J. Polym. Sci.* **16**, 451 (1955).

123. Wiloth, F., *Makromol. Chem.* **14**, 156 (1954).

124. Flory, P. J., *J. Am. Chem. Soc.* **64**, 2205 (1942).

125. Korshak, V. V., and T. M. Frunze, *Synthetic Hetero-Chain Polyamides*, Daniel Davey and Co., Inc., New York 1964, pp. 87–92.

126. Beste, L. F., and R. C. Houtz, *J. Polym. Sci.* **8**, 395 (1952).

127. Brubaker, M. M., D. D. Coffman, and F. C. McGrew (to E. I. du Pont de Nemours and Co.), U.S. Patent 2,339,237 (Jan. 18, 1944).

128. Same as Ref. 125, pp. 123-130.

129. Charles, J., J. Colonge, and G. Descotes, *Compt. Rend.* **256**, 2934 (1963).

130. Wiloth, F., *Z. Physik. Chem.* **11**, 78 (1957).

131. Myagkov, V.A., and A. B. Pakshver, *Kolloid. Zhur.* **14**, 172 (1952); *C.A.* **46**, 8484a (1952).

132. Taylor, G.B., *J. Am. Chem. Soc.* **69**, 635 (1947).

133. Korshak, V. V., and S. V. Rogozhin, *Dokl. Akad. Nauk SSSR* **76**, 539 (1951); *C.A.* **45**, 8455f (1951). Also, *Izv. Akad. Nauk. SSSR, Otd. Khim. Nauk.* **1952**, *531*.

134. Achhammer, B.G., F.W. Reinhart, and G.M. Kline, *J. Res. Nat. Bur. Std.* **46**, 391 (1951); *J. Appl. Chem.* **1**, 301 **(1951)**.

135. Hasselstrom, T., H. W. Coles, C. E. Balmer, M. V. Hannigan, M.M. Keeler, and R. J. Brown, *J. Tex Res.* **22**, 742 (1952).

136. Hopff, H., *Kunststoffe* **42**, 423 (1952).

137. Liquori, A. M., A. Mele, and V. Carelli, *J. Polym. Sci.* **10**, 510 (1953).

138. Korshak, V. V., and S. V. Rogozhin, *Izv. Akad. Nauk. SSSR, Otd. Khim. Nauk.* **1954**, 541.

139. Meacock, G., *J. Appl. Chem.* **4**, 172 (1954).

140. Hill, Rowland, *Chem. and Ind.* **36**, 1083 (1954).

141. Goodman, I., *J. Polym. Sci.* **13**, 175 (1954).

142. Goodman, I., *J. Polym. Sci.* **17**, 587 (1955).

143. Edgar, O. B., and D. H. Johnson, *J. Chem. Soc.* **1958**, 3925.

144. Straus, S., and L. A. Wall, *J. Res. Nat. Bur. Std.* **60**, 39 (1958); *ibid.* **63A**, 269 (1959).

145. Korshak, V. V., G. L. Slonimskii, and E. S. Krongauz, *Izv. Akad. Nauk. SSSR, Otd. Khim. Nauk.* **1958**, 211; *C.A.* **53**, 12803f (1958).

146. Bailey, W. J., and C. N. Bird, *J. Org. Chem.* **23**, 996 (1958).

147. Bailey, W. J., *Polym. Eng. Sci.* **5**, 59 (1965).

148. Smith, S., *J. Polym. Sci.* **30**, 459 (1958).

149. Katorzhnov, N. D., and A. A. Strepikheev, *Zhur. Priklad. Khim.* **32**, 625 (1959); *ibid.* **32**, 1363 (1959); *C.A.* **53**, 11881b and 18537h (1959).

150. Cawthon, T. M., and E. C. Smith, *Am. Chem. Soc.*, Div. Polym. Chem., Papers presented at New York City Meeting, Sept., 1960, p. 96.

151. Ogata, N., Bull. Chem. Soc. Japan **34**, 1201 (1961).

152. Fukumoto, O., *Kobunshi Kagaku* **18**, 19 (1961); *C.A.* **56**, 1585a (1962).

153. Kamerbeek, B., G. H. Kroes, and W. Grolle, in *Thermal Degradation of Polymers*, S.C.I. Monograph No. 13, London, 1961, pp. 357–391.

154. Rafikov, S. R., G. N. Chelnokova, and R. A. Sorokina, *Vysokomolekul. Soedin.* **4** (11), 1639 (1962); *C.A.* **59**, 6531a (1963).

155. Knappe, W., and G. Kress, *Kunststoffe* **53**, 346 (1963).

156. Rafikov, S. R., G. N. Chelnokova, V. V. Rode, I. V. Zhuravleva, and R. A. Sorokina, *Vysokomolokul. Soedin.* **6** (4), 652 (1964); C.A. **61**, 5801b (1964).

157. Chelnokova, G.N., and S. R. Rafikov, *Vysokomolekul. Soedin.* **6**, (4) 710, (1964); *C.A.* **61, 5801c (1964)**.

158. Dine-Hart, R. A., B. J. C. Moore, and W. W. Wright, *J. Polym. Sci.* **28**, 369 (1964).

159. Indest, H. (to Vereinigte Glanzstoff-Fabriken A.-G.), U.S. Patent 3,121,763 (Feb. 18, 1964).

160. Krasnov, E. P., and L. B. Sokolov, *Vysokomolekul. Soedin., Khim. Svoistva; Modifikatsiya Polimerov, Sb. State;* 1964, 275; *C.A.* **62**, 659h (1965).

161. Krasnov, E. P., V. M. Savinov, L. B. Sokolov, V. I. Logunova, V. K. Belyakov, and T. A. Polyakova, *Vysokomolekul. Soedin.* **8**, 380 (1966).

162. Bruck, S. D., *Polymer* **6**, 483 (1965).

163. Bruck, S. D., *Polymer* **7**, 231 (1966).

164. Bruck, S. D., *Polymer* **7**, 321 (1966).

165. Bruck, S. D., and A. A. Levi, *J. Macromol. Sci.* **A1**, 1095 (1967).

166. Dobrokhotova, M. K., B. M. Kovarskoya, and S. R. Rafikov, *Plast. Massy.* 1966 (10), 14; *C.A.* **68**, 50878s (1968).

167. Chevychelov, A. D., *Mekh. Polim.* 1967 (1), 8; *C.A.* **67**, 33100u (1967).

168. Goldfarb, I. J., and A. C. Meeks, AFML-TR-68-347, Part I (1969).

169. Krasnov, E. P., V. P. Aksenova, and S. N. Kharkov, *Vysokomolekul. Soedin.* **11A** (9), 1930 (1969).

170. Bruck, S. D., *Polymer* **10**, 939 (1969).

171. Mortimer, C. T., *Reaction Heats and Bond Strengths*, Pergamon Press, New York, 1962, pp. 129 and 141.

172. Cottrell, T. L., *The Strengths of Chemical Bonds*, Butterworths Scientific Publications, London, 1958, 2nd ed., pp. 274 and 275.

173. Coffman, D. D., G. J. Berchet, W. R. Peterson, and E. W. Spanagel, *J. Polym. Sci.* **2**, 306 (1947).

174. Reimschuessel, H. K., *J. Polym. Sci.* **41**, 457 (1959).

175. Stamatoff, G. S., and N. K. J. Symons (to E. I. du Pont de Nemours and Co.), U.S. Patent 3,247,168 (Apr. 19, 1966).

176. Graf, R., G. Lohaus, K. Boerner, E. Schmidt, and H. Bestian, *Angew. Chem. Int. Ed. Engl.* **1**, 481 (1962).

177. Kuznetsov, G. A., V. D. Gerasimov, L. N. Fomenko, A. I. Maklakov, G. G. Pimenov, and L. B. Sokolov, *Vysokomolekul. Soedin.* 7, 1592 (1965).

178. Dachs, K., and E. Schwartz, *Angew. Chem. Int. Ed. Engl.* 1, 430 (1962).

179. Markley, K. S., in *Fatty Acids,* K. S. Markley, Ed., Interscience Publishers, Inc., New York, 1960, 2nd ed., Part 1, p. 96.

180. Magat, E. E., and I. A. David, publication in preparation.

181. Cannon, C. G., *Spectrochim. Acta* 16, 302 (1960).

182. Trifan, D. S., and J. F. Terenzi, *J. Polym. Sci.* 28, 443 (1958).

183. Starkweather, H. W., Jr., G. E. Moore, J. E. Hansen, T. M. Roder, and R. E. Brooks, *J. Polym. Sci.* 21, 189 (1956).

184. Meyersen, K., in *Polymer Handbook,* J. Brandrup and E. H. Immergut, Eds., Interscience Publishers, a division of John Wiley and Sons, Inc., New York, 1966, p. IV-216.

185. Same as Ref. 125, pp. 315-325.

186. Weiske, C. D., *Kunststoffe* 54, 626 (1964).

187. Ravens, D.A.S., and Mrs. J. E. Sisley, in *Chemical Reactions of Polymers,* E. M. Fettes, Ed., Interscience Publishers, a division of John Wiley and Sons, New York, 1964, pp. 561-564.

188. Frey, H. J., and J. R. Knox in *Analytical Chemistry of Polymers,* G. M. Kline, Ed., Interscience Publishers, Inc., New York 1959, Part 1, p. 277.

189. Haslam, J., and S. D. Smith, *Analyst* 79, 82 (1954).

190. Dunn, P., and G. F. Sansom, *J. Appl. Polym. Sci.* 13, 1641, 1657, and 1673 (1969).

191. Harding, G. W., and B. J. MacNulty, in *Thermal Degradation of Polymers,* S.C.I. Monograph No. 13, London, 1961, pp. 392-412.

192. Anon., *'Zytel' Design and Engineering Data,* Plastics Department, E. I. du Pont de Nemours and Co., Wilmington, Del.

193. Brooks, R. E., and J. F. Cogdell, Jr. (to E. I. du Pont de Nemours and Co.), Brit. Patent 890,437 (Feb. 28, 1962).

194. Gabler, R., H. Muller, G. E. Ashby, E. R. Agouri, H.-R. Meyer, and G. Kabas, *Chimia* 21 (2), 65 (1967).

195. Mikolajewski, E., J. E. Swallow, and M. W. Wabb, *J. Appl. Polym. Sci.* 8, 2067 (1964).

196. Levantovskay, I. I., M. P. Yazvikova, M. K. Dobrokhotova, B. M. Kovarskaya, and K. N. Vlasova, *Plast. Massy* 1963 (3), 19; *C. A.* 58, 14206b (1963).

197. Goldstein, H. E., *J. Macromol. Sci.* A3, 694 (1969).

198. Valk, G., H. Druessmann, and P. Diehl, *Makromol. Chem.* 107, 158 (1967).

199. Valko, E. I., and C. K. Chiklis, *J. Appl. Polym. Sci.* 9, 2855 (1965).

200. Neiman, M. B., *Aging and Stabilization of Polymers,* Consultants Bureau, New York, 1965, pp. 238-247.

201. Schwenker, R. J., Jr., and L. R. Beck, Jr., *Text. Res. J.* 30, 624 (1960).

202. Meschke, R. W., personal communication.

203. Moore, R. F.. *Polymer* 4, 493 (1963).

204. Sharkey, W. H., and W. E. Mochel, *J. Am. Chem. Soc.* 81, 3000 (1959).

205. Guillet, J. E., J. Dahnraj, F. J. Golemba, and G. H. Hartley, in *Stabilization of Polymers and Stabilizer Processes,* Adv. in Chem. Series No. 85, Am. Chem. Soc., 1968, pp. 272-286.

206. Stephenson, C. V., B. C. Moses, and W. S. Wilcox, *J. Polym. Sci.* **55**, 451 (1961).

207. Stephenson, C. V., B. C. Moses, R. E. Burks, Jr., W. C. Coburn, Jr., and W. S. Wilcox, *J. Polym. Sci.* **55**, 465 (1961).

208. Stephenson, C. V., J. C. Lacey, Jr., and W. S. Wilcox, *J. Polym. Sci.* **55**, 477 (1961).

209. Taylor, H. A., W. C. Tincher, and W. F. Hamner, *J. Appl. Polym. Sci.* **14**, 141 (1970).

210. Anton, Anthony, *J. Appl. Polym. Sci.* **9**, 1631 (1965).

211. Same as Ref. 200, pp. 251-261.

212. Heuvel, H. M., and K. C. J. B. Lind, *J. Polym. Sci., Part A-2,* **8**, 401 (1970).

213. Zimmerman, J., in "Radiation Chemistry of Macromolecules" M. Dole, Ed., Vol. 2, Academic Press, New York, 1972.

214. Flory, P. J., *J. Am. Chem. Soc.* **58**, 1877 (1936).

215. Reimschuessel, H. K., and G. J. Dege, *J. Polym. Sci.* **8**, Part A-1, 3265 (1970).

216. Peebles, L. H., Jr., and M. W. Huffman, *J. Polym. Sci.* **9**, Part A-1, 1807 (1971).

217. Wiloth, F., *Makromol. Chemie* **144**, 283 (1971).

218. Reimschuessel, H. K., and K. Nagasubramanian, *Polym. Eng. Sci.* **12**, 179 (1972).

219. Anon., *Chem. Eng. News* **51** (15), 14 (April 9, 1973).

220. Anon., *Chem. Week* **108** (19), 32 (May 12, 1971); Anon., *Chem. Eng. News* **49** (17), 30 (April 26, 1971).

CHAPTER 3

Characterization of Nylons

EDMUND M. LACEY

INTRODUCTION

This chapter discusses tests used to characterize either nylon granules or fabricated parts. These tests can be classified into one of the following groups.

1. Examination, definition, and description of the granulation.
2. Qualitative and quantitative identification and analysis of the polymer.
3. Characterization of the specific composition by molecular weight, moisture content, monomer content, and so on.
4. Identification and estimation of additives such as plasticizers, nucleating agents, glass fiber, and so on.

Since the introduction of nylon in 1938 the literature has been replete with procedures for characterization. This chapter does not attempt to classify this literature or to provide exhaustive references. Instead, selections have been made with the objective of drawing attention to procedures most widely in use today. Also, some tests have been included because they appear to merit further consideration and development.

The analytical techniques are considered in general terms. Discussion centers on usefulness and limitations. The reader should consult the original references for details.

Tests range from the simple to the complex. The research man characterizing a new nylon requires different techniques than the quality control engineer examining an incoming shipment of molded parts. The student or instructor at the university may have more time available for his analysis than the industrial control chemist. It is difficult to select the optimum test for a specific situation. There are a number of factors, however, that should be considered in making this choice: time required, reliability, equipment, complexity, need for technical personnel, capability of automation, and cost.

In brief, our intent in this chapter is to discuss the ways in which the diverse needs for definition of nylon composition are met. Studies of the microscopic and submicroscopic character of nylon in the fabricated part are important to an understanding of the interrelationships of processing conditions, thermal and stress history, and properties. These problems of crystal structure and polymer morphology are described in Chapter 8.

GRANULATION

Not only the chemical character of a nylon composition determines its usefulness for the manufacture of plastic parts. Such simple factors as the size and shape of the original granules, uniformity of the granulation, and inadvertent contamination may also affect processing and properties sufficiently to make a specific composition unsatisfactory while another, nominally similar composition, proves adequate. This section considers tests that assess this physical aspect of nylon characterization.

Size, Shape, Packing, and Flow

Nylon molding or extrusion grade granules are cut from extruded beading or sheets into the form of cylinders or cubes. Particle sizes vary somewhat from manufacturer to manufacturer, but sides, lengths, and diameters range normally from 60 to 120 mils or 1.5 to 3.0 mm. No hard and fast rules for an optimum granulation are possible. Some fabricating techniques are insensitive to variation in granulation, and the cost factor of maximizing use of rework dominates. The quality of the granulation may impose narrow operating limits in some extruders and molding machines although this is sometimes alleviated by suitable modification of equipment, particularly screw design.

Nylon is, of course, granulated by the processor in utilizing runners and sprues that would otherwise be wasted. A variety of size reduction machines is available for this purpose. These vary considerably in capability of cutting particles of equal size and uniformity. It is prudent to remove fines by screening before recycling rework for processing.

Examination of the granulation is a basic part of product quality control. Granulation variables include size and size distribution, the quality of the cut, apparent density, bulk factor, pourability, and angle of repose.

The use of screens for analysis of size distribution is especially useful in examining recut material having gross size variation. For virgin powder, which is much more nearly uniform, variation in the weight and/or dimensions of a number of individual particles provides a measure of uniformity. Examination of several granules with low-power magnification is useful for assessing the quality of the cut. Particles with flattened edges indicate a dull cutting knife. Standards rated for these edge effects are useful inspection aids. Thin edges and sharp corners with marked irregularities in thickness may result in high local moisture absorption during short-time atmospheric exposure and cause weak spots in processed items. Also, irregularities may interfere with constant flow in the hopper.

ASTM D1895-67 provides a method for expressing quantitatively the quality of a molding powder with respect to handling during fabrication (1). This is particularly useful in examining recut material where variability in density or flow must be minimized because of narrow tolerances on the dimensions of molded parts. The quantitites measured are apparent density, bulk factor, and pourability. The apparent density is determined by weighing a standard cup that has been filled with molding powder under controlled conditions. The bulk factor is the ratio of the volume of plastic material to the volume of the same quantity of material after processing. The pourability is used to indicate expected uniformity of flow during processing and is related to dimensional variability in molded parts. Roughly cut and jagged pellets with a range of geometries and sizes do not pour uniformly.

Another test that can serve as an index of feeding behavior is measurement of the angle of repose. In one procedure a conical funnel filled with the molding powder is turned upside down on a clean glass plate. The funnel is then slowly raised to about an inch above the top of the forming polymer cone. The angle of repose is then determined by measuring the angle between the side and base of the cone of polymer.

Color

Apart from those nylons containing colorants, the so-called natural grades may exhibit some degree of color that depends on the type of nylon and the presence or absence of additives. For example, the presence of heat-stabilization systems (antioxidants, see Chapter 11) based on copper salts results in a faint green or blue hue, and some nucleating systems may result in suppression of yellowness.

The measurement of yellowness by ASTM D1925-70 (2) is based upon tristimulus values calculated from spectrophotometric data. Instruments suitable for this determination have been described (3). The yellowness index correlates reasonably well with visual perception of yellowness under daylight illumination. In such measurements geometry is an important factor. Adaptation of the ASTM procedure to comparisons between samples of differing thickness or surface smoothness may not be valid because of the varying ratio of transmitted light to reflected light. The yellowness index measurement may be used to measure the increase in color resulting from oven drying or from exposure to other environmental conditions.

Contamination

Several procedures have been utilized for examining virgin molding powder and recut runners and sprues for contamination. Although the details vary, certain basic elements are common to all procedures. The sampling should be representative of the production unit or system or the lot of recut material. Standard sampling techniques are used, and a sample size of 5 to 10 gal (19 to 38 liters) of material is recommended for thoroughly blended material. If blending is inadequate, a composite is taken of the entire unit.

A specimen from the large standard sample is spread in a layer one cube deep on a white background, and the number of contaminated particles is counted. Grading of the contamination is based on size and type; that is, carbon, gel, metal, wood, and so on. Photographs of the contamination are taken and given numerical ratings which can be used for future evaluations.

IDENTIFICATION OF TYPE OF NYLON

Rapid Identification Using Fisher-Johns Melting, Specific Gravity, and Solubility

The identification of the most common chemical types of nylon is frequently done by determining the Fisher-Johns "melting point" (ASTM D-789), specific gravity, and solubility behavior (8). This simple approach has been used successfully for many years and is still employed today although the proliferation of many new copolymers and mixtures and the periodic introduction of new chemical types can complicate the identification. When results are questionable or when a more precise analysis is required, other analytical procedures described below are used.

The easiest way to distinguish nylon from other plastics is to place a small piece in a Pyrex tube and heat the end of the tube gently with a small flame until the polymer has melted partially and undergone some decomposition. The odor of "scorched" nylon is similar to that of burning hair and is unmistakable.

The melting points, specific gravities, and solubilities of the common commercial nylon homopolymers are summarized in Table 3-1. The Fisher-Johns melting points (ASTM D-789-66) given in Table 3-1 are generally preferred to other types of melting points for industrial use and in specifications for automotive, appliance, and military component parts because of low equipment costs ($200 to $300) and simplicity of operation. A particle in contact with silicone oil that forms a meniscus with a coverglass is heated rapidly to within 20°C of the expected melting point and then at 2 (+1, -0) degrees per minute until the melting point, defined by movement of the meniscus when the cover glass is probed gently with a dissecting needle, is reached. With experience, an operator can achieve reproducibility within a few degrees.

The specific gravities of Table 3-1 are for compositions that contain only minor amounts of modifiers and have been molded or extruded under "average" conditions. Very unusual processing conditions are required to yield material outside the ranges given. The specific gravity of molded parts can be measured by a hydrostatic technique involving the weighing of a specimen in air and in a solvent not rapidly absorbed. Specimens cut preferably to 0.25 X 0.75 X 1.0 in. (0.64 X 1.9 X 2.5 cm) can be used. Saw marks should be removed with fine emery paper or sand paper in order to prevent surface roughness from entrapping air when the specimen is immersed in the liquid. Densities of small chips removed from molded parts are determined by a density-gradient technique (ASTM D-1505) (4). This is useful in studying the effect of processing conditions or examining variations within a part.

Solubility tests used for rapid discrimination of nylon types have been

Table 3.1. Properties useful in rapid identifications of nylons

Nylon	Melting point[a] (°C)	(°F)	Specific gravity	Solubility in 4.2 molar hydrochloric acid Room temp.	Boiling	Solubility in 90% formic acid
66	250-260	482-500	1.13-1.16	Insoluble	Soluble	Soluble
Modified 66	250-260	482-500	1.08-1.10	Insoluble	Partly soluble	Partly soluble
6 (Monomer extracted)	210-225	410-436	1.12-1.14	Soluble	Soluble	Soluble
6 (Monomer not extracted)	195-225	383-436	1.12-1.14	Soluble	Soluble	Soluble
Modified 6	195-225	383-436	1.08-1.09	Partly soluble	Partly soluble	Partly soluble
610	208-220	406-428	1.07-1.09	Insoluble	Insoluble	Soluble
612	206-215	403-419	1.06-1.08	Insoluble	Insoluble	Insoluble[b]
11	180-190	356-374	1.04-1.05	Insoluble	Insoluble	Insoluble[b]
12	175-180[c]	347-356[c]	1.01-1.02	Insoluble	Insoluble	Insoluble[b]

[a] Fisher-Johns (ASTM D789).

[b] These as well as the more polar nylons dissolve in conc. sulfuric acid or *meta*-cresol.

[c] Koeffler Hot Stage (ASTM D2117).

outlined (8). As expected, the more polar solvents dissolve the nylons richer in amide group concentration but not those lower in amide groups (see Chapter 2, pp. 68 to 70). For example, nylons -6, -66, and -610 but not nylons -612, -11, and -12 are soluble in 90% formic acid at room temperature. Crystallinity and morphology also affect solubility. A 4.2 molar solution of hydrochloric acid at room temperature readily dissolves nylon-6 and the low-melting terpolymers but has no effect upon nylons -66, -610, -11, and -612. This test is often used to differentiate nylons -6 and -66. At boil, the polar character of the nylon is dominant, and hot 4.2 molar hydrochloric acid dissolves nylon-66 but not nylons -612, -610, -11, and -12.

The glass-reinforced nylons show solubility behavior characterisitics of unreinforced resin. (The insolubility of the glass fibers does not cause confusion in judging the solubility of the resin.) Plastic parts that have been exposed to unusually severe environmental conditions may undergo molecular changes such as cross linking that alter solubility characteristics and thus make identification more difficult.

Another method (5) identifies nylon by solubility in hot polyhydric alcohols (Table 3-2). Depending upon the nylon and the alcohol, swelling, loss of form, or solution may occur at the boiling point of the alcohol. Cooling of the solution results in a white precipitate of nylon. When the nylon and alcohol are reheated slowly in an oil bath, the temperature at which a clear solution is obtained is characteristic of the nylon and solvent. Incomplete solubility or premature precipitation indicates another component.

Table 3-2. Solubility behavior of nylons in polyhydric alcohols, temperature of redissolution $^\circ$C

	Ethylene glycol		Propylene glycol		Glycerine	
	($^\circ$C)	($^\circ$F)	($^\circ$C)	($^\circ$F)	($^\circ$C)	($^\circ$F)
Boiling point	198	388	189	372	290	554
Nylon-6	135	275	129	264	168	334
Nylon-8	149	300	133	271	Insoluble	
Nylon-11	Insoluble		145	293	Insoluble	
Nylon-66	153	307	153	307	195	383
Nylon-610	156	313	139	282	Insoluble	

Melting Point

Many procedures are used to estimate the melting points as well as the Fisher-Johns method described above. One technique (ASTM D2117) yields a so-called crystalline melting point, the temperature at which the last sign of birefringence disappears when a sample is observed through crossed polaroids on a microscope hot stage. The crystalline melting point, which for nylons ordinarily ranges from 5 to 10°C (9 to 18°F) higher than the Fisher-Johns melting point, correlates more closely to the minimum processing temperature, probably because it corresponds to the temperature at which complete fluidity of the polymer occurs. A melt indexer has also been used (Chapter 16, p. 543) for defining minimum processing temperature.

Differential thermal analysis (DTA) is widely used (Chapter 16, p. 537) for characterizing the melting behavior of nylons and is sometimes included in industrial specifications for nylon component parts because the thermograms provide more information than simple melting points. An obvious advantage to the DTA procedure is the small sample size required, only 2 to 5 mg. The melting point is affected by not only such variables as heating rate and thermal history, but also by variation in the criterion for melting. The temperature at the peak, onset, or end of the endotherm has been used. DTA is useful for differentiating mixtures of homopolymers (Chapter 2, p. 45) from copolymers of similar chemical composition (6). The mixture shows the endotherms of both homopolymers; the copolymer shows one. Repeated melting or prolonged heating of mixtures results in interpolymerization with gradual blending of the endotherms. A single pass of a mixture of homopolymers through a molding machine operated under average conditions will result in only slight interpolymerization and little change in the thermogram. DTA is especially useful for determining freezing points utilizing the cooling cycle of the thermogram. The slope of the exotherm as well as the spread between melting and freezing points has been related to nucleation.

Differential scanning calorimetry (DSC) is also applied to study the melting and freezing behavior of nylons. DSC has also been used for measurement of the heat of fusion (13). The simple observation of melting in a capillary tube as is commonly done for monomeric compounds has been used but is now less common because of the advent of the other techniques already noted.

Additional discussion of the melting points of nylon will be found in Chapter 9, which concerns transitions and relaxations in nylons. Table 9-1 provides a comparison of five different techniques for determination of the melting points of four homopolymers.

Infrared Analysis (with M.I. Kohan)

Infrared analysis is a familiar technique for identification of chemical groups and specific chemical compositions. Absorption bands reported for nylons are obviously those corresponding to the C-C, C-H, N-H, and C=O bands. The C-C stretching vibrations produce weak absorptions in the 8.0 μ (1250 cm^{-1}) to 12.5 μ (800 cm^{-1}) region. This is the lower frequency, lower energy part of the normal infrared absorption spectrum (2 to 15 μ, 5000 to 670 cm^{-1}) that is affected by modes of vibration characteristic of large chemical groupings and is generally less useful for identification of specific chemical bonds. However, this is the region which, because of its association with large groupings, serves as the "finger printing" part of the spectrum and is very useful for identifying specific compositions (9).

Strong absorptions related to the C-H bond appear at 3.4 and 3.5 μ (2940 and 2860 cm-1) and a weaker one at 6.85 μ (1460 cm-1). The N-H and C=O absorptions are strong ones, characteristic of the amide function, and appear at 3.0 μ and 6.1 μ (3330 and 1640 cm-1); another strong N-H absorption is seen at 6.47 μ (1545 cm-1), and a well-defined but weaker absorption occurs at 3.2 μ (3125 cm-1). The relative intensities of the C-H absorption and those of the CONH group would be expected to vary as the concentration of amide groups in linear, aliphatic nylons varies. Over the range of commercial nylon plastics, however, the variation in relative intensities is small.

It is the absorptions at 3.0 and 6.1 μ (3330 and 1640 cm^{-1}) that indicate a nylon. There is some observable change in the ratio of absorbancies of the C-H and N-H bands as the concentration of amide group decreases. This is illustrated in Figure 3-1, which shows the spectra of nylon-6 and nylon-12. Differences occur also in the longer wavelength region where the absorptions are associated with the crystallinity of the nylons and become less structured as crystallinity decreases. This is shown in Figure 3-1 where nylon-12 is less crystalline than nylon-6 and also in Figure 3-2 where crystallinity decreases from a nylon-66 homopolymer to a nylon -66/6 copolymer and to a nylon -66/610/6 terpolymer.

The dependence of the spectrum on crystallinity suggests a complexity arising from different crystalline orders for the same nylon. Thus, the infrared spectrum of nylon-6 depends on the proportion of alpha, beta, and gamma crystalline modifications and the delta amorphous form (Chapter 8). The spectra of nylon-6 with relatively more alpha and delta forms are shown in Figure 3-3. The level of crystallinity may change for nylons as a result, for example, of changing thermal history, and this also affects the infrared spectrum. Crystalline and amorphous bands have been used to provide a measure of the percent crystallinity in 66 and 610 by Starkweather and Moynihan (10). The specific bands employed are shown in Table 3-3.

Table 3-3. Crystalline and amorphous bands in nylons-66 and -610

Nylon	Crystalline band	Amorphous band
66	10.68 μ (936 cm-1)	8.78 μ (1139 cm-1)
610	11.73 μ (853 cm-1)	8.88 μ (1126 cm-1)

The discussion above emphasizes the need to know and to report the method of preparation of the samples used in obtaining the spectra. This is especially true when attempting to differentiate a homopolymer from a copolymer in which the

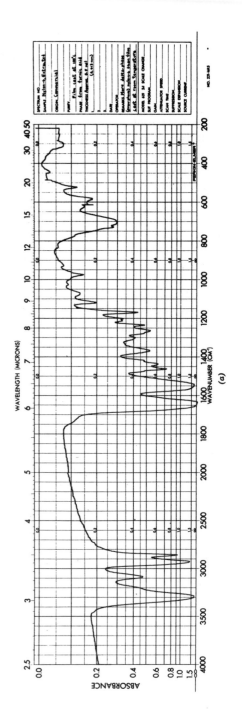

(a)

92

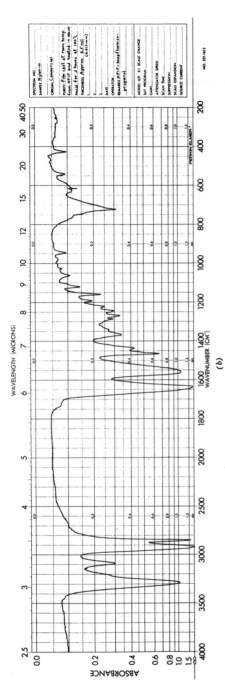

Fig. 3-1. Infrared spectra of (*a*) nylon-6 and (*b*) nylon-12.

93

94

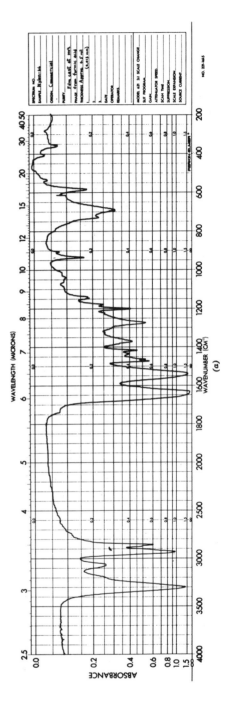

Fig. 3-2a

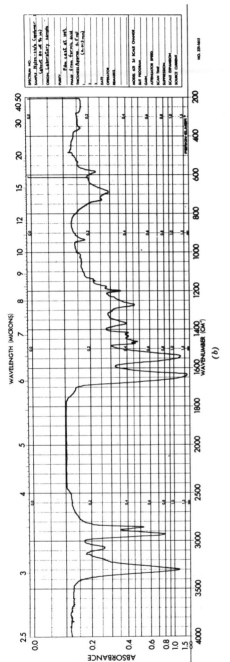

Fig. 3-2b

95

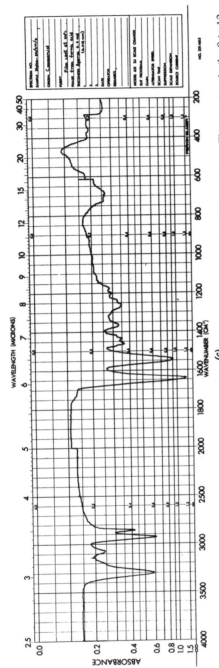

Fig. 3-2. Infrared spectra of (*a*) nylon-66, (*b*) a nylon-66/6 copolymer, and (*c*) a nylon-66/610/6 terpolymer. The structure in the 8 to 12 micron region shows less definition as crystallinity decreases from the homopolymer to the copolymer to the terpolymer.

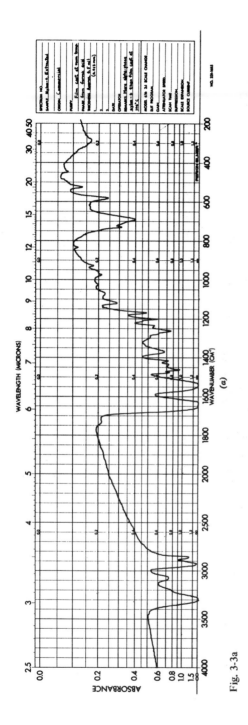

Fig. 3-3a

97

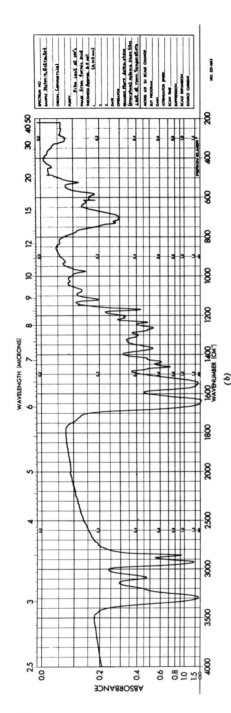

Fig. 3-3.   Infrared spectra of nylon-6 of varying proportions of alpha-phase crystallinity: (a) more and (b) less.

98

principal ingredients are the same as in the homopolymer. Hummel (12a) has discussed the effect of polymorphism on the spectra of nylons and provided references to basic studies.

Additives to nylon compositions are most often present in low concentrations and cannot be identified by the infrared spectra of the polymers but must be isolated first. Solvent extraction and vacuum distillation have been used successfully to remove plasticizers and antioxidants as discussed below.

Infrared absorption at 11.5 μ (870 cm^{-1}) can be used to determine the amount of caprolactam monomer in pressed film of nylon-6 (Ref. 14, p. 128). Obviously, care must be taken to avoid loss of monomer in the preparation of specimens. IR has also been used in measuring the extent of hydrogen bonding in various nylons (11).

The presence of rings or other groupings not found in the linear aliphatic nylons can be detected by IR. For example, a new nylon is poly(trimethyl-hexamethylene terephthalamide). Absorption at 7.3 μ (1370 cm^{-1}) is indicative of the CH_3 groups; absorption at 6.7 and 13.7 μ (1490 and 730 cm^{-1}) is typical of a benzene ring with conjugated substituents at the para positions. Also, a low-melting terpolymer containing segments from bis(4-aminocyclohexyl)-methane shows an absorption at 11.1 μ (900 cm^{-1}), which is attributed to the C-H bond of the aliphatic ring.

Attenuated Total Reflection (ATR) is utilized for measuring the absorption characteristics of a surface layer. ATR requires optical contact with the surface, but melt fabricated parts normally are sufficiently smooth for this technique. By use of microscopic methods, surface areas as small as one square millimeter can be used. Microscopic techniques have also been used to measure the IR transmittance of small samples.

There are numerous published sources of infrared scans of polyamides and additives commonly employed; these include Sadtler (7), Hummel (12), Kagarise and Weinberger (9), and Haslam and Willis (14).

Hydrolysis to Monomer and Analysis

Many industrial acceptance tests are based upon the previously described rapid procedures and infrared scans for identification. It is occasionally necessary, however, particularly with copolymers, to hydrolyze the nylon to its component diacids, diamines, and/or amino acids and to analyze the hydrolyzate.

Stuhlen and Horn (15) hydrolyzed homopolymers and copolymers based on nylons-6, -66, -610, and -11 by heating 18 hr at 100°C in 6N (20 wt %) hydrochloric acid. Sealed, thick-walled glass tubes were used to avoid volatiliza-tion of HCl. Nylons -6 and -66 were completely hydrolyzed, but nylons -610 and -11 were not and required longer periods of heating. Clasper and Haslam (16) also used 6N hydrochloric acid but heated under reflux for 40 hr; they obtained similar results. Increased resistance to hydrolysis with decreasing amide group concentration is to be expected (Chapter 2, p. 70). The hydrolysis of nylon-610 was improved by using a finely dispersed powder obtained by solution in *m*-cresol and reprecipitation with methanol. A summary of procedural details can be found in Haslam and Willis (14).

The dibasic acids found commonly in commercial nylons, that is, adipic, sebacic, and dodecanedioic acids, either precipitate directly from the acid hydrolyzate or are readily caused to precipitate by cooling or dilution with cold water. These can be extracted quantitatively with diethyl either using a continuous liquid/liquid extractor for 6 hr (14). Diacids obtained from homopolymers are readily identified. For convenience, the melting points of the diacids of Table 1-4 are given in Table 3-4. Mixtures from blends of homopolymers or from copolymers may pose a problem.

Table 3-4. Melting points of acids, amine hydrochlorides, and amino acid hydrochlorides

Name	Carbon atoms	Melting point $^\circ$C
A. Diacids		
Adipic acid	6	149
Azelaic acid	9	106
Sebacic acid	10	132
Dodecanedioic acid	12	128
Brassylic acid	13	113
Terephthalic acid	Ring	425 (sealed tube)
B. Amine hydrochlorides		
Hexamethylenediamine • 2 HCl	6	249
C. Aminoacid hydrochlorides		
ϵ-aminocaproic acid • HCl	6	125

If the identities of the dibasic acids are known, the ratio can be determined from the acid numbers (mg KOH per gram of acid mixture) or from the equivalent weights. In general, for binary mixtures:

$$W_A = \frac{100 E_A (a E_B - b)}{b(E_B - E_A)}$$

$$M_A = \frac{100 (a E_B - b)}{a(E_B - E_A)}$$

where

W_A = wt % A in acid mixture

M_A = mole % A in mixture (and in polymer)

E_A = gram equivalent weight of A

E_B = gram equivalent weight of B

a = gram equivalents of base

b = grams of acid mixture

(56, 100/acid number = equivalent weight)

In the special but common case of adipic (A) and sebacic (B) acids where E_A is 73.07 and E_B is 101.12:

$$W_A = \text{wt \% adipic acid} \quad = \frac{7307\,(101.12\,a-b)}{b(28.05)} = \frac{260\,(101.12\,a-b)}{b}$$

$$M_A = \text{mole \% adipic acid} \quad = \frac{100\,(101.12\,a-b)}{a(28.05)} = \frac{3.56\,(101.12\,a-b)}{a}$$

The bishydrochlorides of hexamethylenediamine and the other, relatively infrequently used diamines such as bis(4-aminocyclohexyl)methane are soluble in the hydrolyzate. These are readily isolated after separation of the diacids by evaporation of the excess HCl and water on a steam bath.

Amino acid hydrochlorides in the hydrolyzate may or may not be soluble. That of ε-aminocaproic acid is soluble, but the hydrochloride of the higher molecular weight 11-aminoundecanoic acid precipitates upon cooling and is easily isolated for identification of the parent homopolymer by filtration, washing, and drying.

Mixtures of diamines and diacids from blends of homopolymers or from copolymers can be analyzed conveniently by the gas chromatographic method of Anton (18). Small sample sizes of only 0.05 to 0.2 g are required. For quantitative analysis two samples of known weight can be independently hydrolyzed to avoid the need to evaporate the hydrolyzate to dryness. Alternatively, as shown in Figure 3-4, the hydrolyzate residue can be divided into weighed fractions. The amine analysis involves isolation of the amine via extraction of the saponified residue and a GC analysis of the dried extract. The acid analysis involves esterification with $BF_3 . CH_3 OH$, extraction of the methyl esters, and GC of an aliquot of the extract. Azelaic acid is added in known amount to the GC sample to provide an internal standard. If a single amino acid monomer is known to be present, it can be estimated by difference.

Older techniques for identification of hydrolyzates have been described that make use of ion exchange columns (Ref. 14, p. 117). Amino acid hydrochlorides can sometimes be titrated in the presence of amine hydrochlorides because of the lower basicity of the amine in the amino acid. An indicator such as methyl

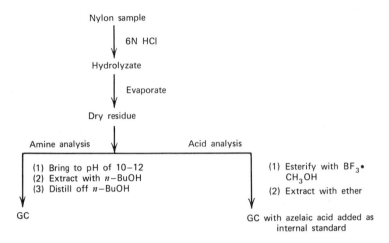

Fig. 3-4.   Anton's method for analyzing complex hydrolyzates (18).

orange, which changes color in the pH 3.1 to 4.4 range, can be used in such instances (Ref. 14, p. 116).

CHARACTERIZATION OF SPECIFIC COMPOSITION

Water Content

Knowledge of the water content in a nylon composition is important to understanding of processing behavior or physical properties as discussed in Chapter 2, p. 33, Chapter 9, pp. 316 and 319, and in the processing chapters (Chapters 5 and 6). The amount of water in any given nylon sample will depend upon the specific nylon; its physical structure (Chapter 8); time, temperature, and relative humidity during exposure; and thickness, if not equilibrated to the environment. In sampling for moisture analysis, consideration must be paid to conditions of exposure and possible variation from surface to core or section to section within a part.

The most commonly used method for moisture analysis is based upon vacuum distillation, condensation of the moisture, and titration with Karl Fischer Reagent (see ASTM Nylon Specifications D789 for details on equipment and procedure). A sample size of 10 to 30 g is normally used with a 30-min distillation and a temperature of 260°C (500°F) for nylon-66 and 220°C (428°F) for nylons -6, -610, and -612. The method refers to its use for moisture contents under 0.28%, but it can be used for moisture levels up to saturation if appropriate reduction in sample size is made. The cited temperatures and times normally avoid complications due to decomposition, polymerization, or hydrolysis. The procedure is employed as a reference method in industrial specifications where the larger sample size is preferred to minimize sampling error.

Another procedure for moisture involves heating a 1- or 10-g sample, depending upon the expected moisture level, in a special device known as a "Moisture Meter" (19). The water vapor generated is indicated by displacement of a piston and is quantified via calibration with samples of known moisture contents. The equipment is commercially available (20) in a compact, portable form that can be used at the processing site. The simplicity of operation enables the analysis to be run with little training.

Another technique involves solution in *meta*-cresol and titration with Karl Fischer Reagent (KFR). Electrometric titrations are required for colored compositions in which visual end points are obscured. The procedure is satisfactory for nylons with high or low moisture contents. Most organic additives do not interfere, although materials easily oxidized by KFR may cause a fading end point (8).

A simple and direct way to determine moisture is to measure the weight loss after drying in an oven, and this is used industrially but only to a limited extent and for moisture levels above 0.4%. A ¼-in. bed of granules is dried in a vacuum oven at 80°C for 24 hr. The possibility of weight loss due to volatilization of low-molecular-weight material must be considered, particularly with nylon-6.

The CEC moisture analyzer is used to determine water content in nylons. The basic element is a Du Pont cell in which moisture carried from the heated sample is absorbed on a phosphorous pentoxide skin and electrolyzed. Readout of the equipment is in terms of micrograms of water electrolyzed. The equipment is commercially available (21). The analysis is simple and satisfactory when representative samples of small size (about 0.1 g) are available.

Gas chromatography has also been used for determining moisture in nylon. Rapidity of analysis is high here also; answers are available in a matter of minutes. Sampling may be critical because of the very small size sample (normally about 0.01 g).

Monomers and Oligomers

All nylons contain at least small quantities of the monomers from which they are made and oligomers (low-molecular-weight material arising from very low degrees of polymerization) as expected from the molecular weight distribution (Chapter 2, p. 44). All nylons contain accordingly an extractable fraction. As also discussed in the previous chapter (pp. 19, 24, 29, and 63), only nylon-6 contains more than 2 wt % of water extractables. This is about 10% as made, but most of the nylon-6 used commercially is extracted before sale to yield products containing less than 2% of monomer and oligomers. Occasionally, low molecular fractions appear in melt processing as condensates; identification can be complex, but these are usually the lower cyclic oligomers many of which have been prepared and characterized (22).

The amount of extractables depends not only on the individual nylon but also on the technique. Different solvents, particle sizes, temperatures, and extraction times have all been used. Hot water is the most commonly used extractant and removes monomer and some oligomers. Adjustment must be made for the moisture content at the start and at the conclusion of the test. Errors in the moisture analyses may amount to about 0.2% absolute and will obviously contribute to uncertainty if the extractables content is small. Other extractants used for this purpose include methanol, glacial acetic acid, and dimethyl sulfoxide (24).

A procedure (23) for determining caprolactam monomer in nylon-6 is based on vacuum extraction, dilution of the distillate to a known volume with water, and comparison of the refractive index with solutions of known concentration. The monomer amounts to about 80% of the water extractables. The monomer analysis holds the advantage of rapidity – about 3 hr versus over 24 hr for the extraction technique.

Another method (24) for the determination of caprolactam involves solution of nylon-6 in o-chlorophenol and measurement of the absorbance at 892 cm^{-1}. The procedure was used primarily for following the early stages of polymerization in which monomer contents ranged from 50 to 90%. The reproducibility of the method is given as ±1.4% absolute. The method is therefore limited to unextracted nylon-6.

Molecular Weight (M.I. Kohan)

Control and determination of molecular weight are common problems in polymer science and industry because of the implication of molecular weight for processing behavior and toughness (Chapter 4, p. 142, and Chapter 11, p. 412). Three levels of assessment can be identified:

a. definition of the detailed distribution of molecular weights which, for nylons, is usually approximated by the "most probable distribution" (Chapter 2, p. 41) characterized by a $\bar{M}_w/\bar{M}_n$ ratio of 2;

b. measurement of molecular weight parameters such as weight-average ($\bar{M}_w$), number-average ($\bar{M}_n$), viscosity-average ($\bar{M}_v$), or other; and

c. establishing a precise, accurate, and readily accessible index of molecular weight that can be correlated with the processing or toughness property in question.

Early work on distribution involved laborious and difficult bulk fractionation procedures which yielded information on the constituent molecular weights. The introduction of gel permeation chromotography (GPC) has provided a technique for relatively rapid acquisition of a distribution curve that, however, requires calibration for translation into an actual molecular weight distribution. Calibration is still a problem, but the changes in GPC-derived curves are useful in characterizing the effect on distribution of different polymerization or degradation histories (25). A recommended practice for GPC terminology has been established (26) (ASTM D3016), but a recommended practice for GPC procedure is still under study by an ASTM committee, D20.70.04.

Each year brings to light new work that makes use of new developments in technique to define more carefully molecular weight averages and their interrelationships or to evaluate new approaches to these parameters. A recent example (27) reports on studies on the oldest of the nylons, that is, nylon-66, via end-group analysis, solution viscosity, osmotic pressure, light scattering, vapor pressure osmometry, and ultracentrifugation. $\bar{M}_n$ ranged from 900 to 31,000. On the average, for linear products of melt polymerization, $\bar{M}_w/\bar{M}_n$ = 1.85 and $\bar{M}_v/\bar{M}_w$ = 0.96. Excessively long polymerization times led to broader distributions.

Of the absolute methods for measuring a molecular weight average, the one most widely used for nylons is end group analysis for $\bar{M}_n$. This is because most nylons are linear so that the number of molecules is exactly one-half the sum of ends and because the analytical techniques are relatively simple. The amine ends can be determined by conductometric titration of a phenol-water-ethanol solution with 0.1N HCl in ethanol (27, 28, 8), but potentiometric titration of a m-cresol solution with 0.5N perchloric acid in methanol is preferred (29). The acids end are analyzed by titration of a hot benzyl alcohol solution with NaOH or KOH, 0.1N in benzyl alcohol, to a phenolphthalein end-point (28). Copolymers soluble in aqueous ethanol can be titrated at room temperature with 0.1N NaOH or 0.1N HCl in aqueous ethanol (28). End-group concentrations are commonly expressed in equivalents per million grams of polymer so that $\bar{M}_n$ = 2 $\times$ 10^6/sum of ends. Compensation must be made for nontitratable ends (Chapter 2, p. 38).

The technique most commonly used to estimate molecular weight is solution viscosity. The basic empirical quantity is the relative viscosity (η_r), which is the absolute viscosity of the polymer solution divided by that of the solvent. In dilute solutions (usually less than 1% polymer) the difference in density of solution and solvent is negligible, and the same capillary viscometer is used to measure the drainage time of both so that η_r = $t_{solution}/t_{solvent}$. Control of temperature and selection of a suitable capillary diameter to avoid a kinetic-energy correction (30) as well as carefully purified solvents are necessary to achieve good reproducibility. Formic acid, $meta$-cresol, and concentrated sulfuric acid are solvents commonly employed for nylons; Cannon-Fenske viscometer sizes 100, 200, and 300 are, in order, convenient for 0.5 g/100 ml solutions.

The intrinsic viscosity or limiting viscosity number (31) ($[\eta]$) is often cited in the literature (32) because it is a viscosity function with a finite value at infinite dilution that yields a straight line on a log-log plot versus molecular weight according to the equation, $[\eta] = KM^a$. It is defined as follows:

$$[\eta] = \left(\frac{\eta_{sp}}{c}\right)_{c=0} = \left(\frac{\ln \eta_r}{c}\right)_{c=0}$$

where

$$\eta_{sp} = \eta_r - 1$$

The identity is valid because $\ln X = X - 1$ if X is only slightly greater than unity. Usually η_{sp}/c or $\ln \eta_r/c$ is plotted against c, and the intercept at $c = 0$ yields $[\eta]$. The Huggins equation (33) has been the most common expression for relating $[\eta]$ to η_{sp}/c:

$$\frac{\eta_{sp}}{c} = [\eta] + k'[\eta]^2 c$$

$$\frac{\ln \eta_r}{c} = [\eta] - k''[\eta]^2 c$$

However, the Schulz-Blaschke equation has been said to be more accurate and to give somewhat higher values (34, 35).

$$\frac{\eta_{sp}}{c} = [\eta] + k[\eta]\eta_{sp}$$

For nylon-6 in 1:1 phenol/tetrachloroethane (36) $[\eta]$ (Huggins) is about 10% below that of $[\eta]$ (Schulz-Blaschke); in 95.6% H_2SO_4, the difference varies from negligible to 10% as the $[\eta]$ increases from 0.2 to 2.0 (in this section, c is always in g/100 ml).

For a polymer with a distribution of molecular weights M should be $\bar{M}_v$ (Chapter 2, p. 45), but the more accessible and readily interpreted $\bar{M}_n$ is often used. Applicability of the $[\eta]$-$\bar{M}_n$ equation then rests on the assumption of an unchanging molecular weight distribution and is best limited to the $\bar{M}_n$ range studied. Table 3-5 indicates the $[\eta]$-M relationships for some nylons in a variety of solvents. A salt has been added occasionally to formic acid solutions to suppress the polyelectrolyte effect which causes curvature in the viscosity-concentration curves, but satisfactory results can be obtained without the salt with sufficient data at low concentrations (27).

Table 3-5. Limiting viscosity number – molecular weight relationship for some nylons[a]

$$[\eta] = K M^a \qquad \text{or} \qquad M = K^{-\frac{1}{a}} [\eta]^{\frac{1}{a}}$$

Nylon	Solvent	°C	$K \times 10^4$	a	$K^{-\frac{1}{a}} \times 10^{-4}$	$1/a$	Range in $M \times 10^{-3}$	Ref.	Comment
				$M = \overline{M}_n$					
66	90% HCOOH	25	11	0.72	1.30	1.39	5.6-24.4	37	
66	90% HOOOH	25	5.55	0.786	1.38	1.27	2.0-31.0	27	b
66	90%HCOOH, 0.1M HCOONa	25	5.15	0.746	2.54	1.34	—	38	6 polymerized in
6(b=1)	conc. H_2SO_4	25	6.3	0.764	1.57	1.31	2.0-14.3	39	presence of acidic
6(b=2)	conc. H_2SO_4	25	4.17	0.794	1.82	1.26	1.9-22.6	39	compound where
6(b=4)	conc. H_2SO_4	25	5.5	0.736	2.70	1.36	2.4-18.4	39	acid groups/cpd = b
6(b=8)	conc. H_2SO_4	25	1.35	0.857	3.37	1.17	4.2-25.7	39	
6	m-cresol	–	5.26	0.745	2.47	1.34	$\overline{M}_w$ = 9-350 for	40	
6	2,2,3-tetrafluoropropanol, 10% H_2O, 0.1M LiCl	–	3.42	0.76	3.76	1.32	fractionated polymer	40	
6	m-cresol	25	18	0.654	1.58	1.53	3.2-31.8	51	
66/6 copolym.	m-cresol	20	10.6	0.71	1.55	1.41	1.1-36.7 (fractions) 1.4-11.5 (whole polymers)	41	
610	m-cresol	25	1.35	0.96	1.07	1.04	8.2-23.8	42	
MX D-6	m-cresol	30	15.7	0.646	2.23	1.55	17.5-46.0	43	
				$M = \overline{M}_v$					
66	90% HCOOH	25	3.53	0.786	(A = 0.025)		$\overline{M}_n$ = 2.0 31.0	27	
66	90% HCOOH	25	1.32	0.873	(A = 0.010)		0.15-50	44	$[\eta] = A + K\,\overline{M}_v^{\,a}$
66	90%HCOOH, 0.1M HCOONa	25	5.16	0.687	(A = 0.010)		0.15-50	44	
66	m-cresol	25	3.53	0.792	(A = 0.015)		0.15-50	44	

[a]Some of the constants are calculated from those given in the references. All are adjusted to c in g/100 ml.

[b]Calculated from reference equation with $\overline{M}_v$ using distribution data of reference.

107

Another viscosity function, the K-value, is sometimes encountered, particularly in the German literature. It is defined by the equations that follow (45) and was designed to provide a characteristic quantity independent of concentration:

$$\log \eta_r = \left(\frac{75k^2}{1 + 15kc} + k \right) c; \qquad K = 1000k$$

It is now recognized that the validity of this expression is limited, and K has been shown not to be constant for nylon-66 in 90% formic acid (46). For nylon-6 the following approximate equations can be derived from the data of Ref. 45b for solutions of 1.0-g polymer in 100 ml sufuric acid at $25°C$:

$$
\left.
\begin{aligned}
\bar{M}_n &= 5,000 - 14,500 \\
K &= 38 - 66
\end{aligned}
\right\} \quad \log \bar{M}_n = 1.9 \log K + 0.70
$$

$$
\left.
\begin{aligned}
\bar{M}_n &= 14,500 - 40,000 \\
K &= 66 - 93
\end{aligned}
\right\} \quad \log \bar{M}_n = 3.0 \log K - 1.30
$$

Correlations that involve a viscosity determined at only a single specific concentration are sometimes useful because of convenience and a linear relationship with molecular weight on a log-log plot for a reasonable range of molecular weights. In freshly distilled, 98% m-cresol at $25.0°C$ for $\bar{M}_n = 10,000$ to 30,000:

$$\bar{M}_n = 1.56 \times 10^4 \left(\frac{\ln \eta_r}{c} \right)^{1.49}_{c=0.5} \qquad \text{for extracted nylon-6(47)}$$

$$\bar{M}_n = 1.30 \times 10^4 \left(\frac{\ln \eta_r}{c} \right)^{1.54}_{c=0.5} \qquad \text{for nylon-66(48)}$$

For nylons -6, -66, -610, and their copolymers, an 8.4 wt % solution in 90% formic acid (11 g per 100 ml) is much used in industry and is described in ASTM D-789. The effects of minor changes in the concentration of polymer, solvent composition, and temperature have been discussed for nylon-66 (49). This concentrated η_r does not yield a linear log-log relationship over a broad range of molecular weights, but it is more sensitive to a change in molecular weight than a dilute η_r. This explains its appeal to industry where a more accurate index of molecular weight is desired. Solution density, solvent density, solvent viscosity, and viscometer calibrations are required so that this procedure appears more arduous, but they need be determined only once for a given polymer and solvent. A suitably calibrated Brookfield viscometer (Brookfield Engineering Laboratories, Stoughton, Mass.), which measures the drag on a rotating spindle, can be used in place of the capillary viscometer to minimize the time required to obtain the viscosity of the polymer solution, but no standardized procedure for solutions of nylons has as yet been published.

The following expression (48) can be used to estimate $\overline{M}_n$ within 10% from the viscosity in 90% formic acid of nylon-66 of normal distribution and $\overline{M}_n$ = 13,000 to 30,000:

$$\log M_n = 0.59 \log \left(\eta_r \right)_{c=11} + 3.23$$

It is consistent with Taylor's $[\eta]$ -M_n equation (37) and the viscosity interrelationship that can be calculated from his data for $[\eta]$ = 0.6 - 1.8:

$$\log \left(\eta_r \right)_{c=11} = 2.36 \log [\eta] + 1.50.$$

It appears that $\left(\dfrac{\eta_{sp}}{c} \right)_{c=0.5}$ and $\left(\eta_r \right)_{c=11}$ will be adopted as inter-

national standards using formic acid as the solvent for nylons -66, -6, -610, and co-polymers and m-cresol for the formic acid insoluble nylons (612, 11, 12). In the International Standards Organization c is in grams per milliliter so that

$$\left(\frac{\eta_{sp}}{c} \right)_{ISO} = 100 \times \left(\frac{\eta_{sp}}{c} \right)_{c=0.5}$$

It should be emphasized that contamination, significant concentrations of moisture or additives in the polymer, changing quality of the solvent (the formic acid should be of constant conductance as well as meet the other purity criteria of ASTM D789), and changes in distribution arising from degradation and anionic polymerization of lactams, all must be considered before attempting to apply any of the above equations.

Additives

Additives in nylon include a variety of organic, inorganic, and polymeric materials to improve processing behavior, increase resistance to chemical attack, or modify physical properties (Chapter 11). The multiplicity, complexity, and frequently low level of additives as well as the steady appearance of new compositions make it impractical to provide specific analytical procedures for every additive in use. Nor is it possible to describe each procedure employed in all of the nylon plastics industry. Our purpose here is to outline some general analytical approaches used to identify and quantify additives in nylon.

In the case of organic additives, removal and isolation can frequently be accomplished by extraction, providing a suitable solvent can be obtained that will dissolve the additive but not the nylon. In general, hot methanol and chloroform are good first choices although a small amount of low-molecular-weight material may also be removed. Grinding the nylon to pass a 20-mesh screen will usually shorten the extraction time from about 1 day for commercial granules to 3 to 4 hr. The additive is isolated by distilling off the solvent and then identified by standard techniques, usually and most simply an ultraviolet or infrared analysis. Compilations of the spectra of common additives are available (7, 12, 14). Quantitative analysis, once the additive is known, may involve solution of the whole polymer in sulfuric acid and measurement of the absorbance at a specific wavelength in the ultraviolet region, or it may require isolation of the additive and subsequent solution or mulling as required for infrared analysis. Obviously, the preferred technique depends on both the specific additive and its concentration. Other methods are employed but are less common. The vacuum distillation of caprolactam followed by measurement of the refractive index of an aqueous solution is described in the section on monomer and olgimers above. Some plasticizers may also be conveniently isolated via vacuum distillation.

It should be recognized that most metal salts of fatty acids that are used as lubricants or mold release agents are soluble in hot alcohol, water, or benzene and can also be isolated via solvent extraction although zinc salts are an exception. The question of surface coated additives also deserves mention in connection with extraction technique. The reflux action often causes transfer of the coating to the liquid even if the coating is insoluble. Filtration removes both the granular polymer and the suspended coating. When isolation of the coating is desired, however, it is possible to select a coarse filter that will pass the suspended coating but not the polymer. Also subject to extraction are water-soluble inorganic additives such as the halide salts used in some antioxidant formulations.

Few dyes are used in molding or extrusion powders as colorants due to instability during processing. Induline and nigrosine are exceptions in black compositions. These can be extracted with boiling methanol and are recognized by the color of the solvent, blue for induline and red to purple for nigrosine.

Inorganic additives include pigments, thermal stabilizers, nucleating agents, fillers, and fire retardants. Emission spectroscopy provides a quick reference on metals present as well as relative concentrations. Frequently this information is adequate for definition of the problem and may indicate the next analytical step. The emission spectrogram can be run directly on the polymer, but concentration of the additives by ashing is advisable. A 2- to 5-g sample can be suitably ashed in a muffle furnace at a temperature of about $550°C$ to avoid loss or decomposition of the more volatile materials. Quantitative analysis of metals is frequently accomplished by x-ray fluorescence or by atomic absorption. X-ray fluorescence is generally the preferred procedure, and it requires only a flat surface of the kind obtained by machining or molding. For atomic absorption, solution of the metal is required, which is difficult for materials such as titanium dioxide. For example, the copper level in certain compositions stabilized against thermal oxidation has been analyzed both ways but more commonly by x-ray fluorescence. Consideration must be given to sources of copper other than the stabilization system such as copper phthalocyanine, a pigment sometimes used in nylon.

Identification of the inorganic additive may require its isolation. This can be done by ignition of the polymer, separation of the additive from a formic acid or *meta*-cresol solution of the nylon, or by extraction with hot water or acid as indicated above. Glass fibers, pigments, asbestos, and other materials unaffected by the nylon solvent can be removed by filtration or centrifugation of formic acid or *meta*-cresol solutions. Carbon black used for protecting against ultraviolet attack is quantitatively determined by filtration from a hydrochloric acid solution of the nylon. The absorbance of a solution in phenol has been used in specifications to indicate that the nylon will effectively prevent transmission of ultraviolet light (17). Pigments used in nylons can usually be isolated by filtration from solution of the nylon in formic acid or *meta*-cresol and can be identified by standard procedures. Identification of the inorganic filler, however isolated, can be done by standard chemical or physical means (52, 54).

Ignition of the nylon and determination of the filler content by the weight of the ash is frequently used for materials such as glass fiber, talc, titanium dioxide, and some salts. Many inorganic additives are partially decomposed during ignition; molybdenum disulfide, for example, is converted to the oxide, and appropriate correction must be made. The uncolored nylons without additives usually contain less than 0.05% ash. Many of the reconstituted nylons derived from textile scrap, however, will contain small amounts of titanium dioxide used in yarn as a delusterant, and its ash content will be correspondingly higher.

The determination of the presence and effectiveness of a nucleating agent is sometimes more important than knowledge of its structure. Although the use of molding tests are best, some measure of the degree of nucleation can be determined by DTA using the slope of the freezing exotherm or the spread between melting and freezing peaks. In comparing the thermogram of virgin polymer with that of a fabricated part, it should be recognized that some slight nucleation can occur during processing.

REFERENCES

1. ASTM General Methods of Testing, Part 27, D1895-69, "Apparent Density, Bulk Factor and Pourability of Plastic Materials."

2. ASTM General Methods of Testing, Part 27, D1925-70, "Yellowness Index of Plastics."

3. ASTM General Methods of Testing, Part 30, "Recommended Practice for Spectrophotometry and Description of Color in CIE 1931 System."

4. ASTM General Methods of Testing, Part 27, D1505-68, "Density of Plastics by the Density-Gradient Technique."

5. Johnson, F.R., and E. Weadon, *J. Text. Inst. Trans.* **55**, T162 (1964).

6. Ke, B., in *Newer Methods of Polymer Characterization*, B. Ke, Ed., Interscience Publishers, Inc., New York, 1964, Vol. 6, Chapter IX.

7. Sadtlers Research Laboratories, Inc., Philadelphia, Pennsylvania, 19104, "Infrared Spectra on Monomers and Polymers."

8. Frey, H.J., and J.R. Knox in *Analytical Chemistry of Polymers,* G.M. Kline, Ed., Interscience Publishers, Inc., New York, 1959, Part 1, Chapter X.

9. Kagarise, R.E., and L.H. Weinberger, "Infrared Spectra of Plastics and Resins," Naval Research Laboratory, Washington, D.C., NRL Report 4369, BP111438, May 26, 1954.

10. Starkweather, H.W., and R.E. Moynihan, *J. Polym. Sci.* **22**, 353 (1956).

11. Trifan. D.S., and J.F. Terenzi, *J. Polym. Sci.* **28**, 443 (1958).

12. Hummel, D.O., *Infrared Analysis of Polymers, Resins and Additives,* Wiley-Interscience, New York, 1969, Vol. 1, Part 2.

12a. Hummel, D.O., *Infrared Spectra of Polymers,* Interscience Publishers, a division of John Wiley and Sons, New York, 1966, pp. 61-64.

13. Liberti, F.N., and B. Wunderlich, *J. Polym. Sci., Part A-2,* **6**, 833 (1968).

14. Haslam, J., and H.A. Willis, *Identification and Analysis of Plastics,* D. Van Nostrand Co., Inc., Princeton, N.J., 1965.

15. Stuhlen, F., and H. Horn, *Kunstoffe* **46**, 63 (1956).

16. Clasper, M., and J. Haslam, *Analyst* **74**, 224 (1949).

17. Military Specification MIL-M-20693A, 8, April 1960.

18. Anton, A., *Anal. Chem.* **40**, 1116 (1968).

19. Symons, N.K.J., and E.C. McKannan, *Anal. Chem.* **31**, 1990 (1959).

20. Manufacturers Engineering and Equipment Corp., Hatboro, Pa.

21. E.I. duPont deNemours & Co., Instrument Products Division, Monrovia, Calif, 91016.

22. Zahn, H., and G.B. Gleitsman, *Agnew Chem. Int. Engl. Ed.* **12**, 410 (1963).

23. Schenker, H.C.C.L. Casto, and P.W. Mullen, *Anal. Chem.* **29**, 825 (1957).

24. Mori, S., and K. Obazaki, *J. Polym. Sci., Part A-1,* **5**, 231 (1967).

25. Goldfarb, I.J., and A.C. Meeks, AFML-TR-68-347, Part 1 (1969).

26. Bly, D.D., K.A. Boni, M.J.R. Cantow, J. Cazes, D.J. Harmon, J.N. Little, and E.D. Weir, *J. Polym. Sci.* **9B**, 401 (1971).

27. Burke, J.J., and T.A. Orofino, *J. Polym. Sci., Part A-2,* **7**, 1 (1969).

28. Waltz, J.E., and G.B. Taylor, *Anal. Chem.* **19**, 448 (1947).

29. Hardy, R. C., Natl. Bur. Standards Monograph **55**, Dec. 26, 1962.

30. Cannon, M.R., Private Report No. 1 to ASTM Subcommittee on Viscosity, Feb. 1950; and Cannon, M.R., and R.E. Manning, Private Report No. 2 to ASTM Subcommittee on Viscosity, Apr., 1959. The latter two are available from The Cannon Instrument Co., P.O. Box 16, State College, Pa. 16801.

31. International Union of Pure and Applied Chemistry, *J. Polym. Sci.* **8**, 257 (1952).

32. Kurata, M., M. Iwama, and K. Kamada, in *Polymer Handbook,* J. Brandrup and E.H. Immergut, Eds., Interscience Publishers, a division of John Wiley and Sons, New York, 1966, Section IV.

33. Huggins, M.L., *J. Am. Chem. Soc.* **64,** 2716 (1942).

34. Schulz, G.V., and F. Blaschke, *J. Prakt. Chem.* **158,** 130 (1941).

35. Ibrahim, F.W., *J. Polym. Sci.* 3A, 169 (1965).

36. Rafler, G., and G. Reinisch, *Faserforsch. Textiltech.* **21,** 412 (1970).

37. Taylor, G.B., *J. Am. Chem. Soc.* **69,** 635 (1947).

38. Howard, G.J., *J. Polym. Sci.* **37,** 310 (1959).

39. Schaefgen, J.R., and P.J. Flory, *J. Am. Chem. Soc.* **70,** 2709 (1948).

40. Tuzar, Z., and P. Kratochvil, *J. Polym. Sci.* 3B, 17 (1965).

41. Batzer, H., and A. Moschle, *Makromol. Chem.* **22,** 195 (1957).

42. Morgan, P.W., and S.L. Kwolek, *J. Polym. Sci.* 1A, 1147 (1963).

43. Nagaoka, T., M. Nakajima, Z. Makita, and M. Mitsuyoshi, *J. Chem. Soc. Japan, Ind. Chem. Soc.* **74,** 786 (1971).

44. Elias, H.-G., and R. Schumacher, *Makromol. Chem.* **76,** 23 (1964).

45. (a) Fikentscher, H., *Cellul. Chem.* **13,** 58 (1932); (b) Anon. BASF Bulletin, "Ultramid, A. Material for the Engineer," Nov., 1961, p. 13.

46. Heim, E., *Faserforsch. Textiltech.* **11,** 513 (1960).

47. Kohan, M.I., in *Macromolecular Syntheses,* J.R. Elliott, Ed., John Wiley and Sons, New York, 1966, Vol. 2, p. 7.

48. Kohan, M.I., unpublished data.

49. Smith, A.S., *J. Tex. Inst.* **48,** T86 (1957).

50. Webber, A.C., personal communication.

51. Reimschuessel, H.K., and G.J. Dege, *J. Polym. Sci., Part A-1,* 9, 2343 (1971).

52. Ryland, A.L., *J. Chem. Educ.* 35 (2), 80 (Feb. 1958).

53. Military Specification MIL-M-20693A, 8, April, 1960.

54. Voter, R.C., *J. Chem. Educ.* 35 (2), 83 (Feb., 1958).

CHAPTER 4

Behavior of Molten Nylons

M.I. Kohan

INTRODUCTION

The viscosity of a molten polymer affects flow in molds, cycle times, power requirements, extruder performance, extrudate shape stability, and is, in brief, an important characteristic of the polymer in all melt processes. In this chapter we will first review the theory of viscosity and techniques for its measurement with emphasis on applicability to molten nylons. The significance of chemical problems in the determination of the melt viscosity of nylons will also be examined before discussing the available data on specific nylons.

A detailed analysis of processing capability involves other important characteristics of a semicrystalline polymer: thermal diffusivity (Chapter 10), heat of fusion (Chapter 9), effect of pressure on melting point (Chapter 9), freezing point (Chapter 9), kinetics of crystallization (Chapter 8), and the effect of processing conditions on morphology (Chapter 8). As indicated, each of these properties is discussed in other chapters; their significance for melt fabrication is dealt with in the chapters on injection molding and extrusion (Chapters 5 and 6). Of course, the polymer is only one of two critical variables. The other is the processing device. A good example of the detailed analysis of the interaction of these variables as applied to extruder design is provided in the book of Tadmor and Klein (1).

THEORY AND MEASUREMENT OF VISCOSITY

Viscosity, Elasticity, and Rheology

The study of viscosity is in the province of rheology, which is concerned with the flow and deformation of matter. Reiner (2) has provided a graphic illustration of what rheology is all about: If one drops a pencil or a piece of putty or pours some water, the laws of mechanics enable us to say that each has the same acceleration due to gravity, but they say nothing about what happens when each strikes a surface – rebound, form a deformed blob, or flow over the surface. Rheology attempts to account for these kinds of variations in behavior: elastic-rebound, plastic deformation, and fluid flow. A sufficiently careful study would show that all materials exhibit elasticity, deform, and flow under suitable, albeit sometimes extreme, conditions. Some materials, polymers in particular, exhibit both elasticity and flow relatively easily, with no change in phase, and even at the same time; these are described as viscoelastic. For example, a solid polymer, when dropped, behaves like the pencil, but under steady load it changes shape or creeps (flows). The swelling of a molten polymer, a fluid, upon extrusion from a die is an example of elastic recovery.

Our principal concern in the typical molding or extruding of nylons is viscous flow, but it should be remembered that nylon melts are viscoelastic. Situations can occur wherein the elasticity of the melt affects processing performance and are more likely at very high levels of viscosity (3) or shear.

Viscosity and Laminar Flow

Viscosity is a measure of resistance to flow and is the reciprocal of fluidity. Flow in a channel may be laminar or turbulent. Laminar flow implies that there is a regular gradient in velocity, from zero in the layer clinging to the confining wall to a maximum at a point remote from the wall. It implicitly excludes motion other than in smooth streamlines. Turbulence describes the flow situation characterized by eddies or motion not in the principal direction of flow. It is normally assumed that the flow pattern in the actual processing of polymers is essentially laminar. The reason for the assumption is that the formulas used in the analysis of flow behavior are valid only for laminar flow. The justification for the assumption is that the calculations do generate guidelines helpful in solving processing problems.

The Reynolds number is a dimensionless quantity used to describe the point at which laminar flow ceases and turbulence occurs. The appropriate formula for the Reynolds number depends on the channel and, for a circular cross-section, is $D\bar{v}\rho/\mu$ where D is the diameter, $\bar{v}$ is the average linear velocity, ρ is density, and μ is viscosity. Assume values that will make this quantity large and that are not totally unreal: $D = 1$ cm (0.4 in.), $v = 1000$ cm sec^{-1} (394 in. sec^{-1}), $\rho = 1.0$ g cm^{-3} (62.4 lb ft^{-3}), and $\mu = 100$ poise (0.0015 lb (wt) sec in.$^{-2}$). In this case, the Reynolds number is 10, well below 2000, the minimum value generally considered necessary for turbulence in a round tube. A viscosity of 100 poise for a molten polymer would be very low, indeed, and is more likely to be higher by a factor of 10 to 1000. The assumption of laminar flow therefore appears reasonable, but a precautionary note is in order. Design factors such as sharp corners in a mold or crosshead and the development of excessive elastic strain can lead to turbulence regardless of very low Reynolds numbers.

In laminar flow velocity increases with distance from a confining surface (Figure 4-1). Viscosity is the property of the liquid that opposes the slipping of one layer past another. A high viscosity therefore means that at a given point from the wall, the velocity will be less than in the case of a lower viscosity. That is, the viscosity and the velocity gradient, the change in velocity with distance across the cross section, are inversely related. Because the applied force is inducing a slipping or shearing action, we call it a shearing force. Consider a cylindrical channel which might, for example, correspond to the runner of a mold used in injection molding, and visualize a volume of flowing material within a small cylinder of radius $= r$ concentric with the channel cylinder of radius $= R$ (Figure 4-2). The force responsible for flow is the difference in pressure multiplied by the cross-sectional area, $(P_1 - P_2)\,\pi r^2$ or $(\Delta P)\pi r^2$. The slower-moving annulus of material which surrounds the cylinder of fluid exerts a restraining influence proportional to the area of contact between the cylinder and the annulus, that is, $2\pi rL$ where L is the length of the cylinder. The shear stress is the shearing force divided by the area of the sliding surface:

$$\text{shear stress} = \frac{(\Delta P)\pi r^2}{2\pi rL} = \frac{(\Delta P)r}{2L} = \frac{(\Delta P)R}{2L} \text{ at the wall} \qquad (4\text{-}1)$$

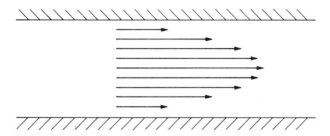

Fig. 4-1.   Laminar flow; length of arrow represents the velocity of the liquid layer.

Clearly, the velocity gradient increases as the applied pressure increases, and Newton postulated that the velocity gradient is directly proportional to the shear stress:

$$\text{velocity gradient} = \text{constant} \times \text{shear stress}$$

Because the velocity gradient is inversely related to viscosity, we may now write the following expression where μ is the coefficient of viscosity or, simply, the viscosity:

$$\text{velocity gradient} = \frac{-dv}{dr} = \frac{\text{shear stress}}{\mu}$$

where v is the velocity and the negative sign is used because r is measured from the center line, and v decreases as r increases, or

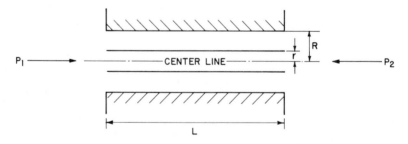

P_1 = PRESSURE DRIVING FLUID FROM LEFT TO RIGHT

P_2 = DOWNSTREAM PRESSURE

$\Delta P = P_1 - P_2$

NET DRIVING FORCE ON FLUID IN CYLINDER OF RADIUS $r = (\Delta P)\pi r^2$

AREA OF CONTACT BETWEEN CYLINDER AND SURROUNDING MATERIAL = $2\pi rL$

SHEAR STRESS = FORCE /AREA OF SHEARING SURFACE = $(\Delta P) r/2L$

Fig. 4-2.   Shear stress in tube.

$$\mu = \frac{(\Delta P) r / 2L}{-dv/dr} \quad \text{in the cylindrical channel} \tag{4-2}$$

The velocity gradient is also called the shear rate and has the units of distance/(time × distance) or, simply, reciprocal time. From Eq. (4-2) it is clear that viscosity has the units of pressure × time. As shown in Table 4-1, this is dyn sec cm^{-2} or, commonly, poise.

Table 4-1. Rheological units

	Common symbol	Recommended[a] units	Engineering units
Shear stress	τ	dyn cm^{-2} [b]	lb (wt) in.$^{-2}$ [c]
Shear rate	$\dot{\gamma}$ [d]	sec^{-1}	sec^{-1}
Viscosity	μ [e]$,\eta$ [f]	poise	lb (wt) sec in.$^{-2}$

[a] ASTM D1703-62.

[b] 1 lb (wt) in.$^{-2}$ = 6.9 × 10^4 dyn cm^{-2}

[c] Lack of uniformity exists not only with respect to units but also abbreviations. Employed here are those recommended by the American Chemical Society (Ref. 4).

[d] The dot is used to indicate that we are dealing with a first derivative, a rate of change, and not an absolute magnitude.

[e] Engineering practice.

[f] Common usage by rheologists, physicists, chemists.

The viscosity as defined above and expressed in poise is the dynamic or absolute viscosity, and we shall refer to it simply as viscosity.

Direct measurement of the velocity gradient is usually not feasible, but it is possible to relate the gradient or shear rate at the wall with the amount of material discharged per unit time. Imagine a thin annular ring, dr in width, at a distance, r, from the center line and assume a velocity at this radial position of v_r. Then

$$Q = \text{volumetric flow rate} = \int_0^R v_r \, 2\pi r \, dr \tag{4-3}$$

By integration of Eq. (4-2) between the limits of v_r and v_R, where v_R is the velocity at the wall and is zero, it is possible to express v_r in terms of r:

$$v_r = \frac{(\Delta P) R^2}{4L\mu} \left(1 - \frac{r^2}{R^2} \right) \tag{4-4}$$

Substitution of Eq. (4-4) in Eq. (4-3) and integration lead to the Hagen-Poiseuille law for laminar flow in tubes:

$$Q = \frac{\pi(\Delta P)R^4}{8L\mu} \tag{4-5}$$

Solving Eq. (4-5) for $\Delta P/\mu$ and using this result in Eq. (4-2) give the shear rate at any radial position:

$$\dot{\gamma}_r = \frac{-dv}{dr} = \frac{4}{\pi}\frac{Qr}{R^4} \tag{4-6}$$

At the wall, where $r = R$:

$$\dot{\gamma}_w = \frac{4Q}{\pi R^3} \tag{4-7}$$

We can now relate viscosity to meaningful and experimentally accessible quantities, the shear stress and shear rate at the wall:

$$\mu = \frac{\tau_w}{\dot{\gamma}_w} = \frac{(\Delta P)R/2L}{4Q/\pi R^3} \tag{4-8}$$

Equation (4-4) says that the velocity changes with the square of radial position; that is, the velocity profile is that of a parabola. On the other hand, both shear stress, Eq. (4-1), and shear rate, Eq. (4-6), vary directly with radius and, unlike velocity, are maximum at the wall.

Viscosity of Real Materials

If the shear stress/shear rate ratio is, in fact, constant, as was implied above, so that the viscosity is independent of the level of shear, then the fluid is said to be pure viscous or Newtonian. This has been said to be generally characteristic of pure liquids or solutions whose molecules contain fewer than 10^2 to 10^3 atoms (5). A liquid whose shear rate increases faster than the shear stress is said to be pseudoplastic or shear thinning; one whose shear rate increases more slowly than the shear stress, dilatant or shear thickening (Figure 4-3). It is possible for a material to exhibit no velocity gradient until a specific level of stress is exceeded. A material with Newtonian behavior above this yield value is called a Bingham plastic. The existence of a yield value is associated with emulsions or slurries

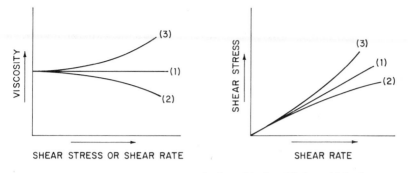

Fig. 4-3.   Flow curves: (1) Newtonian; (2) shear thinning; (3) shear thickening.

rather than single-phase melts or solutions. In each of the above cases a steady-state condition with no dependence of shear rate on time has been tacitly assumed. Gels and sols may exhibit thixotropy where the shear stress required for a given shear rate decreases with time, or rheopexy where the shear stress required to maintain a given shear rate increases with time.

Polymer melts are typically shear thinning because of improved chain alignment and decreased chain entanglement at higher levels of shear. With the increasing advent of polyphase systems such as rubber or glass reinforced polymers, polymer blends, and other modifications of polymers which may introduce a second phase, it is increasingly necessary to examine melt viscosity over a range of shear and be wary of anomalous behavior. A discussion of alternative flow curves and the materials to which they apply is provided in a book by Van Wazer and co-workers (5).

The shear stress versus shear rate behavior of polymer melts is sometimes approximated by a power law over a reasonable range of shear:

$$\tau = k\dot{\gamma}^{n} \tag{4-9}$$

The symbol k is used instead of μ to avoid assumption of Newtonian behavior. An apparent problem in the dimensions of k can be resolved by use of a reference state so that (24)

$$k = k_0 \dot{\gamma}_0^{n-1} \quad \text{and} \quad \tau = k_0 \left| \frac{\dot{\gamma}}{\dot{\gamma}_0} \right|^{n-1} \dot{\gamma}$$

The power-law exponent n is also called the flow-behavior index or flow index, and k is also known as the consistency index (15). However, preferred practice, especially with nylons, is to obtain melt data over a range of shear and make use of the apparent melt viscosity, μ_A, which corresponds to the value of the shear stress/apparent shear rate ratio at the specific level of shear (stress or rate) in question. An apparent shear rate is involved because the expression for $\dot{\gamma}_w$, Eq. (4-7), requires an integration in which the constancy of viscosity with changing shear, that is, Newtonian behavior, is assumed. Most nylons exhibit a region of essentially constant melt viscosity, which occurs below a shear stress of 5×10^5 to 1×10^6 dyn cm^{-2} (7.25 to 14.5 lb (wt) in.$^{-2}$), followed by a non-Newtonian region wherein the viscosity decreases at an accelerating rate as the shear stress increases (Figure 4-4).

Measurement of Viscosity

A device which measures viscosity is called a viscometer or, less commonly, a viscosimeter, and there are many such devices (5). Some viscometers yield results which lend themselves more readily than others to theoretical analysis. Some yield data which serve as useful indices of processing behavior although theoretical analysis is highly complicated. Some are routinely used for quality control or even in-line process control. However, for nylon melts as well as many other molten thermoplastics, capillary viscometers have usually proved to be the device of choice, and our discussion is limited accordingly.

Capillary viscometers come in a variety of sizes and shapes, but, as used to measure the melt viscosity of polymers, consist essentially of a cylindrical chamber from which the molten polymer is caused to flow through a die of smaller radius which is the capillary or, as it is often called, orifice (Figure 4-5). These devices are also referred to as rheometers or plastometers. To avoid confusion with glass viscometers which are used for low-viscosity liquids or solutions, are very different in design, and depend on gravity flow, the term rheometer will be used here. Equations (4-1) through (4-8) apply to Newtonian flow in a capillary rheometer.

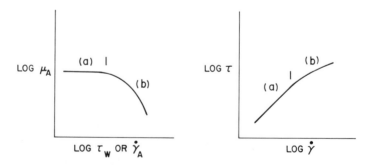

Fig. 4-4.   Common rheological curves for molten nylons: (a) Newtonian region; (b) non-Newtonian region.

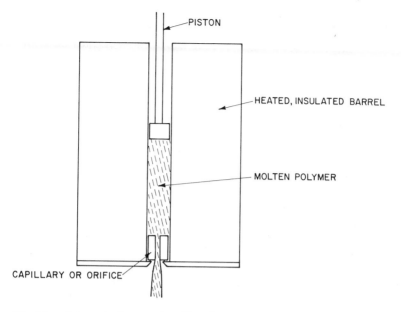

Fig. 4-5.   Schematic diagram of capillary rheometer.

A rheometer may be operated at constant load or constant rate. The constant load can be obtained via a piston supporting a weight, and the pressure needed for use in Eq. (4-8) is simply the load, piston weight included, divided by the average area of the piston face and chamber cross section. The constant load can also be obtained via gas pressure, and this is of particular help in achieving low loads unaffected by friction losses between the piston and barrel wall. The effluent is collected over a period of time and weighed. This can be converted to the volumetric flow rate needed for Eq. (4-8) as long as the melt density is known. It is also possible to time the period necessary to provide a specific piston displacement, l, and calculate the volume of discharge from $l \times A$, where A is the average cross section of the piston face and cylinder. The two techniques can obviously be combined to determine the melt density.

The melt densities of nylons are summarized in Table 4-2.

Variations in pressure and moisture content probably account for the discrepancies observed with nylon-66 and -6. The differences are small, however, and the assumption of a melt density of 1.0 g cm^{-3} is almost always an adequate approximation for these nylons. The melt density, like the room-temperature density, decreases as the hydrocarbon character of the nylon increases. The melt density at minimum processing temperature appears to approximate 0.87 times the maximum room-temperature density.

Table 4-2. Melt densities of nylons

Nylon	Pressure		Temp. range		Density		Ref.
	(lb (wt) in.$^{-2}$)	(kg (wt) cm^{-2})	($^\circ$C)	($^\circ$F)	(g cm^{-3})	(lb ft^{-3})	
66	--	--	270-300	518-572	0.97-0.95	61-59	7
66	2,000	140	274-299	525-570	0.99-0.975	61.6-60.9	6
66	10,000	700	288-299	550-570	1.01-1.00	62.9-62.5	6
610	--	--	230-300	446-572	0.94-0.91	59-57	8
6	--	--	230-300	446-572	0.99-0.95	62-59	7
6	1,000	70	232-288	450-550	0.996-.97	62.1-60.5	6
8	--	--	210-250	410-482	0.94-.92	59-57	7
12	--	--	190-250	374-482	0.90-.87	56-54	7

The dead-weight, constant-load rheometer described in ASTM D1238-65T is the well-known "melt indexer" first introduced to provide "flow numbers" for polyethylene. This instrument has the advantages of relatively low cost and simple operation, but it has a limited range in shear stress and yields a time-averaged result. Standard test conditions of temperature and load are cited for nylons in ASTM D1238-65T, but these have found only restricted application because experimental difficulties preclude use of melt flow for specification purposes for nylons. This is not peculiar to nylons but exists with condensation polymers in general where the dynamics of reversible polymerization-hydrolysis (Chapter 2, p. 33) apply.

A constant-rate rheometer measures continuously the force necessary to drive the molten polymer out of an orifice at a specified rate of travel of the piston which gives the volumetric flow rate ($A \times l/t$) directly. An Instron machine can be adapted to serve this purpose, and one such device has been described by Merz and Colwell (9). One advantage is operability at either low or high volumetric flow rates because there is no need for independent measurements of time and piston displacement or weight of extrudate. The continuous record of load simplifies prompt recognition of steady-state conditions or anomalous behavior. This also makes it easier to make determinations at different levels of shear from a single charge of polymer. Cost is the principal disadvantage.

A capillary rheometer has been developed which records continuously both load and volumetric flow rate and can be operated at either constant stress or rate. It is a modification of a design introduced by Sieglaff and McKelvey (10) for the determination of the melt viscosity of poly(vinyl chloride). Rapidity of measurement, an important factor in melt unstable polymers, and the use of small amounts of polymer, important in developmental work, are claimed as advantages.

A list of capillary rheometers commercially available for high-temperature work with molten polymers is provided in Table 4-3. A maximum temperature of not less than 300°C (572°F) is recommended for general use with nylons.

Problems in Capillary Rheometry

A problem often cited with respect to the determination of shear stress involves end effects. The flow pattern at the capillary exit and entrance differs from that in the capillary. Also, compaction of the fluid at the capillary entrance may be opposed by elastic forces manifest as noted earlier in the swelling of the extrudate. Bagley (11) showed that even for a non-Newtonian, molten polymer such as polyethylene, end effects could be compensated for by assuming a reduction in shear stress which was equivalent to that induced by an imaginary increase in the length of the capillary:

$$(\tau_w)_{corr.} = \tau_{wc} = \frac{(\Delta P)R}{2(L + L_i)} \qquad \text{or} \qquad \tau_{wc} = \frac{\Delta P}{2(L/R + e)} \qquad (4\text{-}10)$$

where L_i = imaginary increase in length of capillary and e = imaginary increment in the ratio L/R or L_i/R.

Assuming that constant shear rate reasonably approximates a constant τ_{wc}, Eq. (4-10) says that a plot of ΔP versus L/R at a fixed shear rate should be linear and should intercept the L/R axis ($\Delta P = 0$) at $-e$, and this is what Bagley found. Greater shear rate means increased throughput for a given capillary and less time to compensate for end effects. An increase in e with increasing shear rate is intuitively reasonable and is observed (Figure 4-6). Note that one can use the e-value appropriate to different shear rates, calculate the corresponding τ_{wc}'s,

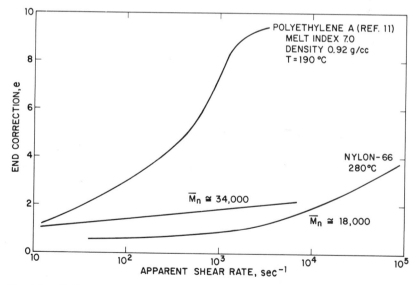

Fig. 4-6.   End correction versus apparent shear rate.

Table 4-3. Commercially available capillary rheometers

		Approx. cost[a] ($)	Type	Max. temp. (°C)	Approximate range in shear[b]		Comment
					Rate (sec^{-1})	Stress (dyn cm^{-2})	
Standard Melt Indexer	F. F. Slocomb Corp. 1400 Poplar St., P.O. Box 1591 Wilmington, Del. 19899	965-1335	Const. load, dead wt.	200	1-5 × 10^2	5 × 10^3-10^6	
Flow Rater	F. F. Slocomb Corp. 1400 Poplar St., P.O. Box 1591 Wilmington, Del. 19899	1145-1510	Const. load, dead wt.	265	1-5 × 10^2	5 × 10^3-10^6	
High Melt Indexer	F. F. Slocomb Corp. 1400 Poplar St., P.O. Box 1591 Wilmington, Del. 19899	1570-1935	Const. load, dead wt.	400	1-5 × 10^2	5 × 10^3-10^6	
Melt Index Apparatus Model CS-127E	Custom Scientific Instruments, Inc. 13 Wing Dr., P.O. Box A Whippany, N. J. 07981	1060	Const. load, dead wt.	400	1-5 × 10^2	5 × 10^3-10^6	
Micro-Flow Melt Indexer, Model CS-127MF	Custom Scientific Instruments, Inc. 13 Wing Dr., P.O. Box A Whippany, N. J. 07981	1090	Const. load, dead wt.	400	1-50	5 × 10^3-10^5	As low as 0.02-gram charge
High Shear Melt Index Apparatus, Model CS-127HS	Custom Scientific Instruments, Inc. 13 Wing Dr., P.O. Box A Whippany, N. J. 07981	2650	Const. load, gas press. (max 650 psig)	400	1-5 × 10^2	5 × 10^3-10^6	
High Shear	Pressure Products Industries,	3375	Const. load, gas	300	1-5 × 10^2	5 × 10^3-5 × 10^6	

Instrument	Manufacturer	Type	Price				Comments
Viscometer HSV-1	Division of the Duriron Co., Inc. Hatboro, Pa. 19040	press. (max 3000 psig) (higher PT avail. on request)					
Extrusion Plastometer (Melt Indexer)	Tinius Olson Testing Machine Co., Inc. Easton Road Willow Grove, Pa. 19090	Const. load, dead wt.	1400-1500	300	$1-5 \times 10^2$	5×10^3-10^6	
Sieglaff-McKelvey Rheometer	Tinius Olson Testing Machine Co., Inc. Easton Road Willow Grove, Pa. 19090	Const. load or const. rate	11,640	426	1-10^4	10^3-10^7	2-gram charge, continuous record of both load and rate
Capillary Rheometer	Instron Corp. 2500 Washington Street Canton, Mass. 02021	Const. rate	5935 (rheometer only) Add $13,500 for 5000-lb floor model Instron tester	340	10^{-1}-10^5	10^3-10^7	Continuous record of load
			Add $8,750 for modified 1000-lb table model Instron tester		10^{-1}-10^5	10^3-2×10^6	

aAs of December,1969. The price range reflects a choice in specific pieces of equipment.
bBased on a reasonable variation in capillary dimensions.

and arrive at a constant μ only if the fluid is Newtonian. For shear thinning polymers such as nylons, the end-correction technique is simply a way of getting a shear-dependent viscosity which does not also depend on capillary goemetry.

Lupton and Regester (12) obtained straight line extrapolations for nylon-66 of two different molecular weights using L/R = 2, 8, and 32 and flat capillary entrances and exits. The end correction differs by a factor almost proportional to the ratio of molecular weights up to about 3×10^3 sec-1 (Figure 4-6). This may be only coincidental; additional data are needed. The L/R in ASTM D1238-65T is 7.64 so that the correction (e = 0.6 to 3.7) varies in this case from about 8 to 48%. A reduction in the orifice diameter from 2.09 to 0.89 mm (0.0825 to 0.035 in.) increases L/R to 18 and decreases the error to less than 10% below 10^3 sec-1 for a $\bar{M}_n$ of 34,000 and below 10^4 sec-1 for a $\bar{M}_n$ of 18,000. For a high-viscosity, low-flow polymer a larger diameter may be necessary to ensure an extrudate in sufficient amount for accurate weighing. If L is increased proportionately to maintain L/R, the flow, assuming Newtonian behavior, will increase as the third power of the ratio in diameters because the flow depends on the fourth power of the radius and inversely on the length according to Poiseuille's law, Eq. (4-5). For example, changing the diameter from 0.89 to 1.27 mm (0.035 to 0.050 in.) and L from 8.0 to 11.4 mm (0.315 to 0.450 in.) involves an increase in both R and L by a factor of 1.43, which leads to an increase in flow of 1.43^3 or almost a factor of 3. The end correction for nylons does not appear to be significantly affected by changing the radius as well as the length of the capillary at least within the interval $R = 0.25$ to 0.625 mm (0.010 to 0.025 in.).

Over virtually all of the range in shear rate (10 to 10^5 sec-1) the end correction is much less for the nylons than for the low-density, medium-flow polyethylene studied by Bagley (Figure 4-6). The unsuitability of the ASTM constant-load plastometer for rheological calculations has often been cited, but the above data show that over a considerable range of shear, particularly if an orifice of moderately higher L/R (18/1) is used, the ability of the ASTM device to provide reasonable estimates of melt viscosity at 280°C (536°F) for nylon-66 is not handicapped by a large end correction. Applicability to other aliphatic nylons at least within this molecular weight range and at roughly similar temperatures is to be expected.

It was noted earlier that $4Q/\pi R^3$ corresponds to the maximum shear rate at the wall, $\dot{\gamma}_w$ in Eq. (4-7), only for Newtonian fluids. It is, however, possible to obtain readily a true maximum shear rate for even non-Newtonian fluids. All that is needed is the curve of $4Q/\pi R^3$ versus shear stress and Eq. (4-11), which was derived by Rabinowitsch (13):

$$\dot{\gamma}_T = \dot{\gamma}_A \left(\frac{3+N}{4}\right) \tag{4-11}$$

where

$\dot{\gamma}_T$ = true shear rate at wall

$\dot{\gamma}_A$ = apparent shear rate = $4Q/\pi R^3$

N = $\dfrac{d \ln \dot{\gamma}_A}{d \ln \tau_w}$ = slope of curve of log $\dot{\gamma}_A$ versus log τ_w at $\dot{\gamma}_A$ or τ_w of interest

ln = natural logarithm

log = common logarithm

This is also written as in Eq. (4-12) if the log of the shear stress is plotted against the log of the shear rate:

$$\dot{\gamma}_T = \dot{\gamma}_A \left(\frac{3N' + 1}{4N'} \right) \qquad (4\text{-}12)$$

where

$N' = 1/N$ = the power-law exponent, n, in Eq. (4-9).

The true melt viscosity is similarly obtained from Eq. (4-13):

$$\mu_T = \mu_A \left(\frac{4}{4 - m} \right) \qquad (4\text{-}13)$$

where

$m = \dfrac{d \ln \mu_A}{d \ln \tau_w}$ = slope of curve of log μ_A versus log τ_w at μ_A or τ_w of interest

For shear thinning polymers such as nylons, N is greater than 1, N' or n is less than 1, and m has a negative value. The true shear rate is, therefore, higher than the Newtonian $4Q/\pi R^3$, and the true melt viscosity is lower than the apparent value. The correction factor involved here for nylons is almost always less than 15% and usually less than 5%. In most practical situations this correction is not important.

The development of Eq. (4-11) includes an expression which shows that the apparent shear rate, $4Q/\pi R^3$, is independent of tube radius at constant stress,

$(\Delta P)R/2L$. This says that even for non-Newtonian fluids, the shear rate information obtained in a capillary instrument of one size can be applied to flow in practical situations involving tubes of different size as long as the shear stress is the same. Computer programs (14) have been written to apply both the end correction of Eq. (4-10) and the shear-rate correction of Eq. (4-11).

The problems discussed above are those most often cited as contributing to errors in the determination of the melt viscosity of polymers. One other frequently mentioned in capillary viscometry is the kinetic-energy effect. The fluid gains in velocity as it moves from the wide chamber into the narrow capillary; this means an increase in kinetic energy, which has to come about at the expense of the applied pressure. The pressure drop due to this kinetic-energy effect, ΔP_{KE}, is

$$\Delta P_{KE} = \frac{KE}{tQ} = \frac{\rho Q^2}{\pi^2 R^4} \qquad \text{(Newtonian fluid)} \qquad (4\text{-}14)$$

ΔP_{KE} for non-Newtonian fluids that obey the power law of Eq. (4-9) can be calculated by dividing Eq. (4-14) by a numerical factor defined by the exponent, n (15).

Assume $\rho = 1$ g cm^{-3} (62.4 lb ft^{-3}) and an unlikely combination of a small R (0.03 cm = 0.014 in.) and a large Q (6 cm^3 min^{-1} = 0.1 cm^3 sec^{-1}); then ΔP_{KE} is approximately $10^{-2}/(10 \times 81 \times 10^{-8}) = 1.1 \times 10^3$ dyn cm^{-2}. Assuming even a small L/R of 8/1, this translates to a $(\Delta P_{KE})R/2L$ of 700 dyn cm^{-2} (0.01 lb(wt)in.$^{-2}$). Correction for non-Newtonian behavior is small for nylons. Because rheometer measurements for molten nylons normally involve shear stresses in the range of 10^4 to 5×10^6 dyn cm^{-2}, the kinetic-energy effect is most often ignored.

The end effects, non-Newtonian behavior, and the kinetic-energy effect all serve to make the observed melt viscosity higher than the true value. Nonetheless, the measurement of the melt viscosity of nylons in a rheometer of adequately high L/R does not normally involve a significant error that can be ascribed to these physical factors. On the other hand, low L/R and high shear rate as may apply in molding or extrusion, can make difficult estimating the viscosity in actual processing.

The literature deals with many other concerns in capillary rheometry: the effect of the absolute level of pressure on melt viscosity (16), piston leakage and friction (17), pressure drop in the reservoir chamber which has been treated as a second capillary in series with the smaller one (17), heat generation (18), the critical stress for melt fracture (19), and slippage (20) where the melt does not adhere to the confining wall. The references cited are only a selected few, but none of this literature deals with nylons. The linearity of the extrapolations of Lupton and Regester noted above and the relatively low end corrections shown in Fig. 4-6 suggest that these are not significant concerns in melt rheometry for many nylons.

Middleman (Ref. 22, pp. 32 and 33) has provided a convenient nomograph for estimating the maximum temperature rise at an adiabatic capillary wall due to viscous heating. The thermal conductivity and diffusivity of nylons in the melt are required in cgs units. Reasonable values are, respectively, 2×10^4 erg cm^{-1} sec^{-1} °C^{-1} and 7×10^4 cm^2 sec^{-1}. A melt viscosity of 1700 poise (0.025 lb(wt)sec in.$^{-2}$) at 10^3 sec^{-1} is reasonable for a nylon (see Figure 4-14 below) and yields a heat rise of 3°C, corresponding to less than a 10% change in viscosity, in a capillary with $L/D = 10/1$ and $D = 0.1$ cm (0.039 in.). Because heat would be lost to the wall and rheometer capillaries have small diameters, the actual temperature rise would be less. Heat rise is therefore not normally a factor in rheometry with nylons. Heat generation can be important, however, in actual processing, particularly where high pressures are needed to overcome high melt viscosities and the large channels compared to capillary diameters make heat transfer more difficult. Heat generation means, of course, lower viscosity and opposes the physical factors cited earlier.

For those interested in more extensive remarks on the phenomenological, macroscopic analysis of viscosity and its measurement, Refs. 5 and 21 are helpful. These provide additional references which explore the subject in greater depth. References 22 and 23 are for those interested in recent, mathematically sophisticated analyses.

Flow in Channels of Different Shapes

Application of viscosity information to real situations requires consideration of flow in channels other than tubes. It can be shown (Figure 4-7) that in a slit die the shear stress on a rectangular box of fluid within the die is $(\Delta P)h/L$ where h is the half-height above or below the center line and L is the land length. The width does not enter the expression as long as it is much greater than the height of the box, $2h$. The maximum shear stress at the wall is

$$\tau_w = \frac{(\Delta P)H}{2L} \qquad \text{where } H \text{ is the die opening} \qquad (4\text{-}15)$$

Note the similarity to Eq. (4-1) which involves, however, the tube radius whereas Eq. (4-15) involves the die opening rather than half the die opening. We ignore the effect of the sides and assume that the only velocity gradient that is important is $-dv/dh$ so we can write as in Eq. (4-2).

$$\mu = \frac{(\Delta P)h/L}{-dv/dh} \quad \text{or } -dv = \frac{\Delta P}{L\mu} h \, dh \qquad (4\text{-}16)$$

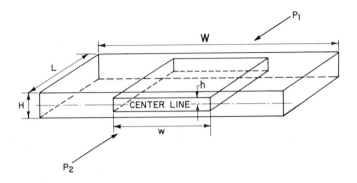

P_1 = PRESSURE ON REAR RECTANGULAR FACE DRIVING FLUID FORWARD

P_2 = DOWNSTREAM PRESSURE ON FRONT RECTANGULAR FACE

$\Delta P = P_1 - P_2$

NET DRIVING FORCE ON FLUID IN BOX OF DIMENSIONS, 2h × w × L, = $(\Delta P)(2h)w$

AREA OF CONTACT BETWEEN BOX AND SURROUNDING MATERIAL = $2(2h)L + 2wL$

SHEAR STRESS = FORCE / AREA OF SHEARING SURFACE = $(\Delta P)hw/L(2h + w)$

$$= \frac{(\Delta P)h}{L} \text{ IF } w \gg 2h$$

Fig. 4-7.   Shear stress in slit die.

Integrating Eq. (4-16) between the limits of v_h and v_w corresponding to the limits of h and $H/2$ yields the slit analog of Eq. (4-4) for a Newtonian fluid:

$$v_h = \frac{(\Delta P)H^2}{2L\mu} \left(\frac{1}{4} - \frac{h^2}{H^2} \right) \tag{4-17}$$

Consider two thin rectangular layers as shown in Figure 4-8. Then

$$Q = \text{volumetric flow rate} = 2 \int_0^{H/2} v_h \, W dh \tag{4-18}$$

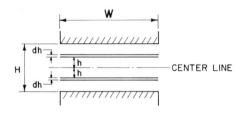

Fig. 4-8.   Two thin layers in slit die.

Substitution of Eq. (4-17) in Eq. (4-18) and integration lead to the slit analog of Eq. (4-5):

$$Q = \frac{(\Delta P)WH^3}{12L\mu} \tag{4-19}$$

Solving Eq. (4-19) for $(\Delta P)/\mu$ and using this result in Eq. (4-16) yield the expression for shear rate at any distance from the center line:

$$\dot{\gamma}_h = \frac{-dv}{dh} = \frac{12Qh}{WH^3} \tag{4-20}$$

At the wall where $h = H/2$,

$$\dot{\gamma}_w = \frac{6Q}{WH^2} \tag{4-21}$$

In analogy to Eq. (4-8) and for a Newtonian fluid, using Eqs. (4-15) and (4-21),

$$\mu = \frac{\tau_w}{\dot{\gamma}_w} = \frac{(\Delta P)H/2L}{6Q/WH^2} \tag{4-22}$$

Flow through an annulus which is thin so that the circumferences of the inner and outer surfaces are nearly the same can be approximated using Eqs. (4-15) through (4-22) in which W is the circumference of the mid-line of the annulus. The reader is referred to McKelvey's book (24) for a power-law treatment of flow through slits and narrow annuli as well as for flow analysis involving other shapes. A generalized equation for Newtonian flow in channels of different geometry together with the graph of the flow coefficient versus aspect ratio, the minimum dimension divided by the maximum dimension, has been provided by Lahti (25) and is especially convenient (Figure 4-9).

MELT VISCOSITY OF SPECIFIC NYLONS

Studies of the melt viscosity of nylons prior to the advent of commercial capillary rheometers were few but varied. Glass capillary viscometers involving undefined but low shear rates (27,28), falling ball viscometers (29-32), rotational and torsional techniques (33,34), and the damping of a vibrating reed (35) were all used. Much of this work involves polycaprolactam and notes effects of molecular weight, the extractable low-molecular-weight fraction, moisture content, and temperature, but the information necessary to ensure adequate control of variables leading to reliable correlations is mostly lacking. Some results reported, such as the dependence of melt viscosity on the presence of a small amount of acetamide end groups, probably reflect inadequate control leading to questionable conclusions.

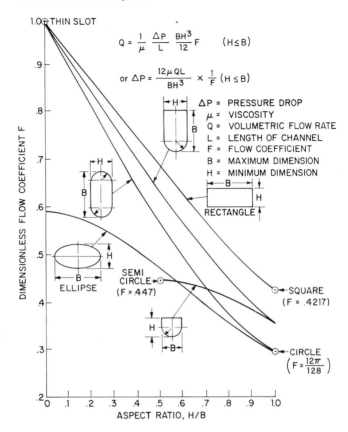

Fig. 4-9. Flow coefficient versus aspect ratio for various channel cross sections (25).

Importance of Chemistry

Nylons obey the equilibrium expressed in Eq. (2-1) (Chapter 2, p. 33):

$$K_c = \frac{[NHCO]\ [H_2O]}{[NH_2]\ [COOH]}$$

This says that if the water content is sufficiently high, hydrolysis to a lower molecular weight will occur; and if it is sufficiently low, polymerization to a higher molecular weight will occur:

$$\underline{\hspace{1cm}} COOH + H_2N \underline{\hspace{1cm}} \underset{\text{high } H_2O}{\overset{\text{low } H_2O}{\rightleftharpoons}} \underline{\hspace{1cm}} CONH \underline{\hspace{1cm}} + H_2O$$

A change in temperature must affect K_c and $[H_2O]$ in exactly the same way if the product of ends, $[NH_2] \times [COOH] = EP$, is to remain unchanged. The product of ends determines the molecular weight of the polymer:

$$\overline{M}_n = \frac{2 \times 10^6}{(\Delta^2 + 4EP)^{1/2} + U} \qquad (4\text{-}23)$$

where
Δ = imbalance = $[COOH] - [NH_2]$, equiv./10^6 g
U = unreactive, nontitratable ends, equiv./10^6 g

Raising temperature lowers K_c; maintaining molecular weight at a higher processing temperature therefore requires a lower moisture content in the polymer or a lower hold-up time. Time is obviously a factor since the effect of a too dry or too wet polymer depends on how fast the hydrolytic and polymerization reactions occur (Figure 4-10). Summing up, the molecular weight of a processed nylon depends on the following:

1. the molecular weight of the virgin polymer,
2. the water content of the virgin polymer,
3. the processing temperature,
4. the hold-up time.

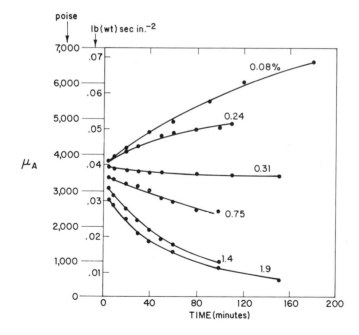

Fig. 4-10.   Effect of moisture content and time on apparent melt viscosity at 10^{-1} sec, 230°C (446°F), of extracted polycaprolactam, $\overline{M}_n$ = 17,600 (26).

The need to maintain molecular weight depends on the process and the product. In many cases small variations in molecular weight are of no significance. Other factors may be controlling; for example, in extrusion of a low-viscosity nylon, a low moisture content may be necessary to avoid undesirable bubbling of the melt, and the accompanying increase in molecular weight during extrusion may not significantly affect extruder output or other process variables.

The greatest need for very precise control of moisture content is in the study of the viscosity of molten nylons as a function of temperature and molecular weight. Here, accurately known moisture contents checked by measuring the molecular weight of the extrudate is essential to guarantee that the observed relation is the desired one free of other effects. It is the dynamics of the temperature dependent, reversible polymerization-hydrolysis that makes determination of the melt viscosity of nylons less precise than that of the more extensively studied polyethylenes or polystyrenes. Some investigators using a constant rate device have measured the force after a 3- to 6-min hold-up at a specified shear rate and then remeasured it after a 10- to 15-min hold-up and used a small change (10%) to argue in favor of reasonable control of molecular weight. It is clear from Figure 4-10 that even the initial observation may vary so that the concept of a "small" change in viscosity over a 10-min period as evidence of molecular weight control is not necessarily valid.

An important distinction is being made and should be understood. Qualitative estimates of melt viscosity can provide useful guidelines in many processing problems as is illustrated in Chapters 5 and 6. A precise analysis of the effect of small changes in variables on melt viscosity requires a sophisticated understanding of the fabrication process, the techniques of measuring viscosity, and the chemistry of nylons. It also requires accurate determination of moisture content, molecular weight, and the rates of polymerization and hydrolysis at the temperatures involved. This kind of analysis is a challenging task of academic interest; it is unlikely, however, that this level of precision is required to cope with the kinds of processing problems encountered in commercial practice.

The above analysis assumes unmodified nylons and deals only with the dynamics of the polymerization-hydrolysis equilibrium. The presence of additives may alter the rate at which changes occur and may even alter the nature of the changes. Nonhydrolytic, thermal degradation is not a factor except in unusual circumstances. Because of concern on the part of all who must contend with molten nylons, the subject of thermal degradation was dealt with in some detail in Chapter 2. Very little of the studies on degradation deal, however, with the initial phases which would be most likely to have implications for behavior in melt processing. On the practical level, the processor must not overlook the possibility that the instrument readings may be higher or lower than actual melt temperatures depending on the location of thermocouples, the

working of the melt, and the throughput rate. Concern for melt temperature
rather than instrument settings can sometimes lead to optimum performance.

Effect of Shear Rate and Shear Stress

A thorough study of the melt viscosity of polycaprolactam is that of Pezzin and
Gechele (26), who made use of a constant-rate rheometer. A total of 35
polymers extracted to a water soluble content of less than 0.8 wt. % and ranging
from a number-average molecular weight ($\bar{M}_n$) of 3500 to 42,500 were examined
over four decades of shear rate (1.7 to 15,000 sec^{-1}) at 230°C (446°F). Control
of moisture content and hold-up times and use of a low temperature minimized
fluctuation in molecular weight and content of extractables. High capillary
length to radius ratios of 67.4 to 271.2 were used. End corrections were not
calculated, but the data obtained appear independent of L/R and would be
expected to be independent based on the experience with nylon-66 cited in
Figure 4-6.

Nearly Newtonian behavior (viscosity constant with changing shear) at low
levels of shear are observed (Figure 4-11). Newtonian behavior requires that a
plot of log shear stress versus log shear rate be a straight line with a slope of
unity. A straight line with a slope less than unity indicates a shear thinning fluid
obeying the power law. Figure 4-12 demonstrates linearity up to a limiting value
of shear rate for several polymers differing in molecular weight. Figure 4-13
plots slope (the power-law exponent) against molecular weight. A linear but
small decrease from unity at a $\bar{M}_n$ of 4000 to 0.83 at a $\bar{M}_n$ of 30,000 is
indicated.

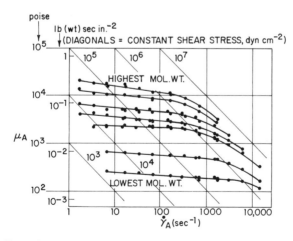

Fig. 4-11. Dependence of apparent melt viscosity on apparent shear rate at 230°C
(446°F) for extracted polycaprolactams, $\bar{M}_n$ = 8050-25,600 (26).

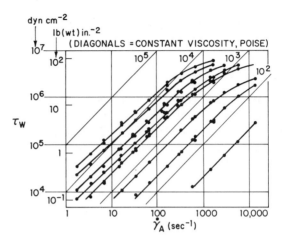

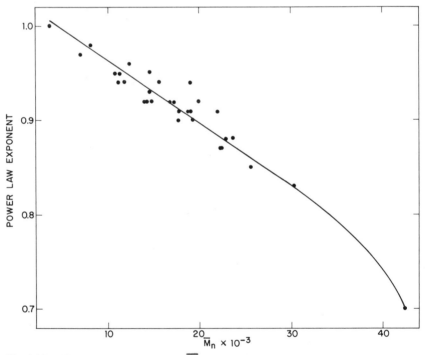

Fig. 4-12. Relation of wall shear stress and apparent shear rate at 230°C (446°F) for extracted polycaprolactams, $\overline{M_n}$ = 3500-25,600 (26).

Fig. 4-13. Power-law exponent versus $\overline{M_n}$ for extracted polycaprolactam (based on data at 230°C in Ref. 26).

Similar behavior is observed with nylon-6 at a higher temperature and with other linear, aliphatic nylons (Figure 4-14). Inclusion in the polymer chain of a ring structure as in the case of TMDT (poly(trimethylhexamethylene terephthalamide), see Table 1-4) appears to increase shear sensitivity (Figure 4-14).

At high rates of shear the rate of decrease in viscosity increases; further, the onset of nonlinearity occurs at lower shear rate as the viscosity (molecular weight) increases (Figure 4-11). On the other hand, it is clear from Figure 4-11, where the diagonals represent lines of constant shear stress, and from Figure 4-12 that the departure from linearity corresponds roughly to constant shear stress (5×10^5 - 1×10^6 dyn cm^{-2} or 7.25-14.5 lb(wt)in.$^{-2}$).

Lamb (45) discusses the application of rheological measurements to injection molding and suggests the ratio of melt viscosities at shear rates of 100 and 1000 sec^{-1} as an index of shear sensitivity. His data, summarized in Table 4-4, indicate that commercial molding compositions of nylon-6, -66, -11, and -610 are significantly less sensitive to shear than other molding compounds.

Unstable flow resulting in rheometer extrudates with a wavy surface has been observed at high shear in many polymers (53). In general, the onset of this kind of physical instability occurs at a true shear stress of about 10^6 dyn cm^{-2} (14.5 lb(wt)sec in.$^{-2}$) (53,54). Thus, the shear rate necessary for instability increases as viscosity decreases. A nylon-66 molding compound at 275°C (527°F), but not a nylon-6 at 240°C (464°F), required about an order of magnitude higher shear stress, about 10^7 dyn cm^{-2} (53). The instability phenomenon is associated with the elastic properties of the melt (54).

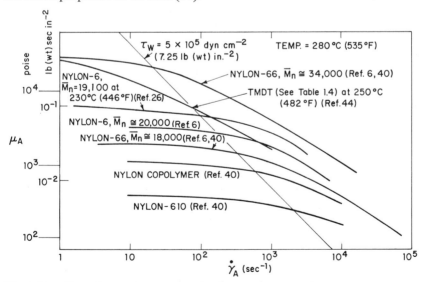

Fig. 4-14. Dependence of apparent melt viscosity on apparent shear rate for various nylons.

Table 4-4. Shear sensitivity of polymers (Ref. 45)

Polymer	Temperature ($^\circ$C)	($^\circ$F)	μ_A at 100 sec^{-1} / μ_A at 1000 sec^{-1}
Acetal copolymer	180	356	2.1
Acrylic	200	392	2.3
Nylon-6	280	536	1.4
Nylon-66	310	590	1.2
Nylon-11	250	482	1.8
Nylon-610	280	536	1.6
Polyethylene	190	374	$\geqslant$ 2.6
Polypropylene	230	446	3.6-4.7
Polystyrene	240	464	3.9
Polycarbonate	270	518	2.7
Poly (vinyl chloride)	190	374	5.0-6.0

Effect of Temperature

Processing temperatures for molten nylons are sufficiently high, that is, at least 100°C above the glass-transition temperature (see Chapter 9), to permit use of an Arrhenius equation (46):

$$\mu = Ae^{-E_v/RT} \tag{4-24}$$

where A is a constant, E_v is the energy of activation of viscous flow, R is the gas constant, and T is absolute temperature, degrees Kelvin. A linear plot of log viscosity versus reciprocal temperature with a slope proportional to E_v is indicated by Eq. (4-24). The E_v for nylon-6 at an apparent shear rate of 10 sec^{-1} over the interval 230 to 250°C, was found to be 9 to 13 kcal mole^{-1} and to be independent of molecular weight (26). A value of 11 kcal mole^{-1} can be calculated for a nylon-6, $\overline{M}_n \cong 20{,}000$, over the interval 232 to 288°C, using the viscosities at 10 sec^{-1} reported by Westover (6). E_v for nylon-66, $\overline{M}_n \cong 18{,}000$, from the melt viscosities (6) at 10 sec^{-1}, 280 to 300°C, is about 15 kcal mole^{-1}, not much different from the E_v for nylon-6.

Within the limits of accuracy of most measurements and within the processing range of most nylons, the log viscosity can, for simplicity in practical application, be related directly to temperature. This was done in the Westover compilation (6) from which Figures 4-15 and 4-16 are derived. Viscosities shown at 10 and 1000 sec^{-1} for nylons-6 and -66 suggest that the rate of change of viscosity with temperature is somewhat less at the higher shear rate.

A viscosity range is plotted against temperature in Figure 4-17. The slopes of the bands suggest a similar sensitivity to temperature for the viscosities of

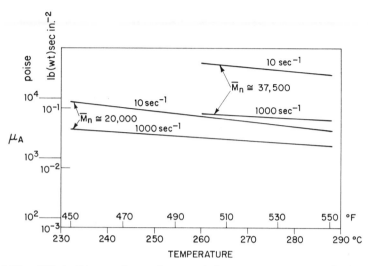

Fig. 4-15. Effect of temperature and apparent shear rate on apparent melt viscosity of nylon-6 (6).

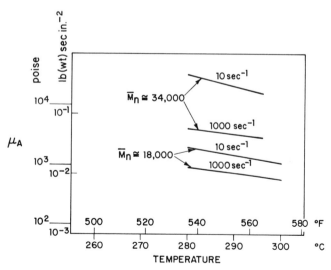

Fig. 4-16. Effect of temperature and apparent shear rate on apparent melt viscosity of nylon-66 (6).

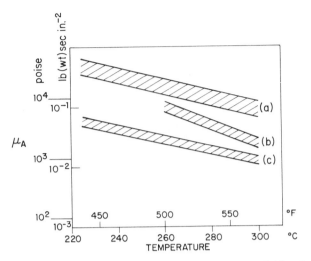

Fig. 4-17. Effect of temperature on apparent melt viscosity of (*a*) nylon-610, $\overline{M}_n \cong$ 19,500; (*b*) nylon-66, unspecified $\overline{M}_n$; and (*c*) nylon-6, $\overline{M}_n = 18,000$ (shear rates not indicated) (8, 43).

nylon-6 and nylon-610 (E_ν calculated on minimum slope = 12 to 13 kcal mole^{-1}) and a higher sensitivity for nylon-66 ($E_\nu \cong 21$ kcal mole^{-1}). Absence of information with respect to water content and extrudate molecular weight casts some doubt on the data because of the structural similarity of these resins.

A rapid change of viscosity with temperature ($E_\nu \cong 33$ kcal mole^{-1}) is suggested for TMDT, poly(trimethylhexamethylene terephthalamide), in Figure 4-18. This is probably real and attributable to the ring structure in TMDT.

Lamb (45) uses the change in viscosity at 1000 sec^{-1} over a 40°C (72°F) interval as an index of temperature sensitivity. The nylons are more sensitive than the polyhydrocarbons, an acetal copolymer, or plasticized poly(vinyl chloride) but generally less sensitive than acrylic or polycarbonate (Table 4-5).

Effect of Molecular Weight

The Newtonian melt viscosity of linear polymers has been related both theoretically (36) and experimentally (37) to the 3.4 to 3.5 power of a molecular weight average:

$$\mu = K\overline{M}_x^{3.5} \tag{4-25}$$

$\overline{M}_x$ approximates $\overline{M}_w$ for narrow distributions ($\overline{M}_w/\overline{M}_n$ less than about 2) and $\overline{M}_z$ (Chapter 2, p. 41) for broader distributions. Equation (4-25) is said to apply if a critical value of molecular weight associated with the onset of chain entanglement is exceeded, and for nylon-6 this is reported to be a $\overline{M}_w$ of 5,230

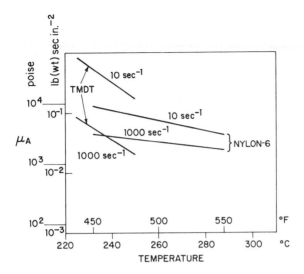

Fig. 4-18.   Effect of temperature and apparent shear rate on apparent melt viscosity of poly (trimethylhexamethylene terephthalamide) (6, 44).

Table 4-5. Temperature sensitivity of polymers (Ref. 45)

Polymer	T_1 (°C)	(°F)	T_2 (°C)	(°F)	$\dfrac{\mu_A, 1000 \text{ sec}^{-1}, T_1}{\mu_A, 1000 \text{ sec}^{-1}, T_2}$
Acetal copolymer	180	356	220	428	1.35
Acrylic	200	392	240	464	4.1
Nylon-6	240	464	280	536	2.2
Nylon-66	270	518	310	590	3.5
Nylon-11	210	410	250	482	2.4
Nylon-610	240	464	280	536	2.0
Polyethylene	150	302	190	374	Usually <2.0
Polypropylene	190	374	230	446	1.3-1.5
Polystyrene	200	392	240	464	1.6
Polycarbonate	230	446	270	518	3.0
Poly (vinyl chloride)	150	302	190	374	1.45-2.0 (flexible-rigid)

(37). Support for a gradual change, with increasing molecular weight, from an exponent of unity to exponents in excess of 3.4 has recently been offered, but no data on nylons are included (55).

In their study of the melt viscosity of extracted polycaprolactam, Pezzin and Gechele (26) reported moisture contents that for most samples were below 0.15

wt. % and hold-up times that were usually 10 to 11 min including 4 min to achieve thermal equilibrium. As expected, little change in molecular weight at the low extrusion temperature of 230°C (446°F) was observed although changes did occur at higher temperatures (Figure 4-19). The viscosity at the essentially Newtonian shear rate of 10 sec^{-1} yielded on a log-log plot against $\overline{M}_n$ a pair of parallel lines, each of slope 3.4 (Figure 4-20). $\overline{M}_n$ was used because it was available from end-group analysis and was expected to be in constant ratio to $\overline{M}_w$. This assumed a most probable distribution, $\overline{M}_w/\overline{M}_n = 2.0$. The polymers

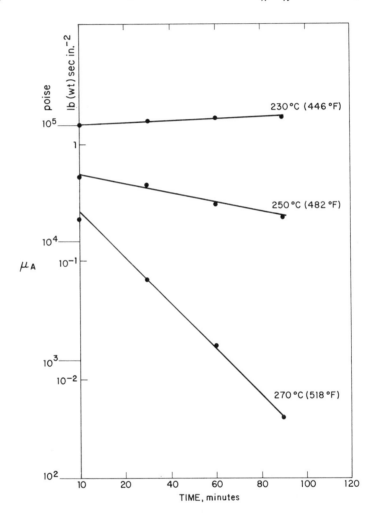

Fig. 4-19.   Effect of time and temperature on apparent melt viscosity at 10 sec^{-1} of extracted polycaprolactam, $\overline{M}_n = 42,500$ and less than 0.15 wt % water (26).

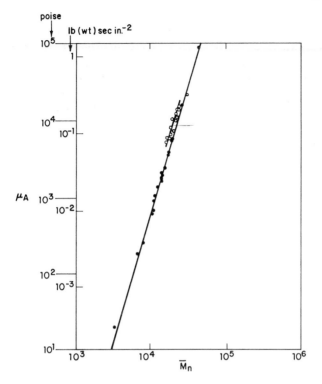

Fig. 4-20.   Dependence of apparent melt viscosity at 10 sec^{-1}, 230°C (446° F), on $\overline{M}_n$ of extracted polycaprolactam (26).

finished under reduced pressure (open circles in Figure 4-20) rather than at atmospheric pressure define the shorter line displaced to a higher viscosity at a given $\overline{M}_n$. It was postulated that a difference in molecular weight distribution caused by the difference in polymerization procedure might be responsible for the two lines.

Schaefgen and Flory (27) prepared mono, di, tetra, and octachain polymers with different molecular weight distributions by polymerizing caprolactam in the presence of a mono, di, tetra, or octafunctional acid. A polymer molecule corresponds in structure to the acid because the polymer molecule can contain only one of the acid molecules as long as interaction of the acid end groups does not occur:

$$(CH_2)_5 \genfrac{}{}{0pt}{}{C{=}O}{NH} \ + \ R(COOH)_n \longrightarrow R \left(CO \left[NH(CH_2)_5 \ CO \right]_x OH \right)_n$$

$$n = 1, 2, 4, \text{ or } 8$$

The theoretically minimum $\overline{M}_w/\overline{M}_n$ ratio at complete reaction was shown to be $1 + n^{-1}$ or 2, 1.5, 1.25, and 1.125. Calculations based on end-group analysis to define actual extents of reaction indicated the following $\overline{M}_w/\overline{M}_n$ ratios: 1.94 to 1.99 for $n = 1$, 1.51 to 1.54 for $n = 2$, 1.30 to 1.37 for $n = 4$, and 1.18 to 1.26 for $n = 8$. The log viscosity-log $\overline{M}_w$ plots indicate slopes which approximate 3.5 in all cases, but the critical molecular weight which must be exceeded increases as n increases (Figure 4-21). The displacement of the curve for the octachain polymer can be ascribed to structural complexity, but the tetrachain line falls

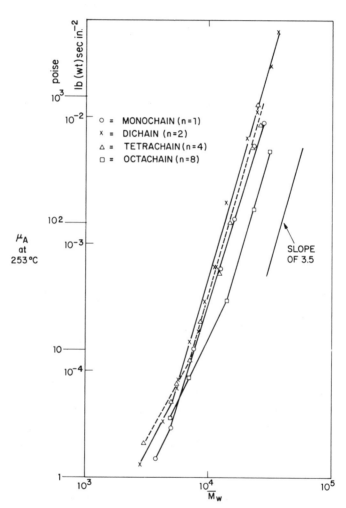

Fig. 4-21. Dependence of apparent melt viscosity on $\overline{M}_w$ for nylon-6 containing R(COOH)$_n$, where n = 1, 2, 4, or 8 (data from Ref. 27).

right between those of the monochain and dichain polymers. Further, the monochain and dichain polycaprolactams are both linear polymers, but they do not fall on the same curve. This suggests questioning the use of the weight average. Longworth and Busse (39), working with mixtures of polyethylenes and a low-molecular-weight paraffin wax, found a linear relationship of log melt viscosity and log $\overline{M}_\nu$ (a solution viscosity average, $\overline{M}_\nu < \overline{M}_w$, Chapter 2, p. 45). However, Schaefgen and Flory (27) showed that the octachain polymers did not follow the same line as the other polymers on a log-log plot of intrinsic viscosity versus $\overline{M}_\nu$ and suggested that the same factors affected both melt viscosity and dilute solution viscosity. It is of interest that the melt viscosities of the caprolactam polymers do fall on a single curve on a log-log plot against intrinsic viscosity (Figure 4-22) and describe a linear log-log relationship for $[\eta]$ > 0.3.

To summarize, the studies on nylon-6 indicate agreement with theory for linear polymers (μ_A depends on 3.5 power of $\overline{M}_w$) and suggest the use of intrinsic viscosity to predict melt viscosity where question exists with respect to structural complexity and/or molecular weight distribution.

Insufficient data are available to assess precisely the dependence of melt viscosity on molecular weight for other nylons. The same dependence on the 3.5 power of molecular weight with possibly some displacement as a function of the $CH_2/CONH$ ratio would be expected for the linear, aliphatic nylons. The useful generalization is made, however, that for such nylons the melt viscosity, to a first approximation, depends on the molecular weight and temperature and is independent of proximity to the melting point.

Nylons with ring structures which make for stiffer chains have, as expected, higher melt viscosities at low shear than the non-ring polymers of comparable molecular weight. For example, the poly(trimethylhexamethylene terephthalamide) of Figure 4-18 with μ_A of about 10,000 poise (0.14 lb(wt)sec in.$^{-2}$) at 10 sec^{-1} and 250°C (482°F) is estimated to have a $\overline{M}_n$ of 12,500. Figure 4-15 shows that at the same conditions a nylon-6 with the much higher $\overline{M}_n$ of 20,000 has a lower melt viscosity ($\mu_A \cong 6000$ poise or 0.09 lb(wt)sec in.$^{-2}$). Most probable distributions are assumed to apply in both cases.

Effect of Additives

Additives may or may not affect the melt viscosity of nylons. Nylon-66 modified for color stability, heat stability, rapid crystallization, or hydrolysis resistance is cited (40) as having the same relative flow in a snake flow mold as an unmodified polymer at the same melt and mold temperature. On the other hand, a composition containing carbon black for weather stability has 0.75 times the flow of the unmodified polymer (40). Melting point and rate of crystallization affect performance in a snake flow mold, so correlation with melt viscosity requires that these factors be eliminated.

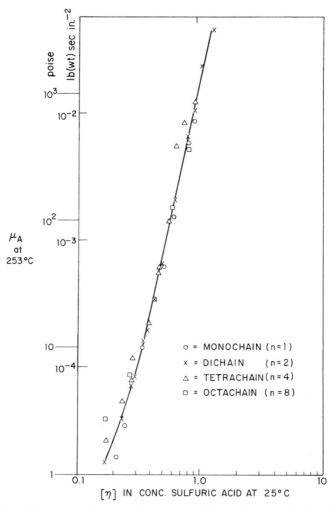

Fig. 4-22. Dependence of apparent melt viscosity on intrinsic viscosity for nylon-6 containing R(COOH)$_n$, where n = 1, 2, 4, or 8 (data from Ref. 27).

Fillers and reinforcing agents tend to increase melt viscosity. The use of short glass fibers as a reinforcing agent increases both melt viscosity and shear sensitivity according to Figure 4-23 so that the difference diminishes as the shear increases. Addition of 10 wt % of 17 mμ silica to nylon-6 increases melt viscosity at 270°C (518°F) and a shear stress of 5.4 × 10^6 dyn cm^{-2} (78.4 lb(wt)in.$^{-2}$) by a factor of 24 which is only reduced to 12 if corrected for the difference in $\overline{M}_n$ (Ref. 56; Chapter 11, p. 429). Polymerizing caprolactam in the presence of ethyl orthosilicate also reduces flow (60).

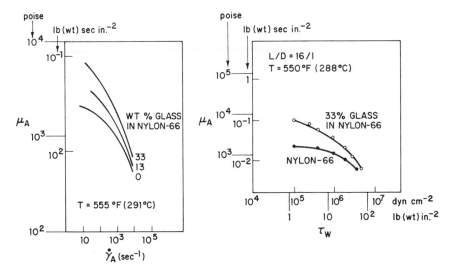

Fig. 4-23. Dependence of apparent melt viscosity on apparent shear rate and wall shear stress for nylon-66 containing short glass fibers (41, 42).

Polymer combinations affect viscosity. Increasing the amount of polyethylene grafted to nylon-6 from 10 to 30 to 50 wt % lowers the melt flow in a rheometer, in units of g/5 min at 235°C (455°F), from 8.9 to 6.2 to 2.3 (Ref. 57; Chapter 11, p. 421). Decreased flow is also observed in nylon-6 compositions containing ethylene/alkyl acrylate copolymers (58).

Additives are employed specifically for their thickening effect on molten nylons. These include compounds containing at least three acid groups or their derivatives such as anhydrides, esters, amides, and salts (59), di- and tri-alkyl and aryl phosphites (47), similar chloroalkyl and chloroaryl compounds (47), diepoxides (48), diisocyanates (49), polylactam derivatives of polyfunctional acids (50), haloalkylxylenes (51), and 4-ketopimelic acid (52). Supporting evidence has usually been based on arbitrary flow measurements or improved facility for extrusion of profiles but without determination of melt viscosity or its shear dependence.

NOMENCLATURE

A	=	area
D	=	diameter
e	=	natural logarithm base or imaginary increment in L/R
EP	=	product of ends = $[COOH]\ [NH_2]$
h	=	half-height above and below center line in slit die

H	=	opening of slit die
k	=	consistency index = proportionality constant in flow power law
K_c	=	amidation equilibrium constant
KE	=	kinetic energy
l	=	piston displacement
ln	=	natural logarithm
log	=	common logarithm
L	=	land length
L_i	=	imaginary increment in L in capillary rheometry
m	=	slope of log μ_A versus log τ_w
$\overline{M}_n$	=	number-average molecular weight
$\overline{M}_w$	=	weight-average molecular weight
$\overline{M}_v$	=	viscosity-average molecular weight
n	=	flow index = shear rate exponent in flow power law
N	=	slope of log $\dot{\gamma}_A$ versus log τ_w
N'	=	reciprocal of N = slope of log τ_w versus log $\dot{\gamma}_A$
P	=	pressure
ΔP	=	pressure drop
ΔP_{KE}	=	pressure drop due to increase in kinetic energy
Q	=	volumetric flow rate
r	=	radial position
R	=	radius
t	=	time
U	=	unreactive, nontitratable ends
v	=	linear velocity
w	=	width of section of fluid in slit die
W	=	width of slit die
[COOH]	=	concentration of acid ends in polymer
[NH$_2$]	=	concentration of amine ends in polymer
[CONH]	=	concentration of amide groups in polymer
Δ	=	end-group imbalance = [COOH] — [NH$_2$]
$\dot{\gamma}$	=	shear rate
$\dot{\gamma}_A$	=	apparent shear rate
$\dot{\gamma}_T$	=	true shear rate
$\dot{\gamma}_w$	=	shear rate at the wall
τ	=	shear stress
τ_w	=	shear stress at the wall
τ_{wc}	=	shear stress at the wall corrected for end effects
μ	=	coefficient of viscosity = viscosity
μ_A	=	apparent viscosity
μ_T	=	true viscosity
ρ	=	density

REFERENCES

1. Tadmor, Z. and I. Klein, *Engineering Principles of Plasticating Extrusion,* Van Nostrand Reinhold Book Co., New York, 1970.
2. Reiner, M., *Deformation, Strain, and Flow: An Elementary Introduction to Rheology,* 2nd ed., Interscience Publishers, Inc., New York, 1960.
3. Westover, R. F., *Polym. Eng. and Sci.* **6,** 83 (1966).
4. Anon., *Handbook for Authors,* Amer. Chem. Soc., Washington, D.C., 1967.
5. Van Wazer, J. R., J. W. Lyons, K. Y. Kim, and R. E. Colwell, *Viscosity and Flow Measurement: A Laboratory Handbook of Rheology,* Interscience Publishers, a division of John Wiley and Sons, New York, 1963.
6. Westover, R. F., in *Processing of Thermoplastic Materials,* E. C. Bernhardt, Ed., Reinhold Publishing Corp., New York, 1959.
7. Weiske, C.-D., *SPE Tech. Pap.* **13,** 676 (1967).
8. Anon., BASF Tech. Bulletin, " 'Ultramid' S Processing Properties," July, 1969.
9. Merz, E. H., and R. E. Colwell, *ASTM Bull.*, No. 232, 63 (1958).
10. Sieglaff, C. L., *SPE Trans.* **4,** 129 (1964); Anon., *Chem. Eng. News* **41,** 54 (May 27, 1963).
11. Bagley, E. B., *J. Appl. Phys.* **28,** 624 (1957).
12. Lupton, J. M., and J. W. Regester, Plastics Department, E. I. du Pont de Nemours and Co., Inc., personal communication.
13. Rabinowitsch, B., *Z. Physik. Chem.* **A145,** 1 (1929).
14. Klein, I., and D. I. Marshall, *Computer Programs for Plastics Engineers,* Reinhold Book Corp., New York, 1968.
15. Metzner, A. B., in *Advances in Chemical Engineering,* T. B. Drew and J. W. Hoopes, Jr., Eds., Academic Press Inc., New York, 1956, Vol. I.
16. Westover, R. F., *SPE Tech. Pap.* **6,** 80 (1960); Metzner, A. B., E. L. Carley, and I. K. Park, *Mod. Plast.* **37** (11), 133 (July, 1960).
17. Rogers, M. G., and C. McLuckie, *J. Appl. Plym. Sci.* **13,** 1060 (1969); Marshall, D. I., and D. W. Riley, *ibid.* **6,** S46 (1962).
18. Lupton, J. M., and W. H. Hale, *Polym. Eng. Sci.* **5,** 244 (1965).
19. Dennison, M. T., *Trans. J. Plast. Inst.* **1967,** 803.
20. Lupton, J. M., and J. W. Regester, *Polym. Eng. Sci.* **5,** 235 (1965).
21. Lyons, J. W., in *Treatise on Analytical Chemistry,* John Wiley and Sons, Inc., New York, 1967, Part I, Vol. 7, Chap. 83.
22. Middleman, S., *The Flow of High Polymers,* Interscience Publishers, a division of John Wiley and Sons, New York, 1968.
23. Coleman, B.D., H. Markovitz, and W. Noll, *Viscometric Flows of Non-Newtonian Fluids,* Springer-Verlag, Inc., New York, 1966.
24. McKelvey, J. M., *Polymer Processing,* John Wiley and Sons, Inc., New York, 1962.
25. Lahti, C. P., *SPE J.* **19,** 619 (1963).
26. Pezzin, G., and G. B. Gechele, *J. Appl. Polym. Sci.* **8,** 2145 (1964).
27. Schaefgen, J. R., and P. J. Flory, *J. Am. Chem. Soc.* **70,** 2709 (1948).
28. Dumanskii, I. A., L. V. Khailenko, and L. V. Prokopenko, *Colloid J. USSR* **25,** 542 (1963).

29. Kokhomskaya, T. N., and A. B. Pakshver, *Colloid J. USSR* **18**, 179 (1956).

30. Steffens, H., *Z. Phys. Chem.* (*Leipzig*) **216**, 356 (1961); *C. A.* **55**, 25334i (1961).

31. Albrecht, W., *Abh. Deut. Akad. Wiss. Berlin, Kl. Chem., Geol. Biol.* **1963** (1), 393; *C. A.* **60**, 14624f (1964).

32. Vessereau, J., and C. Mermoud, *Bull. Soc. Chim. Fr.* **1961**, 1096.

33. Mukouyama, E., and A. Takagawa, *Chem. High Polym.* *(Japan)* **18**, 323 (1956); *C. A.* **51**, 18616g (1957).

34. Morávec, J., *Chem. prumysl* **10**, 657 (1960); *C. A.* **55**, 21643f (1961).

35. Chatain, M., G. Héry, and J. Prévot, *Ind. Plast. Mod.* **10** (9), 32 (1958).

36. Bueche, F., *J. Polym. Sci.* **43**, 527 (1960).

37. Fox, T. G., and S. Loshaek, *J. Appl. Phys.* **26**, 1080 (1955); Fox, T. G., S. Gratch, and S. Loshaek, in *Rheology*, F. R. Eirich, Ed., Academic Press Inc., New York, 1956, Vol. I, Chap. 12.

38. Smith, S., *J. Polym. Sci.* **30**, 459 (1958).

39. Longworth, R., and W. F. Busse, *Trans. Soc. Rheol.* **6**, 179 (1962); also, *J. Polym. Sci.* **58**, 49 (1962).

40. Anon., Plastics Department, E. I. du Pont de Nemours and Co., Tech. Bulletin, "Molding Du Pont Zytel® Nylon Resins . . . a Handbook for the 70's," 1970.

41. Anon., Plastics Department, E. I. du Pont de Nemours and Co., Tech. Bulletin, "Zytel® Glass Reinforced Nylon Resins Molding Manual," 1968.

42. Williams, J. C. L., D. W. Wood, I. F. Bodycot, and B. N. Epstein, Soc. Plast. Ind., Reinf. Plast. Div., 23rd Ann. Conf., Feb., 1968, Sec. 2-C.

43. Anon., BASF Tech. Bulletins, " 'Ultramid' A Processing Properties" and " 'Ultramid' B Processing Properties," July, 1969.

44. Gabler, R., H. Müller, G. E. Ashby, E. R. Agouri, H.-R. Meyer, and G. Kabas, *Chimia* **21** (2), 65 (1967).

45. Lamb, P., in *Advances in Polymer Science and Technology*, S.C.I. Monograph No. 26, London, 1967, pp. 296-312.

46. Miller, M. L., *The Structure of Polymers*, Reinhold Book Corp., New York, 1966, p. 219.

47. Kessler, J. C. F., and H. R. Spreeuwers (to N. V. Onderzoekingsinstituut Research), U.S. Patent 2,959,570 (Nov. 8, 1960).

48. Anon. (to Farbenfabriken Bayer A. G.), Brit. Patent 831,207 (March. 23, 1960); Anon. (to N. V. Onderzoekingsinstituut Research), Brit. Patent 807,217 (Jan. 7, 1959); Reichold, E., and W. Böckmann (to Farbenfabriken Bayer, A. G.), U.S. Patent 3,458,481 (July 29, 1969).

49. Hechelhammer, W., and H. Streib (to Farbenfabriken Bayer A. G.), Ger. Patent 1,027,398 (Mar. 15, 1962).

50. Anon. (to Wingfoot Corp.), Brit. Patent 693,645 (July 1, 1953).

51. Anon. (to N. V. Onderzoekingsinstituut Research), Brit. Patent 807,215 (Jan. 7, 1959).

52. Ellery, E., B. J. Habgood, and R. J. W. Reynolds (to Imperial Chemical Industries, Ltd.), Brit. Patent 709,705 (June 2, 1954).

53. Tordella, J. P., *Rheol. Acta* **1**, 216 (1958).

54. Tordella, J. P., in *Rheology*, F. R. Eirich, Ed., Academic Press, New York, 1969, Vol. 5, Chap. 2.

55. Cross, M. M., *Polymer* 11, 238 (1970).
56. Symons, N. K. J. (to E. I. du Pont de Nemours and Co.), U.S. Patent 2,874,139 (Feb. 17, 1959).
57. Craubner, H., and G. Illing (to H. Roemmler G.m.b.H.), U.S. Patent 3,261,885 (July 19, 1966).
58. Anspon, H. D., and H. E. Robb (to Gulf Oil Corp.), U.S. Patent 3,472,916 (Oct. 14, 1969).
59. Walker, I. F. (to E. I. du Pont de Nemours and Co.), U.S. Patent 2,557,808 (June 19, 1951).
60. Hayes, R. A. (to The Firestone Tire and Rubber Co.), U.S. Patent 3,461,107 (Aug. 12, 1969).

CHAPTER 5 ⎯⎯⎯⎯⎯⎯⎯⎯⎯⎯⎯⎯⎯⎯

Injection Molding of Nylons

C. M. BARANANO

INTRODUCTION

Nylons have been injection molded for over a quarter of a century. From a modest beginning, nylons have proliferated to a variety of products that differ in processing characteristics and physical properties. Concurrent with the growth of nylons and other thermoplastics has come a steady growth in the injection molding industry. This latter growth has motivated an increasing desire for understanding the process, for measuring and controlling basic variables, and for applying sound engineering principles to the design, construction, and operation of molding equipment. The net result is the emergence of systems engineering in injection molding.

This chapter reviews material handling, equipment, and machine operation relevant to the injection molding of nylons and emphasizes the need to adopt an integrated systems approach from raw material to finished product in order to optimize quality and productivity.

MATERIAL HANDLING

Efficient material handling is basic to operating a nylon molding system because of the sensitivity of nylons to moisture and contamination. A well-engineered system will do the following:

1. Maintain a uniform and low level of moisture.
2. Avoid contamination.

3. Maintain a uniform and adequate temperature.
4. Minimize waste and spillage.
5. Maintain a uniform feed to the molding system in terms of virgin to regrind ratio, particle size, and material level in hopper.

Nylons absorb water at different rates and to different levels as shown in Table 5-1 and as discussed in Chapters 9, 10, and 17.

Table 5-1. Absorption of water by nylons (wt % dry basis)[a]

Nylon type	24 hr (0.125 in. or 0.32 cm) ASTM D-570	Equilibrium with 50% R.H.	Saturation
6	1.6	2.7	9.5
66	1.5	2.5	8.0
610	0.4	1.5	3.5
612	0.4	1.3	3.0
11	0.25	0.8	1.9–2.9[b]
12	0.25	0.7	1.4–2.5 est.[b]

[a] Data from trade literature.
[b] Varies with temperature.

The implications of varying water contents for chemical transformations in different nylons have been treated in Chapter 4 (p. 134).

Control of Moisture

Control of moisture requires consideration of the plant environment, the polymer storage, and the effects of varying degrees of exposure in handling and processing. Our emphasis here is on control factors other than drying, which is of course very important and is discussed in a separate section that follows.

Any system that controls the total plant environment offers the best safeguards against moisture-related problems (splay, blisters, drooling) and hence decreases down-time and improves quality. Plants can control their environment by installing air conditioning and dehumidifying facilities. The objective is to keep relative humidity below 35% and temperature below 75°F (24°C) even in the worst of summer days. Geographical location of the plant must be kept in mind, as Table 5-2 clearly shows. Humidity is the controlling factor, but temperature and other reasons for variability at the specific site contribute to the reported moisture absorptions based on the average annual relative humidities available from statistical tables.

Table 5-2. Effect of location on moisture content of nylon-6

Cities	Mean moisture content(1) in nylon-6 granules after 1 yr exposure	Annual(1a) ave. % R.H. 1:00 p.m.
Boston, Massachusetts	1.96%	55
Queens, New York	1.98	56
Pittsburgh, Pennsylvania	2.50	57
Rochester, New York	2.17	––
Atlanta, Georgia	2.35	57
Miami, Florida	3.83	61
Houston, Texas	2.55	60
New Orleans, Louisiana	3.02	63
Detroit, Michigan	2.21	58
Chicago, Illinois	2.04	58
Minneapolis, Minnesota	2.63	61
Cleveland, Ohio	2.50	62
Cincinnati, Ohio	2.68	57
San Francisco, California	2.23	69
Denver, Colorado	1.70	40
Los Angeles, California	2.26	61
Seattle, Washington	3.40	75

Partial exclusion of moisture in the vicinity of the machine can be accomplished with automatic, closed material-handling systems equipped with dehumidified hopper dryers. Total exclusion of the environment can be achieved by introducing inert gases. Such systems have to be leak-proof to avoid losses of the relatively expensive inert gas. Inert gas blanketing can also help to reduce oxidation of nylons in dryers where temperatures above 150°F (65°C) and long residence times are required to dry the material.

Nylons should be stored in dry places at temperatures near that of the operating area. Cold material can condense moisture and carry it into the molding machine. When storage in a cold location cannot be avoided, the container must be allowed to come to the temperature of the molding area before opening. Storage should allow a "first in, first out" inventory policy.

As noted in Chapter 2 (p. 18), nylons are customarily supplied in containers ranging from 25-lb cans or 50-lb bags to 1200-lb cartons. It is obviously desirable to select a size of package compatible with the scale of operation, and it sometimes is convenient to transfer the contents of several cans or bags into a container sized for a specific molding job. Because the upper layers of resin serve as a desiccant for the lower layers and transfer to the machine hopper occurs via a suction line that is immersed in the bed of granules, it is usually safe to assume that the moisture pick-up corresponds to that expected for half the time it takes to empty the container, that is, 0.5 times the size of the container divided by the rate of consumption of resin. The moisture absorption can be estimated for both virgin and regrind using absorption curves such as those in Figures 5-1 and 5-2. It is good practice to dry material whenever either virgin or regrind contains over 0.3% water, and processes that involve high temperatures or long cycles may require moisture contents below 0.2%. The moisture absorption prior to grinding of material to be reused should not be overlooked. The combined absorption of reworked resin before and after grinding is the most likely source of excessive moisture.

In general, exposure times of polymer at the loading station, through tubes, and in conveyors cannot be controlled unless there is uniform demand and accurate metering. These systems are open to the atmosphere and unless the environment is controlled, material is vulnerable to them. The optimum situation is naturally total and prompt feedback of the runner system and selection of the container size that will be consumed in an adequate time.

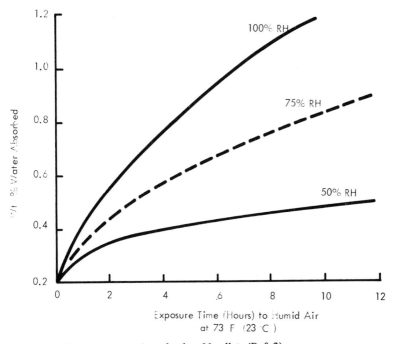

Fig. 5-1.   Moisture absorption of nylon-66 pellets (Ref. 2).

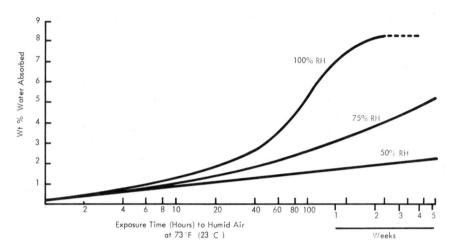

Fig. 5-2.   Moisture absorption of nylon-66 ground with 0.25 in. (0.63 cm) screens (Ref. 2).

Control of Contamination

Factors in contamination are housekeeping, storage, conveying systems, grinders, temperature control, and spillage.

After a material change, the hopper, conveyors, grinding equipment, screw, nozzle, nozzle valve, and barrel should be carefully examined and cleaned if necessary. The distance between auxiliary equipment, its location, and baffling are important to prevent cross-contamination. Grease, oil, and rust from the molding machine are other sources of contamination. Valves, manifold surfaces, fittings, and O-rings are usually the cause of oil problems; and over-lubrication, the common cause of grease. Condensation in chilled molds can be a major source of moisture and rust.

The use of plastics in biomedical applications, aerospace, computers, and communication equipment is creating the need for stringent control of contaminants in plastic manufacturing. Application of aspects of the "cleanroom" technology developed by the communication and aerospace industries is to be expected. Literature on the design and construction of such facilities is available from United States government agencies (3,4,5).

Containers used to store other plastics should not be used to store nylons unless they have been cleaned thoroughly. "Fines" and "shavings" of other materials left in the walls and bottoms of such containers adhere by either static charges or moisture and are hard to remove. Empty containers should be kept closed to prevent collecting dust. When opening bags, a clean cut, rather than a tearing cut, should be made to prevent paper from falling in the material.

There is no substitute for periodic inspection of the raw material, runners, and parts at all stages of the process. Regrind obviously carries more contaminants than virgin material because it has been exposed longer to the environment and equipment and should be checked periodically.

Careful engineering of automatic handling systems is necessary to avoid contamination. Conveyors (6) should be of the dry-lubrication type with belts that withstand abrasion, oil, grease, and localized high temperatures (up to 200°F, 93°C). The transport problem is aggravated by the variety of runners and parts encountered. Cardboard improvisation around conveyors is a poor solution to the problem of runner and part containment. Covered conveyors can be used to keep dust and contaminants out. The best approach to transport of runners and parts is via pneumatic and vibratory systems.

A serious source of contamination and spillage is the emission of fines through seams and gaskets of automatic loading equipment, dehumidified hopper dryers, and hopper feed ports. Relief openings supplied with filters, better structural rigidity, gasketing, and separate screening of fines minimize this problem.

Grinders also contribute to contamination. The feed chute should accept the runners easily and baffle the fly back of material. It is better to oversize than to undersize the feed throat. Fly back can be cut down with augur grinders. Good fits, adequate clearances, and streamlining and gasketing of the collection bins and rotor axle reduce spillage from grinders.

Hopper magnets remove most of the adventitious ferrous metal contaminants (such as from chipped grinder knives) that can cause serious machine damage, poor check ring operation, plugging of gates, and process interruptions. Sprues that stick should be removed with plastic instead of brass rods to avoid nonmagnetic metal particles. The use of compressed air is being discouraged for reason of safety.

Minimizing exposure to elevated temperatures in material handling systems is important to avoid contamination from oxidation as well as to control moisture content. There is no incentive for cooling the rework; instead, care must be taken to reduce recycle hold-up time. With the finished product, however, removing the heat carried into packaging can be important to avoid changes in shape or dimensions as well as prevent discoloration.

The temperature of dehumidified hopper driers should be controlled. There is rarely a need for temperatures above 175°F (80°C) to accomplish drying. Residence times in excess of 4 hr at 175°F may cause discoloration.

Grinders that are overloaded, or where clearances between cutting knives and housing are tight, can cause localized high temperatures with damage to the material and grinder.

Control of Feed

Control of feed involves control of the ratio of virgin to regrind resin, of particle size, and of the level of material in the hopper.

Changes in processing characteristics (melt viscosity due to a change in molecular weight, melting rate because of variation in particle size, and freezing point via nucleation by chance contamination) can be caused by random introduction of virgin and regrind materials. These changes can result in quality fluctuations in the final product. In general, 50% regrind levels will not impair the performance of engineering parts made of nylon. It is only in very specific situations, such as the molding of glass-reinforced nylons, critical appearance applications, or tight dimensional control molding, that it may be desirable to keep the regrind ratio below this figure.

Grinding equipment should be selected and maintained to produce regrind with an average particle size close to that of the virgin nylon. Sometimes it is necessary to remove fines and shavings with vibrating screens or cyclone separators. Fines and shavings impair the function of automatic loaders, particularly by plugging filters. They absorb moisture and melt faster and can cause discoloration and loss of properties.

A constant level of material in a hopper insures uniformity of environmental exposure and of feed. Level sensing devices in a continuously fed hopper are available, and it is desirable to install audiovisual alarms to warn if material level falls due to malfunction or lack of feed material.

Hoppers should be streamlined to allow free flow of the material into the machine barrel. Sometimes it is helpful to bias the hopper discharge in the direction of screw rotation. The hopper capacity should be such as to allow at least 1 hr of machine production.

Automatic Handling Systems

The guidelines discussed above apply equally to manual or automatic handling. The specific advantages of automatic systems over manual are several:

1. Ability to handle high production rates.
2. Reduction in labor costs.
3. Reduction in human error.
4. Constancy of feed into the hopper.
5. Elimination of machine shutdowns because of lack of feed material.
6. Reduction of contamination level.
7. Reduction of exposure to a wet environment.
8. Improvements in yield.

Realization of these advantages requires a well-engineered and a well-maintained system. Poorly designed or operated systems can be a deficit rather than an asset. Figure 5-3 provides a sketch of an integrated, automated handling

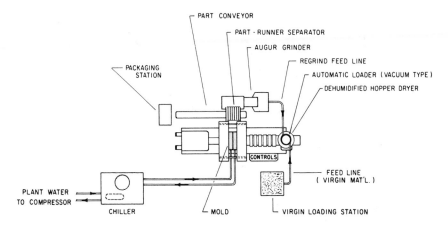

Fig. 5-3.   Sketch (top view) of automated system for handling, drying, and cooling.

system including conveyors, part-runner separators, grinders, vacuum tubes, classifiers, and hopper proportioning devices. A good analysis of pneumatic conveying systems has been provided (7).

The author's experience suggests special mention of the problems of containment of the molded shot, separation of the part and runner, and layerization in hopper loaders. Confinement of the molded shot in the drop area to minimize spillage requires consideration of mold design, relative size of mold to platens, baffling techniques, machine door design, drop area space, amount of overlap between machine and conveyors, skimming of parts off conveyors, height of side guards in transfer and separation equipment, and overlap between conveyors and grinder throat chute. Part-runner separators that can be phased into the system without need for synchronization with the machine cycle are to be preferred over synchronous separators.

The mode of sequencing from regrind to virgin to regrind in hopper loaders can lead to layerization and fluctuations in molding. It is sometimes desirable to install a suitable mixing device in the hopper or loader.

DRYING

The time required to dry a nylon depends on the initial moisture content, the goal level of moisture, the thickness of the polymer particles, and the temperature and humidity in the drying device. Bed thickness in drying trays is normally not a factor if kept below 1 in. (2.5 cm). The type of nylon affects the initial moisture level and drying rate and hence the drying time, but this problem

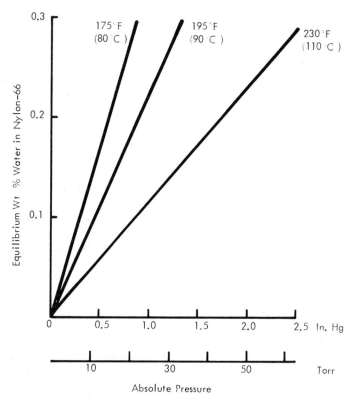

Fig. 5-4.   Effect of drying temperature and vacuum on equilibrium moisture content of nylon-66 (Ref. 2).

has not been carefully quantified because the degree of variation has not been a significant factor with the currently available aliphatic nylons.

The moisture content can be estimated from manufacturer's data like that in Figures 5-1 and 5-2 or determined by analysis as described in Chapter 3 (p. 102). Proper use of ovens requires elimination of leaky ducts, which can be determined by comparison of inlet and exit dew points of an empty oven. Application of a psychometric chart to estimate drying temperature is described in trade literature (for example, Ref. 2) which also provide curves of moisture content versus time for specific temperatures, initial moisture levels, and dew points. The use of desiccants permits more consistent performance and is discussed further below in connection with hopper dryers. The effect of vacuum in an oven depends on the temperature and is illustrated for nylon-66 in Figure 5-4.

Hopper dryers are the trend in the molding industry because they permit

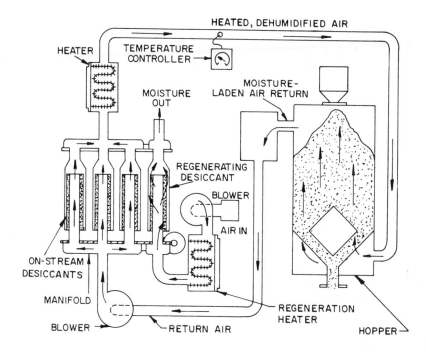

HEATED, DEHUMIDIFIED AIR

HEATER

TEMPERATURE
CONTROLLER

MOISTURE
OUT

MOISTURE-
LADEN AIR RETURN

MOISTURE-
LADEN AIR RETURN

REGENERATING
DESICCANT

BLOWER

AIR IN

ON-STREAM
DESICCANTS

MANIFOLD

BLOWER

RETURN AIR

REGENERATION
HEATER

HOPPER

Fig. 5-5.   Diagram of carrousel-type hopper dryer (Ref. 8).

automated systems. Hopper dryers are taken to include both hoppers positioned at the feed throat of the molding machine and plenums located away from the machine. Although a static bed holds no advantage in drying time over ovens operated at the same temperature and humidity, the hopper dryer permits countercurrent flow so that dry polymer exits from the bottom while moist resin is added at the top. An adequate capacity and flow pattern of resin in the hopper must be assured. Figure 5-5 (8) diagrams a four-cartridge, carrousel-type, desiccant system using recycled air. Molecular sieves are commonly employed as the desiccant. Regeneration of the sieves is accomplished by purging with air heated to about 400°F (205°C).

Table 5-3 shows one manufacturer's recommended drying times and air velocities at 160°F (71°C) for several types of nylon, drying hopper sizes, and drying system outputs for those dryers that completely isolate process air from ambient air (8). When selecting a drying system for any application enough heat energy should be available for the required processing rates. This is as important as making sure a conveyor's tube capacity is high enough to keep the hopper filled. Standard engineering approximations can be applied:

Table 5-3. Drying times and air velocities for carrousel-type dryers at $160°F$ ($71°C$)(8)

Nylon	% Water	Time (hr)	Air velocity			
			$120 \text{ ft}^3 \text{ min}^{-1}$ ($0.057 \text{ m}^3 \text{ sec}^{-1}$)		$240 \text{ ft}^3 \text{ min}^{-1}$ ($0.113 \text{ m}^3 \text{ sec}^{-1}$)	
			Hopper capacity	Output	Hopper capacity	Output
11 or 12	1	2	260 lb (118 kg)	130 lb hr^{-1} (58 kg hr^{-1})	525 lb (239 kg)	260 lb hr^{-1} (116 kg hr^{-1})
6, 66, or 610	1	3	525 lb (239 kg)	190 lb hr^{-1} (86 kg hr^{-1})	1085 lb (493 kg)	380 lb hr^{-1} (173 kg hr^{-1})
6, 66, or 610	2	4	370 lb (168 kg)	95 lb hr^{-1} (43 kg hr^{-1})	800 lb (364 kg)	190 lb hr^{-1} (86 kg hr^{-1})

$$Q = mC_p (T_d - T_o) + m(W_o - W_d) \Delta H$$

where

Q = total heat required per unit time

m = weight of dry nylon per unit time

C_p = specific heat of nylon

T_d = mean drying temperature

T_o = initial temperature of nylon

W_o = initial moisture content per unit weight of dry nylon

W_d = final moisture content per unit weight of dry nylon

ΔH = heat of vaporization of water per unit weight of water

Heat losses must be accounted for and can vary from 10% for small L/D hoppers at low temperatures to 60% for large L/D hoppers at high temperatures. Table 5-4 (8) indicates the requirements for drying 100 lb hr^{-1} (45 kg hr^{-1}) of nylon at 160°F (71°C) assuming a 30% heat loss and very high initial moisture contents. These are obviously conservative values.

Table 5-4. Requirements for drying nylons at 160°F (71°C) assuming 30% heat loss(8)

| Nylon | Time (hr) | Initial % water | To dry 100 lb hr^{-1} (45 kg hr^{-1}) | | | |
| | | | Power | | Air velocity | |
			(kw)	(Btu hr^{-1})	(ft^3 min^{-1})	(m^3 sec^{-1})
6	3-4	9.0	4.5	15,400	180	0.085
66	3-4	7.5	3.8	13,000	152	0.043
610	3-4	3.0	2.7	9,200	108	0.031
11	2	1.8	2.2	7,500	88	0.025
12	2	1.5	1.4	4,800	56	0.016

The mass flow rate of air that a dryer blower produces must be high enough to transfer heat energy to the pellets efficiently. It must maintain the desired temperature and overcome the resistance of the dryer lines and the packed bed of granular material. Recommended air velocities are also given in Table 5-4. These change in direct proportion to the weight of nylon dried per unit time.

If the blower is undersized, not enough air flow is available and the required heat energy cannot be transferred at the drying temperature. Increasing the air temperature to overcome the reduced air flow causes two problems. First, the heat losses in the connecting hose and hopper will greatly increase, which eliminates most of the gain in temperature. Second, if the temperature is raised excessively, there is danger of the pellets sticking in the lower portion of the hopper or the feed throat of the machine.

Both the size and the design of the dryer hopper are important to the system. The material must be exposed to the dry air for a long enough period of time to allow internal moisture to be released. The residence time for different nylons varies generally from 2 to 4 hr (Table 5-4). The volume, that is, usable capacity, of the hopper must be sufficiently large to hold an amount of material equal to the processing rate multiplied by the recommended residence time. For example, 200 lb hr^{-1} of a nylon with a residence time of 4 hr would need a hopper that will hold 800 lbs.

Because the heat energy must be transferred equally to all pellets, the air must flow uniformly across them. For each pellet to have the same residence time the flow of material down through the hopper must also be uniform. A spreader cone in the bottom of the hopper (see Fig. 5-5) helps accomplish this and also helps to spread the air flow evenly around the pellets.

There are two opposing factors involved in the slenderness ratio (height/diameter) of hoppers. For ideal material flow and air distribution, a tall, narrow hopper would be best. However, from the standpoint of heat losses and ceiling height, a hopper with its diameter equal to its height would be best. The most commonly used hopper shape is one with the height equal to twice the diameter. Regardless of shape, a hopper with insulated sides is well worth its cost in terms of power savings and is to be recommended.

Often ignored is the drying of additives such as lubricants and colorants. Sometimes the colorants used are polyethylene concentrates. In this case care should be taken not to heat above 150°F (66°C) prior to entering the injection cylinder because the polyethylene concentrate may "cake-up" or agglomerate.

EQUIPMENT FOR MOLDING NYLONS

In this section attention centers on the molding machine and the mold. The question of controls will be discussed in a later section on the injection molding operation.

The Molding Machine

Machines for molding nylons must take into account the high heat requirements of these resins, their ease of flow, the possibility of degradation, and the need for precision and quality normally associated with nylon applications.

The Injection End

The injection end of a molding machine for nylons must have adequate power to melt and inject the nylon. Heater bands normally have a watt density of 25 to 35 W in.$^{-2}$ of outside barrel surface (3.9 to 5.4 W cm^{-2}), and the screw drive should have 1 hp (746 W) for every 7 to 8 lb hr^{-1} (3.2 to 3.6 kg hr^{-1}) of melt output based on continuous rotation of the screw. An efficient screw is required to transfer the energy from the machine to the material. Temperatures, pressures, displacements, and times have to be controlled accurately.

Melt temperature in the nozzle depends on heater band regulation, back pressure, and screw design. To achieve heater band control it is necessary to:

1. Divide heater bands into zones (preferably four).
2. Locate properly the temperature sensing elements with respect to the zones.
3. Select the proper depth for the sensing elements in the wall of the barrel and insure contact.
4. Calibrate regularly the sensing elements and controllers.

Figure 5-6 shows the energy supply arrangement for a molding machine with

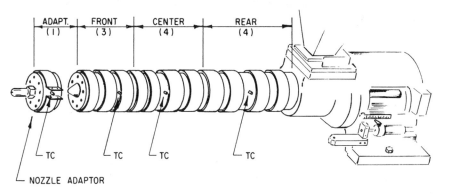

(1) All heater bands of a given zone are connected in parallel.

Rear:	4 x 1400 W	
Center:	4 x 1400 W	Total = 16,340 watts
Front:	3 x 1400 W	
Adapter:	1 x 940 W	

(2) Outside surface area = 640 in.^{2} = 4130 cm^2
      Watt density   = 25.5 W in.$^{-2}$ = 3.96 W cm^{-2}

(3) Thermocouples are located as shown, and each is connected to its own controller.

(4) Screw drive = 20 horsepower = 14,920 W.

(5) Melting capacity = 140–160 lb hr^{-1} = 63–73 kg hr^{-1}.

Fig. 5-6.   Typical heater band arrangement for injection molding of nylons (Ref. 9).

the capacity to melt about 150 lb hr^{-1} (68 kg hr^{-1}) of nylon. Continuity in production is improved by adding meters or light signals to detect heater band failures. Cooling of the feed throat is good practice to prevent excessive temperatures during interruptions that can lead to "bridging" and erratic screw operation.

Uniformity of screw retraction depends on screw design and material and operating factors. Material and operating factors are discussed below in the section on operation.

Screw injection machines generate significantly more heat by viscous work than do ram injection machines. Mixing is obviously better in screw than in ram machines. Nonetheless, ram machines are preferred in specific instances such as the molding of patterns.

Because nylons involve a high enthalpy of processing and are low viscosity materials, their melting rate is slower than polystyrene and many other plastic materials processed in a screw machine. Ratings given for screw machines are usually given in terms of the melting rate of polystyrene and should therefore be multiplied by about two-thirds for application to nylons.

Screw design can be very important in nylons. One polymer manufacturer's suggestions are outlined in Figure 5-7. Theoretical limitations to screw parameters (defined in Figure 5-7) can be obtained and have been verified by experiment (11). One formula (28) is

$$13 \geq \frac{n_m + (1/2)n_t}{(h_m)^2 (\cos^2 \phi) N} \geq 8 \qquad \begin{array}{l} n_m, n_t, h_m, \phi \text{ as in Figure 5-7} \\ N = \text{maximum operating screw} \\ \text{speed} \end{array}$$

This expression holds true for nylons if the L/D of the screw is 10 to 24, the diameter is 1.5 to 3.5 in. (3.8 to 8.4 cm), and 40 to 60% of the screw turns are allocated to the feed section. Below a value of 8, unmelt results at high rates unless excessive back pressure is used; above 13, overheating of the resin results. The author's experience suggests a feed zone depth of not less than 0.29 in. or 0.74 cm even in the smallest diameter screws to avoid variation in screw retraction.

The injection cylinder must have the capacity to inject the molten nylon at high pressures and rates. For molds difficult to fill (long, thin sections in the flow system), injection pressure capability may have to be as high as 20,000 lb (wt) in.$^{-2}$ (1760 kg (wt) cm^{-2}). Volumetric displacement rates of 3 in.^{3} sec^{-1} (49 cm^3 sec^{-1}) are typical in molding nylons. Only infrequently is it required to inject at rates higher than 5 in.^{3} sec^{-1} (82 cm^3 sec^{-1}), and even the smallest of modern molding machines will inject at these rates.

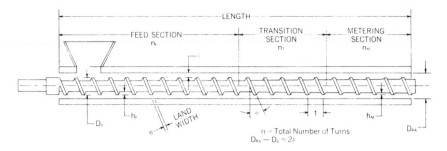

	LOW VISCOSITY RESINS					HIGH VISCOSITY RESINS			
	Screw Dia		Feed Depth		Metering Depth		Feed Depth		Metering Depth
	(in.)	(cm)	h_F(in.)	(cm)	h_M(in.)	(cm)	h_F (in.) (cm)		h_M(in.) (cm)
20 L/D									
Square pitch	1.5	3.8	.300	.762	.060	.152	.300	.762	.080	.203
screws	2.0	5.1	.310	.787	.065	.165	.320	.813	.090	.229
n_F = 10	2.5	6.4	.320	.813	.075	.190	.380	.965	.100	.254
n_T = 5 (4)*									
n_M= 5 (6)*									
16 L/D									
Square pitch	1.5	3.8	.300	.762	.055	.140	.300	.762	.075	.190
screws	2.0	5.1	.310	.787	.060	.152	.320	.813	.085	.216
n_F = 7.5	2.5	6.4	.320	.813	.070	.178	.380	.965	.095	.241
n_T= 3.5									
n_M = 5.0									

General practice in the industry is to have the land width (e) = 1/10 the distance between the flights (t) and the radial clearance (δ) one-thousandth of the diameter of the screw (Ds).

*ALTERNATE DESIGN FOR HIGHER OUTPUT RATES.

Fig. 5-7.   Screw design for nylons as suggested by one manufacturer (Ref. 2).

Control of shot size is important in the molding of nylons. Good nonreturn valve sealing; proper clearances between screw, check ring, and barrel; and uniformity of screw retraction are necessary for optimum shot metering. Devices that stop the movement of the screw after recovery are often desirable to improve shot uniformity with electric drives.

Nozzle valves prevent drooling, allow rotation of the screw while the mold opens and closes, and are of great help in fast molding. Whether the spring-pin type, positive slide, or ball check type is used, is a matter of preference. Valves should be streamlined for flow. Preheating at least one-half hour prior to start up insures proper operation.

The low viscosity of most common grades of nylons requires that special attention be paid to nozzles and nozzle tips (12). The nozzle tip should be of the reverse taper type to facilitate sprue extraction and minimize drooling. The land length and bore diameter best suited to an operation cannot be specified without considering mass and heat transfer and cycle. To avoid drooling, the melt must partially cool in the nozzle tip. In a fast cycle operation, the flow of hot melt is so frequent that partial cooling is harder to achieve and longer landed bores are preferred. A sample nozzle is illustrated in Figure 5-8.

The Clamp End

Platen support, strength, and parallelism are important to avoid deflection at high injection pressures. In particular the front platen can be troublesome because it is inherently weaker. Platen parallelism and tie bar stretching should be checked periodically, and with some machines care should be taken not to hang extremely small molds that will act as concentrated loads and deform the platens. Tie rods should be provided with sufficient bushing area to minimize tilting and wear of platens. The high fluidity of nylons makes these precautions imperative to prevent flash and obtain good dimensional control.

The clamp should provide adequate tonnage to maintain the mold closed under the forces of injection. For nylons this requirement ranges usually from 2.0 to 5.0 tons in.$^{-2}$ (280-700 kg(wt)cm^{-2}) of projected cavity and runner area depending on the pressure required to fill the mold. It is most often between 3.5 and 4.5 tons in.$^{-2}$ (490 to 630 kg(wt)cm^{-2}). The sequences of mold opening and closing should be readily programmed and reproducible. Opening and closing stroke should be cushioned (or decelerated), and knock-out should be adjustable in position, speed, and sequence (multiple knock-out). It is highly advisable in molding nylons to reproduce a set-up with respect to tonnage, strokes, and accelerations.

The Mold

A mold must be designed to minimize resistance to flow; it must remove heat quickly; and it must consistently produce desired shapes within specified tolerances. The problems of flow and heat transfer are considered in a

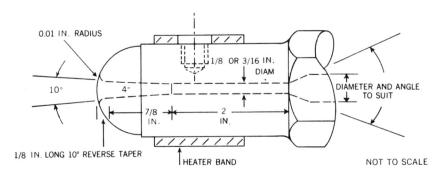

Fig. 5-8.   Sample nozzle for injection molding of nylons (Ref. 2).

subsequent section. Adequate recommendations for gating, layout, undercuts, and the like, are available in resin manufacturers' technical literature for nylons. Comments here are confined to areas of mold design and operation that are not frequently mentioned but are important in injection molding of nylons.

Sprue-Runner Systems

Two factors must be kept in balance in a nylon sprue-runner system of the conventional type. Large-diameter runners are favored to provide an adequate cross section for flow, to balance flow in multicavity molds, and to avoid deformation when the ejector system acts. Small-diameter runners are preferred to minimize heating or cooling of the mold and reduce the amount of regrind.

The T-junction at the base of the sprue is usually the thickest part of the layout and is frequently the source of ejection difficulties. This problem can limit cycle. Separate cooling of this area can be difficult but should be considered along with sprue puller design. Sprueless molding is feasible with nylons, and recommendations for such systems appear in the literature(13).

It is excellent practice to cut the runner network as a separate block of metal and to harden it. Many problems are caused by flash in a deteriorated runner parting plane. This eventually leads to poor fits and dimensions which can cause breakage of ejector pins and distortion of mold plates.

Unbalanced flow layouts may be balanced by adjusting secondary runners through flow equations. Except in extreme cases, balancing through gates is undesirable since it affects gate sealing times. The layout should also be balanced for projected area to avoid uneven stresses on the mold surface and ejector system.

Gates

Gates serve to throttle flow into the cavities (via control of cross section), balance flow (via adjustment of location), and limit (via thickness and seal-off time) the time the cavity is under the influence of cylinder pressure. Pressure-time relations in the cavity can be crucial to the properties and reproducibility of nylon molded parts. Figures 5-9 and 5-10 illustrate for nylon-66 the effect of gate thickness or diameter on maximum cavity fill rate and gate seal-off time for specific melt and mold temperatures. Gates are not normally responsible for a major part of the pressure drop in a mold layout (see p. 183), but gate sizing is critical to the molding of nylons to avoid premature freezing. Small gates can also present problems in flow speed (jetting) and generation of high temperatures through high shear rates. This can affect the surface quality and properties of molded nylons.

With the current emphasis on automatic operations and fast molding, small gates and, in particular, tunnel gates have gained in frequency of use. The design and construction of such gates require care. For example, adequate thickness of metal must be provided between tunnel gates and cavity walls; also, tunnel gating into thin walls can lead to distortion and sticking of parts during ejection.

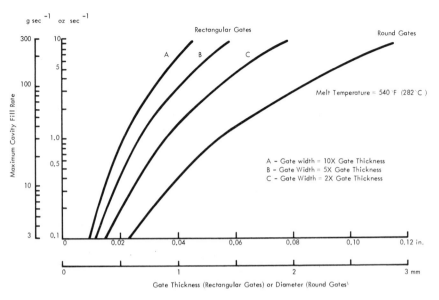

Fig. 5-9.   Maximum fill rate for a specific nylon-66 versus gate thickness or diameter (Ref. 2).

Cavities

The versatility of nylons in injection molding is exemplified by the extent to which they are molded successfully in complex cavity designs. Nylons are outstanding in stripping from undercuts without deformation or breakage, and they release better than the average thermoplastic from cavities with thin ribs, deep thin walls, and deep-draw cores.

Because nylons have a high heat content, good heat transfer away from parts that have thick walls and areas inaccessible to coolant must be provided by proper selection of materials of construction. The injection of hot nylon melts into relatively cold molds places a stress on the mold via the cyclic thermal expansion and contraction of the metal. This means care must be taken to maintain good mating of mold surfaces to avoid excessive tool wear and scoring.

In many molds nylons flash easily at the viscosities typical of injection grade resins. For this reason the contact areas around cavity edges should withstand clamp pressures, be ground flat and parallel, and be hardened (50 – 62 Rockwell C). In general, venting depth should not exceed 0.001 in. (0.025 mm).

Cavity sizing for nylons is as much an art as a science because of such complexities as wide variation in cavity design, differences among nylons in rates of crystallization, moisture induced growth, deformation problems during ejection, and anisotropy in the molded part. In some instances, resin suppliers have worked out guidelines. For example, one supplier provides a nomograph

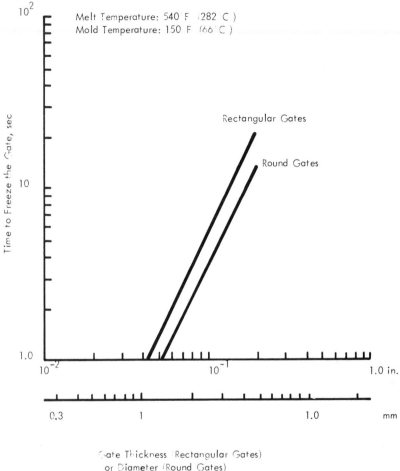

Fig. 5-10.   Effect of gate thickness or diameter on gate seal time for nylon-66 (Ref. 2).

that relates gate and part dimensions and process variables to mold shrinkage for a variety of compositions(2). The type of nylon, tolerance requirements, coolant demand, and technical and cost factors in mold construction determine the optimum number of cavities per mold.

Ejector Systems

Every part requires a certain amount of force to be ejected from a mold. The force required depends on the total drag area of walls, ribs, and bosses, and the coefficient of friction between the cavity metal and the nylon. Packing pressures in the cavity affect drag forces. Lubricants can reduce significantly the ejection force required for nylons.

Balancing the ejector forces and distributing them over the part by proper spacing and correct selection of ejector area is of great importance to secure smooth ejection. For consistent performance of the ejection system and to minimize part distortion, it is preferred to use guide pins and sleeve bushings to support and actuate the ejector plate. In automatic operation it is advisable to return the ejector assembly prior to closing, and this normally means hydraulic ejection. Of particular advantage is the facility thereby provided for multiple knock-out sequencing.

Cooling Systems

As cycles become faster in the processing of nylons the need for high coolant capacity and efficiency becomes more stringent. Mold heaters are being used less, and chillers are increasing in use. Many problems associated with processing and cavity-to-cavity fluctuations have to do with poor control and quality of the coolant. Water treatment systems are available to maintain the quality of water necessary to minimize scale build-up in mold passages, molding machine heat transfer equipment and chillers, and temperature zone controllers. Piping networks in a plant should be able to handle high flow rates while minimizing pressure drops to each part of the coolant system.

Obviously, higher cooling rates can be achieved by decreasing the temperature or by increasing the rate of circulation of the coolant. When using low coolant temperatures, care must be taken to avoid moisture condensation ("mold sweating").

Extensive use of refrigeration systems for cooling raises questions of system reliability and product quality. Chiller malfunction must be signaled promptly or a large number of bad parts may be produced. A variety of temperature control equipment of great accuracy and compact size is now available and includes devices that permit simultaneous, multiple mixing of plant water with the chilled water to allow independent control of individual molds.

FLUID FLOW AND HEAT TRANSFER

The injection molding process involves at its core two engineering disciplines – fluid flow and heat transfer. This section briefly reviews these two areas as they concern the molding of nylons.

Fluid Flow   (with M. I. Kohan)

The flow of nylons into an injection mold is a complex operation that involves non-Newtonian behavior and nonisothermal conditions. Viscosity varies with changing shear and temperature as the melt travels to the cavities. Discussions of such complexities are available(14) and are beyond the scope of this book.

Two kinds of calculations are generally of interest: (1) that of flow rate or shear rate, and (2) that of shear stress and related pressure drops. Shear rate is important because flow irregularities in the injection molding of nylons have been associated with excessively high rates (above 10,000 sec^{-1}). Knowledge of pressure drops is necessary in mold design and flow analysis.

The calculations outlined in this section ignore the heat transfer part of the problem and are only approximations, but they have proved to be of value in many actual molding situations. The development of the generally useful flow equations and their applicability to nylons were considered in Chapter 4 where it was pointed out that flow is normally laminar. The following simplifications for calculation of flow in a mold are made:

1. Flow is isothermal at the temperature developed in the cylinder, that is, the temperature of the melt as it exits the nozzle.

2. Only average flow rates are considered. Variation in flow rate within channels and turns is ignored. In some molds turn corrections are necessary and can be obtained empirically only. The mold is assumed to be filled at a constant injection rate, and the total time of fill is taken to be equal to the ram-in-motion time.

3. Resistance to flow due to compression of air is not considered because adequate venting is assumed to exist.

4. Flow cross sections are taken to be those cut into the mold. No allowance is made for the reduction in cross section caused by partial solidification of the melt.

Analysis begins with estimation of flow in the sprue, the runner system, and the gates. The mass flow through the nozzle is obviously equal to the total shot weight divided by the fill time. Everything that passes through the nozzles goes through the sprue so that the mass flow rate in the sprue must be the same as in the nozzle and therefore is also equal to the shot weight divided by the fill time. Assuming a symmetrical runner system, the mass flow in each of the runners depends on the number of such runners. The flow rate through each gate is simply the flow rate through the sprue divided by the number of gates:

$$M_g = \frac{1}{n} \left(\frac{W_T}{\text{RIM}} \right) \tag{5-1}$$

where

$$M_g = \text{mass flow rate through one gate}$$
$$n = \text{number of gates}$$
$$W_T = \text{total shot weight}$$
$$\text{RIM} = \text{ram-in-motion time}$$

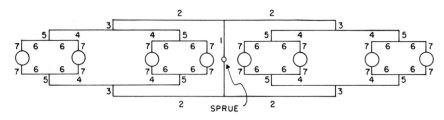

SECTION	CROSS-SECTION	MAXIMUM AND MINIMUM DIAMETER OR WIDTH		THICKNESS		LENGTH	
		IN.	CM	IN.	CM	IN.	CM
SPRUE	CIRCULAR (TAPERED)	285, .160	725, .41	–		2.5	6.35
RUNNERS							
1	TRAPEZOIDAL	.460, .325	1.17, .825	.285	.725	2.0	5.1
2	"	.365, .270	.93, .685	.225	.57	4.0	10.2
3	"	.365, .255	.93, .65	.220	.56	.5	1.3
4	"	.330, .200	.84, .51	.190	.48	1.8	4.6
5	"	.330, .200	.84, .51	.190	.48	0.5	1.3
6	"	.240, .165	.61, .42	.165	.42	1.0	2.5
7		.195, .130	.495, .33	.140	.355	.5	1.3
GATE	RECTANGULAR	.050	.13	.040	.10	.060	.15

Fig. 5-11.   Initial layout for sample gear mold.

The mass flow rates are converted to volumetric flow rates by dividing by the density at the melt temperature in question (Chapter 4, p. 124, Table 4-2).

Correction factors for different channel cross sections are provided in Chapter 4 (p. 134, Fig. 4-9). The problem of gate freezing prior to fill was noted previously (see Fig. 5-10). The important relationship between fill time, cavity dimensions, and pressure as dictated by the freezing characteristics of the nylon bears mention here also. This relationship has been discussed at length for nylon-66 by Filbert(15a).

Our analysis now continues and employs the example of an eight-cavity gear mold to be filled with a standard grade of nylon-66 having a $\bar{M}_n$ of 18,000 to illustrate application of flow calculations to a mold layout problem. The initial layout is depicted in Figure 5-11, and additional pertinent information follows:

Shot weight	111 g
Part weight	6 g
Melt temperature	536°F (280°C)
Fill time	6 sec

Thus, the mass flow rate in the sprue is 111 g/6 sec or 18.5 g sec^{-1}, and the volumetric flow rate (assuming a density of 0.96 g cm^{-3} for nylon-66 at 280°C) is 18.5/0.96 or 19.3 cm^3 sec^{-1} (1.18 in.^{3} sec^{-1}). The flow rates in the runners depend only on the number of streams into which the melt divides and are shown below:

Runners	Flow rate
1	(Sprue flow rate)/2
2 and 3	(SFR)/4
4 and 5	(SFR)/8
6 and 7	(SFR)/16

The flow rate in the gate is the same as in runners 6 and 7.

The shear rate in the sprue is calculated using Eq. (4-7), which applies to a circular channel of uniform cross section:

$$\dot{\gamma}_w = \frac{4Q}{\pi R^3}$$

where

Q = volumetric flow rate

R = radius of channel

$\dot{\gamma}_w$ = maximum shear rate at wall

The average of the minimum and maximum cross-section radii is used for the conically shaped sprue (from Figure 5-11, 0.111 in. or 0.283 cm). If the runners are circular in cross section, Eq. (4-7) would again apply, but the trapezoidal shape of the runners introduces a complexity worth explanation because it shows how to use the information made available in Chapter 4. It is now helpful to refer to Figure 4-9. The trapezoidal shape more closely resembles the rectangle than any other cross-section geometry in Figure 4-9. We now assume that the trapezoid can be equated to a rectangle with the same thickness and with a width equal to the average width of the trapezoid. It can be shown that this is a very good approximation. The ratio of thickness to average width for the data given in Figure 5-11 is about 0.715 for runners 1 through 5 and 0.84 for runners 6 and 7. Figure 4-9 indicates that the corresponding dimensionless flow coefficients are 0.56 and 0.50. These coefficients indicate that Q, the volumetric flow rate, when *calculated* for a given pressure drop (ΔP) is less than that determined from the thin-slit equation, Eq. (4-22). This is because the sides of the trapezoid or rectangle contribute restraints to flow that can no longer be ignored. The conversion of an *actually measured* flow rate to a shear rate is accomplished with only minor error (see p. 183) by using the slit shear-rate equation, Eq. (4-21):

$$\dot{\gamma}_w = \frac{6Q}{WH^2}$$

where

$$Q \;=\; \text{volumetric flow rate}$$
$$W \;=\; \text{width}$$
$$H \;=\; \text{thickness}$$
$$\dot{\gamma}_w \;=\; \text{maximum shear rate at wall}$$

The flow rates and calculated shear rates that correspond to the situation in Figure 5-11 are shown in Table 5-5.

Table 5-5.

	Mass flow rate		Volumetric flow rate		Shear rate
	(g sec^{-1})	(oz sec^{-1})	(cm^3 sec^{-1})	(in.^{3} sec^{-1})	(sec^{-1})
Sprue	18.5	0.653	19.3	1.18	1100
Runners					
1	9.25	0.327	9.65	0.59	111
2	4.63	0.163	4.82	0.30	112
3	4.63	0.163	4.82	0.30	120
4, 5	2.31	0.082	2.41	0.15	94
6	1.16	0.041	1.22	0.075	82
7	1.16	0.041	1.22	0.075	141
Gate	1.16	0.041	1.22	0.075	5620

The flow rate in the cavity is not of concern in this case; however, the pressure drop in cavities is often considerable and can constitute a major resistance to flow(16).

Calculation of the pressure drop requires use of the melt viscosity which is determined for the nylon in question, a 66 with a $\overline{M}_n$ of 18,000, and for each shear rate from Fig. 4-14 (Chapter 4, p. 139). The pressure drop in the sprue is readily obtained from the circular cross-section expression, Eq. (4-8), which appears below in rearranged form:

$$\Delta P = \frac{8QL\mu}{\pi R^4}$$

where

ΔP = pressure drop

Q = volumetric flow rate

L = length of channel

μ = melt viscosity

R = radius

Alternatively, to make use of the previously calculated shear rate:

$$\Delta P = \dot\gamma_w \left(\frac{2L\mu}{R}\right) \qquad (5\text{-}2)$$

Thus,

$$\Delta P_{sprue} = 1100 \text{ sec}^{-1} \left(\frac{2 \times 6.35 \text{ cm} \times 1000 \text{ poise}}{0.283 \text{ cm}}\right)$$

$$= 4.94 \times 10^7 \text{ dyn cm}^{-2} = 715 \text{ lb(wt)in.}^{-2}$$

$$= 50.3 \text{ kg(wt)cm}^{-2}$$

Care must be taken to use consistent units (see Table 4-1). The shear stress can then be determined from Eq. (4-1) or Eq. (4-8):

$$\tau_w = \frac{(\Delta P) R}{2L}$$

where τ_w = maximum shear stress at wall, or

$$\tau_w = \mu \dot\gamma_w$$

For the runners, the slit equation cannot be used directly because it does not take into account the restraints of the side walls as discussed above. It can be applied, however, with a correction factor which, as noted above, is the flow coefficient (F) obtained from Fig. 4-9. Thus, Eq. (4-22) takes the form

$$\Delta P = \frac{12QL\mu}{WH^3} \times \frac{1}{F}$$

Using the shear rate yields a form analogous to Eq. (5-2):

$$\Delta P = \dot{\gamma}_w \left(\frac{2L\mu}{H} \right) \frac{1}{F} \qquad (5\text{-}3)$$

Thus,

$$\Delta P_{\text{runner 1}} = 111 \text{ sec}^{-1} \left(\frac{2 \times 2.0 \text{ in.} \times 0.0245 \text{ lb(wt)sec in.}^{-2}}{0.285 \text{ in.}} \right) \frac{1}{0.56}$$

$$= 61.5 \text{ lb(wt)in.}^{-2} = 4.32 \text{ kg(wt)cm}^{-2}$$

The shear stress can be obtained directly from the shear rate and the viscosity: $\tau_w = \mu \, \dot{\gamma}_w$. To calculate the shear stress from the pressure drop, it is necessary to go back to the basic relationship that the shear stress is the pressure drop multiplied by the cross-sectional area divided by the product of the cross-sectional perimeter and the channel length. For the rectangular approximation of the trapezoid, this is (compare Figure 4-7):

$$\tau_w = \frac{(\Delta P)WH}{2L(W + H)} \qquad (5\text{-}4)$$

$$= \frac{(\Delta P)H}{2L} \times \frac{1}{1.715} \qquad \text{for } \frac{H}{W} = 0.715$$

$$= \frac{(\Delta P)H}{2L} \times \frac{1}{1.84} \qquad \text{for } \frac{H}{W} = 0.84$$

Because $(\Delta P)H/2L$ is the slit expression for shear stress (see Eq. (4-22)), the actual shear stress is the slit shear stress multiplied by a geometrical correction factor which in the above cases are $1/1.715$ or 0.584 and $1/1.84$ or 0.544. For consistency, the shear-stress correction factor divided by a shear-rate correction factor must equal the flow coefficient obtained from Figure 4-9. The flow coefficients were shown (p. 179) to be 0.56 and 0.50, respectively. Thus,

$$\text{shear rate factor} = \frac{\text{shear stress factor}}{\text{flow coefficient}}$$

$$= \frac{0.584}{0.56} = 1.04 \text{ in one case and}$$

$$= \frac{0.544}{0.56} = 1.09 \text{ in the other case.}$$

It is the proximity of the shear-rate factor to unity that permits direct use of the slit expression for shear rate as was done above. The validity of these conclusions can be established on a more theoretical basis(17).

The viscosities, pressure drops, and shear stresses that apply to the situation in Figure 5-11 are summarized in Table 5-6.

Table 5-6.

	Apparent melt viscosity		Pressure drop		Shear stress	
	(lb(wt)sec in.$^{-2}$)	(poise)	(lb(wt)in.$^{-2}$)	(kg(wt)cm^{-2})	(lb(wt)in.$^{-2}$)	(kg(wt)cm^{-2})
Sprue	0.0145	1000	715	50	15.9	1.12
Runners						
1	0.025	1700	63	4.4	2.7	0.19
2	0.025	1700	160	12	2.7	0.19
3	0.025	1700	25	1.7	2.9	0.20
4	0.025	1700	78	5.5	2.4	0.17
5	0.025	1700	21	1.5	2.4	0.17
6	0.0255	1750	47	3.3	2.1	0.15
7	0.024	1650	45	3.2	3.4	0.24
Gate	0.0075	520	210	15	42.2	2.97
			1364	96.6		

The length to thickness ratio is small in runners 3, 5, and 7 and in the gate, but the only major reduction in channel dimension is in entering the gate. There is very likely an end effect in the gate which the above, simplified treatment also overlooks. Nonetheless, it is clear that the total pressure drop (or resistance to flow by the layout) is small relative to the machine pressure available, 15,000 lb(wt)in.$^{-2}$ or 1050 kg(wt)cm^{-2}. The largest resistance to flow is in the sprue followed by the gate and the second runner. Although the shear rate in the gate is high, it appears to be well below the critical level, about 10,000 sec^{-1}, for flow irregularities.

The preceding flow and pressure-drop calculations indicate that the runner layout is unnecessary and can be considerably reduced without sacrificing mold filling. This is highly desirable because the layout in question generates 57% regrind, and part quality may be improved by reducing the amount of regrind. Also, the gates may be decreased in number. If necessary, the gates could be

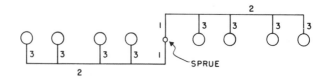

SECTION	CROSS - SECTION	DIAMETER OR WIDTH		THICKNESS		LENGTH	
		IN.	CM	IN.	CM	IN.	CM
SPRUE	CIRCULAR (TAPERED)	.285, .160	.725, . 41	—		1.0	6.35
RUNNERS							
1	TRAPEZOIDAL	.460, .325	1.17, .825	.285	.725	1.0	2.5
2	"	.225, .180	.57, .46	.165	.42	6.7	17.0
3	"	.200, .100	.51, .25	.135	.34	.5	1.3
GATE	RECTANGULAR	.050	.13	.040	.10	.060	.15

Fig. 5-12.   Revised layout for gear mold.

increased in size to reduce the shear rate (reducing the rate of fill further was not considered advisable because the rate of fill was already slow). The layout was changed to that shown in Figure 5-12; sprue and gate dimensions were not altered. The new layout reduced the shot weight from 111 to 68 g and decreased the amount of regrind from 57 to 29%. The color of the gears improved noticeably because of the use of less regrind. Shear in the gates did not prove to be excessive.

In a layout for which pressure drops to individual cavities can be shown to differ, it is generally advisable with nylons to check dimensional variation if the difference exceeds 1000 lb(wt)in.$^{-2}$ (70 kg(wt) cm^{-2}). In situations that are especially critical with respect to toughness, a performance check on parts from different cavities is recommended if the difference in pressure drop exceeds 500 lb(wt)in.$^{-2}$ (35 kg(wt) cm^{-2}).

Heat Transfer

The heat transfer problem is of special importance in nylons because of their generally high melting points and the sensitivity of their physical structure and properties to thermal history (see Chapter 8, pp. 282 and 299). The energy requirements of the injection end were mentioned earlier (p. 169).

Metzner, Vaughn, and Houghton[18] have dealt with the problem of heat transfer in a flowing liquid under laminar conditions in tubes, and Skelland (Ref. 14, p. 383) has given an excellent review of the subject. Knudsen and Katz[19] have provided an explanation of the physical significance of the dimensionless groups that appear in the Metzner equation.

At typical conditions for molding nylons (radial temperature gradients of 536 to 50°F or 280 to 10°C in a conventional runner(20), a 1.3 viscosity ratio between the bulk stream and the wall conditions, shear rates at the wall of 1000 sec^{-1}, and a 0.250-in.-diameter runner, 10 in. long), Graetz numbers (N_{Gz}) of 6.0 to 7.0 $\times$ 10^5 and Nusselt numbers (N_{Nu}) of 300 to 400 result. This leads to a heat transfer coefficient h of about 2240 Btu hr^{-1} ft^{-2} °F^{-1} (1075 cal hr^{-1} cm^{-2} °C^{-1}). Assuming that all heat transfer is radial from the bulk of the stream to the boundary layer, this leads to a heat transfer rate of 6000 Btu hr^{-1} (1500 kcal hr^{-1}). However, this type of heat transfer is not continuous because mold filling takes a small percentage of the total cycle time in the average situation. Furthermore, the boundary of solidifying plastic usually acts as the controlling resistance to heat transfer(21). The net result is a considerable reduction (50- to 100-fold) in the heat transfer rate of the flowing melt as calculated by the Metzner equation. The most important part of heat transfer occurs, therefore, after fill, and it is with this aspect of the problem that the rest of this discussion deals.

The calculations that follow are presented in the form of a specific mold problem. Details are provided with respect to cooling of the filled cavity, and a similar analysis can be applied to the sprue and runner system. Figure 5-13 shows the cavity design planned for a bushing to be made of a nylon-66 with a $\bar{M}_n$ of 18,000. Other pertinent data follow:

No. of cavities	:	64
Estimated cycle	:	12 sec
Cycle rate	:	300 cycles hr^{-1}
Melt temperature	:	536°F (280°C)
Ejection temp.	:	250°F (121°C)
Room temperature	:	70°F (21°C)
Part weight	:	0.07 oz (2.0 g)
Runner weight	:	1.81 oz (51.2 g)
Shot weight	:	6.3? oz (179.2 g)
Total part wt.	:	4.52 oz (128 g)

Heat Transfer Requirements

The enthalpy of processing is, using Figure 5-14;

$$\Delta H_{in} = H_{536°F} - H_{70°F} = 300 \text{ Btu lb}^{-1} (167 \text{ cal g}^{-1}). \qquad (5\text{-}5)$$

Thermodynamic data have been published on nylon-66(22,24), nylon-610 (22,25), and nylon-6(23) and are available from trade literature. The enthalpy leaving with the parts and runners is

$$\Delta H_{out} = H_{250°F} - H_{70°F} = 80 \text{ Btu lb}^{-1} (45 \text{ cal g}^{-1}). \qquad (5\text{-}6)$$

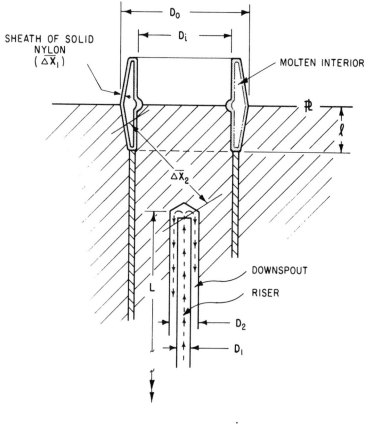

		in.	cm
Bubbler Length	(L)	12.00	30.5
Riser ID (final)	(D_1)	0.25	0.635
Downspout OD (final)	(D_2)	0.3125	0.794
Half Cavity Height	(ℓ)	0.225	0.570
Cavity OD	(D_0)	0.600	1.524
Cavity ID	(D_i)	0.350	0.889
Sheath Thickness	($\Delta \overline{X}_1$)	0.042	0.107
Core Thickness	($\Delta \overline{X}_2$)	0.400	1.016

Fig. 5-13.   Cavity design for nylon bushing.

The mass production rate (m_p), from the above data, is

$$300 \times 6.33 \text{ or } 1900 \text{ oz hr}^{-1} \text{ (119 lb hr}^{-1}, 54.5 \text{ kg hr}^{-1}).$$

Thus, the net enthalpy to be removed by the mold cooling system is

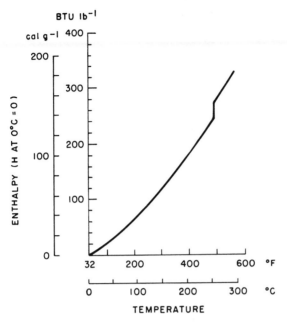

Fig. 5-14.   Enthalpy of nylon-66 (Ref. 22).

$$\Delta H_{net} = \Delta H_{in} - \Delta H_{out} = 119\,(300\text{-}80) = 26,200 \text{ Btu hr}^{-1} \qquad (5\text{-}7)$$
$$(6600 \text{ kcal hr}^{-1})$$

The convection and radiation losses of mold and machine should be subtracted, but these losses are usually not high. Calculation of these losses is postponed until a later section.

The enthalpy to be removed per cavity is

$$\Delta H_c = (\text{wt of one part/wt of shot})\ \Delta H_{net} \qquad (5\text{-}8)$$

$$= (2/179) \times 26,200 = 293 \text{ Btu hr}^{-1} \ (74 \text{ kcal hr}^{-1}).$$

Because of the part geometry, the cavities are to be placed symmetrically into the front and rear cavity plates. Each cavity plate will have its own cooling system. Hence, we can confine the problem to a half-cavity:

$$\Delta H_{c/2} = 0.5 \times \Delta H_c = 147 \text{ Btu hr}^{-1} \ (37 \text{ kcal hr}^{-1}). \qquad (5\text{-}9)$$

Minimum Rate of Coolant Circulation
It is helpful to make a preliminary estimate of the required rate of coolant circulation assuming that all the heat is removed efficiently. Good practice

dictates that the coolant temperature rise be small – preferably about 7°F (4°C) or less (26).

If the coolant is water circulating through one side of the mold at a rate of 2 gal min^{-1}, the mass flow rate of water is

$$m_w = 2 \times 8.35 \times 60 = 1000 \text{ lb hr}^{-1} \text{ } (454 \text{ kg hr}^{-1})$$

Because the coolant must remove all of the heat from one half of the mold:

$$m_w \times \text{sp. heat} \times \Delta t = \Delta H_{net}/2$$

Hence

$$\Delta t = \Delta H_{net}/2 \times m_w \times \text{sp. heat} = 26,200/2000 = 13.1°F \text{ } (7.3° \text{ C})$$

Increasing the rate of circulation to 4 gal min^{-1} (15.2 liter min^{-1}) reduces Δt to an acceptable value.

At this point, it should be kept in mind that we have not yet considered whether this rate of water circulation can be achieved. This depends on the diameter of coolant passages, the size of the pump, and the distance that the coolant must travel.

A critical question concerns the implication in the above calculations that a completely solid part is removed from the mold. The next section considers this question.

Conduction from the Part to the Cavity Wall

The first resistance to heat transfer encountered is the sheath of plastic that forms immediately as the molten nylon hits the cavity. The sheath grows with time in a nonlinear fashion (27), and hence the heat transfer through the growing sheath decays also in a nonlinear fashion. This is obviously a complex situation, but we can estimate the time required for cooling the part. Assuming that heat removal is adequate to maintain boundary conditions, one can calculate, with the aid of Heyman's equations (27), the time required to freeze the part to the centerline. For a melt temperature of 536°F, a wall temperature of 250°F, and a slab of half-thickness equal to 0.0625 in., this turns out to be 11.5 sec.

In the given problem with an over-all cycle of 12 sec, the actual time of the part in the mold would be about 8 sec. The ratio of 8 sec cooling to 11.5 sec for complete solidification yields an estimate of sheath thickness equal to two-thirds of the half-thickness of the part or one-third of the average part thickness. This corresponds to 0.042 in. (0.107 cm). Experience has shown that for standard nylon-66 moldings not over 2 in. (5.1 cm) in the long dimension and not over

0.2 in. (0.51 cm) in width, ejection will not cause part distortion if this sheath size, that is, one-third of the average part thickness, is obtained. Fast molding resins with less tendency to stick in the mold require even less of a solidified sheath for ejection.

Consider the amount of heat that must be conducted per hour (q_s) in order to achieve this sheath size. The area over which heat transfer occurs for each half-cavity can be approximated as follows:

$$A_s = \pi \ell (D_O + D_i) + \pi (D_O^2 - D_i^2)/4$$

$$= 3.14 \times 0.225 (0.6 + 0.35) + 3.14 (0.36 - 0.12)/4$$

$$= 0.86 \text{ in.}^2 (5.6 \text{ cm}^2)$$

Application of the integrated form of Fourier's conduction formula (Ref. 10, p. 7) yields

$$q_s = k_n A_s \frac{\Delta t}{\Delta \bar{X}_1} \tag{5-10}$$

where

$$k_n = \text{thermal conductivity of nylon sheath}$$
$$\Delta t = \text{temperature difference}$$
$$\Delta \bar{X}_1 = \text{sheath thickness}$$

For a melt temperature of 536°F, a wall temperature of 250°F, and k_n equal to 1.7 Btu hr^{-1} ft^{-2} °F^{-1} in. (5.8×10^{-4} cal sec^{-1} cm^{-2} °C^{-1} cm)

$$q_s = 1.7 \times \left(\frac{0.86}{144}\right) \left(\frac{286}{0.042}\right) = 69 \text{ Btu hr}^{-1} (17.4 \text{ kcal hr}^{-1})$$

Equation (5-9) indicates that the requirement ($\Delta H_c/2$) is 147 Btu hr^{-1}. Examination of the heat transfer equation reveals that the only parameter over which we have control is the temperature gradient. Improving heat transfer requires a lower cavity wall temperature. This can be accomplished by bringing the bubbler closer. A reduction to 100°F (37.8°C) exterior sheath temperature would improve the situation by 436/286 or 1.53 fold. The rate of heat transfer through the sheath becomes 1.53 × 69 or 105 Btu hr^{-1} (26.4 kcal hr^{-1}). This still falls short of the requirement for complete solidification, but the real need is to achieve sufficient rigidity to permit knocking the part out of the mold without distortion. Thus, the 12-sec overall cycle appears feasible if more than 69 Btu hr^{-1} can be removed from the surface of the cavity.

As nylon parts cool in the cavity, the possibility of shrinkage away from the walls arises. Air or vacuum gaps that result would introduce another, conceivably controlling resistance to heat transfer. The obvious friction between cavity walls and nylon parts and evidence with respect to residual mold pressure just before mold opening suggest that this does not normally occur.

Conduction from the Cavity Wall to the Coolant

Before advancing into this section we must estimate the approximate temperature gradient that will develop in the cavity core, that is, from the cavity wall to the coolant. This gradient will depend on the material of construction of the core, the water temperature in the coolant passage (a bubbler in our case), and the cavity wall temperature. For symmetrical cavities, the problem can be approached by using the two-dimensional relaxation approach outlined by McAdams (Ref. 10, pp. 19 to 24). Figure 5-15 provides a two-dimensional map of isotherms (heat flux lines would form a second two-dimensional map orthogonal to the isotherms) assuming a wall temperature of 200°F (93°C) and a coolant temperature of 50°F (10°C) at the centerline of the bubbler.

From the temperature map and the equation for conduction we can estimate the heat transferred through the cavity core:

$$q_c = \frac{k_c A_c \Delta t_c}{\overline{\Delta X_2}}$$

The thermal conductivity, k_c for 440 stainless steel, the material of construction of the core, is 150 Btu hr^{-1} ft^{-2} $°F^{-1}$ in. (5.1 $\times$ 10^{-2} cal sec^{-1} cm^{-2} $°C^{-1}$ cm). Figure 5-15 shows a bubbler external wall temperature that varies from 60°F (15.6°C) to 77°F (26°C). Using 70°F as the average external bubbler wall temperature, a cavity wall temperature of 200°F (94.3°C), an area of 0.86 in.^{2} for the half-cavity, and an average cavity wall to coolant distance (ΔX_2) of 0.40 in. (1.0 cm),

$$q_c = 150 \left(\frac{0.86}{144}\right) \frac{130}{0.4} = 292 \text{ Btu } hr^{-1} \text{ (73.6 kcal } hr^{-1})$$

This is more than adequate for even the total solidification requirement of 147 Btu hr^{-1} (37 kcal hr^{-1}). The rate of heat transfer obviously increases with increasing temperature difference (for example, a cavity wall temperature of 250°F) so that conduction from the cavity wall to the coolant is not limiting for a centerline coolant temperature of 50°F.

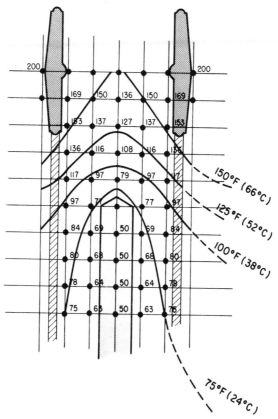

Fig. 5-15. Temperature profiles in the core of the bushing cavity. A wall temperature of 200°F (93°C) and a coolant temperature of 50°F (10°C) at the center line of the bubbler are assumed.

Heat Transfer in the Bubbler

Before a forced convection heat transfer coefficient can be estimated, flow calculations must be performed. The heat transfer coefficient will then be determined for the downspout part of the bubbler (see Figure 5-13) because this is the side in contact with the half-cavity core.

Consider first the internal riser tube in the bubbler. The integrated form of the Fanning equation (Ref. 10, p. 145) can be applied:

$$\Delta P = \frac{f_m W^2 v_m L}{2 g_c r_h} \tag{5-11}$$

where

ΔP = pressure drop

f_m = dimensionless friction factor

W = mass velocity per unit area of cross-section

v_m = specific volume

L = length

g_c = dimensional conversion factor from Newton's law of motion

r_h = hydraulic radius = area of cross-section divided by wetted perimeter

The friction factor, f_m, depends on the Reynolds number, $N_{Re} = 4Wr_h/\mu$. In the cgs system of units, ΔP is in dyn cm^{-2}, g_c is unity, and μ is in poise. For ΔP to be in kg (wt) cm^{-2}, g_c must be 980,000 dyn kg(wt)$^{-1}$. In the English system, ΔP is in lb(wt)in.$^{-2}$, g_c is 386 lb in. sec^{-2} lb(wt)$^{-1}$, and μ must be in lb sec^{-1} in.$^{-1}$ so that $\mu = \mu$ (in lb(wt)sec in.$^{-2}$) $\times$ g_c or μ (in centipoise) $\times$ 5.60 $\times$ 10^{-5}. Because the viscosity of water changes significantly with temperature and affects N_{Re}, f_m varies with temperature.

The downspout of the bubbler is the annular region that is in direct contact with the cavity core. Equation (5-11) applies also to the downspout via use of $(D_2 - D_1)/4$ for the hydraulic radius:

$$r_h = \frac{(\pi/4)(D_2{}^2 - D_1{}^2)}{\pi(D_2 + D_1)} = \frac{D_2 - D_1}{4}$$

Tables 5-7 and 5-8 summarize calculations for one resin and one downspout. Note the very high Reynolds numbers which indicate turbulent flow. The pressure drop for 4 gal min^{-1} flow rate exceeds the pressure normally available in cooling equipment — in particular because the mold layout of 64 cavities will require that several bubblers be connected in series. Otherwise, there would be a maze of water hoses dangling from the mold, an obviously impractical situation considering space limitations and operating problems. Increasing the diameter of the bubbler and decreasing the rate of coolant circulation are indicated to obtain a lower pressure drop. Increasing pumping capacity is an obvious alternative.

At the larger diameters and lower coolant flow rates, the pressure drops become more practical, permit connection of about sixteen half cavities in series, and lead to a simpler hose arrangement.

Now it remains to be found whether the lower coolant flow rate will accomplish the necessary heat removal. McAdams (Ref. 10, p. 242) gives the following equation for cooling on outside walls of annular tubes where the ratio of diameters is less than 10 to 1 (turbulent flow):

Table 5-7. Pressure drops for one riser[a]

State	Temp. (°F)	Temp. (°C)	Circulation (gal min⁻¹)	Circulation (cm³ sec⁻¹)	Dia. (D_1) (in.)	Dia. (D_1) (cm)	W (lb sec⁻¹ in.⁻²)	W (g sec⁻¹ cm⁻²)	r_h (in.)	r_h (cm)	μ (lb sec⁻¹ in.⁻¹) ×10⁵	μ (cp)	N_{Re}	f_m	ΔP (lb(wt)in.⁻²)	ΔP (kg(wt)cm⁻²)
1	100	37.8	4	252	0.125	0.318	45.2	3180	0.0312	0.0795	3.83	0.684	147,000	0.0051	150	10.5
2	100	37.8	2	126	0.125	0.318	22.6	1590	0.0312	0.0795	3.83	0.684	73,500	0.0057	42	3.0
3	50	10.0	2	126	0.125	0.318	22.6	1590	0.0312	0.0795	7.33	1.308	38,400	0.0065	47	3.3
4	100	37.8	2	126	0.188	0.476	10.0	703	0.0469	0.119	3.83	0.684	49,000	0.0062	6.0	0.42
5	50	10.0	2	126	0.188	0.476	10.0	703	0.0469	0.119	7.33	1.308	25,600	0.0070	6.7	0.47
6	50	10.0	2	126	0.250	0.635	5.64	396	0.0625	0.159	7.33	1.308	34,200	0.0063	1.4	0.10

[a] L = 12.0 in. (30.5 cm); $v_m \cong$ 27.8 in.³ lb⁻¹ (1.00 cm³ g⁻¹); 1 gal. = 231 in.³ = 3785 cm³.

Table 5-8. Pressure drops for one downspout

State	Temp. (°F)	Temp. (°C)	Circulation (gal min⁻¹)	Circulation (cm³ sec⁻¹)	Dia. (D_2) (in.)	Dia. (D_2) (cm)	D_1 (in.)	W (lb sec⁻¹ in.⁻²)	W (g sec⁻¹ cm⁻²)	r_h (in.)	r_h (cm)	μ ×10⁵ (lb sec⁻¹ in.⁻¹)	μ (cp)	N_{Re}	f_m	ΔP (lb(wt)in.⁻²)	ΔP (kg(wt)cm⁻²)
1	100	37.8	4	252	0.250	0.635	0.188	25.8	1810	0.0156	0.0396	3.83	0.684	42,100	0.0064	118	8.3
2	100	37.8	2	126	0.250	0.635	0.188	12.9	905	0.0156	0.0396	3.83	0.684	21,000	0.0073	36	2.5
3	50	10.0	2	126	0.250	0.635	0.188	12.9	905	0.0156	0.0396	7.33	1.308	11,000	0.0089	39	2.7
4	100	37.8	2	126	0.313	0.795	0.250	10.0	703	0.0156	0.0396	3.83	0.684	16,300	0.0079	22	1.6
5	50	10.0	2	126	0.313	0.795	0.250	10.0	703	0.0156	0.0396	7.33	1.308	8,500	0.0089	24	1.7

$$\frac{1}{h_L} = \frac{(N_{Re})^{0.2} \, (N_{Pr})^{2/3}}{0.023 \, C_p W} \left(\frac{\mu_w}{\mu_b} \right)^{0.14} \tag{5-12}$$

where

h_L = convection heat transfer coefficient

N_{Re} = Reynolds number

N_{Pr} = Prandtl number = $C_p \, \mu/k$ where k is thermal conductivity

C_p = specific heat

W = mass flow per unit cross-section

μ_w = viscosity at wall temperature

μ_b = viscosity at bulk temperature

The calculations that follow make use of data provided in Table 5-6 and assume that the wall temperature is 20°F (11.1°C) higher than the bulk temperature.

State 4:

$$\frac{1}{h_L} = \frac{(16,300)^{0.2} \, (2.42 \times 0.684/0.363)^{2/3}}{0.023 \, (1.0)(10.0)} \left(\frac{0.559}{0.684} \right)^{0.14}$$

$$= \frac{6.96 \, (2.73)}{0.23} (0.76) = 63$$

$$h_L = \frac{1}{63} \text{Btu sec}^{-1} \text{ in.}^{-2} \,^{\circ}\text{F}^{-1} = 8200 \text{ Btu hr}^{-1} \text{ ft}^{-2} \,^{\circ}\text{F}^{-1}$$
$$(1.1 \text{ cal sec}^{-1} \text{ cm}^{-2} \,^{\circ}\text{C}^{-1})$$

State 5:

$$\frac{1}{h_L} = \frac{(8,500)^{0.2} \, (2.42 \times 1.308/0.348)^{2/3}}{0.023 \, (1.0)(10.0)} \left(\frac{0.979}{1.308} \right)^{0.14}$$

$$= \frac{6.1 \, (4.35)}{0.23} (0.67) = 77$$

$$h_L = \frac{1}{77} \text{ Btu sec}^{-1} \text{ in.}^{-2} \,^{\circ}\text{F}^{-1} = 6700 \text{ Btu hr}^{-1} \text{ ft}^{-2} \,^{\circ}\text{F}^{-1}$$
$$(0.91 \text{ cal sec}^{-1} \text{ cm}^{-2} \,^{\circ}\text{C}^{-1})$$

Doubek (29) used another approach to determine film heat transfer coefficients for glycol-water coolant systems. Calculation of h_L via this method yields a value for State 4 of 7500 Btu hr^{-1} ft^{-2} $°F^{-1}$ (1.0 cal sec^{-1} cm^{-2} $°C^{-1}$).

The rate of heat transfer, q_w, is $h_L A_w \Delta t$. If h_L is taken to be 6700 Btu hr^{-1} ft^{-2} $°F^{-1}$, if A_w is the outside surface area of the bubbler (0.202 in.^{2}) calculated for 0.2 in. of length, and if Δt is the temperature difference (20°F) between the outside and inside walls of the bubbler, then q_w is 188 Btu hr^{-1} (47.4 kcal hr^{-1}). This is obviously a high value in spite of the use of only 0.2 in. of bubbler length in estimating A_w.

Convection and Radiation Losses

McAdams (Ref. 10, p. 173) gives for natural convection of short vertical plates a heat transfer coefficient (h_v) of 0.9 Btu hr^{-1} ft^{-2} $°F^{-1}$ (1.2 × 10^{-4} cal sec^{-1} cm^{-2} $°C^{-1}$). Based on a vertical mold area (A_v) of 2 ft^2 and a temperature difference of 80°F (44.4°C) between the mold surface and the environment, q_v is (0.9) (2) (80) or 144 Btu hr^{-1} (36.3 kcal hr^{-1}).

For horizontal plate losses to air McAdams (Ref. 10, p. 180) indicates that h_h equals 0.22 $(\Delta t)^{1/3}$ which, for an 80°F difference, is also 0.9 Btu hr^{-1} ft^{-2} $°F^{-1}$. For a horizontal mold area (A_h) of 3 ft^2, q_h is 216 Btu hr^{-1} (54.5 kcal hr^{-1}).

Estimating radiation loss from the sides requires compensation for radiation from the machine doors and is accomplished via Eq. (5-13) (Ref. 10, p. 63):

$$q_{\text{sides}} = A_v k_r (T_M^4 - T_D^4) \frac{1}{1/E_1 + 1/E_2} \qquad (5-13)$$

where

k_r = Stefan-Boltzmann constant

T_M = absolute temperature of mold

T_D = absolute temperature of doors

E_1 = emissivity for mild steel, unpolished = 0.2

E_2 = emissivity for painted doors = 0.9

The value of k_r is 4.88 × 10^{-12} kcal hr^{-1} $°K^{-4}$ cm^{-2}, and T is in degrees Kelvin; or, in English units, k_r is 1.71 × 10^{-9} Btu hr^{-1} $°R^{-4}$ ft^{-2} and T is in degrees Rankine. Emissivity is dimensionless. Assuming T_M is 150°F and T_d is 70°F,

$$q_{\text{sides}} = (2)(1.71 \times 10^{-9})(610^4 - 530^4) \left(\frac{1}{1/0.2 + 1/0.9} \right)$$

$$= 3.42 \times 10^{-9} (5.95 \times 10^{10}) \left(\frac{1}{6.11} \right)$$

$$= 33.3 \text{ Btu hr}^{-1} \ (8.4 \text{ kcal hr}^{-1})$$

The same expression can be applied to the bottom of the mold and the painted surface below. The area is 1.5 ft^2, so q_{bottom} is $(1.5/2)$ (33.3) or 25.0 Btu hr^{-1} (6.3 kcal hr^{-1}). A maximum value for radiation from the top of the mold is obtained by ignoring reflection from ceiling, roof, or other confining surface:

$$q_{top} = A_{top} E_1 k_r T_m^4 = (1.5)(0.2)(1.71 \times 10^{-9})(610^4)$$

$$= 71.1 \text{ Btu hr}^{-1} \ (17.9 \text{ kcal hr}^{-1})$$

The total convection and radiation loss, $q_v + q_h + q_{sides} + q_{bottom} + q_{top}$, is $144 + 216 + 33 + 25 + 71$ or 489 Btu hr^{-1} (123 kcal hr^{-1}). As shown in Eq. (5-8), the heat to be removed is 293 Btu hr^{-1} per cavity or $18{,}800$ Btu hr^{-1} (4740 kcal hr^{-1}) for the 64-cavity mold. Even if the assumption is made that other losses through the machine and clamp areas double the total convection and radiation loss, these losses constitute about 5% of the heat removal requirement and can be ignored in estimating demand on the cooling system.

Summary

The multiple resistances to heat transfer through the sheath of plastic and the metal core into the bubbler can be expressed in terms of a suitably averaged heat transfer coefficient, an appropriate area, and the over-all temperature gradient:

$$\frac{1}{U_O A_s} = \frac{\overline{\Delta X_1}}{k_n A_s} + \frac{\overline{\Delta X_2}}{k_c A_c} + \frac{1}{h_L A_w} \tag{5-14}$$

U_O is the overall heat transfer coefficient defined in terms of the half-sheath area, and the other terms are as before. Because $A_s = A_c$, Eq. (5-14) can be simplified to

$$\frac{1}{U_O} = \frac{\overline{\Delta X_1}}{k_n} + \frac{\overline{\Delta X_2}}{k_c} + \frac{1}{h_L}\left(\frac{A_s}{A_w}\right)$$

$$\frac{1}{U_O} = \frac{0.042}{1.7} + \frac{0.40}{150} + \frac{1}{6700}\left(\frac{0.86}{0.202}\right) = 0.0280$$

$$U_O = 35.7 \text{ Btu hr}^{-1} \text{ ft}^{-2} \ {}^\circ\text{F}^{-1} \ (4.85 \times 10^{-3} \text{ cal sec}^{-1} \text{ cm}^{-2} \ {}^\circ\text{C}^{-1})$$

The overall rate of heat removal, q_O, is $U_O A_s \Delta t_O$ where Δt_O is the difference in temperature between the melt and the coolant. Thus, for a coolant temperature of 50°F:

$$q_O = (35.7) \left(\frac{0.86}{144} \right) (486) = 104 \text{ Btu hr}^{-1} (26.2 \text{ kcal hr}^{-1})$$

This is only 70% of $\Delta H_{c/2}$ of Eq. (5-9) which corresponds to total solidification with a surface temperature of 250°F. Nonetheless, it is 1.5 times the 69 Btu hr^{-1} which results in a sheath thickness equal to one-third of the part thickness. Because this sheath thickness normally allows knockout without part distortion (see p. 189), the indicated 12-sec cycle should not be limited by heat transfer.

It is clear that transfer through the nylon sheath is limiting. Thermal conductivity and part design are not subject to change; so, still faster cycles will require use of lower coolant temperatures or faster coolant flow and, possibly, some sacrifice in coolant efficiency. Another possibility is to use a resin modified for faster molding. With a standard nylon-66, bubbler diameters of 0.250 in. for D_1 and 0.313 in. for D_2, the top of the bubbler 0.200 in. from the parting plane, a circulation rate of 2 gal min^{-1}, and a coolant temperature of 30°F, a 10.2-sec cycle was actually obtained.

THE INJECTION MOLDING OPERATION

Any injection molding operation can be reduced to the control of four key process variables, the properties of the material being molded, and the mold. The process variables are: melt temperature, melt pressure, melt displacement, and time. Polymer variables include susceptibility to change in viscosity because of polymerization, depolymerization, degradation, temperature, and shear; rate of crystallization; melting point and freezing point; and thermal properties. These polymer properties are considered in Chapters 4, 8, 9, and 10, and special attention has been given to the heat transfer problem in the preceding section. The discussion in this section focuses on the effect of operating conditions on the performance of nylons in the injection molding machine and mold. Our first concern is how the process variables are measured. In general, advances in solid state instrumentation have significantly simplified and improved the methods of measurement and control.

Instrumentation

Control of temperature, pressure, displacement, and time are obviously limited by the instrumentation employed, their sensitivity, accuracy, and speed of response.

Temperature Control

The temperature of the cylinder wall and the mold cavity are commonly measured, but it is best to determine also the melt temperature. It is important to measure and control the temperature of the hydraulic fluids in the machine because this too can significantly affect consistency of operation.

Fast-response thermocouples of rugged construction, connected to fast-response controllers or recorders, can be used for cylinder temperature measurement and control. Also, platinum wire resistance pyrometry can be used to eliminate the problem of compensating for cold junctions or variable lead wire length in thermocouples. Discussions of temperature controlling devices are available (30, 31, 32). It is desirable to have continuous read-outs of temperature rather than depend on periodic inspection.

For best control, the average melt temperature should be determined for a specific screw under different conditions of rotation, back pressure, and cycle. Fine tuning of the melt temperature can be achieved by adjustment of the front and adapter sections of the cylinder. With good feedback and adequate heater band power, control to $\pm 2°F$ or $\pm 1°C$ is feasible.

Mold temperature regulation can be obtained by monitoring inlet and outlet temperatures although the effect of circulation rates and the inlet and outlet pressures of the coolant cannot be overlooked. Again, $\pm 2°F$ is feasible and desirable. Because of the importance of gate freeze-off and cycle reduction in the molding of nylons, the control of temperature in the mold can be critical to the molding operation.

Pressure Control

Variation in hydraulic pressure can be critical and should be measured and controlled. The time in first and second stage pressure can be adjusted in various machines by timers, position switches, or pressure level controllers. It is also desirable to monitor pressure in the nozzle and mold cavities. Pressure level in the cavities and its dependence on time and position determines the dimensions and properties of a molded nylon product.

Pressure transducers and related signal conditioners that measure up to 25,000 lb (wt) in.$^{-2}$ (1760 kg (wt) cm^{-2}) and 1000°F (538°C) with automatic balancing circuits and digital read outs are available. These permit the reading and recording of pressures in the nozzle, runners, and cavities. Ideally, in molding nylons no more than 5% variation of the pressure in the cavities should occur for optimum control. This may require sophisticated feedback instrumentation.

Displacement and Shot Control

Measurement of displacement is best accomplished by following movement of the screw (during both injection and retraction) with potentiometers or other position transducers. Equipment is available to do this and, in addition, plot displacement versus time. Control of feed and displacement, that is, shot control, can be a problem in screw machines because of leakage, slow speed of response (33), wear, poor clearances, and malfunction of check rings. Unmelted material lodging between the check ring and screw seat can be the cause of shot to shot variations. Timer fluctuations can induce similar variability.

Devices to prevent overrotation of the screw contribute to better control of screw retraction.

Time Control

Solid-state timers are preferred because they can reproduce signals with 0.1 sec whereas electromechanical timers fluctuate by as much as 0.5 sec. The improvement in performance gained by use of solid-state timers more than compensates for the additional cost involved.

Accurate timing of a molding cycle involves understanding of all of the events that occur, and these are shown in Table 5-9.

Operating Conditions

There is no foolproof way of predicting the best operating conditions for a particular nylon composition in a specific injection molding process. A careful program to define satisfactory operating limits is necessary. Resin suppliers provide much useful information that helps narrow these limits, for example, the dependence of melt viscosity on shear and temperature, gate seal-off times, recommended gate fill rates, and mold design factors. A recent series of articles has reviewed these recommendations for one manufacturer's line of nylon-66 resins (15).

Broad limits of melt temperature, cavity pressure, and shear rate for various nylons are outlined in Table 5-10. Instrument readings, whether on the barrel or nozzle, should not be confused with melt temperature, which may be measured by collecting air shots (maintaining cycle) and inserting into the melt a needle probe attached to a pyrometer. Normal process fluctuations due to a change in the resin (moisture content, regrind ratio, switching to a different lot, and so on) or inherent variability in machine performance (mold temperature change because of radical change in room temperature or controller response lag, injection pressure change because of change in oil temperature, and so on) should be clearly distinguishable from equipment breakdown. Automated processing of nylons is becoming more common to take advantage of fast molding characteristics, but high productivity often lies in attention to the maintenance details such as regular calibration of instruments; checking oil lines for dirt, water, and air; watching for loss of ring seal integrity, nozzle build-up, tie bar stretching, platen parallelism, and so on.

Table 5-9. Injection molding events in sequence[a]

Event	Explanation
1. Ram-in-motion	Time from start of injection until the screw stops moving forward
2. Ram forward hold[b]	Time from end of screw forward motion until the screw starts to move back
(First stage pressure[c]	Time under first, usually higher pressure)
(Second stage pressure[c]	Time from end of first stage pressure until end of ram forward)
3. Forward screw decompression or "screw delay"	Time from end of ram forward until start of screw rotation
4. Screw rotation	Time from start to end of screw rotation
5. Rear screw decompression	Time from end of screw rotation until the end of all rearward motion of the screw
6. Screw dead-back	Time from end of screw movement to mold opening
7. Mold opening	Time from start to end of mold opening movement
8. Mold open delay	Time from end of mold opening to start of mold closing
(Ejection[d]	Time from start to end of ejector action)
9. Mold closing	Time from start to end of mold closing motion

[a] Events in parentheses do not add to the total cycle.

[b] Just "ram forward" is commonly interpreted to include ram-in-motion and ram forward hold because one timer controls both events and a special effort is necessary to determine ram-in-motion alone.

[c] First and second stage pressures apply during ram-in-motion and ram forward hold and add to the same total time.

[d] Ejection can occur during mold opening or mold open delay or both.

Table 5-10. Normal operating limits for injection molding of nylons

Nylon	Melt temp. range ($^{\circ}$F)	($^{\circ}$C)	Cavity press. range (lb(wt)in.$^{-2}$)	(kg(wt)cm^{-2})	Maximum[a] shear rate (sec^{-1})
66[b]	525-580	275-305	2500-CLP[c]	175-CLP[c]	10,000
66[d]	540-610	280-320	2500-CLP	175-CLP	2,000
610	450-550	230-290	2000-CLP	140-CLP	10,000
612	450-550	230-290	2000-CLP	140-CLP	10,000
6[b]	450-550	230-290	2000-CLP	140-CLP	10,000
6[d]	480-610	250-320	2000-CLP	140-CLP	2,000
11	390-550	200-290	2000-CLP	140-CLP	10,000
12	370-550	190-290	2000-CLP	140-CLP	10,000

[a] These are approximate values that depend on melt viscosity. Unsteady flow occurs at a roughly constant shear stress(34) of about 10^7 dyn cm^{-2} or 145 lb(wt)in.$^{-2}$.

[b] Normal molding grade.

[c] Clamp limiting pressure is that cavity pressure which, when multiplied by the total mold projected area, does not exceed the available clamping force.

[d] Extrusion grade.

In general, it is advisable to operate at the lowest possible melt temperature consistent with goal properties and dimensions in order to minimize the rate at which the nylon might hydrolyze or polymerize. Temperature profile across the barrel can be adjusted as needed. For example, increasing the rear zone temperature may eliminate variability in screw retraction; however, the best solution to this problem normally lies in increasing the depth of the feed zone in the screw. The addition of small amounts of a lubricant may also help in limiting variations in screw retraction.

Experience suggests a minimum pressure of 2000 lb (wt) in.$^{-2}$ (140 kg (wt) cm^{-2}) in the cavity to obtain a good level of properties. Thus, first and second stage hydraulic pressures in the injection cylinder should generate at the nozzle a melt pressure such that, when pressure losses due to flow are subtracted, at least 2000 lb (wt) in.$^{-2}$ is available in the cavity. The time that the pressure is available in the cavity is also important. This time is controlled by the time it takes the nylon in the gate to freeze. The total injection time (ram-in-motion plus ram forward hold) should be greater than the gate seal time. In selecting cycle and injection pressure, one should determine the effect of injection time and pressure on part weight. There usually is a noticeable decrease in weight below a critical screw forward time which increases with decreasing pressure. Melt and mold temperatures will also affect this critical gate seal time. The pattern of pressure development (pressure versus time) can be important and involves different ratios of ram-in-motion and ram forward hold times for the same or different total injection times and different timing for switching from first to second stage pressure.

Control of mold temperature is essential in molding nylons. Many of the properties in the final product are influenced by the thermal history of the part both in and outside of the mold. It is good practice to monitor part temperature at a precise time after the mold opens and not rely on just the mold surface or coolant temperature as an index of thermal conditions in the mold.

Definition of Limits

The many factors that have to be considered in the injection molding of nylons have already been discussed, but the variety of available compositions and the range of molding operations in which they are used mean that specific molding limits have to be established for any specific job. Assuming that the mold design, mold temperature, overall cycle, and approximate melt temperature to achieve a full shot are known, a limit defining procedure such as that outlined in Figure 5-16 can be followed. Upper and lower pressure limits corresponding to flash and surface defects such as wrinkles, splay, or small pits are first determined. The pressure range is narrowed further by checking the dimensions on parts from every cavity for pressures within these upper and lower boundaries. The pressure domain for dimensions that meet tolerance requirements are thus obtained. Similar scans for small changes in temperature, fill rate, and so on, are made and lead to precise definition of the required control limits.

Fast Cycles

Small amounts of processing modifiers such as mold release agents or nucleating agents (see Chapter 11, p. 423) have made possible very fast cycles in the injection molding of nylons. Realization of the potential gains in productivity most often requires a carefully integrated and automated system such as that described in the earlier section dealing with the handling of nylons. Quality as well as productivity can benefit from automation where in-line controls reduce fluctuations in machine performance, constancy of feed is improved, and contamination from rework is minimized.

Assuming that the operating limits for a particular molding job have already been defined for a standard nylon resin, the following steps to achieve cycle reduction with a fast molding resin have proved effective:

1. Reduce the mold closed time by a small interval, about two seconds, without changing the total injection time (events 1 and 2 of Table 5-9).

2. Lower coolant temperature or increase circulation rate to achieve the same part temperature that existed before cycle reduction.

3. Increase rate of screw rotation only if screw retraction is limiting.

4. Make necessary adjustments in cylinder temperatures and back pressure to

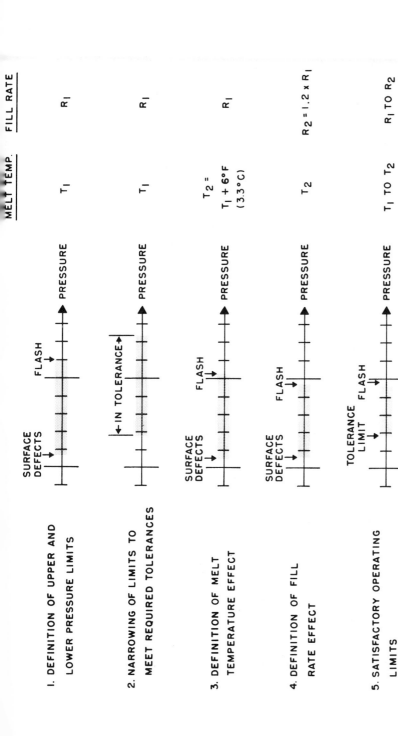

Fig. 5-16. Sample procedure for definition of molding limits. A specific mold temperature and overall cycle are assumed. The fill rate is approximately the part weight divided by the ram in motion time.

compensate for the above changes and return the melt temperature to the original level. It may be necessary to change screws, heater bands, or even the injection molding machine. Screw design and energy requirements for nylons are discussed above in the equipment section.

5. Repeat the above procedure as many times as possible or until there is insufficient time for screw retraction.

6. Incorporate a shut-off valve to permit screw rotation while the mold is open if screw retraction time is limiting.

In summary, the key to cycle reduction in nylons is to define the injection rate, melt pressure, melt temperature, and part ejection temperature necessary to produce parts to specification. From there on the task becomes one of increasing the production rate while adjusting heat transfer so that the key injection, melt, and part parameters are not affected. Heat transfer can be altered not only by changes in design of the coolant passage or flow rate and temperature of traditional coolants but also by use of carbon dioxide or "Freon" as the coolant.

REFERENCES

1. Anon., Plastics Division, Allied Chemical Corp., "Plaskon" Nylon Sales Bulletin, p. 18.

1a. Anon., Statistical Abstracts of the United States, Ed. 92, U.S. Department of Commerce, Washington, D.C., p. 183, 1971.

2. Anon., Plastics Department, E.I. du Pont de Nemours and Co., Tech. Bulletin, "Molding Du Pont Zytel® Nylon Resins . . . a Handbook for the 70's," 1970.

3. Anon., NASA Bulletin NHB5340.2, "NASA Standard for Clean Rooms and Work Stations for the Microbially Controlled Environment," Aug., 1967.

4. Ballard, D.W., Sandia Laboratories, Non-Destructive Testing Division, Bulletin SC-M-70-549, "Check List of Good Contamination Control Practices from a Manufacturing Viewpoint," Apr., 1971.

5. Anon., U.S. Government Federal Standard No. 209a, "Clean Room and Work Station Requirements, Controlled Environment," Aug., 1966.

6. Reiter, P., *Automation* 18 (12), 63 (Dec., 1971).

7. Lyons, A.L., *Automation* 18 (1), 48 (Jan., 1971).

8. Conair Corp., personal communication.

9. Reed-Prentice Division, Package Machinery Co.

10. McAdams, H. W., *Heat Transmission,* 3rd Ed., McGraw-Hill Book Co., New York, 1954.

11. Squires, P. H., and C. F. W. Wolf, *SPE J.* 27 (4), 68 (Apr., 1971).

12. Morse, A. R., *Plast. Design and Process.* 8 (6), 23 (June, 1968).

13. Reichelt, W., in BASF Technical Bulletin, "Injection Molding Technology and Related Problems," Part 3.

14. Skelland, A. H. P., *Non-Newtonian Flow and Heat Transfer,* John Wiley and Sons, New York, 1967.

15. (a) Filbert, W. C., Jr., *Plast. Tech.* 17 (6), 35 (June, 1971); (b) *ibid.* 17 (9), 48 (Sep., 1971); and (c) *ibid.* 17 (11), 36 (Nov. 1971).

16. Barrie, I. T., *SPE J.* 27 (8), 64 (Aug., 1971).

17. Miller, C., *Ind. Eng. Chem. Fundam.* **11** (4), 524 (1972).
18. Metzner, A. B., R. D. Vaughan, and G. L. Houghton, *Am. Inst. Chem. Eng. J.* **3**, 92 (1957).
19. Knudsen, J. G., and D. L. Katz, *Fluid Dynamics and Heat Transfer,* McGraw-Hill Book Co., New York, 1958, p. 357.
20. Nanigian, J., Nanmac Corp., personal communication.
21. Skelland, A. H. P., *SPE Tech. Pap.* **17**, 1 (1971); *J. Eng. Physics (USSR),* 1970.
22. Wilhoit, R. C., and M. Dole, *J. Phys. Chem.* **57**, 14 (1953).
23. Marx, P., C. W. Smith, A. E. Worthington, and M. Dole, *J. Phys. Chem.* **59**, 1015 (1955).
24. Griskey, R. G., and J. K. P. Shou, *Mod. Plast.* **45** (10), 138 (June, 1968).
25. Griskey, R. G., M. W. Din, and C. A. Gellner, *Mod. Plast.* **44** (3), 129 (Nov., 1966).
26. Waters, C. E., *Mod. Plast.* **46** (4), 147 (April, 1969).
27. Heyman, E., *SPE Tech. Pap.* **13**, 848 (1967).
28. Baranano, C.M., unpublished information.
29. Doubek, O., *SPE J.* **25** (6), 47 (June, 1969).
30. Willer, A., West Instrument Division, Gulton Industries, Inc., Tech. Bull., "How to Understand Plastics Temperature Control Instruments," 1970.
31. Emich, K., *SPE J.* **20** (4), 363 (Apr., 1964).
32. Nanigian, J., *SPE J.* **27** (2), 51 (Feb., 1971).
33. Schiedrum, H.O., *SPE J.* **27** (1), 31 (Jan., 1971).
34. Tordella, J. P., in *Rheology,* F. R. Eirich, Ed., Academic Press, New York, 1969, Vol. 5, Chap. 2.

Extrusion of Nylons

R. M. BONNER

INTRODUCTION

In the early years of the nylon plastic industry, the compositions available were limited to low-viscosity nylon-66 and nylon-610 resins containing as much as 0.3 wt % water. These factors and the lack of adequately heated extrusion equipment limited the use of nylon in extrusion applications. Today, through the introduction of new homopolymers, copolymerization, plasticization, nucleation, variation in molecular weight, and other techniques, extrusion resins (1-8) to suit almost any purpose are available. Furthermore, the improved heating systems of modern extrusion equipment have provided the capability of handling high-melting polymers. Where early nylon extrusions were limited to the production of filament and rod or bar stock, nylon resins are now available for extrusion into blown and cast film, blow molding, tubing, pipe, profiles, and various coating applications. Co-extrusion of nylon with other polymers to achieve desired combinations of properties is also a commercial reality.

The multitude of nylon resins on the market, for example, nylon-6, -66, -610, -612, -11, -12, and their copolymers may cause confusion in specifying a resin for an application. To make a suitable resin choice, one must consider its physical and chemical properties, processibility, and cost. The final choice is usually a compromise. The physical and chemical property requirements for a particular application will normally limit the number of nylon candidates. For example, if an electronic component requires good stiffness and high-temperature resistance, then a heat stabilized nylon-66 should be considered. If weatherability is a further requirement, then carbon black is generally added to promote weather stability. Another example is a nylon film application in which good heat sealing at a minimum temperature is required. In this case, one may prefer a lower-melting resin such as nylon-6, -610, -612, -11 or -12, or a copolymer.

The definition of processibility in nylon extrusion is not an easy one and involves such factors as melt viscosity, melting point, and rate of crystallization. Melt viscosity may have to be high ($\geqslant$ 50,000 poise) or low ($\leqslant$ 20,000 poise) depending upon the application. High-viscosity resins (that is, low melt flow) are preferred in blown film, blow molding, and free extrusion of rod and profiles. Low-melt-viscosity resins, on the other hand, are usually more suitable for wire coating and monofilament extrusion. Intermediate viscosity resins are generally desired for extrusion molding of rod and slab (that is, forming box method), coating applications, and the manufacture of tubing. A given resin may sometimes be used for more than one kind of extrusion by changing the processing temperature to achieve the desired level of melt viscosity.

The processibility of a nylon resin is not necessarily a direct function of its melting point. Low-melting resins are easier to liquify, but crystallize more slowly, particularly if the low melting point is due to copolymerization. Slow crystallization is advantageous where shrinkage and distortion due to non-uniform cooling are problems as in free extrusion of profiles. It is disadvantageous where sticking to a cold metal surface is a problem as in blow molding or making cast film. Furthermore, extrusions (for example, tubing) of some slow-crystallizing resins must be conveyed and handled with extreme care to prevent permanent distortion of the product.

In some applications, where regrind must be used, processibility must take into account the extrudability of the reground resin. A resin which maintains its melt viscosity and which requires a minimum amount of drying on regrinding is usually easier to process.

In summary, the production cost of a nylon extrusion will depend upon the raw material cost, production rate, yield, and the ease of reprocessing regrind.

MATERIALS HANDLING

Handling of nylon extrusion resins prior to their being processed is very important from the standpoint of cost, physical properties, and processing. Any loss due to mishandling will obviously add to the manufacturing cost of the extruded product. Proper inventory control is necessary to determine the amount of raw material, regrind, and finished product so that realistic yields can be determined. Too many times, the true cost of manufacture is not known because there is no record of how many times the resin has been processed before the order is filled.

Mishandling of the extrusion resins can lead to excessive water pickup or crosscontamination of the feed stock. Excess moisture in the resin will cause a loss of melt viscosity (Chapter 4) and difficulty in maintaining constant dimensions. In some cases, excess moisture may also cause bubbling of the melt and a splayed surface. Contamination by airborne dust or other thermoplastic resins should be avoided. Contamination may lead to discoloration and a loss in toughness. Separate handling systems for virgin and regrind (Figure 6-1) minimizes the contamination hazard.

In general, the discussion on handling nylons in Chapter 5 also applies to nylon extrusion. One difference in handling nylon resins for extrusion is that

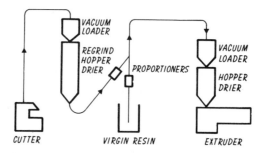

Fig. 6-1.   Materials handling system for nylon extrusion resins.

more care must be taken to insure a water content below about 0.15 wt %. Higher water contents can be tolerated in the injection molding process because the moisture remains in solution when the resin is molded under pressure.

The discussion that follows emphasizes those aspects of material handling that are important to the extrusion of nylon resins.

Virgin Resin

Nylon extrusion resins are shipped in various types of containers which include cans, bags, drums, boxes, and tote bins. Regardless of the size container used in a particular extrusion operation, there are a few fundamental rules to which one should adhere. Incoming resins should be sorted by lot numbers and the lot numbers recorded. The lots should be used on a first in, first out basis. Lot numbers relate to the time and the conditions under which the resin was produced, and, therefore, all resin of one lot number should have similar, if not identical, extrusion and physical properties. Using resin from a single lot increases the probability of maintaining a constant melt viscosity, melt elasticity (swelling), drawability, and obtaining a more uniform extruded product. Sometimes, minor extruder temperature changes will have to be made when a lot change takes place. At this time, the extrusion should be monitored closely to make sure tolerances are held within specifications.

It is important that the incoming extrusion resins be stored at a temperature above the dew point so that moisture will not condense on the resin when the container is opened and cause the aforementioned processing problems. The containers should be wiped free of dust, and so on, prior to opening to prevent contamination of the virgin material.

Drying

Properly packaged nylon extrusion resins do not require drying prior to processing. On the other hand, drying may be necessary if nylon molding resins are to be extruded, if extrusion resins are overexposed to atmospheric water, or if reground resin is too wet to be processed.

The most automated method for drying nylon extrusion resins utilizes a hopper dryer. Preferably, a closed system is used in which heated dehumidified air is circulated through the resin in a hopper dryer and the air flow is then recycled through the dehumidifier and back to the hopper dryer. This system minimizes contamination by keeping the manual handling to a minimum. The use of hopper dryers is most suitable when one type of material is to be dried since clean-out of the system requires considerable down time. A schematic drawing of a materials handling system including hopper dryers is shown in Figure 6-1. Note the separate dryer for the regrind. The resin is preheated as well as dried in the hopper dryer. This means that more resin can be melted in the extruder per unit time and output is increased.

A molecular sieve type desiccant is preferred for drying the air since it is very efficient and does not require a refrigerator system for regeneration. Although not as efficient, silica gel dessicant and mechanical dehumidifiers have also been used somewhat successfully in drying nylon extrusion resins. A recently developed humidity monitor (9) is very worthwhile because it makes possible prompt recognition of improper dehumidifier operation. A drying temperature of about $180°F$ ($82°C$) is satisfactory for most commercial nylon resins. Drying above $180°F$ in air may cause excessive yellowing due to surface oxidation. Drying times will depend upon the initial moisture content in the resin, the relative humidity and velocity of the drying air, and the air temperature. Details of the drying rates for several drying temperatures may be found in the injection molding chapter.

Dehumidified circulating air ovens or vacuum ovens may also be employed for drying nylon extrusion resins. However, these methods are only satisfactory if the oven capacity is sufficient to keep up with the extrusion rates and if strict cleanliness procedures are maintained.

Still another method for drying nylon extrusion resins is to make use of a vented extruder. This procedure in combination with a hopper dryer is very efficient. Extremely wet resin (for example, containing 8 wt % water) can be dried to less than 0.15 wt % water if proper procedures are used. Single screw extruders require a two-stage screw in this type of extrusion. Vacuum is applied at the vent on the extruder barrel to increase the rate of water removal, to decrease discoloration by air oxidation, and to eliminate bubbling of the extrudate due to air entrapment. The operation and some limitations of the vented extrusion process for nylon resins are given in a subsequent section.

Hopper Loading

The method of hopper loading will depend upon the size of the extruder and the extrusion rates involved. Manual loading is satisfactory for extruders up to 2.5

in. (63.5 mm) in diameter and for outputs no greater than 100 lb hr^{-1} (45.5 kg hr^{-1}). Because the output varies with hopper level in some extruders, it is a good policy to maintain as constant a hopper level as possible. In low-output extrusions such as those used in thin wire coatings, one must be careful not to expose the resin too long to the atmosphere. A good solution to this problem is to open the container at one corner and place the open end in the feed throat. In this way, the resin first out of the container flows directly into the extruder, and air exposure is minimized. If resin is poured from the container directly into the hopper, the hopper lid should be closed and a small counter-current flow of inert gas should be passed into the feed throat as indicated in Figure 6-2. This inert gas will provide a positive pressure and will keep moist air out of the system. Figure 6-3 illustrates how tubing dimensions can be improved by using the inert gas system.

If the extrusion rate is greater than 100 pph (45.5 kg hr^{-1}) then it is more expedient to use an automatic hopper loader. Vacuum loaders have been used very successfully in nylon extrusion operations. If this system is used, care must be taken to insure that the vacuum system shuts off when the hopper is filled. Otherwise, moist atmospheric air will be drawn into the hopper. Screw conveyors have also been used advantageously in loading hoppers, especially when the fines content is rather high, as may be possible in the case of reground nylon. Screw conveyors also minimize exposure of the resin to moist air. When large quantities of regrind are blended with virgin resin, it is best to have one hopper for the virgin and one for the regrind and to make use of a proportioning system. The regrind is then automatically fed at a pre-set rate by means of the proportioning unit. This prevents mixing large quantities of virgin resin with contaminated regrind. It is good practice to use a magnet in the feed hopper of the extruder to pick up tramp metal that might accidentally be carried into the system.

Coloring

Nylon extrusion resins can be colored by dry blending with a pigment or dye (Chapter 11), but care must be taken to select a colorant system which is stable to the nylon and the extrusion process. Color concentrates are often used instead of dry coloring because they are cleaner and they do not affect the feeding characteristics of the granules. The color concentrate may be added to the virgin resin by tumble blending or by using a metering device such as that shown in Figure 6-4. An intensely colored molding or extrusion resin can often be blended with a natural color resin to give the desired color. For example, a resin containing carbon black for weatherability may be added to natural colored resins in ratios up to one part black to 100 parts of natural and still yield a satisfactory black color. Of course, the depth of color depends upon the

SEALED HOPPER

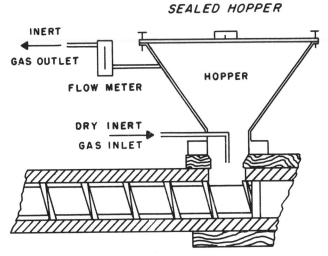

Fig. 6-2.   Schematic drawing of a sealed hopper design for nylon extrusion (Ref. 6).

thickness of the extruded part and the ratio of the black resin. Black extrusion resins made by this dilution process can no longer classify as weatherable resins.

Extruded nylon may also be dyed (Chapter 17). Dyeing offers some advantages in extrusion processing. There is no waste due to purging the extruder from one color to another, and because no blending operation is required, there is less chance of moisture pickup in the feed material. Also, the extrusion characteristics of the nylon resin is constant since there is no surface coated binder or pigment to disrupt the feeding of the granules.

Nylon wire jackets or tubing, and the like may be printed (Chapter 17) or striped using commercial inks designed for nylon. The resin manufacturers and

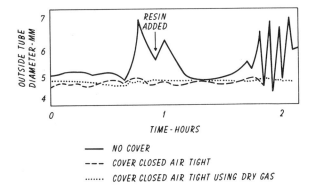

Fig. 6-3.   Tubing dimensions versus type of hopper cover (Ref. 6).

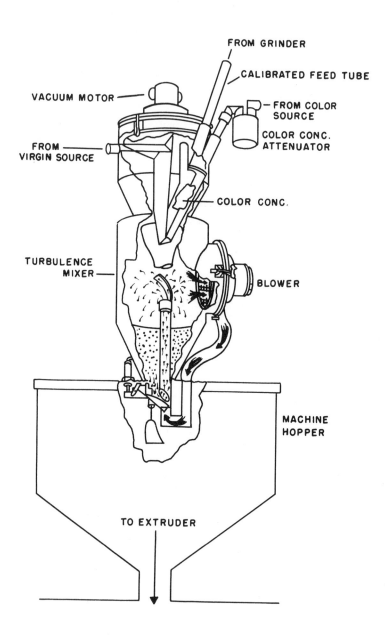

FROM GRINDER

CALIBRATED FEED TUBE

VACUUM MOTOR

FROM COLOR
SOURCE

COLOR CONC.
ATTENUATOR

FROM
VIRGIN SOURCE

COLOR CONC.

TURBULENCE
MIXER

BLOWER

MACHINE
HOPPER

TO EXTRUDER

Fig. 6-4. Color blender for extrusion resins (Courtesy Conair, Inc.).

the manufacturers of printing equipment and inks should be consulted for specific printing needs. Care must be taken that neither the ink nor printing equipment affects the final quality of the nylon surface. Conventional physical tests, expecially those mentioned in specifications, should be run with samples before and after printing or striping.

Rework

Nylon extrusions may be reworked in many different ways depending upon the resin, the the process involved, and the tolerances which have to be held. For example, air or chill roll quenched nylon extrusions may be immediately reground and re-extruded since the moisture level should be low. The thickness of the extrudate should be greater than about 30 mils (0.76 mm) for producing regrind which can be blended with virgin resin and readily re-extruded. Thinner regrind such as that made from film will generally have to be repelletized before it can be adequately extruded. Water-quenched extrusions can be reused if the extrudate exit the quench is hot enough to evaporate the surface water and if the material is reground immediately. The regrind is best blanketed with an inert gas such as dry nitrogen.

Most general-purpose fly knife cutters designed for grinding plastic will be suitable for nylon. The cutter blades should be checked periodically to be sure they are sharp so that excessive fines will not be developed during the cutting operation. Care must be taken to choose a screen which will give a particle size suitable for feeding in the extruder. A 0.375-in. (9.5-mm) hole in the screen is normally satisfactory. Smaller screens will cause overheating and will produce too many fines in the reground resin. Larger screens may yield long particles which will cause erratic feeding in the extruder.

Contamination is a problem which is ever present in the handling of regrind (10). For best protection of nylon regrind from contamination, the extrudate should be reground immediately and stored in drums lined with polyethylene film. In many extrusion processes, 100% regrind may be extruded directly or it may be blended with virgin resin in any proportion as long as the ratio of the regrind is constant. The amount of regrind which can be handled in any extrusion will be determined by the tolerances required, color, toughness, and other quality requirements. On many occasions, regrind is repelletized to insure a uniform feed material.

EXTRUSION EQUIPMENT

The broad class of nylon resins can be readily extruded in conventional equipment, providing some basic machine requirements are fulfilled. Probably, the most important factors to consider are screw design and temperature control. Many nylon resins have been processed in equipment designed for extruding polyethylene providing a proper metering screw is used and the extruder components are sufficiently heated. The following sections discuss the various components of a nylon extrusion line and emphasize the factors which are important to the nylon extrusion process.

Extruder Design

Nonvented, single-screw extruders like that shown in Figure 6-5 are predominantly used for the extrusion of nylon. Except when noted, the majority of this chapter will deal with nonvented screw extruders equipped with a single-stage screw.

Vented single screw extruders are used in some specific applications such as reclaiming reground nylon (11). In this operation, regrind can be dried and repelletized in one operation. Vented extrusion is not normally used for converting regrind directly into a finished product because the output variation is not easily controlled to a satisfactory level. Holdup of material in the vent section is another difficulty. Resin that collects in the vent zone may carbonize and cause black specks. Excessive loss of volatile additives such as plasticizer can also occur.

Twin screw extruders are useful in compounding additives such as fillers, plasticizers, colorants, and so on into various nylon compositions. Compounding twin-screw extruders are designed to provide extensive shearing action so the additives will be incorporated as uniformly as possible. The high shear rates generated by these machines lead to melt temperatures which may be excessive for extruding nylon directly into an end use product. Although twin-screw extruders can be designed to process nylon at moderate melt temperatures, there appears to be no need to make use of these machines in present commercial nylon extrusions.

When choosing an extruder for a particular nylon extrusion job, one should consider the output in pounds per hour which is required to make the run profitable. The extruder choice is based upon the assumption the extruder will be operating in the range of 50 to 75% maximum output. This is usually the optimum output range for good output uniformity and a reasonably low holdup time in the extruder. Table 6-1 illustrates typical extrusion rates for various extruder diameters. Because there is much overlap of the size extruders used for various extrusion operations, Table 6-1 should be considered only as a rough guide. Furthermore, the production rate is based on a single die. If a manifold system with a number of dies is attached, then larger extruders can be used effectively. Blow molding and rod extrusion are two processes which make use of multiple dies.

Although extruders having barrel lengths of only 10 diameters have been used successfully in some nylon extrusions, it is generally accepted that longer barrels

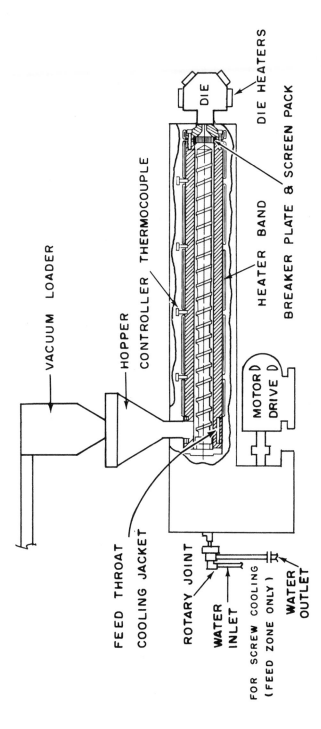

Fig. 6-5. Single screw extruder — one possible setup for nylon extrusion.

Table 6-1. Typical production extruders (24/1 L/D barrel)

Barrel, I.D.		Output range		Types of extrusion
(in.)	(mm)	(lb hr^{-1})	(kg hr^{-1})	
1.25-1.5	31.8-38	15-30	6.8-13.6	Rod, co-extrusion (film, wire coating)
2	51	40-60	16.2-27.2	Monofilament, co-extrusion, slab, rod, wire coating
2.5	63.5	80-150	32.4-68.0	Tubing, film, sheet, wire coating
3.5	89.0	175-300	79.5-136.0	Film, extrusion coating, pipe
4.5	114.0	325-500	147.2-227.0	Film, extrusion coating

yield higher outputs and smoother extrusion performance (12-14). For this reason, present day nylon extruders have a barrel length of 24 diameters. A barrel length of at least 24 diameters is almost mandatory for vented extrusion.

Toll (14) has shown that overhead or tangential feed throat designs are satisfactory for processing nylon-66. Both rectangular and circular feed throat cross sections have performed equally well. Undercut feed throats such as those used in the rubber industry are not satisfactory for extruding nylon. Since nylon is very tough and ductile, the extrusion granules will wedge between the flight of the screw and the barrel wall if the feed throat is undercut or the screw clearance is too large. This will cause erratic feeding of the granules and will result in an extruder surging problem.

Motor Drives

No special drive requirements are necessary for processing nylon as compared to other thermoplastic resins. Many successful nylon extrusions have been made using a mechanical drive. Mechanical speed variators use a split sheave to vary the sheave ratio, which, in turn, controls the screw speed. These drives are generally used on the smaller machines, where the motor load is under 50 hp. AC motors with a magnetic eddy current clutch are also suitable for extruding nylon resins. These drives are typically used for machines requiring 25 hp or more.

Direct-current speed control drives have been used successfully in many nylon extrusion applications. Although AC-DC motor generator sets have been used in commercial nylon extrusion plants, the solid state silicon controlled rectifier (SCR) speed control systems are becoming popular because they are efficient and quiet.

The horsepower rating of the motor drives should be sufficiently high so that adequate output can be obtained without overloading the motor. Table 6-2 shows typical horsepower ratings of modern extruders used for processing nylon.

Table 6-2. Drive horsepower for nylon extruders

Extruder size		Horsepower requirement
(in.)	(mm)	
1.25-1.5	31.8-38	10-20
2	51	20-40
2.5	63.5	40-60
3.5	89.0	75-125
4.5	114	150-250

The motor drives are normally monitored by an ammeter and voltmeter. The ammeter will give a good indication of the performance of the extruder. If the ammeter needle remains steady, the operation is proceeding under control. When the ammeter indicates a low or fluctuating reading, the resin is probably not feeding properly. When a bridge occurs on the screw and no resin is feeding, the ammeter will read the same as a "no load" condition. High ammeter readings indicate that the rear barrel temperature is too low or possibly that resin particles are wedging between the screw flights and the barrel surface. The motor ammeter should not be used as a substitute for a melt pressure gauge. Many times the ammeter will read normally and yet extremely high melt pressures can be developed.

Temperature Control

Extruder components including adaptors and dies must be adequately heated or nylon resins cannot be properly processed. In fact, insufficiently heated machines can be a safety hazard since excessive melt pressures may be generated if the die and its components are not adequately heated and the resin freezes in these areas. This may result in "blowing off" a head and causing ruptured bolts to travel at speeds high enough to cause serious injury. Extruders having large flanges or gates to which dies are attached are frequently not sufficiently heated to maintain nylon above its melting point. Some blow molding machines also suffer from inadequate heating capacity. A good rule of thumb to provide adequate heating is to use a watt density of 30 W in.$^{-2}$ (4.6 W cm^{-2}). Thus, if the outer surface area of the barrel adaptor and die is 400 in.^{2} (2580 cm^2) then the heater band capacity should be 12,000 W.

Control of metal temperatures by use of suitable controllers (15, 16) and by proper location of thermocouples is also very important in preventing either freeze-offs or excessive melt temperatures. Because the melt viscosity is affected by modest changes in melt temperature, the extrudate quality (that is, melt viscosity, dimensional control, and so on) is very dependent upon the maintenance of constant temperatures. For this reason, the better controllers, such as the SCR type, are used in the die and adaptor sections. On some extruders, it is also necessary to have a sophisticated controller on the rear barrel zone or section next to the hopper. Good rear barrel temperature control is required since the feeding characteristics of nylon resins are dependent upon the temperature of this zone. Temperature cycling in the rear barrel section can cause erratic feeding of the extrusion granules which, in turn, will be reflected as an output variation or surging of the extrudate. Less sophisticated controllers are required in the center and front barrel sections where heat is emitted because of the mechanical working of the screw. In these zones, the controller remains off (that is, overriding occurs) during a large percentage of extrusion time.

Placement of controller thermocouples is very important in maintaining suitably constant temperatures for nylon extrusion. Best temperature control is usually obtained by using a shallow well thermocouple close to the heat source. Spring-loaded thermocouples which are about 0.375 in. (9.5 mm) deep usually perform very satisfactorily. Spring loading is required to maintain constant contact of the thermocouple with the metal being heated. Again, since constant heat control is mandatory the various sections of the extruder should be heated independently. Thus, the barrel usually is divided into at least 4 separate zones and each extruder attachment (for example adaptor, neck, die, and so on) are individually heated. Very few nylon extrusions are run with barrel cooling as part of the temperature control system.

Adequate temperature control requires continuous monitoring of melt and metal temperatures. Stock thermocouples (17) are widely used to determine the melt temperature of nylons. These instruments are usually located in the adapter section between the extruder and the die. For rapid response, an unshielded thermocouple is used for this measurement. The melt temperature may be recorded, read out on a "Student" potentiometer, or an unused controller may be used to indicate the temperature reading. A better method for measuring melt temperature is the infrared thermometer technique (18). This instrument makes use of the 3.4 μ carbon-hydrogen absorption band of nylon resins. One simply aims the instrument at the nylon extrudate and reads the melt temperature directly. This method offers the following advantages over the melt probe technique:

1. The extrudate temperature immediately exit the die is measured and thus the true temperature of the melt is known.

2. The melt is not touched by the instrument, and therefore, the extrusion process is not disturbed.

3. The molten web from film or sheeting dies can be scanned for melt

temperature gradients (providing the film thickness is relatively uniform).

4. The cooling rate of the extrudate can be measured.

Multipoint recorders are invaluable for evaluating the performance of the temperature controller system. Since recorders operate with a standard cell, they compensate for ambient air temperature changes and they indicate a true temperature reading. Ideally, each temperature control zone is monitored on the recorder. To save drilling and tapping extra holes in the extruder barrel, adaptor, and so on, one may use a dual thermocouple in each control zone. One thermocouple lead is installed in the control panel and the other in the recorder. By periodic scanning of the recorder chart, the extruder operator can readily determine if all zones are heating properly. More importantly, he can make an immediate correction if a control zone is malfunctioning.

Pressure Control

Nylon extrusion resins are similar to other thermoplastic materials in that they normally extrude more uniformly when the extruder screw operates against an optimum head pressure. This pressure may range from 500 to 3000 lb (wt)in.$^{-2}$ (19.5-117 kg cm^{-2}), depending upon the screw design, melt viscosity, extrusion rate, die design, and many other factors. In some cases, the melt pressure may be below or above these values and still yield a commercial product.

Of the numerous ways to control the melt pressure in the head of the extruder, a screen pack is probably the best known and most widely used. Screen packs also serve as a filtering medium to eliminate dirt, metal, and the like, from the extrudate. The number of screens and the mesh size in a screen pack have to be determined for each individual case. Many successful nylon extrusions have been produced using a combination of two 80- and two 120-mesh screens. The 120-mesh screens are generally supported between larger screens. Control of head pressure by means of a screen pack may have more limitations than advantages. First, the screens may clog with foreign material, and the resultant increase in resistance to flow creates an increase in head pressure. Second, the screens may puncture due to excessive pressures at the start-up operation, causing much of the melt to short circuit the screens. Third, the melt pressure is not adjustable or predictable. One simply accepts the melt pressure which is attained at any set of extrusion conditions. The use of a screen changer could overcome some of these difficulties, but the expense of this type equipment is normally not warranted for most commercial nylon extrusions.

Another method for controlling the operating melt pressure is to use a restricted orifice in the adapter or in place of the breaker plate. One may estimate the orifice dimensions required for the desired head pressure by using the well-known volumetric flow equation for Newtonian flow. Thus, as discussed in Chapter 4,

$$\text{Volumetric flow} = Q = \frac{\pi R^4 \, \Delta P}{8\mu L}$$

By rearranging the equation and solving for "R", one can calculate the radius of the orifice that is needed. Thus,

$$R = \sqrt[4]{\frac{8\mu L Q}{\pi \Delta P}}$$

Since an output Q, pressure drop ΔP, melt viscosity μ, and land length L can be assigned estimated values one may solve for the radius. Generally, the orifice radius is made somewhat smaller than the calculated value. If the head pressure proves to be too high in actual use, then the orifice can easily be machined to a larger size.

A more elegant way to control head pressure is to use an infinitely variable valving system. The valve may be within the extruder or it may be located in the adapter. Internal valves generally make use of a movable screw. The delivery end of the screw is designed with a conical shape such that it matches a conical adapter piece. The screw may be moved axially by means of a mechanical or hydraulic system. This axial movement increases the head pressure as the screw moves forward and decreases it if the screw is moved rearward. The hydraulic method is more versatile since melt pressure may be varied while the extruder screw is rotating. Mechanical types require an interruption in the extrusion to make the pressure adjustment.

Valving in the adapter is satisfactory providing the system is well heated and a streamlined gate valve is used. The valve should have adequate land length so melt pressures can be controlled with small displacement of the stem. Needle valves generally are not satisfactory since they provide very little latitude in melt pressure control. Regardless of the type valve used, it is mandatory that the valve be designed such that the melt flow will never be cut off entirely when the valve is fully closed. This will prevent the possibility of excessive melt pressure being inadvertently developed. For added safety, a pressure relief system (17) (that is, rupture disc) may be installed upstream from the valve. When operating in a nylon extrusion line in which a valve is used, it should be general practice to open the valve to its maximum aperture at startup and at shutdown. This will help insure that excessive melt pressures do not develop at these critical times.

A melt pressure gauge is almost mandatory for the safe operation of a nylon extruder. This is especially true when one uses the higher melting resins such as nylon-66. Melt pressure gauges are invaluable at startup to protect the operator and the extrusion equipment. If the front end of the extruder (for example, head or die) is not sufficiently heated at start-up, then resin will freeze off in this zone and excessive pressures can develop. Proper monitoring of a pressure gauge will eliminate this problem.

The least expensive and most frequently used pressure gauge is the Bourdon gauge. The stem of the gauge which is screwed into the adapter section is filled with high-temperature silicone grease. Since this grease will melt and flow, the instrument should not be inserted until just before the extruder is started. Furthermore, the gauge should be pointed downward so that a minimum amount of grease flows out of the stem. If the grease flows out and is replaced by molten nylon, the nylon may freeze and therefore melt pressure cannot be transmitted to the gauge. To determine if the stem is plugged with solid nylon, silicone grease is pumped into the stem of the gauge. If the gauge indicates an increase of pressure, the stem is plugged. The plug may be removed by carefully flaming the stem with a propane torch. Silicone grease is immediately pumped into the stem to force out any remaining nylon. Because silicone grease adversely affects the adhesion of nylon to other materials, the Bourdon gauge should not be used when the nylon extrudate has to be bonded to a substrate, or the like. Instead, one may use a pressure transducer (19) to indicate the melt pressure. This instrument has a metal diaphragm which contacts the molten nylon. As pressure is developed, the nylon melt works against the diaphragm and causes a deflection in a strain gauge. This, in turn, causes a change in electrical current which is transformed to a pressure reading. Transducers of this type may be direct reading or they may be connected to a recorder. Melt pressure recordings have been very helpful in evaluation of screws (20) and barrel temperature profiles (19) for nylon extrusion resins.

High-speed electronic recorders have been shown (19) to be very useful in nylon extrusion evaluations and especially in trouble shooting when problems arise. By using suitable pre-amplifiers one may measure changes in melt pressure in the range of one lb (wt) in^{-2}. Similarly, temperature variations of less than $0.5°F$ ($0.28°C$) can be determined. The recorder speed can be adjusted so that response times are less than 0.1 sec. By correlating the short- and long-term pressure and temperature variations with machine conditions one can optimize a nylon extrusion operation.

Screw Design

In the plastics extrusion industry, a nylon screw is typically thought of as a rapid transition metering type screw as indicated in Figure 6-6a. Note that the metering length is about 25% of the total length of the screw and that the transition zone occupies only about one-half of a flight. Although this so-called nylon screw design has been used successfully in many nylon extrusions, it has been shown (12,20) that the combination of short metering and rapid transition sections is not always suitable or desirable for nylon. Screws of this type will

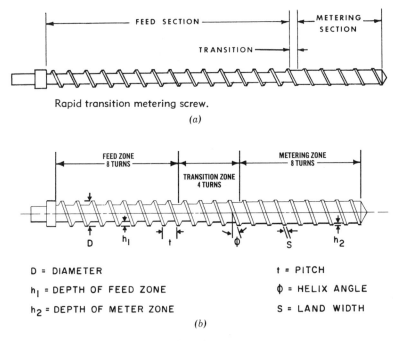

Rapid transition metering screw.

(a)

D = DIAMETER   t = PITCH

h_1 = DEPTH OF FEED ZONE ϕ = HELIX ANGLE

h_2 = DEPTH OF METER ZONE   S = LAND WIDTH

(b)

Fig. 6-6.   Typical screw designs for nylon resins.

tend to exhibit surging, air entrapment, poor plastication, and so on, especially at higher extrusion rates. Short metering screws can be used for nylons if an adequate transition length is incorporated in the screw configuration. Transition lengths of five to ten flights usually improve the performance (20). A more satisfactory design is to lengthen the metering section to at least eight flights (21, 22) (Fig. 6-6b). This is done regardless of the barrel length to diameter ratio, that is, whether it is 16 to 1 or larger. The longer metering section performs well for both high and low melt viscosity resins (that is, extrusion or molding resins).

The metering depth depends on the resin as well as the process. Thus, a high-melt-viscosity resin requires metering depths which are 15 to 30 mils (0.38 to 0.76 mm) deeper than those used for resins of lower melt viscosities. The deeper flights permit high-viscosity resins to be extruded without restricted output and at reasonably controlled melt temperatures. High-output high-melt-temperature extrusion processes such as paper coating will require a somewhat shallower metering depth as compared to blown film operations where low melt temperatures and high viscosities are mandatory.

The feed section design is as important as the metering zone for satisfactory extrusion. Both the length and depth of the feed sections must have adequate dimensions to convey nylon extrusion granules at a rate which is uniform and yet fast enough to maintain a constant supply of melt in the metering zone. For feeding cubes or cylinders about 100 to 125 mils (2.54 to 3.18 mm) in dimension, a good rule of thumb is to design the feed zone channel depth at least 0.300 in. (7.6 mm) deep regardless of the extruder size. Extruders as small as 1.0 in. (25.4 mm) diameter have been successfully run with this feed depth. When the nylon particle size is smaller than about 100 mils (2.54 mm) then more shallow feed depths can be used.

For most nylon screws(12,21), another rule of thumb is to design the feed zone depth so it is about 3 to 4 times that of the channel depth of the metering section. As in the case with other extrusion resins, when one is designing a screw for a new nylon extrusion application, it is usually desirable to make the feed and metering sections perhaps 5 to 10% more shallow than the proposed design. After a trial run, the screw can be machined if necessary.

The transition section of typical screws used in nylon extrusion does not appear to be critical once a properly designed feed and metering section are incorporated in the screw. However, modified transition sections (or mixing zones) may be included in the screw to improve mixing or blending of colorants, plasticizers, and the like. Some of the mixing devices used in thermoplastics extrusion have been published(23). The extruder manufacturer and the material supplier should be contacted for advice on mixing screw devices for specific applications.

A recent development in the concept of screw design is based upon the theory of melting proposed by Tadmor et al (24). In this screw design, the feed section is rather conventional, but the transition zone, designated the "Barr" Section after R. Barr who designed the screw, is double flighted. One flight transports solid cubes and as the polymer melts, it is conveyed back to the second flight or melt reservoir. As the material proceeds down the extruder, the solids conveying flight becomes progressively more shallow and finally it no longer exists. At this point, the resin is completely melted and is in the deep melt reservoir. The melt is then conveyed to a metering zone, and it is then pumped from the extruder. This "Barr" transition design has yielded high outputs and a minimum of surging with nylon resins of greatly different melting points and melt viscosities. One limitation of this type of screw may be its tendency to generate high melt temperatures. More work will be required to determine its suitability for the various types of nylon extrusion operations.

It should be apparent that there is no universal screw design for processing nylon resins. As in the case with other thermoplastic resins, screws for nylon have to be designed for a specific process. Factors that must be considered are machine type, extrusion rate, melt temperature, resin, extrusion process, and so on. The effects of barrel temperature profiles, surface lubricants, plasticizers, and so on further complicate the procedure for designing screws for nylon extrusions.

Although screw design is complicated, to say the least, there are a number of screw designs which have processed the various types of nylon resins satisfactorily. This includes nylon-6, -66, -610, -612, -11, and -12 as well as their many copolymers. Table 6-3 lists some screw designs which have been used in processing tubing, rod, film, coated paper, and other extrusions as well.

Two-stage screws are primarily used in vented extrusion for removing volatiles such as water, solvents, and so on. The two-stage screw may be considered as two screws in series in which the first section conveys solids, melts the material, plasticates, and conveys the resin to the vent section. The first stage in the screw may be designed like a conventional single stage screw as regards feed and metering depths. Since a limited number of flights are available one generally divides this section into feed and metering zones using a minimum transition length (that is, 0.5 to 1.0 flight).

The second stage of the screw consisting of the vent and pump zone is conventionally designed such that the vent depth is equal to the depth of the feed zone of the first section or first stage. Again, conventionally, the pump zone depth is about 50% deeper than that of the metering zone of the first stage of the screw.

A typical two-stage screw for a 24/1 L/D extruder might have nine flights in the feed section, four meter, five vent, and five flights in the pump section. All the transition sections would be rapid (that is, 0.5 flight).

Extruder Attachments

In this section, some general concepts concerning the design of the attachments to the extruder will be considered. Specific die designs will be discussed in connection with specific processes (p. 234 et seq.). The adapter, breaker plate and die are usually fabricated from metals which are satisfactory for the extrusion of polyethylene. Many successful dies have been prepared from 4140 steel. Components made from this metal can be cleaned by careful burnout procedures without causing ill effects to the part. If the parts are chromed, then the burnout procedure cannot be used and the components must be cleaned by scraping with an aluminum or soft brass spatula. Brass screening (Chore Girl®) may then be used to remove the final particles of nylon. A purge material such as a high-density low-melt-index polyethylene (for example, Alathon® 6700 polyethylene resin) may be used to remove the majority of the nylon from the die prior to cleaning with Chore Girl®. Other clean-out procedures are given in the extruder operation section.

Table 6-3. Typical single stage screw designs for nylon extrusion resins

Extruder ID (in.)	(mm)	Feed depth (in.)	(mm)	Meter Depth (in.)	(mm)	20/1L/D Feed	Trans.	Meter	24/1L/D Feed	Trans.	Meter
1.5	38.0	0.30	7.6	0.080	2.0	11	1	∞	12	4	∞
		(0.30)a	(7.6)	(0.070)a	(1.8)	11	1	∞	12	4	∞
2.0	50.8	0.360	9.1	0.090	2.3	8	4	∞	8	8	∞
		(0.320)	(8.1)	(0.080)	(2.0)	8	4	∞	8	8	∞
2.5	63.5	0.400	10.2	0.100	2.5	8	4	∞	8	8	∞
		(0.340)	(8.6)	(0.085)	(2.2)	8	4	∞	8	8	∞
3.5	88.8	0.460	11.7	0.115	2.9	8	4	∞	8	8	∞
		(0.380)	(9.7)	(0.095)	(2.4)	8	4	∞	8	8	∞
4.5	114.0	0.500	12.7	0.125	3.4	8	4	∞	8	8	∞
		(0.420)	(10.7)	(0.105)	(2.7)	8	4	∞	8	8	∞

Number of flights

aValues in parentheses are depths for low-viscosity nylon resins, for example, $<$10,000 poise.

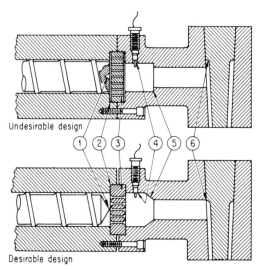

Undesirable design

Desirable design

1. Screw centering holes should be filed or eliminated.
2. Breaker plates should fit tight and act as sealing surfaces.
3. There should be no dead holes in the breaker plate.
4. Stock thermocouples should seat properly and dead space should be eliminated.
5. Shoulders should be streamlined.
6. The mandrel should meet in the crosshead properly and dead space eliminated.

Fig. 6-7.   Streamlining of extruder components.

Because the degradation of nylon increases with hold-up time (Chapter 2, p. 46 *et seq.,* and Chapter 4, p. 134), the breaker plate, adapter and die should be streamlined (25) (Figure 6-7). The breaker plate should be counterbored on the up and downstream side. The adapter should have a minimal orifice diameter so the shear rate in this section is sufficient to maintain adequate flow at the region of the wall. Spider sections which hold a mandrel in position perform best if they are streamlined on the up and downstream side. If these precautions are followed and there is sufficient melt pressure to promote welding of the melt streams, then there will be a minimum problem with so-called die lines in the extrudate.

A positive displacement pump (for example, gear pump) is sometimes placed between the extruder and die to insure that a constant output is maintained. A melt pump can be beneficial to processes such as monofilament production(26) where a minimum of output variation or surging is required to prevent strand breakage. Melt pumps must be carefully engineered and well heated to operate properly. When correctly installed and used, they can be an important adjunct to the extruder.

Quenching Systems

The cooling method chosen for nylon extrudates must allow for the melt characteristics (for example, melt viscosity, sticking tendency, and so on) and for the rate of crystallization of the particular nylon. This means that different quenching techniques may be used depending upon the type of nylon being extruded. To maintain good control of dimensions and physical properties, uniform cooling is a primary requisite. The more crystalline and the higher the melting point of the nylon being extruded, the more care must be taken in the cooling process. For example, a profile extruded of nylon-66 will warp and distort more readily than a low-melting copolymer. In some extrusions, such as rod, slab, and the like, slow cooling must be used to prevent the formation of shrinkage voids in the part. Shrinkage voids come about because the outer surface solidifies, while the inside is still in a molten state. As the inside cools, it shrinks due to thermal contraction and crystallization. Because the rigid outer surface cannot contract as the inner volume diminishes, a shrinkage void is formed. Any method which will keep the outer surface hot and pliable while the inner surface is setting up minimizes or eliminates void formation.

The cooling method chosen for nylon extrusions will usually depend upon the solid-state physical properties required, the dimensional tolerance needed, and for economic reasons, the maximum rate of extrusion which can be attained.

When possible, vertical downward extrusion into water is preferred for cooling because the distance between the die and the water surface can be easily and accurately controlled. Also, there is no problem of water running up the extrudate and cooling the die as in the horizontal process. Vertical downward extrusion can be used for extrusions which are not distorted by the guide rolls which direct the extrudate out of the quench tank. Thus, monofilament, film, and narrow sheet (strip) have been successfully produced by this method.

When cross-sectional distortion occurs around the guide rolls or if a straight extrusion is required, then a horizontal water quench is used. To maintain a constant freeze line (and therefore constant dimensions) and to prevent water from the quench tank from running up the extrudate to the die, it is best to apply a water spray on the extrudate prior to its entry into the quench tank. The water spray should impinge the extrudate uniformly so equal cooling is attained.

It is good practice to maintain a constant water quench temperature so that a constant crystallization rate and uniform dimensions are obtained. Ambient tap water (50 to 70°F, 10 to 21°C) is often used in water quenching, but some extrusions (27) require high water temperatures (for example, 180°F, 48°C) to obtain good physical properties.

The quenched surface of some copolymers becomes sticky when water cooling is used. In these cases, the surface may have to be coated with talc or a lubricant to prevent sticking in windup as when tubing is being reeled. Another way of solving this problem is, simply, to cut the product to lengths and not coil. It is sometimes possible to avoid water quenching by extruding vertically upward and air quenching as in the blown film process.

Air rather than water quenching may be useful in a number of other processes. For example, in some profile extrusions, air quenching allows one to perferentially cool individual sections of an extrudate so that warpage can be minimized. Air quenching also eliminates the problem of surface water marking caused by fluctuating or nonstable water flow. Air quenching requires a constant air flow and a number of flexible (but easily positioned) air jets. The air flow to each position should be monitored by a flow meter so constant cooling conditions are maintained.

Chill roll cooling is used primarily for film, sheet, and extrusion coating. The inlet temperature of the fluid flowing through the chill rolls is normally set as high as possible commensurate with optical properties and easy release of the film or sheet from the roll. High roll temperatures in the range of $200 \pm 50°F$ ($93 \pm 28°C$) are common. This minimizes buildup of condensed vapors (for example, oligomers, monomers, and the like) on the cooling surface of the roll. Condensate on the chill roll may be in the form of a clear liquid or a waxy solid. If chill roll buildup is excessive, it will tend to be picked off by the film or sheet and cause splotches on the extrudate.

Some extrusion processes require the extrudate to be cooled on or inside a cooled metal surface. The metal forming device is of the desired shape and is usually fabricated of brass or aluminum to facilitate heat transfer. The forming device must be cooled uniformly if distortion of the extrudate is to be avoided. The surface in contact with the nylon extrudate may have to be coated with one to two mils of a tetrafluoroethylene resin to prevent excessive sticking. The TFE coating may be applied by spraying a suitable TFE dispersion onto the surface and then baking (28). The thin TFE layer does not appreciably affect the heat transfer of the forming box and greatly improves the lubricity of the surface in contact with the extruded nylon.

Nylon rod and slab is generally produced by forcing the extruded nylon under pressure into a water cooled forming box. The take-off and extrusion rates are adjusted to produce void-free extrusions. This process is a continuous extrusion molding operation. Hollow shapes such as tubing and pipe may also be made using a pressure type forming box. In this case, the extruded tube is forced to conform to the geometry of the cylindrical forming box by means of an inert gas which is passed through the hollow mandrel and inside the tube.

Nylon pipe or tubing may also be manufactured by use of vacuum to force the extrudate to the shape of the forming device. This process offers the advantage that tubing can be cut to any length without causing change in the tubing dimensions. If internal air is used, as in the pressure forming technique, then dimensional control is disturbed every time a length of tube is cut.

The vacuum forming box technique has been supplanted by the differential pressure calibration (DPC) method for many tubing and pipe applications. The DPC method, described on p. 238, offers the advantage of maintaining vacuum control of the extrudate for a much longer period of time.

EXTRUDER OPERATION

Start-Up

From a safety standpoint, it is very important to make sure that the extruder attachments (for example, gate, adapter, and die) are heated well above the melting point of the resin being processed. This will ensure that resin will not freeze off in the system and cause dangerously high head pressures. Before starting up, each section should be checked out by means of a touch pyrometer because experience has shown that a controller can be faulty.

Important to the success of start-up is the method for heating the barrel. Care must be taken not to heat the feed zone of the screw above the melting temperatures of the resin to avoid adhesion of granules to this section of the screw root. When this occurs near the feed throat, incoming pellets will be blocked and a bridge will form. This results in extruder surging, which is indicated by a loss in motor load, and then a stoppage of production. If a bridge has developed, it is best to clean the extruder and start over again because nylon that has solidified on the root of the screw is difficult to remove. Removing the bridge by increasing the rear barrel temperature to 600°F (315°C) or higher and by feeding solid chunks of nylon (for example, rod) into the machine is sometimes attempted. This can be time-consuming and usually does not remove all of the nylon clinging to the screw, and bridging often reoccurs in a few hours after production is started. Bridging should not be a problem if proper start-up procedures are used.

In order to melt the resin in the front end of the extruder while preventing overheating of the feed section of the screw, an increasing ("uphill") temperature profile on the extruder barrel is recommended. That is, the rear barrel temperature (hopper end) is set 10 to 25°F (6 to 15°C) below the melting point of the resin while the front end is set 30 to 50°F (17 to 28°C) above the melting point. To insure that the bulky extruder components are heated sufficiently at start-up, it is best to turn on the controllers for the front barrel, adapter, die, and so on. When these zones are up to control temperature, the remaining barrel controllers are turned on and when all control temperatures are reached, the extrusion should be started within half an hour or so. After start-up, the rear barrel temperature may be adjusted to the preferred operating temperature based on prior experience. Initially, the screw speed should be low, about 10 rpm, and the melt pressure gauge should be observed continually. It is desirable, but not necessary, to hand-feed the resin until it exits the die. In this way, it can be determined whether or not the resin is feeding properly before it emerges from the die.

Operation

The heat distribution along the length of the barrel has a profound effect upon the output and the extrudate quality (19). Because no two extruders are alike with regard to heater band capacity, location of thermocouples, types of controllers, and so on, it is difficult to give any hard and fast rules on barrel temperature settings. In general, however, a low-melting resin performs better using an uphill profile. In some cases, an uphill temperature profile does not provide enough heat to completely melt the resin at the operating speeds. Extruder surging will result because there is an insufficient melt reservoir feeding into the metering section of the screw. This problem may at times be overcome by using a so-called hump back temperature profile wherein the rear barrel temperature is set low and the next zone is set high. The following zones are set progressively lower until the front barrel temperature adaptor, and die is near the melting point of the resin. The die lips may have to be at a higher temperature (separate heater band) to prevent melt fracture and to obtain a glossy surface.

The higher-melting nylon homopolymers and copolymers usually perform better with a downhill temperature profile. The extruder is started up using an uphill temperature profile, and once extrusion is underway and melt is extruding at a safe pressure, the rear temperature is increased up to 50°F (28°C) above the resin melting point. The front barrel is then reduced as required for minimum degradation or maximum melt viscosity. Barrel temperatures may have to be readjusted if the extrusion rate is changed significantly. Higher barrel temperatures are required for high-speed operations; lower temperatures for low-speed extrusions.

Given a properly sized extruder and good screw design, most nylon resins extrude more uniformly at moderately fast screw speeds, for example, 50 rpm. If the screw speed is low, that is 10 rpm, or lower, then feeding problems can occur because there is insufficient force generated in the feed section to maintain a constant movement of the solids. Erratic feeding or surging may occur and cause the output to vary beyond an acceptable level for production runs. Furthermore, low screw speeds lead to long hold-up times which can cause resin degradation problems.

Many nylon extrusions operate more smoothly as the line speed is increased to an optimum level. Apparently, the faster the nylon extrudate passes through the transition of the melt to solid, the better is the control of the extruded part. This appears to be especially true in operations where sizing or shaping devices are used.

Shutdown

There are a number of procedures for shutting down a nylon extrusion line. These may be classified as short-term, long-term, and shutdowns to clean out. If an extruder is being run at temperatures no greater than 550°F (288°C), it may be shut down for short periods (for example, 10 min) without taking any special precautions. If a longer shutdown is required, then all controller temperatures may be set to the resin melting point, and the extrusion is continued until the motor load (ammeter reading) increases. The extruder is then run until it is empty. At this time, all the heaters may be turned off. Higher melting resins such as nylon-66 may be purged from the extruder with a low melting copolymer or a nylon-6 resin so that low shut-off temperatures can be attained. This will minimize the chance of oxidation or degradation of the nylon in the extruder. Also, on reheating the machine, the low-melting resin remaining in the adapter and die will melt more readily. This will minimize the chance of "blowing off" a head due to insufficient melting of the resin in these zones. Polyethylene may be used as a low-melting material for shutdown providing the system can be easily purged. A method using polyethylene containing a blowing agent has also been described(5) for purging nylon from extrusion systems. This procedure is useful for purging nylon from large extruders equipped with bulky dies (for example, extrusion coating machines), especially when they are not dismantled for cleaning.

When shutting down a nylon extruder to clean out and change over to a new set-up, one may proceed in several ways. First, the system may be purged with a 1.0 melt index polyethylene resin while leaving the adapter, die, and so on, temperatures at the nylon operating point. When the nylon is completely removed from these areas, the temperatures may be dropped to about 350°F (176°C). When the extruder and its components equilibrate to this temperature, rigid PVC is added to the hopper and the system is purged free of polyethylene. Then the extruder is run dry and the heaters are shut off. The PVC is stripped off the parts in the usual way. This procedure alleviates the necessity of burning out the accessory components of the extruder.

Another clean-out procedure consists of purging the extruder with polyethylene as above and then removing the die, adapter, and so on from the extruder. The screw and barrel may then be cleaned by using a cast acrylic purge compound. This material is ground waste which results from the casting of liquid acrylic monomer into sheets, rods, and so on. The resin has a molecular weight in the millions, and it really does not melt under normal extrusion temperatures. The acrylic purge compound is not to be confused with commercial grades of acrylic molding and extrusion resins. The latter materials are not suitable for purging an extruder. The acrylic purge compound may be purchased from various casters of acrylic sheet and rod. The purging procedure consists of running the extruder at a fairly high speed and feeding a couple of scoopsful of acrylic purge compound to the hopper. This is repeated until the extrudate is free of nylon and the acrylic purge extrudes as a powder. The extruder is then run dry and the screw removed. Both screw and barrel are wiped free of any adhering acrylic purge material.

The die and adapter which have been purged with polyethylene may be cleaned by scraping the parts while they are still warm. Frequently, these parts have to be burned out in some manner in order to completely remove the plastic residue. For small parts, one may use a propane torch. Larger dies and so on are more easily cleaned in specially designed burn-out ovens. These ovens (for example Lindberg oven) are designed to operate under an inert or reducing atmosphere so metal oxidation or rusting will not occur. A more recent development is a furnace produced by Procedyne Corp. which cleans extruder parts using heated, fluidized aluminum oxide. The dies and so on may also be immersed in a potassium nitrate/sodium nitrate salt bath and burned out. This procedure is dangerous since highly combustible materials may oxidize explosively. After burn-out, the parts may be liquid sandblasted to remove carbon and other material not removed by the burn-out process. Chromed parts should not be burned out.

SPECIFIC PROCESSES

Tubing

Because of its unique combination of physical and chemical properties, nylon is used in many diversified tubing applications. Unsupported nylon tubing is used in automotive gasoline lines, air control systems, and electrical applications. Nylon tubing is also overbraided with wire or filament and used in brake and hydraulic hose systems, refrigeration lines, and push-pull control cables.

Although nylon tubing has been made from resins of widely different melt viscosities, medium-melt viscosity resins (that is, 20 to 40,000 poise) are generally preferred. A good tubing resin must have enough melt viscosity so that the molten extrudate does not sag, but it must not be so high that the melt cannot be easily drawn. Commercial quality nylon tubing can be made using either an in-line or crosshead die (Figures 6-8 and 6-9). The choice of tubing head depends on the orientation of the extruder relative to the cooling trough. For example, if an extruder is used for both wire coating and tubing extrusion, a crosshead system is indicated because the extruder is normally perpendicular to the quench bath.

Nylon tubing can be manufactured by any of the well-known methods including free extrusion, vacuum calibrating, pressure sizing, and internal calibration. Each procedure has specific advantages and disadvantages which the manufacturer must consider before embarking on a particular process.

Free Extrusion

The simplest and most versatile method of manufacturing nylon tubing, at least from a tooling standpoint, is to free extrude it into a horizontal water

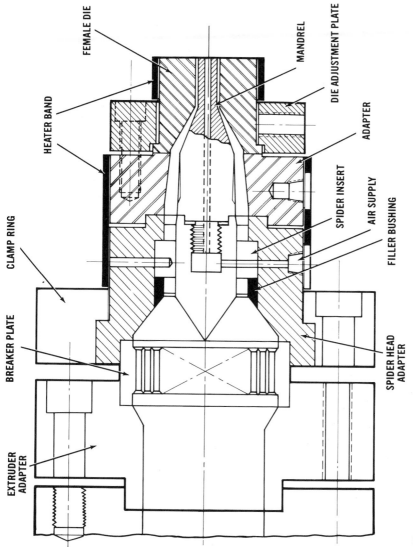

CLAMP RING

BREAKER PLATE

EXTRUDER ADAPTER

HEATER BAND

FEMALE DIE

MANDREL

DIE ADJUSTMENT PLATE

ADAPTER

SPIDER INSERT

AIR SUPPLY

FILLER BUSHING

SPIDER HEAD ADAPTER

Fig. 6-8. In-line head for tubing extrusion.

235

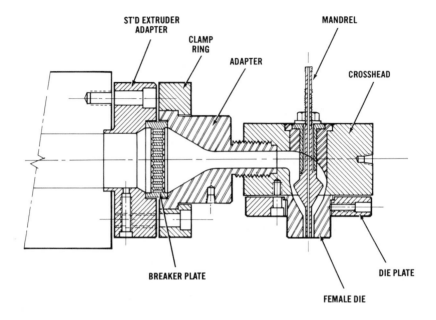

Fig. 6-9.   Cross-head for tubing extrusion.

quench tank (Figure 6-10). This procedure requires a fairly inexpensive die and mandrel setup and noncomplicated sizing plates in the cooling trough (29). Numerous tubing sizes can be made from one die and mandrel combination by varying the screw speed, take-off rate, or air pressure inside the tube. Fairly round tubing (for example, ±2 mils or ± 0.08-mm OD variation) can be made up to tubing diameters of about 0.25 in. (6.4 mm) and extrusion take-off rates in the neighborhood of 200 ft min^{-1} (60 m min^{-1}) are not uncommon.

Larger diameter tubing tends to float in the water quench bath, and

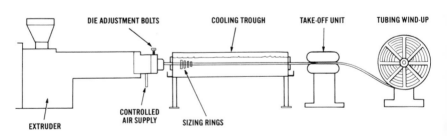

Fig. 6-10.   Typical free extrusion line for nylon tubing.

cross-sectional roundness is more difficult to control. More positive sizing methods are desirable.

The die and mandrel used for free extruding nylon tubing are designed somewhat oversized and the tubing is drawn down to the desired dimensions. Some tubing producers size the die by simply multiplying the tubing OD by a factor of 2 or more. The mandrel dimensions are obtained by using a similar ratio between the die annulus and the tubing thickness. These relationships may be described as follows:

$$ID_d = X\,OD_t \qquad \text{and} \qquad \frac{ID_d - OD_m}{OD_t - ID_t} = X$$

where

ID_d = die inside diameter

OD_m = mandrel outside diameter

ID_t = tube inside diameter

OD_t = tube outside diameter

X = drawdown ratio

Another method used for calculating the tooling dimensions is based on the cross-sectional area of the tubing and the die annulus. The drawdown ratio based on cross-sectional area is usually in the range of 4 to 10/1. Empirically it has been found convenient to make the mandrel 60% larger than the tubing OD. The die ID is then calculated using a cross-sectional area drawdown ratio in the range of four to ten to one. The mathematical relationships are the following:

$$OD_m = 1.6\,OD_t$$

$$\frac{ID_d^2 - OD_m^2}{OD_t^2 - ID_t^2} = Y$$

where Y = cross-sectional area drawdown ratio.

Table 6-4 reveals that both methods lead to closely related, if not identical, tooling dimensions when appropriate drawdown ratios are used.

Table 6-4. Tooling for nylon tubing extrusions

		Calculated Tooling Dimensions			
Tubing dimensions		Cross-sectional area basis 4/1		$ID_d/OD_t = 2/1$	
OD_t	ID_t	Die	Mandrel	Die	Mandrel
(mils)(mm)	(mils)(mm)	(mils)(mm)	(mils)(mm)	(mils)(mm)	(mils)(mm)
200 7.9	100 3.9	470 18.5	320 12.6	400 15.8	200 7.9
300 11.8	200 7.9	655 25.8	480 18.9	600 23.6	400 15.8
400 15.8	300 11.8	830 32.6	640 25.2	800 31.5	600 23.6
500 19.7	400 15.8	1000 39.0	800 31.5	1000 39.0	800 31.5

Nylon tubing is best made using a melt temperature close to the polymer melting point. After the extrusion is underway and the desired machine temperatures are reached, the die is centered relative to the mandrel. To determine if the die is centered, one simply cuts the extrudate with a brass knife at the face of the die. If the extrudate extrudes in a straight line, then the die is well centered. If the extrudate curves away from the machine direction, the die is not centered. To correct this situation, one moves the die toward the direction in which the tube curves or towards the thin side. Once centering is achieved, the tubing is "strung up" through the sizing plates in the water bath and through the take-off. The centering is rechecked by marking the top of the tube at the die face with a colored grease pencil, cutting the marked sample as it leaves the take-off, and measuring the wall thickness with a tubing micrometer. If necessary, the die is adjusted relative to the mark on the top of the tube. When the wall thickness is uniform, the desired tubing dimensions can be obtained by adjusting the screw speed, take-off, and internal gas pressure. The tubing is wound on a reel until the desired length is attained. Before cutting and changing wind-up reels, the tube must be sealed or pinched off to prevent loss of internal gas pressure. For this reason, it is best to run free extruded tubing with no internal pressure – if satisfactory dimensions can be maintained.

Vacuum Quench Tank

Many tubing applications require cross-sectional roundness control, which is difficult if not impossible to attain by the free extrusion process. Tubing in the range of 0.375 in. (9.5 mm) ID and larger tends to float in the water trough, and the cross section becomes oval by the force of the fixtures holding the tube under the water. For this reason, much nylon tubing is produced by a process known as the differential pressure calibrating or vacuum quench method. A schematic drawing of this process is shown in Figure 6-11. Extruded tubing is passed through a series of sizing plates or a sizing cylinder which extend into a

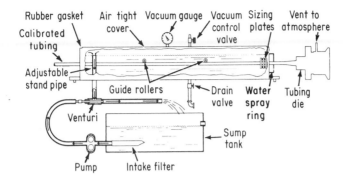

Fig. 6-11.   Differential pressure tube sizing device for nylon.

water quench tank that is under vacuum. The purpose of the vacuum is to reduce the external pressure exerted on the outer surface of the tube (that is, air and water pressure) to a value less than the atmospheric pressure inside the tube. One advantage of this process over the vacuum forming box technique which is described in a following paragraph, is that vacuum is applied for the full length of the quench tank (for example 10 to 20 ft or 3.3 to 6.6 m) instead of only the forming box length (6 to 12 in. or 15 to 30 cm). Thus most nylon compositions will be solidified enough to be handled on the exit side of the vacuum quench tank. Generally, no internal gas is used in this process and, therefore, tubing may be cut to any length without affecting the diametral dimensions.

There are essentially two commercial types of differential pressure tubing calibraters on the market. They differ in the way the tube enters the vacuum quench tank. In one case (Figure 6-11), the tubing passes directly into the vacuum tank through a series of sizing plates or a sizing sleeve. In the other method, tubing is passed into a water-cooled vacuum sizing sleeve and then into a vacuum tank. Both methods have been used successfully in the production of quality nylon tubing.

The following processing details are important to consider when using a differential pressure system for producing nylon tubing.

1. The die ID should be between 1.5 to 3 times larger than the sizing plate ID for tubing up to at least 0.50 in. (12.7 mm) ID. This is to make sure a vacuum seal is attained at the entrance of the sizing plate without the aid of internal air pressure. Larger tubing [for example, 6 in. (15 cm) ID] has been made using a sizing plate ID equal to the die ID. However, internal gas pressure was required to maintain a seal in the vacuum box.

2. The ID of the sizing plates is generally 6 to 13% larger than the OD of the desired tube. This is to allow for shrinkage as the resin passes from the melt to the solid phase. A graph such as that shown in Figure 6-12 is useful to estimate

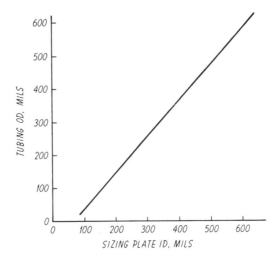

Fig. 6-12.   Effect of sizing plate dimensions on nylon tubing OD.

the required sizing plate dimensions. The sizing plate may be machined slightly undersized and then adjusted to produce the correct tubing dimension.

3. A water spray ring is required at the entrance of the sizing plate to provide lubrication. Otherwise, the tube will stick and slip in the entrance sizing plate and the tubing will have internal concentric rings. The outer surface will usually be unsuitable for commercial use. The water must contact the tubing uniformly to maintain a constant wall thickness.

4. A rubber gasket at the exit end of the quench tank can be made of soft rubber and the hole size must be cut small enough so a good vacuum is attained but not so small that excessive frictional resistance is developed. If the gasket hole is too small, the tube will "chatter" and concentric rings will form on the inside of the tube. If the gasket hole is too large, it will be difficult to maintain a constant vacuum inside the tube. It is best to sandwich the soft flexible gasket between two or more rigid rubber backup discs. These are made with hole diameters about 50% larger than the gasket. The backup discs prevent the flexible gasket from vibrating excessively and causing a fluctuation in the vacuum level.

5. The water quench level of the tank should be adjusted so it is about 2 in. (5 mm) above the tube surface. Higher water levels may cause excessive outside pressure on the tube and will require higher vacuum levels to maintain roundness.

6. The water in the quench bath should be maintained at a constant temperature. This will insure a constant setup or crystallization rate and will aid in maintaining constant tube dimensions.

The tubing diameter will depend upon the dimensions of the sizing plate, the amount of vacuum applied, extrusion rate, and the type of resin being processed. Therefore, some adjustment in tubing OD and wall thickness is possible with one set of sizing plates.

When starting up a nylon tubing extrusion (for example, 0.5-in. or 12.7-mm OD tubing) using the differential pressure sizing method, the water tank is first filled to the level of the sizing plates, then after centering the die, one simply threads the nylon melt through the water spray ring (water on), sizing plates, vacuum tank, water quench bath, and haul-off. This "string-up" operation is usually done at slower speeds until the operator gains experience. Next, the inlet water to the tank and the vacuum pump is turned on and the quench tank lid is secured. The vacuum inside the tank keeps the quench water in the tank and prevents it from running onto the die face and causing a freeze-off. This is true only when the sizing plate and gasket diameters are small enough to maintain a vacuum in the tank. The extruder speed is then gradually increased until a vacuum seal is formed on the sizing plates. If a vacuum seal cannot be readily attained, the vacuum unit is moved toward the die until the seal is formed. Once the seal is formed, the extruder speed and the take-off rate, as well as the vacuum level, is adjusted to give the desired tubing size.

Forming Boxes

The vacuum forming box method is very similar to the differential pressure technique for manufacturing nylon tubing. In both processes, the vacuum is created on the outer surface of the tube so the atmospheric pressure on the inner surface will force the tube to conform to the desired dimensions. The processes differ in that vacuum forming boxes are only about 6 in. (15 mm) long whereas the vacuum tanks may be 20 ft (6 m) or more in length. This means that vacuum tanks offer more time under vacuum during the solidification process. Long forming boxes are not feasible because of the high frictional forces created when nylon is in contact with the metal forming tube. To reduce the frictional forces, the inside surface of the forming box is often coated with polytetrafluoroethylene. This coating is normally 1 to 2 mils thick and provides the necessary lubricity for easy transport of the tube through the forming box. As in the case of the differential pressure sizing method, vacuum forming box extruded tubing can be cut to any length without disturbing dimensional control.

The pressure forming box method utilizes an externally cooled forming tube which is bolted directly to the die. The internal surface of the forming box is normally TFE coated to provide lubricity. It is essential that the connection between the die and the forming box be well heated so that resin will not freeze off in the die. Air or nitrogen pressure is applied to the inside of the tube through the spider and hollow mandrel (in-line system). In this process, the tube end must be plugged or pinched off to obtain constant internal pressure and good dimensional control. This process is not amenable to the production of short lengths of pipe or tubing, that is 20-ft (6-m) lengths, because up to 20% of the tubing near the pinch-off point will have to be reground and re-extruded. A procedure for calculating the required length of the calibrating unit has been developed by Schenkel(30).

The extended mandrel method has been used to produce 1- to 4-in. (2.5 to 10-mm) diameter nylon-66 tubing having nominal walls of 10 to 20 mils (0.25 to 0.5 mm). The extended mandrel is usually made of brass for good heat transfer and it must have an adequate taper (that is, about 10 mils/in.) to release the nylon properly. A TFE coated, extended mandrel is necessary for the more sticky or tacky nylon resins.

Stock Shapes

Nylon rod (circular and polygonal cross section) and slab are produced in stock sizes for use in screw machines and other machining operations. These shapes are prepared in different ways depending upon the size and cross-sectional geometry of the shape and the type of nylon resin used. Nearly all round nylon rod is centerless ground to provide close dimensional control of the product as well as providing a good looking surface. Other extruded nylon stock shapes may also be machined prior to shipment.

The least complicated method for manufacturing rod is to free extrude it into a water quench bath. This process utilizes a circular die which is about 1.7 to 2.0 times larger in diameter than the desired rod. This drawdown ratio is sufficient to maintain tension on the extrudate and help attain roundness. The extrusion is generally carried out horizontally rather than vertically downward to make it easier to produce a round rod. Vertical upward extrusion of rod is also possible if an air quench is used. The latter process avoids the force of gravity which tends to flatten horizontally extruded rod.

The conventional horizontal free extrusion process requires a closely balanced cooling system to obtain a round, void-free rod. Rapid quenching in water enhances roundness but usually induces shrinkage voids. Apparently the outer surface of the rod solidifies and forms a rigid shell which will not retract as the inside crystallizes. As the center of the rod cools and shrinks, there is insufficient resin to maintain a continuous mass and a shrinkage void is formed. Shrinkage voids are usually eliminated by shortening the length of the quench bath, by air quenching, or by resorting to a hot oil bath. To maintain roundness, it is helpful to misalign the take-off somewhat so that a twist or rotation is imparted to the rod as it is being quenched. Although free extrusion is primarily used for making rod around 0.5 in. (12.7 mm) or less in diameter, some larger diameter rod has also been made by this method.

When precise control of diameter and shrinkage voids is desired for rod about 0.375 in. (9.5 mm) diameter or larger, a forming box method is indicated (Figure 6-13). The molten nylon is extruded under pressure through a cylindrical water cooled steel tube. The rod is shaped in the forming box by the pressure

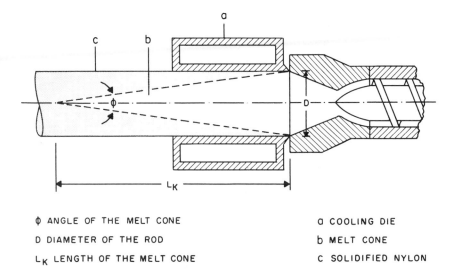

φ ANGLE OF THE MELT CONE   a COOLING DIE

D DIAMETER OF THE ROD   b MELT CONE

L_K LENGTH OF THE MELT CONE   c SOLIDIFIED NYLON

Fig. 6-13. Schematic drawing of forming box for extruding nylon rods (Ref. 31).

generated by the extruder, and it is pulled at a constant rate by a suitable haul-off. Because the forming box procedure is a closed system, it is advisable to install a melt pressure gauge in the adaptor section. Control of melt pressure is important for maintaining a constant movement of the extruded rod. If the rod sticks in the forming box, the melt pressure will increase. Generally, the rod sticks, then slips and so moves in a stepwise manner. If the rod sticks too long and the pressure increases above a critical value, the rod may then emit as a projectile traveling at very high and dangerous speeds. With a pressure gauge in the system, one can turn off the extruder before excessive pressures occur. It is also advisable to include a rupture disc (17) in this system.

A special melt pressure control device has been described (31) for producing nylon rods. It utilizes a spring-loaded valving arrangement that controls melt flow into the forming box. Care should be taken to see that this system is adequately heated for the high-melting nylon resins (for example, nylon-66).

To improve the lubricity of the forming box, it is generally coated with polytetrafluoroethylene. TFE is also applied to the flange of the forming box which contacts the hot die so that heat transfer is minimized (that is, to prevent freezing at the die).

Because the output of rod produced by the forming box method is slow, an extruder may be equipped with a manifold and a multiplicity of dies. Some type of valving mechanism is required to maintain equal flow in the series of forming boxes.

Rod and slab produced by the forming box method generally require annealing to remove the extruded-in stresses. These stresses can be so great that the rod may shatter explosively even while in storage. Annealing is most efficient when done in-line. Thus, the heat content of the rod can be used and the annealing times can be shortened. The procedures described in Chapter 17 can be used for out-of-line annealing of nylon rod and slab.

Monofilament

Nylon monofilament is produced in diameters ranging from about 0.0035 in. (0.089 mm) to about 0.100 in. (2.5 mm). Typical uses of nylon monofilament are: fishing line, toothbrushes, paintbrushes, fish nets, and the like. The nylon monofilament extrusion process has been described in some detail by Riggert (26) and others(32,33). Briefly, the process consists of pumping molten nylon through a die having a series of circular holes. The pump may be a single screw extruder or an extruder with a gear pump attached. Gear pumps are commonly used in filament production to insure that a uniform extrusion of melt is obtained. Nonuniform melt flow through the die will result in caliper variation or strand breakage. The extruded strands are normally carried vertically downward into a water bath in which they solidify. With high-viscosity resins, the melt does not sag, and the extrusion may be done horizontally. The quenched strands are pulled through the water bath by means of a series of Godet rolls (pull rolls) which determine the size of the melt-draw filament. It should be noted that little orientation is accomplished during this melt draw-down step. Furthermore, the method of drawing at this point determines whether the filament is level (constant diameter) or tapered. To produce a level filament, the initial drawing step is made using a constant speed pull roll or Godet. When tapered filament is desired, the pull roll speed is varied in a predetermined manner so the diameter will be uniformly variable from a maximum to a minimum value.

The filament is then heated in water or air (for example, infrared heaters) and stretched by means of another series of Godet rolls which are run at a higher speed than the first set. The speed differential is adjusted to attain the desired degree of drawdown or orientation. Typically, the drawdown ratio is in the range of 2 - 6 to 1. This orientation step increases the tenacity (that is, tensile strength) and stiffness but decreases the ultimate elongation. When the filament leaves the second Godet rolls, it is passed through an annealing chamber which again may be a hot air or liquid system. The filament is allowed to relax by adjusting the speed of the third Godet rolls to a value somewhat lower than the second rolls. Annealing increases the elongation and reduces tenacity and stiffness. By adjusting the drawdown ratio and annealing conditions, one may obtain a filament having the specific combination of physical properties required for a particular application. On leaving the annealing system, the filament may be wound up on reels or, if straight hanks are required, it may be wound on a

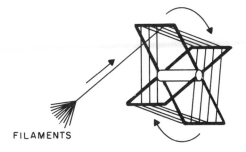

FILAMENTS

Fig. 6-14.   Device to make hanks of nylon filament.

square winding mechanism, such as that shown in Figure 6-14. The hanks of filament between the pins will be cut off and packaged as straight bundles. The filaments near the pins are curved and have to be discarded or re-extruded.

Profiles

Extrusions having shaped cross sections can be manufactured from many nylon extrusion resins. As in the case of other thermoplastic profiles(34), nylon shapes having a uniform thickness are much easier to produce than those with an irregular wall thickness. Variations in wall thickness cause problems in flow distribution in the cross section of the die and bowing or warping of the extruded shape as it is being cooled.

Nylon profiles may be arbitrarily grouped into two general categories: those which can be free extruded and those which must be produced in a forming box. Free extrusion is almost always tried first because it offers more flexibility, it is less complicated, and it yields higher extrusion rates than the forming box technique.

Free extrusion is most successful with symmetrical shapes, especially those having modest dimensions (for example <0.125 in. or 4.9 mm in thickness and <3 in. or 7.5 mm in width). The forming box method is used when the cross section of the profile is such that excessive warpage occurs when the shape is free extruded. The forming box method is essentially the same process as that discussed for stock shapes.

Free extrusion of small and thin profiles (<0.5 in. or 12.7 mm wide and 0.1 in. or 2.5 mm thick) can be accomplished with resins having a melt viscosity in the range of 20,000 to 40,000 poise. Thicker and wider extrusions may require resins having melt viscosities of at least 50,000 poise. The higher melt viscosity is necessary to obtain enough strength and elasticity for retention of the desired configuration. The lower-melting nylon copolymers are easier to process than the higher melting homopolymers because they crystallize more slowly and tend to warp less on solidification.

Profile extrusion dies for free extruding nylon resins are generally made from a mild tool steel. As in the case with other resins, die design is a cut and try process. The die is initially machined to the same shape as the desired extrusion except allowance is made for drawdown. Some drawdown is required to maintain control of the extrudate, that is, to prevent it from sagging, twisting, and so on. A number of nylon shapes have been extruded using a drawdown ratio of about 2 to 1. In other words, the width and height of the die shape and the gap opening is designed with dimensions twice that of the desired part. Extrusions having nonuniform wall thicknesses usually require a differential drawdown from the thick to the thin section. These are much more difficult dies to design and again the cut and try method is used. The new die is checked by running a trial extrusion and then measuring the part dimensions.

If a die modification has to be made, it should be remembered that melt flow rate through the die varies inversely with the land length and directly with the cube of the gap opening (see Chapter 4). Thus, shortening the land by a factor of 2 will increase flow twofold while increasing the gap by a factor of 2 increases the flow by a factor of 8. Minor changes in flow are therefore made by adjusting land length. Longer land dies normally lead to better shape retention. Although no hard and fast rules are available regarding the land dimensions, a land length of 2 in. (5 cm) has performed satisfactorily in a die having a gap opening of 0.140 in. (5.5 mm) and a cross-sectional area of 0.20 in.^{2} (62 mm^2). A longer land die would be required if the gap opening were increased.

There is little information published on specific profile die designs to use in nylon extrusions. However, some die design information for producing narrow width strip (less than 1.0 in. or 25 mm wide) has been reported (35). Nylon strip was extruded vertically downward into water using slit dies of different widths (D_w) and heights (D_h). It was found that the strip width (S_w) and strip height (S_h) were proportional to the die width and height: $D_w/D_h = S_w/S_h$. Working charts can be made by calculating the strip width and thickness assuming the above relationship is true over a range of drawdown ratios. To demonstrate the procedure, let us examine the following specific case given a die width and a die height. The objective is to calculate the strip width and strip height when different drawdown ratios are used.

Let

$$D_w = 1.250 \text{ in.}$$
$$D_h = 0.065 \text{ in.}$$

Then,

$$\text{the die area} = D_w \times D_h = 1.25 \times 0.65 = 0.0813 \text{ in.}^2$$

Since $S_w/S_h = D_w/D_h = 1.25/0.065 = 19.25$, then $S_w = 19.25\ S_h$. The area of the strip cross-section $= S_w \times S_h = 19.25\ S_h{}^2$. Assume the drawdown ratios (cross-sectional area basis) of 4, 10, and 30 to 1. Then,

$$\frac{\text{area die}}{\text{area strip}} = \frac{0.0813\ \text{in.}^2}{19.25\ S_h{}^2} = 4, 10, \text{and } 30$$

$S_h = 0.0324, 0.0205,$ and 0.0118 in. $(0.82, 0.52,$ and 0.30 mm)
$S_w = 0.624, 0.394,$ and 0.229 in. $(15.8, 12.5,$ and 5.8 mm)

Plotting these values (Figure 6-15) facilitates estimating the various combinations of strip sizes that can be obtained from a particular die. Note that the experimental data lie close to the calculated line. The correlation holds true over a reasonable range of outputs and take-off rates.

It is possible that this type of data would be helpful in designing dies for more complicated shapes. This information might be used in Carley's method (36) in which he visualizes complex dies as made up of combinations of simple shapes such as strips, hemispheres, circles, and so on.

Nylon shapes of small cross sections can be extruded vertically downward into a water bath (14). The advantage of this quench method is that the air gap can be controlled quite accurately and there is little chance of water splashing on the die and causing a freeze-off. Air gap control is essential in determining the shape and dimensions of the cross section. The disadvantage of this method is that the extrusion has to be directed around a wheel or shoe to convey the extrudate to the take-off rolls. Bending the extrudate around this shoe may cause flattening or distortion of the shape. Horizontal water quenching offers the advantage that no bending is required. On the other hand, it is more difficult to maintain a constant freeze point since water will tend to run from the quench tank toward the die. In the case of a shape like a U channel, the water will actually run up the extrudate to the die and cause a freeze-off. This problem can be alleviated by placing a water spray ring between the die and the quench bath. Water is sprayed toward the quench bath to provide a constant freeze point on the extrudate and to prevent water from the quench bath from running up the extrudate toward the die. The horizontal water quench method has several disadvantages. In the more complicated profile extrusions, it is difficult to use differential water cooling to preferentially quench the extrudate and prevent warping of the cooled product. This is especially true for the higher melting resins such as nylon-66. Water quenching can cause pitting on the surface of the extrudate due to air in the water. Air bubbles will form on the surface of the extrusion and will insulate that part under the bubble from the water quenching

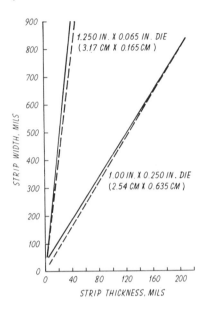

Fig. 6-15.   Extruded nylon strip — width versus thickness (high-viscosity nylon-66, 1-in. air gap).

media. Because the surface under the bubble of air cools more slowly, it shrinks more and causes a pit to form. Another difficulty is water marking of the extrudate surface which is caused by slight variations in the flow of the water stream. In other words, the quench line is not absolutely constant and, therefore, the zone in which solidification takes place varies slightly. The resulting surface imperfections, albeit quite small, can be easily detected visually.

Air quenching is another satisfactory method for cooling many nylon shapes. The primary requirement for this process is to have adequate cooling in the first few feet where drawdown occurs and where most solidification takes place. When the air supply is directed properly on the extrudate, the line may be strung up as easily as if it were a water quench system. Air quenching offers the advantage of differential cooling, eliminates water marking, and minimizes the possibility of a freeze-off of the die. The primary disadvantage of air quenching is that air is a relatively poor heat transfer medium and the cooling rate is relatively slow. A fogging device involving a combination of water and air to cool the surface can overcome this heat transfer problem.

Film

The market for nylon film is expanding rapidly especially in the food and packaging area. The combination of heat resistance, grease resistance, toughness,

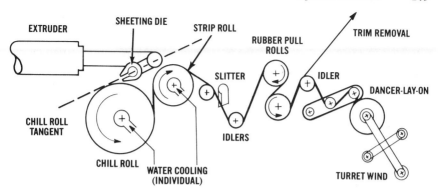

Fig. 6-16.   Typical chill roll unit for nylon film.

clarity, and oxygen impermeability of nylon make it a very desirable packaging material. In addition, nylon film can be laminated, extrusion coated, or co-extruded with polyethylene or ionomer resins to improve the end-use properties. Good quality nylon film has been made by flat film extrusion using chill roll casting or water quenching and by blown film extrusion. Resins used in the production of film must be clean and free of gel particles. Low- to medium-viscosity resins, for example 10 to 20,000 poise, are preferred for flat film; higher-melt-viscosity resins (that is, greater than 25,000 poise) for blown film. Because melt viscosity changes with processing temperature, one resin is sometimes adequate for several film processing methods. The water content in the resin should be low (0.10 wt % or less) and constant (see Chapter 4). The melting point of the resin may be dictated by the end use. For example, if heat resistance is required, a high-melting, heat-stabilized resin is indicated.

Casting of nylon film is best accomplished by using a system such as that shown in Figure 6-16. Note that the film is cast downward onto a roll such that the web is tangential to the chill roll.

Because the film is quenched rapidly in this process, crystallization and the growth of spherulites are arrested (see Chapter 8), and the product is clear. The success of this operation depends upon proper design of the die, good die heat control, and proper annealing of the film prior to windup. These factors will be discussed in some detail in the following paragraphs.

Die design requirements for any nylon film process are:
1. Minimum holdup time in the manifold and pre-land region.
2. Good melt distribution across the die width.
3. Adequate caliper adjustment to compensate for flow differentials.

To insure sufficient shearing action in the possible stagnant areas, a good rule of thumb is to design for a minimum output of three pounds (1.4 kg) per hour per lineal inch (2.5 mm) of die width (37). All the interior die sections are chrome plated to improve the scratch resistance, corrosion resistance, and ease of cleaning. Good die temperature control is an absolute necessity because of the change in melt flow with variations in temperature. Many caliper variations can be directly related to variations in die temperatures. The better type controllers such as the silicon controlled rectifier (SCR) type are recommended for die temperature control.

A further requirement for good die temperature control of a flat film die requires it be divided into equal or symmetrical heating zones across the width of the die. The number of zones required depend upon the die width. For example, a 36-in. (91-cm) wide film die may have three control zones, whereas a 60-in. (1.5-m) wide die will normally have five. The end plates should each be individually heated and controlled.

Nylon film should be fairly well crystallized before it reaches the windup rolls. If a noncrystalline or amorphous film is wound up, crystallization will occur on the roll and puckering may result. Film that has shrunk or puckered on the roll may be used for some packaging applications, but puckering detracts from formability and appearance. To anneal or crystallize the nylon film as it leaves the casting roll, one may set the casting roll temperature about 10 to 20°F (8 to 16°C) below the temperature at which the film will stick to the roll. The high chill roll temperature also minimizes the condensation of monomers or oligomers on the casting roll. If annealing cannot be accomplished on the casting roll (that is, contact time is too short), then another heated roll must be added to the line.

In the operation of the film line, the barrel and die temperature may be set about 25 to 50°F (14 to 28°C) above the melting point of the resin. To prevent edge weaving and excessive neck-in of the film as it contacts the chill roll, it is usually expedient to pin the edges with a jet of air (Figure 6-17). This technique assures a uniform contact across the width of the film with a chill roll. An air knife may also be used to pin the molten web against the casting roll. The air knife must be designed so that a uniform air stream flows across its width. In addition, the air knife must be positioned such that the air flow is directed away from the die.

Nylon film may be slit in-line as is done with films of polyethylene and other resins. The edge trim is usually not reprocessed directly into film because of contamination and feeding problems.

The blown film method (Figure 6-18) offers the advantage of no waste from edge trim and the film can be readily converted to packaging applications. Since the cooling rate of film made in this manner is rather slow, the film is usually not as optically clear as that made by the chill roll method. A bottom fed die, preferably one with a spiral mandrel, is desirable for nylon because hold-up and attendant weld line problems are minimized. Rotating dies are best for the blown film process since better caliper distribution is attained and more uniform

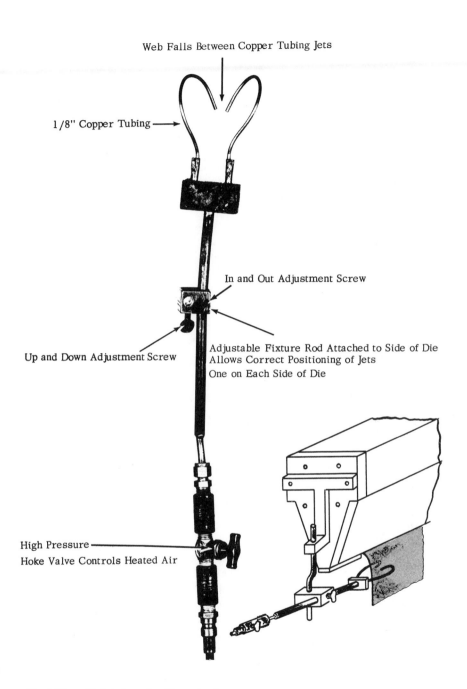

Web Falls Between Copper Tubing Jets

1/8" Copper Tubing

In and Out Adjustment Screw

Up and Down Adjustment Screw

Adjustable Fixture Rod Attached to Side of Die
Allows Correct Positioning of Jets
One on Each Side of Die

High Pressure
Hoke Valve Controls Heated Air

Fig. 6-17.   Air jets for nylon film extrusion.

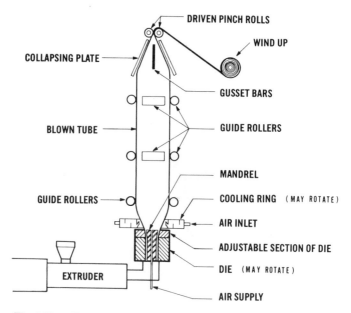

DRIVEN PINCH ROLLS

WIND UP

COLLAPSING PLATE

GUSSET BARS

BLOWN TUBE

GUIDE ROLLERS

MANDREL

GUIDE ROLLERS

COOLING RING (MAY ROTATE)

AIR INLET

ADJUSTABLE SECTION OF DIE

DIE (MAY ROTATE)

EXTRUDER

AIR SUPPLY

Fig. 6-18.   Extrusion of blown film.

and better looking rolls of film can be made. A rotating air ring is also desirable to even out any variations which may occur in the distribution of cooling air. As is the case with flat film, it is imperative to have good temperature control, especially on the die and the adapter sections. Larger blown film dies operate well and safely if the mandrel is heated. Heating of the mandrel decreases the time to heat up the die and helps insure against solidification of melt.

The blow-up ratio (diameter film/die diameter) is usually in the range of 2/1. Although molten nylon can be blown many diameters larger than the die, the maximum blow-up ratio is limited by caliper uniformity and the ability to maintain a stable bubble.

Wrinkling of the blown film as it passes through the nip of the pull rolls is a constant problem with nylon. The wrinkling becomes progressively more severe as one extrudes resins of increasing density. Thus, nylon-66 will tend to wrinkle more than a low density copolymer. It has been suggested (39) that reducing the distance between the nip rolls and the die, so the bubble is pinched while hot, decreases wrinkling. The wrinkling tendency is also minimized by proper alignment of the bubble with the collapsing frame and pull rolls, by good gauge control, and by minimizing the frictional forces between the film and the collapsing frame.

The water quench process shown schematically in Figure 6-19 is not widely

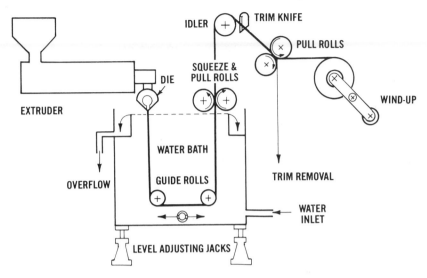

Fig. 6-19.   Water quench bath for nylon film.

used for producing nylon film. However, this method has some interesting attributes. The quench line can be accurately controlled since the film passes vertically downward into a water bath of constant height, and both sides of the film are quenched at equal rates since both sides are in contact with the cooling medium. This is not the case with either of the previously described methods. Also, the volatile monomers are water soluble and, therefore, present no problem (38). A final advantage of the water quench method is that the water accelerates the nylon crystallization process so that there is no puckering problem on winding rolls of film.

One long-standing objection to the water quench process is the problem of water carry over. That is, if the system is not designed properly, the film acts as a pump and transfers water to the take-off and wind-up system. This can be easily taken care of by making sure the pull rolls carrying the film from the water bath are perpendicular to the bath. Even though most of the water is removed by this procedure, the film is moist and must be passed through a drier prior to wind-up.

Nylon film made by water quenching is not highly glossy. Some of the problem is caused by a surface etching on a micro-scale, and part is from the dissolved salts. These difficulties can be eliminated by quenching in an organic medium such as a fluorinated hydrocarbon (40).

Nylon resins have been co-extruded with polyethylene and ionomer resins using the blown film technique (39). Other combinations of nylon with various resins are conceivable with this process or by using a flat film die. In the blown film method, the lower melting resin is usually extruded as the inner layer, and

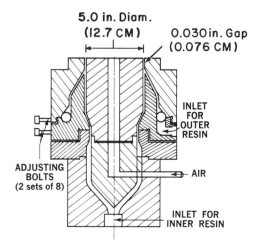

5.0 in. Diam.
(12.7 CM)   0.030 in. Gap
                  (0.076 CM)

INLET
FOR
OUTER
RESIN

ADJUSTING
BOLTS
(2 sets of 8)

AIR

INLET FOR
INNER RESIN

Fig. 6-20.   Blown film co-extrusion die.

the nylon is applied as the outer layer as indicated in Figure 6-20. Since the inner layer is usually strong enough to support the molten bubble, the melt viscosity of the nylon resin need not be as high as is required in blowing nylon film alone.

Nylon Sheet

Nylon sheet is commercially produced in thicknesses ranging from 10 to 300 mils (0.25 to 7.5 mm) and in widths from 12 to 60 inches (30 to 150 cm). The extruded sheet may be slit into smaller widths and used for stamping washers and other flat parts or for thermoforming. Oriented sheet is used in power transmission belting. Thin gauge sheet (that is, 25 mils or 0.6 mm) of some resins may be cast onto a chill roll using the same method as described previously for cast film, but most sheet is manufactured using a three-roll polishing stand (Figure 6-21). In this process, a sheet of uniformly thick melt is cast between the nip of the top and center roll of the polishing stand. The sheet is cooled by S wrapping around the center and bottom roll and by air cooling on the supporting rolls before entering the pull rolls. The die design and temperature control must be such that the flow of the melt is constant and uniform across the width of the die. This is required to maintain a small constant bank of resin in between the nip of the top and the center roll. This bank of resin acts as a reservoir of melt and compensates for minor feeding deviations from the extruder or a minor degree of eccentricity of the polishing rolls. The bank of resin should not be too large or a herringbone pattern will develop on the surface of the sheet. This is caused by oxidation of the melt as it rolls in nip of the rolls.

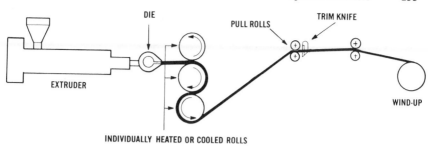

Fig. 6-21. Nylon sheet extrusion line.

If the bank is too small, the sheet will have patterns of thin caliper (usually called puddles, lakes) which will mar the appearance of the product.

Control of the bank becomes very difficult when nylon sheet in the range of 15 to 30 mils thick is produced. Thus, sheeting in this thickness range is the most difficult to manufacture. The casting roll temperatures should be as high as possible (for example 200 to 275°F, 93 to 135°C internal liquid temperature) for nylon 66, but about 10°F (6°C) below the sticking temperature. This will help maintain the casting rolls free of condensed oligomers, and the like, and it will insure good contact between the molten web and the casting roll. At lower roll temperatures, the condensed gases may crystallize as an oil or a white powder and mar the appearance of the sheet. The three polishing rolls will usually have to be run at somewhat different temperatures so that a flat nonbowed sheet is obtained.

Coating

In the extrusion coating process, molten nylon is extruded through a flat film die onto a substrate such as polyethylene film, paper, paperboard, or aluminum foil. The coating of nylon is pressed into the substrate as it passes through the nip of a pressure (elastomeric) roll and a chill roll (polished steel). The nylon extrusion coating process has been described in detail (5, 37, 40), and a typical extrusion coating line is shown in Figure 6-22.

Most of the common types of nylons are suitable for extrusion coating providing the moisture content is low enough to prevent bubbling and the melt viscosity is sufficiently low for rapid drawdown rates. For these reasons, the better coating resins have a water content below 0.1 wt % and a melt viscosity in the region of 5000 to 10,000 poise at 575°F (302°C) and at shear stress = 3×10^5 dyn cm^{-2}.

Extrusion coating equipment normally used for polyethylene resins can be employed for nylon coating providing all the extruder components are adequately heated and good temperature controllers are available. SCR

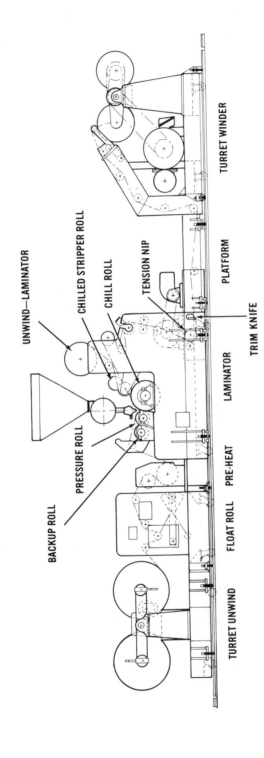

BACKUP ROLL

UNWIND—LAMINATOR

CHILLED STRIPPER ROLL

CHILL ROLL

PRESSURE ROLL

TENSION NIP

TURRET WINDER

PLATFORM

TRIM KNIFE

LAMINATOR

PRE-HEAT

FLOAT ROLL

TURRET UNWIND

Fig. 6-22.   Extrusion coating line.

controllers are well suited for controlling the adaptor and die sections. Long metering screws are recommended (2, 38, 42) for obtaining a constant output and a homogeneous melt for extrusion coating.

Air jets are used (5, 38) (Figure 6-18) to pin the edge of the film to the chill roll and to minimize the neck-in of the molten nylon web. Because air pinning holds the edge of the web in place, it eliminates "edge weaving" and minimizes the bead thickness at the edge of the coating. As the coating speed is increased and the draw rates become faster, there comes a time when the coating thickness varies in an s pattern across the width. This problem, commonly known as "center weaving," occurs because the molten nylon web has insufficient time in which to draw down to the desired gauge. "Center weaving" is eliminated by increasing the gap between the die lips and the nip rolls. This allows more time for the drawing to take place, and thus a uniform coating is obtained.

When coating flexible substrates such as polyethylene film, it is necessary to crystallize the nylon laminate before it is wound up; otherwise, puckering of the roll will occur. The same techniques used for accelerating the crystallization rate in cast film production are employed in extrusion coatings.

Many different nylons may be extrusion coated onto mechanical cable to provide corrosion and abrasion resistance. Nylon coatings are also used for low voltage insulation or more generally as an armor over a primary insulation such as PVC or polyethylene. The general processing equipment and techniques used for coating polyethylene are applicable to the family of nylon resins (41). Perhaps the primary requirement is to provide enough heating capacity so all nylons can be processed. It is especially important to examine the gate or flange area of the extruder to make sure this zone can be heated to at least 550°F (288°C). This will provide a safety factor when processing the higher melting nylon-66 resins.

Self-centered dies [Figure 6-23(c)] have been used with limited success. The problem with this type of die is that there is no adjustment for centering in case the melt flow is not well distributed and the caliper is not uniform. Sometimes the die may be rotated around the horizontal axis to correct a particular flow problem. Two types of adjustable dies are used in wire coating production. For coating over smooth substrates such as single strand wire or polyethylene and PVC insulated wire, a tubing die is used [Figure 6-23(a)]. With this system, the nylon jacket is drawn tightly onto the substrate by vacuum a short distance from the exit of the die. The vacuum is generally applied at the entrance of the guide tip by means of a suitable vacuum pump. The advantage of this type of jacket extrusion is that the wall thickness is constant regardless of the configuration of the wire or cable. A disadvantage of the tubing method is that the structure of thsubstrate is reproduced on the jacket. For example, the jacket over stranded wire will not disguise the lay of the strands and the coating will not appear to be smooth.

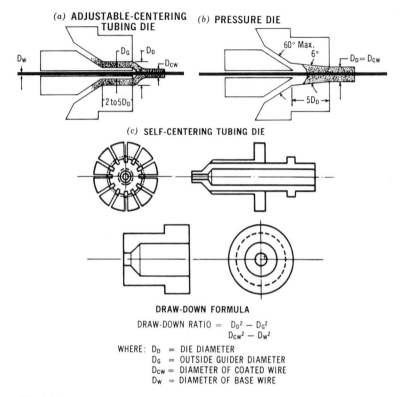

DRAW-DOWN FORMULA

DRAW-DOWN RATIO $= \dfrac{D_D{}^2 - D_G{}^2}{D_{CW}{}^2 - D_W{}^2}$

WHERE: D_D = DIE DIAMETER
D_G = OUTSIDE GUIDER DIAMETER
D_{CW} = DIAMETER OF COATED WIRE
D_W = DIAMETER OF BASE WIRE

Fig. 6-23.   Wire coating dies for nylon.

A pressure die is another adjustable type die for wire and cable coating work. In this process, the guide tip is recessed inside the die [Figure 6-23(b)] and the coating is applied under pressure. The pressure die offers the advantage that the diameter of the coated cable is constant even though the diameter of the substrate may vary. The pressure die is used in applications where good looking, smooth coatings are required and a constant diameter is wanted.

Nylon coated wire and cable is usually quenched in a water bath. In some cases, an air quench is used prior to the bath so marking or striping ink can be applied at the optimum temperature. Thin coatings (5 to 10 mils or 0.1 to 0.25mm) of nylon-610 require (28) a hot rather than cold water bath to develop optimum physical properties, especially resistance to cracking. Water quench temperatures in the range of 180 to 190°F (82 to 88°C) are used to control the crystalline structure of the resin as it solidifies. Preheating the substrate is usually advantageous to improve adhesion and, in the case of nylon-610 and similar resins, to increase the crystallization rate and control the crystalline structure.

Blow Molding

Hollow shapes, such as containers, tanks, carburetor floats, and the like, may be blow molded from various types of nylon resins providing the melt viscosity is sufficiently high to prevent sagging of the parison. Small objects (≤ 8 in. or 20 cm in length) may be blown from general purpose extrusion resins having a melt viscosity of 25 to 50,000 poise. Larger blow moldings require resins of higher melt strength, and a melt viscosity of over 100,000 poise is desirable. Blow molding machines designed for polyethylene may be used for nylon processing, but care must be taken that all systems of the blow molding machine are heated sufficiently to handle the complete family of nylon resins.

Because many nylon resins have insufficient melt strength to prevent sagging of the parison, they are best processed in machines having either an accumulator or a reciprocating screw such as is commonly used in injection moldings. In these machines the melt is extruded at a high rate, and, therefore, there is little time for sagging to take place. Parison extrusion times of about 1 or 2 sec are desirable to minimize the sagging tendency of the resin. With a standard extruder and no accumulator, the parison extrusion time may be 2 or 3 times longer than that attained with the accumulator system.

The die, mandrel, mold, and so on, are designed (13) using principles similar to those for high-density polyethylene. Allowance in tooling design must be made for the differences in the way various nylons swell exit the die, as well as their different shrinkage characteristics.

The appearance of weld lines in blow molded nylon parts is not uncommon. Weld lines may be formed where the polymer melt welds together after it passes the spider which holds the mandrel in place. The weld lines arise because the melt viscosity of the resin in contact with the spider leg is different and usually lower than that of the bulk resin. The low viscosity in the spider region occurs when the head temperature which contains the spider is higher than the bulk resin. Thus, the resin in contact with the spider becomes overheated with a reduction in melt viscosity. The hotter melt travels in streamlines which tend to draw down more than the bulk resin when the parison is blown to shape. To correct the weld line problem, one therefore reduces the temperature of the zone containing the spider section. If weld lines are still pronounced, then the spider may require further streamlining.

High melting resins, such as nylon-66, freeze quickly and shrink away from the mold, due to the high degree of crystallization. Thus, molding cycles can be quite fast with this type of resin. On the other hand, low melting copolymers freeze slowly with minimal shrinkage and are difficult to remove from the mold. Care must be taken with resins that crystallize slowly not to distort the part as it is removed from the mold. Any distortion will remain because these resins have little resilience until crystallization is at least partially complete. The crystallization rate can be speeded up by immersing the blow molding in tap water.

REFERENCES

1. Anon., Badische Anilin and Soda Fabrik AG, "Ultramide B Resins," May, 1969.

2. Anon., Badische Anilin and Soda Fabrik AG, "Ultramide S Resins," June, 1969

3. Anon., Badische Anilin and Soda Fabrik AG, "Ultramide B6 — Instructions for Blow Molding," July, 1969.

4. Anon., Badische Anilin and Soda Fabrik AG, "Ultramide A Resins," July, 1969.

5. Anon., Plastics Department, E. I. du Pont de Nemours & Company, "Du Pont Zytel® Nylon Resins — Extrusion Manual," March, 1970.

6. Anon., Plastics Division, Gulf Oil Corporation, "Moisture Control of Gulf Nylon Before and During Processing," May, 1968.

7. Anon., Plastics Division, Gulf Oil Corporation, "A Report on Nylon 12," June, 1968.

8. Bedarida, J., and Claude Bessard, *SPE Tech. Pap.* **14**, 490 (1968).

9. Wall, W. C., Plastics Department, E. I. du Pont de Nemours & Company, "Drying Oven Humidity Monitor," May, 1970.

10. Scheiner, Lowell L., *Plast. Tech.* **11** (8),37 (August, 1965).

11. Mikhailov, P. A., P.N. Mabysheo, and Ya V. Duplenko, *Soviet Plastics* (English trans.) (1961), p. 42.

12. Pfluger, Richard, *Kunstoffe* **52**, 277 (1962).

13. Cunliffe, S. R., *Plastics* **28**, 65 (April, 1963).

14. Toll, K. G., *SPE J.* **13**, 17 (July, 1957).

15. Scheiner, Lowell L., *Plast. Tech.* **14** (6), 35 (June, 1968).

16. Harris, H. E., *Mod. Plast.* **44** (8), 115 (Aug., 1967.

17. Bernhardt, E.C., *Mod. Plast.* **32** (6), 127 (Feb., 1955).

18. Van Ness, R. T., *Plast. Tech.* **11** (12), 30 (Dec., 1965).

19. Kessler, H. B., *et al.*, *SPE J.* **16** (3), 267 (Mar., 1960).

20. Bonner, R. M., *SPE J.* **19**, 1069 (Oct., 1963).

21. Bonner, R. M., Plastics Department, E. I. du Pont de Nemours & Co., POD TR 1017, April, 1968.

22. Weiske, Claus-Dieter, *SPE Tech. Pap.* **13**, 676 (1967).

23. Schenkel, G., *Plastic Extrusion Technology and Theory*, American Elsevier Publishing Co., New York, English Edition, 1966, p. 32.

24. Tadmor, Z., I. Duvdevani, and I. Klein, *Polym. Eng. Sci.* **7** 198 (1967).

25. Richardson, P. N., *SPE J.* **14**, 40 (August, 1958).

26. Riggert, K., *Kunstoffe* **55**, 788 (1965).

27. Bonner, R. M., E. W. Kjellmark, Jr., and R. E. Shaw, Twelfth Annual Symposium Sponsored by U. S. Signal R & D Lab Dec., 1963, "Tougher Wire Jackets of Nylon."

28. Anon., E. I. du Pont de Nemours & Co., Industrial Finishes Technical Bulletin No. 1, Ed. 13, Teflon® TFE Fluorocarbon Finishes.

29. O'Toole, J. L., H. G. Tinger, and T. R. Von Toerne, *SPE J.* **14** 47 (Oct., 1958).

30. Schenkel, G., *Kunstoffe* **51**, 153 (1961).

31. Voigt, J. L., *Kunstoffe* **51**, 450 (1961).

32. Mink, W., *Grundzuge Der Extrudertechnik,* Rudolf Zechner-Verlog Speyer R. H. (1964).

33. Rodenocker, W. G., *Faserforsch. Textil. Tech.* **18** (9), 433 (1967).
34. Bordner, P. G., G. C. Fulmer, and D. Meadows, *Mach. Des.* **40** (13), 155 (1968).
35. Bonner, R. M., Plastics Dept., E. I. du Pont de Nemours & Co., TR 145, June, 1965.
36. Carley, J. F., *SPE J.* **19**, 977, 1263 (Sept., Dec., 1963).
37. Regan, J . F., 20th Plast. Pap. Conf. Tappi, Detroit, Michigan, 1965.
38. Hamilton, E. F., and J. P. Harrington (to E. I. du Pont de Nemours & Co.), U. S. Patent 3,027,602 (April 3, 1962).
39. Guillotte, J. E., *SPE J.* **25**, 20 (November, 1969).
40. Werner, J. C., and G. D. Murphy, *Mod. Plast.* **42** (11) 128 (July, 1965).
41. Anon., "Wire and Cable Coaters Handbook," Plastics Dept., E. I. du Pont de Nemours & Co., 1968, p. 120.

CHAPTER 7

Processing and Product Quality

EDMUND M. LACEY

INTRODUCTION

An adequate system of quality control is basic to the successful fabrication and use of molded parts. The extent and complexity of this control will vary with product needs. Visual inspection may suffice if part requirements are simple and service failure unlikely. If mechanical property requirements are high, as in components for appliances or automobiles under warranty, more detailed testing may be necessary to insure that processing, contamination, or other factors have not impaired properties. As an engineering material, nylon is used in applications with high performance requirements. Quality control testing of nylon is, therefore, important.

The increasing demand for reliable performance has led to more stringent specifications and, in turn, increased use of the laboratory by the processor. However, laboratory testing can be costly and can involve excessive delay in feedback to the processing operation. To optimize performance there has been expanded use of sophisticated instrument controls such as in-line devices that sense temperature, pressure, and viscosity and provide for immediate process correction. Increased attention has been given also to design to be sure that potential weak spots are recognized and eliminated. A critical factor often is the handling of rework not only to minimize loss but also to assure control of moisure content, cut, and cofeeding with virgin resin so that consistent performance is guaranteed. These aspects of process control have been discussed in the two preceding chapters.

SAMPLING

Sampling practice varies with the application, and even in a single application sampling frequency changes in the event of an emergency. Some shops collect a single specimen from a multicavity mold hourly or every two hours. If this procedure proves inadequate, then a larger number of specimens or more frequent sampling is used. Specifications can include a particular sampling technique.

Although trial and error sampling frequently works out successfully, a more statistical approach is usually preferred. Much has been published on sampling, and basically similar techniques can be used for all plastic parts. Recommended sampling techniques are described in ASTM (1) and military (2) literature as well as in texts (3, 4). An acceptable level of rejects must first be decided upon by processor and end user. Sampling to insure continuance of this level can then be determined. Although this procedure utilizes a mathematical basis for selection, changes in sampling frequency when "advisable" are usually made. This keeps testing to a reasonable and economic level.

Nylon is frequently molded into small mechanical parts, sometimes 2 or 3 g each, using multicavity molds with up to 80 and more cavities. Cavities are often numbered to identify individual parts. This is important in quality control because different parts from the same shot may vary due to differences in flow, temperature in the mold, gate locations, vent size, or other factors. Maintaining cavity identity is important in effective quality control because rejects may not be randomly distributed. Cavities from which a large incidence of rejects are found can sometimes be corrected should examination indicate a flaw such as a sharp radius, small gate, or inadequate vent. When correction cannot be readily effected, the cavity may be blocked off.

When parts cannot be related to cavities, random sampling must be used. The probability of obtaining a sample from each cavity increases with the sample

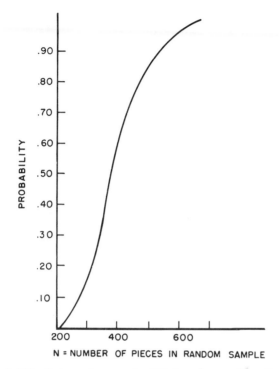

Fig. 7-1. Probability that a random sample of N pieces from an 80-cavity mold contains at least one piece from each cavity versus N (9).

size. Figures 7-1 and 7-2 show the probability that a random sample of N pieces from multicavity molds will contain at least one piece from each cavity. Figure 7-1 shows that a random sampling of an 80 cavity mold requires 380 individual pieces for a 50% probability of obtaining one piece from each cavity and 585 pieces for a 95% probability. Figure 7-2 shows probability versus sample size for 12, 36, 72, and 80 cavity molds. The periodic testing of each cavity is the only alternative to large random sampling if all cavities are to be tested.

Moisture absorption affects dimensions and physical properties of nylon and is therefore a consideration in any sampling technique. Toughness tests, for example, are frequently conducted by the end user on dry, as-molded samples, primarily because this represents the most severe condition that can be anticipated. Samples taken by the molder for testing either on the site or at the end user's location may accordingly be stored in waterproof packaging. Differences in test results at the processing plant and at the end user are frequently due to variations in moisture level. Another source of discrepancy between plant and user is an increase in crystallinity, which is especially noticeable in thin sections from relatively cool molds.

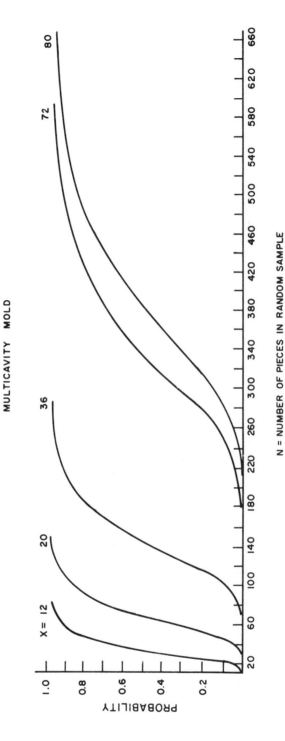

X = NUMBER OF CAVITIES

MULTICAVITY MOLD

Fig. 7-2. Probability that a random sample of N pieces from the output of a multiple-cavity mold contains at least one piece from each cavity versus N (9).

TESTS OF PROCESSING PERFORMANCE

Visual Defects

The quality of fabricated nylons is often based upon visual inspection either by the processing operator or by inspectors at a viewing bench. Inspection is surprisingly effective, especially when certain steps are taken to improve reliability. Simple magnifiers, choice of lighting such as illuminant C (10) for colors, and availability of a powerful light beam equipped with an adjustable iris to examine thick sections improve performance. Standards illustrating acceptable and nonacceptable parts with respect to defects are helpful.

In injection molding, splay, burn marks, weld lines, unmelt, flash, contamination, voids, and part distortion are defects that may be observed in, although not limited to, nylon parts. Splay is a surface phenomenon usually caused by too high a moisture content during processing and consists of small "V" shaped imperfections on the surface that become frothy when severe. Splay may be associated with a loss in molecular weight due to the presence of the excess moisture. For both reasons it is wise to check toughness on parts that splay. Burn marks, which arise from oxidation in an air pocket because of inadequate venting, represent both a physical blemish and a weakness and suggest that such remedial steps as better venting or changes in processing conditions such as a slower ram forward speed. Weld lines are not normally troublesome with nylon but in some instances can cause a reduction in impact strength. The observation of a prominent line alerts the operator to increase melt or mold temperature, reduce injection time, or take other steps to correct the problem. Unmelt refers to incompletely melted pellets that appear as discrete particles of different shade. The discontinuity can represent a physical weakness. Obviously, higher melt temperatures or longer dwell times eliminate unmelt. Flash, a defect due to unsatisfactory mating of the parts of a mold or insufficient clamping pressure, can be a factor in control of dimensions and impact resistance. Contamination and color are readily observed visually, preferably using reference standards. Voids can sometimes be spotted by simple inspection although not in thick or pigmented parts as discussed in the following section. Part distortion refers to warpage or to depressions in the surface that are commonly called sink marks. Knock out pins may cause surface defects also when the cycle is too short or melt temperature too high.

Suppliers of nylon polymers customarily provide summaries of the typical problems encountered in processing their compositions and outline alternative remedies.

Part Weight and Voids

Injection molders use the part weight of large moldings to monitor the uniformity of processing conditions. Part weight is sometimes plotted on a control chart at the machine. Allowable weight limits are based on previous experience with critical properties. Frequently, although not always, weight variation outside the control limits will indicate a change in properties or another flaw such as a void.

Voids in molded parts may impair properties as well as appearance. They are most frequently present in thicker sections, especially when gates are small and freeze-off prevents pressure from being maintained during a large portion of the cooling cycle.

In uncolored nylon, voids can frequently be observed by viewing with a narrow penetrating beam of light. For thick or pigmented parts it is possible to employ sink or float tests which involve a close matching of the density and a fluid that is not absorbed by the nylon. The density is adjusted to allow a void free part or a cut section of a part to sink very slowly when immersed. Parts or pieces therefrom that contain only small voids will float. The technique can be used at the molding machine site, where satisfactory results are obtained as long as temperature and fluid compositions are held constant. A suitable hydrometer is helpful in this type of "at the machine" testing. X-ray or fluoroscopic examination of large specimens such as rod stock is frequently used prior to machining because the added cost of this inspection is insignificant relative to machining costs. Although ordinary medical-type equipment can be used, the higher resolutions of specialized industrial equipment permits determination of smaller voids. Ultrasonic detection of voids can be used for parts of regular geometry such as extruded rod stock. For complex shapes, the need for a variety of transducers has restricted this approach.

Dimensions

Dimensions are checked with micrometers, verniers, or other calipers; ring and plug; thickness, thread, and depth gauges, etc. Optical comparators are used frequently to check shape and contours. The techniques are discussed in Ref. 5.

Conformance of nylon parts to specified dimensional tolerances sometimes involves measurements at both the plant site and the end user's location. At the processing site, critical dimensions may be determined soon after fabrication to minimize production of nonstandard material. Similar measurements at the end user's location, perhaps one month later, may show poor agreement because of dimensional changes due to temperature, moisture absorption, crystallization, and stress relief. Frequently these dimensional changes are unimportant to the end use and do not constitute a problem. When dimensions are critical, one of the following procedures may be used to obtain samples for specification purposes.

a. The processor anneals the parts, measures critical dimensions, and packages them in waterproof containers before shipment. This is the ideal situation because the dimensions are based upon dry, stress-free parts.

b. The processor measures without annealing and packages in dry containers prior to shipment. Stress relief on dry material during a 3 to 4 week period is normally not a problem.

c. The processor conditions in boiling water to a specified moisture level, measures, and packages in a waterproof container. Immersion in boiling water is often practiced where added moisture is desired to enhance toughness, but the sealed package is used only for sampling and not for production because of the possibility of oxidation during prolonged storage at high warehouse temperatures. Dimensional changes may occur during shipment as the moisture distribution becomes more uniform but these changes are not critical.

Internal Stress

Internal stress, which is discussed in Chapter 17, can be determined by either annealing or stress cracking tests, usually with zinc chloride solution.

Toughness

Toughness as a requirement for satisfactory performance in mechanical applications has frequently led to the selection of nylons over other, less expensive materials. Because toughness may be of major importance, many quality control tests have been used to evaluate this property. One test consists of dropping a ball or dart onto the part and using a height and mass that breaks parts of poor quality and approximates the impact likely to be encountered in service. Go/no-go tests are obviously of no value in determining whether the toughness of parts vary in random or nonrandom fashion. It can be important to know of nonrandom variation, which suggests careful examination of weak parts for contamination, weld lines, etc., and of the possible relationship of weak parts to specific cavities that may require modification. The toughness distribution of nylon parts can be obtained by high-speed tensile studies or by breaking a large number of specimens in a pendulum impact machine (6) such as that used for determining Izod impact strength (8). Parts are held in position using rigs designed for each individual geometry. Values can be plotted on probability paper to show the cumulative percent failing at different impact energies. This procedure has been used successfully (6) for determining optimum processing conditions as well as comparing quality of various grades of nylon. This test is discussed further in Chapter 10.

Molecular Weight

The measurement of molecular weight is discussed in Chapter 3. Relative viscosity (concentrated) is frequently used in military and industrial specifications for nylon parts where minimum values are set to insure that degradation by thermal or hydrolytic means has not occurred during processing. A reduction in relative viscosity during processing may be accompanied by a loss in toughness. The absence of a reduction in viscosity is frequently interpreted as indication that toughness is satisfactory, but this is not always true because other factors such as contamination, poor weld line strength, voids, sections of inadequate radius, and the like, can cause poor impact strength. Some end users require a toughness test (on dry, as-molded specimens) for assurance that the toughness is satisfactory and the relative viscosity has not been significantly lowered by processing.

REFERENCES

1. ASTM D1898.
2. Military Standard MIL-STD-105D, "Sampling Procedure and Tables for Inspection by Attributes," 29 April 1963.
3. Statistical Research Group at Columbia University, *Sampling Inspection,* McGraw-Hill, New York, 1948.
4. Davis, O. L., *Statistical Methods in Research and Production,* Hafner Publishing Co., New York, N.Y., 1957.
5. Publication of Society of Plastics Industry, "Proposed Test Methods for Plastic Parts Used in Appliances," Jan. 8, 1965.
6. Heyman, E., *SPE J.* **24** (5), 49 (May, 1968).
7. ASTM D1505.
8. ASTM D256.
9. Dunleavy, M. T., personal communication.
10. Committee on Colorimetry, *J. Opt. Soc. Am.* **34,** 643 (1944).

CHAPTER 8

Physical Structure of Nylons

E. S. CLARK AND F. C. WILSON

INTRODUCTION

Most nylon plastics of commercial interest are partially crystalline, and an understanding of their physical structure requires investigations at several dimensional levels between the molecule and the article-of-commerce. Figure 8-1 illustrates the main structural entities and their size ranges. These can conveniently be separated into three areas: (1) crystal structure, (2) morphology, and (3) macroscopic specimen.

● *Crystal structure* deals with the organization of long chain molecules into some sort of sub-microscopic array. If the array, on the average, is highly ordered in three dimensions, the material is said to have a high crystallinity. If the packing of the chains, averaged over the sample, is not very regular, the crystallinity is low. Nylons are different from many other crystalline polymers such as polyethylene in that the degree of crystallinity of a given nylon can be controlled over a wide range.● High crystallinity requires both parallel alignment of the chains and uniformity in the manner in which hydrogen bonds are

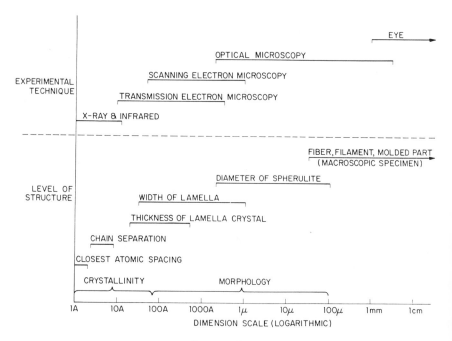

Fig. 8-1.   Experimental techniques and corresponding structural elements.

formed. This question of uniformity with respect to hydrogen bond formation accounts for the wide range of crystallinity and is discussed further below.

Morphology deals with structure on the microscopic level. The simplest concept of morphology treats the molecular aggregations as two distinct phases — a nearly perfect crystal phase imbedded in an amorphous entanglement of chains with many chains joining the two phases to provide strength. This classical concept is useful for providing a simple quantitative description of morphology from which a "percent crystallinity" can be derived from x-ray, infrared, or density data. For the most part, engineering parameters and physical property data have been correlated with percent crystallinity and orientation.

However, the morphology of nylons is far more complicated than this simple model. The molecules fold to form thin ribbon or layer shaped crystals termed "lamellae" and these in turn often aggregate into spherical clusters called "spherulites." Orientation results when operations such as drawing or rolling induce the partial breakup of lamellae and spherulites and form new structures. Unfortunately, the knowledge of these morphological types is limited; identification and analysis of the structures are complex. Nevertheless, an improved understanding of physical properties and their relationship to processing variables requires a detailed knowledge of crystal structure and morphology, and this chapter is concerned with this problem.

CRYSTAL STRUCTURE

The crystal structures of the aliphatic nylons fall into two general categories. One, in which are represented all the commercially important nylons, includes the alpha and beta phases. The second category is the gamma phase which is the characteristic structure of over half the polyamides, but none of commercial importance.

This phase terminology characterizes the general conformation of the polymer chains and their mode of packing, but the details of the structure vary from one polyamide to another. These structural assignments have been derived primarily from x-ray diffraction studies of oriented fibers (1). A word of caution is in order: Different designations exist in the literature; one should check usage before assuming a structure.

Even-Even Nylons

The structures of the even-even polyamides, nylons-66 and -610, were determined by Bunn and Garner (2). The stable structure is the α-phase, which is comprised of planar sheets of hydrogen-bonded molecules stacked upon one another. The molecules are in the fully extended zigzag conformation (Figure 8-2). In crystallographic terminology, the structure is triclinic with one chemical repeat per unit cell. The triclinic structure is of low symmetry with all axes and interaxial angles unequal. The only symmetry element is a center of symmetry, which is possessed by the molecule itself. In both the diamine and the diacid, or their residues in the nylon, there is a center of symmetry at the midpoint of the monomer. This is preserved in the crystal structure and means that only half of a chemical repeat is needed to define the structure.

Figure 8-3a, a perspective drawing representing the structure of nylon-66, shows chemical repeat segments of four molecules and the outline of a unit cell. Each chemical repeat runs along the c-axis of the cell, but since each segment is shared by four unit cells there is only one per cell in the crystal. The molecules are hydrogen bonded (indicated by arrows) in the a-c faces of the cell. The hydrogen-bonded sheets are developed by extension of the a-c faces. Each successive molecule in a hydrogen-bonded sheet is displaced by one chain atom in the c-direction. Examination of Figure 8-2 will show that this is required for strain-free formation of the hydrogen bonds. Each successive sheet is displaced the equivalent of three chain atoms in the c-direction as an accommodation to the packing of the amide groups. Since the packing of the molecules is approximately hexagonal, each sheet is displaced about one-half unit normal to the c-axis in its own plane. It is interesting to note that the more tightly bonded molecules in the hydrogen-bonded sheets are actually farther apart than the nearest molecules in neighboring sheets. The unit cell dimensions, a and b,

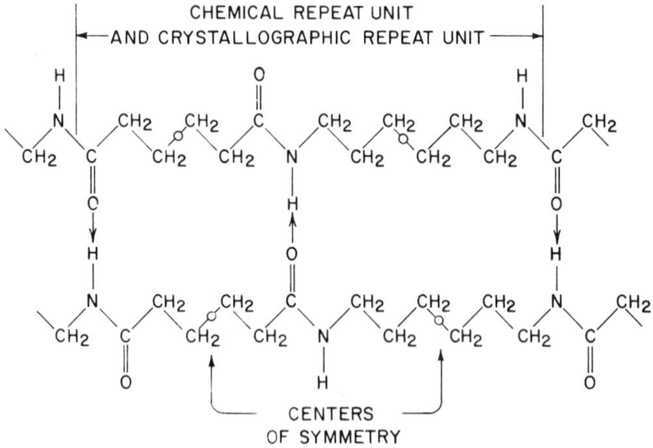

Fig. 8-2.   Schematic representation of nylon-66.

correspond to the distance between like positions on adjacent chains. The angular displacements of adjacent chains are such that the perpendicular distance between hydrogen-bonded sheets is 3.6 A and the perpendicular distance between chains within a hydrogen-bonded sheet is 4.2 A.

Figure 8-3b shows the crystallographic planes of major interest superimposed on the outline of the unit cell of Figure 8-3a. The numbers identifying the planes are the Miller indices. The (010) planes have the direction and spacing of the a-c faces and can be considered to include the hydrogen-bonded sheets. The (100) planes have the direction and spacing of the b-c faces, and the (110) planes cut the unit cell diagonally. The interplanar spacing of the (110) planes is always very close to that of (010), although the intensity from (110) is substantially less. The (001) planes constitute the a-b faces of the unit cell and mark the termination of a crystallographic repeat.

The β phase, which apparently is not a distinct phase in the thermodynamic sense, amounts to a slight perturbation of the α phase. It is recognized by the presence of a meridional spot and layer-line streaks in an x-ray fiber photograph (Figure 8-4). The original interpretation (2) was that successive sheets were staggered up and down instead of always being displaced in the same direction. Another interpretation (3) postulates initial stages of ordering and the presence of unrelated small crystallites. Since the presence of the β phase is based upon the appearance of fiber photographs, the question is somewhat academic for the general case of samples with low levels of orientation. The presence of the β phase cannot be reliably determined except in samples possessing a high degree of axial orientation, and the two phases probably represent stages in a continuum of possible states of order. This point will be returned to in the discussion of measurement of crystallinity.

Other even-even nylons crystallize into closely analogous structures differing essentially only in the length of the c-axis in order to accommodate the varying numbers of methylene groups according to the monomers involved.

Even Nylons

All the even nylons from nylon-8 up generally crystallize into the γ phase (1) which is discussed on p. 278. However, nylon-4 and the commercially important nylon-6 usually crystallize into the α phase (1), although they can be converted to the γ phase by unusual treatments. As is the case with the even-even nylons, the molecules in the α phase are in the fully extended zig-zag conformation and grouped into essentially planar, hydrogen-bonded sheets (Figure 8-5). Details of the intermolecular bonding differ, however, because of the differing symmetry of the molecules. The even-even nylons are characterized by centers of symmetry along the molecule, and there is no sense of direction to the molecule. The even (and odd) polyamides, in contrast, are not centrosymmetric and are

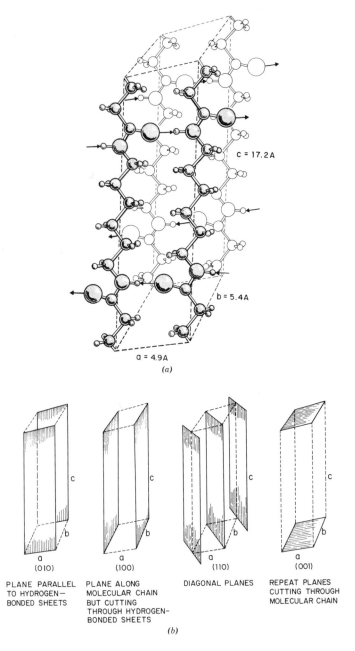

$c = 17.2A$

$b = 5.4A$

$a = 4.9A$

(a)

c

a
(010)

PLANE PARALLEL
TO HYDROGEN—
BONDED SHEETS

c

b

a
(100)

PLANE ALONG
MOLECULAR CHAIN
BUT CUTTING
THROUGH HYDROGEN-
BONDED SHEETS

c

b

a
(110)

DIAGONAL PLANES

c

b

a
(001)

REPEAT PLANES
CUTTING THROUGH
MOLECULAR CHAIN

(b)

Fig. 8-3. (a) Perspective drawing of a unit cell of nylon-66. The viewpoint is 11A up, 10A to the right and 40A back from the lower left corner of the cell. **(b)** Principal crystallographic planes of nylon-66.

276

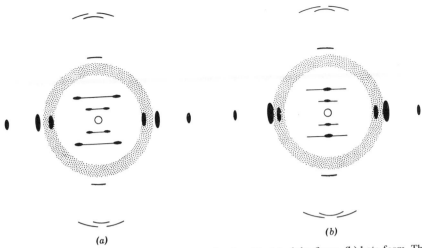

Fig. 8-4. Schematic x-ray fiber diagrams of nylon-66: (a) alpha form, (b) beta form. The fiber axis is vertical.

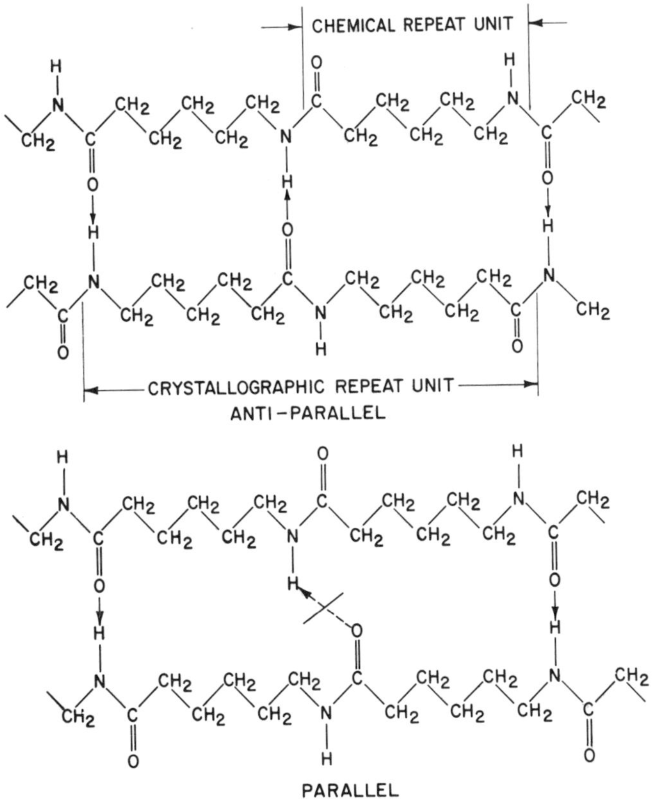

CHEMICAL REPEAT UNIT

CRYSTALLOGRAPHIC REPEAT UNIT

ANTI-PARALLEL

PARALLEL

Fig. 8-5. Schematic representation of nylon-6.

characterized by a directionality to the molecule (NH-CO or CO-NH) such that if a molecule is imagined turned end-for-end it cannot be superimposed upon itself (Figure 8-5). The hydrogen-bonded sheets of the α phase of nylon-6 involve adjacent molecules which have opposite directionality and are said to be antiparallel. Arranged in this manner, all the hydrogen bonds can be formed without strain. If adjacent molecules were of the same direction (parallel), only half of the hydrogen bonds could be formed, and infrared studies show that this is not the case (1, 5, 6). The hydrogen-bonded sheets are also staggered up and down instead of always being displaced in the same direction. The resulting structure is monoclinic instead of triclinic, and there are four molecular segments in a unit cell, where the crystallographic repeat unit consists of two chemical repeat units or monomer residues. Because of crystallographic convention for monoclinic structures, the direction along the molecular chain (the "fiber" axis) is labeled the b-axis rather than the c-axis as is the case for the triclinic nylon-66. Because there are four molecules rather than one in a unit cell, the Miller indices are doubled for the nonfiber axes. Thus, the intense equatorial reflections of nylon-66 are indexed as (100) and the (010, (110) doublet; the equivalent reflections for nylon-6 α phase are (200) and the (002), (202) doublet.

Odd Nylons

The odd nylons, of which nylon-11 is the commercial example, also crystallize into α and β phases (1). As can be seen for nylon-7 (Figure 8-6), the requirement that adjacent chains be antiparallel for formation of all the hydrogen bonds does not hold for the odd polyamides. All the hydrogen bonds can be formed in either the parallel or antiparallel arrangement of molecules in a sheet. There is disagreement among authors as to which represents the structure. The antiparallel arrangement must be considered the more likely. First, it would require much less rearrangement from the random melt. Second, and even more important, it is the inevitable consequence of crystallization with chain folding. This concept was first introduced in 1957 (7), and its generality and implications were not appreciated by many until somewhat later. Since a majority of the structural determinations of polyamides predate 1960, it can be assumed that the consequences of chain folding were not considered. Both the quality and quantity of x-ray data obtainable from polymers are often insufficient to make a determination of this kind from first principles.

Other Nylons – The Gamma Phase

The structure of the even-odd, odd-even, odd-odd, and the even nylons, are different from those discussed above and are generally designated the γ phase.

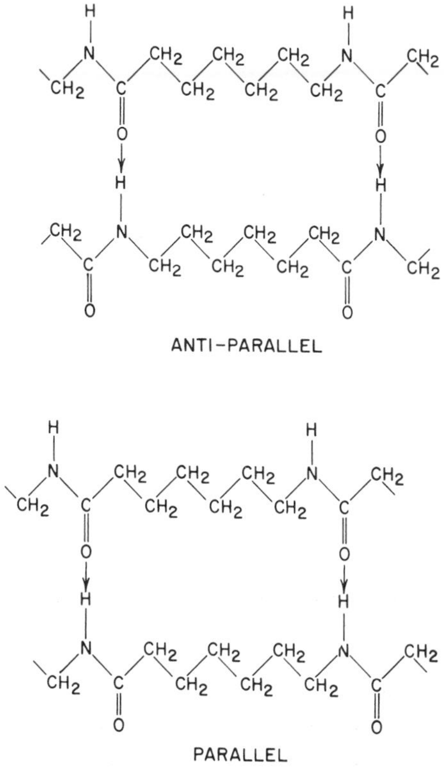

ANTI-PARALLEL

PARALLEL

Fig. 8-6.   Schematic representation of nylon-7.

This is true for all of these nylons even though some, the odd-odd nylons where the acid has two more carbon atoms than the diamine and the even nylons, can theoretically form strain-free intermolecular hydrogen bonds with the chains in the fully extended planar zigzag conformation characteristic of the α phase. Infrared studies (5) show that the hydrogen bonds are essentially complete, and as a consequence the polymer chains accommodate to this situation by rotation about single bonds into a puckered or pleated conformation. This results in a molecular repeat distance which is shorter than that of the fully extended chains and a molecular packing which is pseudohexagonal (1). As noted earlier, nylons-4 and -6 are unusual in that they can crystallize into both the α and γ phases. The obvious difference in the appearance of x-ray fiber photographs from the α phase and the γ phase is illustrated for the case of nylon-6 in Figure 8-7. An x-ray fiber photograph from γ-phase nylons is characterized by a single strong equatorial reflection instead of the pair of strong reflections observed from well-crystallized α-phase nylons. Pseudohexagonal means that the packing of the molecules is such that the location of the molecules can be described by a

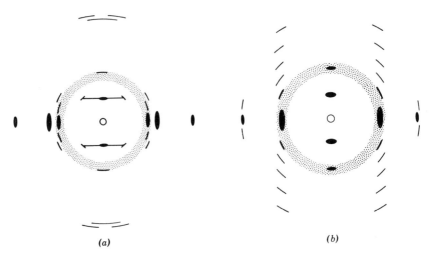

Fig. 8-7.   Schematic x-ray fiber diagrams of nylon-6: (*a*) alpha form, (*b*) gamma form. The fiber axis is vertical.

lattice of hexagonal dimensions but that the structure lacks true hexagonal symmetry. True hexagonal symmetry requires that each element of the structure be exactly duplicated when rotated 60° or 120° (Figure 8-8). A structure such as that on the right of Figure 8-8 is also described as metrically hexagonal. It is the knowledge of the chemical structure of the nylon molecules that leads to these conclusions. The molecule does not possess the required symmetry to exist in a hexagonal cell of the observed dimensions.

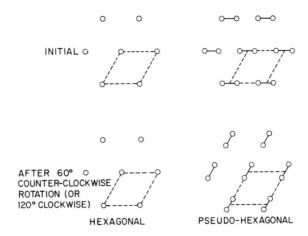

Fig. 8-8.   Hexagonal and pseudo-hexagonal symmetries. The dashed lines outline the basal plane of a hexagonal unit cell.

CRYSTALLINITY

Estimation of Percent Crystallinity

A numerical measure of "crystallinity" is often used as a measure of the degree of crystalline order in a semicrystalline polymer. The term implies the presence of a two-phase system of crystalline and amorphous regions. This was a reasonable interpretation of the diffraction patterns of such polymers as polyethylene and polytetrafluoroethylene. Diffraction from these polymers resembles the superposition of a crystalline diffraction pattern on an amorphous pattern, which in turn appears to be an extrapolation of the diffraction pattern of the melt. It is now generally considered that this interpretation is a gross oversimplification, but the resolution of such patterns into amorphous and crystalline "areas" and the derivation of a degree of crystallinity from such an operation can still be empirically useful.

For all the nylons there is no obvious procedure for resolution of a diffraction pattern into crystalline and amorphous regions and the calculation of even an empirical degree of crystallinity. There is no obvious demarcation between crystalline and amorphous areas (8). An instrument such as the Du Pont 310 curve analyzer can be used to resolve these patterns (9) but the results can be quite arbitrary. Thus for polyamides, x-ray diffraction is not commonly used to derive a measurement of crystallinity.

The usual methods for assessment of degree of crystallinity in nylons are by measurement of density or by infrared techniques. Measurement of density is perhaps most satisfactory because it is rapid, precise, and unaffected by sample orientation and geometry. It is not an absolute method, however, and requires calibration — the assumption of amorphous and crystalline densities. It also requires close control over water content. Ideally, the samples to be compared should be dry. If this is not practical, they should be equilibrated to the same moisture content. However, this can change the degree of crystalline order. The plasticizing effect of water on a dry sample of low crystallinity may increase its crystallinity. It should also be realized that massive samples are typically not uniform in crystallinity. For example, the rapidly cooled surface of an injection molding will be less crystalline than the more slowly cooled interior. Annealing will make the sample more uniform and also more crystalline, but if the sample is subjected to a flash annealing, the surface can become more crystalline than the interior.

Infrared measurement of the intensities of crystalline and amorphous bands has been used as a measure of crystallinity and, when carefully done, correlates well with density measurements (10). For these studies, thin films were used so that the samples could be considered homogeneous.

As mentioned above, x-ray diffraction techniques are not satisfactory for a routine determination of crystallinity of nylons in the usual sense of the term.

However, much useful information can be gained regarding the state of order in a typical α- or β-phase nylon, since the diffraction pattern varies a great deal as a function of sample history. The most intense region of a nylon diffraction pattern arises from interactions among neighboring molecules and consists of the two intense equatorial reflections of a fiber pattern. For a triclinic (even-even) nylon, these are the (100) peak and the (010), (110) doublet. Figure 8-3b shows in heavy outline, superimposed on the unit cell of nylon-66, the planes of particular interest for this discussion. For a well-annealed, highly crystalline nylon, these peaks are well resolved. When the degree of crystalline perfection is lower, the peaks are closer together, and for a sample of low order, they fuse into a single peak. In this case, the structure is often referred to as pseudohexagonal, but it should not be confused with pseudohexagonal γ phase. A measurement of the separation of these two peaks in degrees 2θ or converted to a difference in angstroms can give a very useful measure of the degree of order, the order being higher for a greater degree of separation. When the peaks coalesce to the point where they can no longer be resolved, a measurement of the width of the peak at half-maximum intensity becomes a useful parameter of order.

As these two peaks fuse together with decreasing crystalline perfection, it is actually the 010 peak that moves the most, moving to a lower angle corresponding to a larger spacing. The spacing of the 010 peak is the distance between hydrogen-bonded sheets. As disorder increases and hydrogen bonds are no longer restricted to these sheets, the molecules are forced farther and farther apart until all six nearest neighbor molecules are essentially equidistant in the pseudohexagonal structure. It is from such a structure that only a single intense peak is observed in the diffraction pattern. The determination of peak separation thus is a measure of the degree of perfection of the formation of hydrogen-bonded sheets.

Figure 8-9 shows the change in the appearance of the diffraction pattern of a nylon monofilament with variation in crystallinity achieved by different annealing treatments after cold-drawing. The changes described above are clearly evident. It is in the characterization of drawn samples such as this example that x-ray diffraction is most useful. The test is not disturbed by the adventitious orientation often observed in moldings, and samples are easily prepared for study by either photographic or diffractometric techniques. The degree of orientation can also be measured simultaneously.

Effect of Crystallinity on Properties

• The effect of crystallinity on the properties of nylons is substantially the same as it is for other semicrystalline polymers. Modulus and strength and related properties such as hardness and yield point increase with increasing crystallinity. Measures of toughness such as impact strength decrease, particularly in the

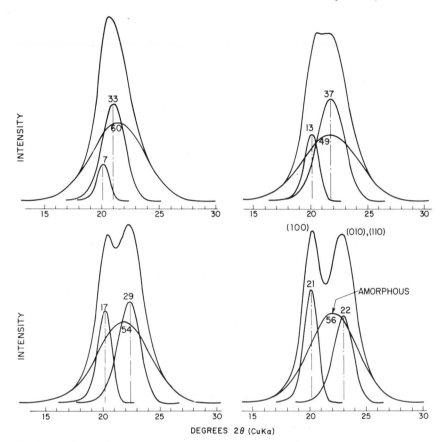

Fig. 8-9. X-ray diffraction scans from nylon monofilaments of differing crystallinites. The curves were resolved with the Du Pont 310 Curve Analyzer with the assumption that each curve was the sum of three Gaussian peaks. The numbers on the curves indicate relative areas.

high-crystallinity range. Table 8-1 gives a more complete list of the crystallinity dependent properties.

However, the effect of crystallinity can hardly be discussed independent of that of water. The properties of polyamides are as dependent on water content as they are upon crystallinity. Not only can water have an effect on crystallinity *per se*, as discussed above, but it also changes physical properties independently. Once a sample has absorbed a given amount of water at any temperature, this water can be removed and replaced at this temperature without noticeable effect on crystallinity (as judged by infrared (13) or x-ray diffraction), but the effects on properties will be substantial. Water acts as a plasticizer for nylons and lowers

Table 8-1. Effect of increasing crystallinity on properties

Property increases	Property decreases
Stiffness or modulus	Impact resistance
Density	Elongation
Yield stress	Thermal expansion
Chemical resistance	Permeability
Electrical properties	Swelling
Tg and Tm	Mechanical damping
Abrasion resistance	
Creep resistance	
Dimensional stability	

the glass transition temperature and the characteristic temperatures of mechanical relaxation (see Chapter 9). In many respects, the addition of water to nylon is equivalent to raising its temperature by a substantial amount. Thus, in order to predict the physical properties of a given sample, one would have to know both the level of crystallinity and the water content. Figure 8-10 (11) demonstrates these effects on the stiffness of nylon-610.

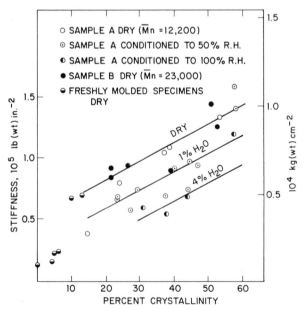

Fig. 8-10. Stiffness versus crystallinity (corrected according to Ref. 10) of nylon-610 films (11).

Rate of Crystallization

Primary Crystallization

The rate of crystallization of a nylon is dependent upon the chemical structure, molecular weight, thermal history, and the presence of nucleating or plasticizing additives. Each of these controlling factors will be discussed below.

Both the amide concentration and molecular symmetry are important. For a given class of nylons, the rate of crystallization will typically rise with increasing amide concentration. At a given amide concentration, the classes with the highest degree of symmetry (e.g., even-even) will have the highest rate of crystallization. These same factors also control the melting points, with higher melting polyamides generally crystallizing more rapidly.

Once a certain minimum molecular weight is attained, the rate of crystallization at a given temperature is an inverse function of molecular weight, because the increase of viscosity with molecular weight decreases the mobility of the polymer chains.

For any particular resin, temperature is the most important parameter for control of the rate of crystallization and resultant level of crystallinity. For any semicrystalline polymer there is a particular temperature or degree of super-cooling at which the rate of crystallization is a maximum. The rate is obviously zero above the equilibrium melting point, and some degree of supercooling is required to initiate crystallization. At sufficiently low temperatures, below the glass transition temperature, the rate is again effectively zero. Some polymers (e.g., polyethylene) crystallize so rapidly over a broad temperature range that they cannot be quenched to a fully amorphous state. Others [e.g., poly(ethylene terephthalate)] can be quenched to a fully amorphous state easily. The nylons are typically intermediate between these extremes, and encompass a consider-able range depending upon chemical structure. The temperature of the maximum rate of crystallization can be substantially below the melting point. Burnett and McDevit (12) observed the highest spherulite growth rate of nylon-6 at $132°C$.

Both the rate of nucleation and the subsequent growth rate of nucleated regions play an important part in the overall crystallization rate. In general terms, the greater the degree of supercooling, the greater is the driving force toward nucleation and subsequent crystallization, but this is counteracted by an increase in viscosity. If a polyamide is quenched to a sufficiently low temperature rapidly enough, very little crystallization will occur, but this is possible only with thin sections because of poor heat transfer owing to the low thermal conductivity of the nylon. As mentioned above, molecular structure has an important effect on rate of crystallization. It is very difficult to prepare and maintain a low-crystallinity specimen of nylon-66 but relatively easy with nylon-610 (10). One reason for the more rapid crystallization rate of nylon-66 is its shorter repeat length. A crystallizing chain segment need not move so far on the surface of a growing crystal to reach a suitable site for formation of hydrogen bonds.

The prior history of a sample can have an effect in many ways. The melt temperature and time in the melt are of importance, with an increase in either tending to lower crystallization rate. The influence is on both the induction time for crystallization and on subsequent growth rate, and the implication of this is that the melt has a memory. It has been observed in hot-stage studies that spherulites will recur at the same place on a microscope slide if the melt history is not too severe. This is a nucleation phenomenon and the effect is lost with prolonged or high temperature heating. The lower subsequent growth rate can be explained on the basis of crystal growth occurring by the addition of preordered units which can be disrupted by an extreme melt history. The presence of moisture in nylon has been reported to increase the rate of crystallization based on an increase in rate of spherulite growth. This is attributed to a plasticizing action which increases chain mobility (13), but this is not a practical means of control because it causes degradation and bubbles. The presence of nucleating agents will also enhance crystallization via an increase in the concentration of nuclei. These are often intentionally added, but impurities from any source can act in a similar manner. In fact, most semicrystalline polymers crystallize via adventitious nucleation. Reworked polymer is typically more highly nucleated than virgin polymer.

Secondary Crystallization

After initial crystallization, a typical nylon is metastable and will undergo further crystallization. The factors affecting this are essentially the same as those controlling the rate of primary crystallization except that the plasticizing effect of water can be much more important for secondary crystallization. Those polymers which initially crystallize most rapidly will also undergo secondary crystallization most readily, particularly if they were initially quenched to a low level of crystallinity. This effect is not typically encountered because after a given crystallization history, a rapidly crystallizing polymer will be more crystalline and more stable than one which crystallizes more slowly. However, when a rapidly crystallizing resin is severely quenched to achieve an equivalent low level of crystallinity, it will undergo secondary crystallization more readily. For any polyamide, the lower the level of crystallinity and the higher the level of absorbed water, the higher will be the rate of secondary crystallization at a given temperature. The ability to absorb water is dependent upon crystallinity and chemical structure, being higher for low-crystallinity samples and usually higher for high amide content, although polyoxamides are an exception.

A prime difference between primary and secondary crystallization is that the latter usually does not involve morphology to any great extent. Secondary crystallization is reorganization on a micro scale and often will not change the appearance of a sample unless it was quenched to a very low level of initial crystallinity with a high degree of transparency. Crystallinity changes can be detected by the usual techniques (density, infrared, x-ray) but not optically. In this connection, it is interesting to note that the majority of the studies of the kinetics of primary crystallization were performed optically, either by measurement of the rate of nucleation and growth of spherulites or by rate of increase of birefringence. Neither of these techniques would give a reliable measure of the rate of secondary crystallization.

Obviously, the degree of crystallinity achieved after both primary and secondary crystallization is an important determinant of the properties of nylon plastics. Given a particular resin, chosen for its chemical structure and molecular weight, the effect of processing variables on thermal history will determine the extent of primary crystallization. Among the factors to consider are melt temperature, hold-up time, part thickness and quench temperature. The effect of additives on crystallization may also have to be considered. Secondary crystallization will be more or less important depending upon the resin and its initial thermal history. The rate of secondary crystallization is enhanced by increasing temperature and level of absorbed water. Final equilibration of thick sections can take months under typical ambient conditions. Part stability can best be achieved by annealing but in principle the properties may differ slightly from those acquired by equilibration at ambient conditions.

LAMELLAE

Single Crystals from Solution

Numerous studies on the morphology of polymers crystallized from solution nave shown that nearly all polymers capable of crystallization precipitate as thin layer crystals termed "lamellae" (7). The interpretation of the conformation of the chains within the lamellae continues to be a controversial topic, but it is generally accepted that these lamella crystals result from folding of chains with the idealized conformation that of "adjacent reentry" illustrated in Figure 8-11.

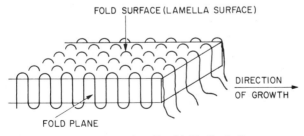

Fig. 8-11.   Illustration of folded-chain lamella with idealized adjacent reentry of chains.

Ideally, a molecule folds back on itself to form a loop which is tight and located with crystallographic regularity on the surface of the lamella. More realistically, depending on the conditions of crystallization, the folds may project with varying degrees of looseness above the surface.

When crystallized from solution, nylons can form single crystals similar to those well known from studies of polyethylene (7) but with certain distinct differences resulting from hydrogen bonding. Nylon-66 crystallized from dilute solution has received the most attention and illustrates most of the features characteristic of nylon single crystals. Nylon single crystals in the form of laterally extended lamellae similar to those of polyethylene have been observed in several studies (7, 14, 15, 16). Frequently observed in nylons are crystals with a flat ribbonlike shape (16, 17, 18). This type of crystal is often called "fibrillar," but it should be understood it is a special type of folded-chain lamella and is not to be confused with drawn fibrils. Evidence from electron microscopy (16) suggests that a laterally extended lamella results from the aggregation of ribbon crystals. A single crystal of nylon-66 grown from solution in 1,4-butanediol is shown in Figure 8-12. From assignment of the infrared 1327- and 1224-cm^{-1} bands to the chain fold itself, it was proposed (17) that chainfolding occurs with regular adjacent reentry (Figure 8-11) and the (010) planes of the unit cell (the plane of the hydrogen-bonded sheets) are the preferred fold surface. A molecule in folding must form hydrogen bonds between its adjacent segments as illustrated by the model in Figure 8-13. On initial precipitation, the folding embodies some irregularity dependent on the degree of supercooling. However, annealing can impart sufficient thermal energy to the chains to transform irregular folds to a regular conformation. Furthermore, the infrared band assignment gives evidence (17) that the chainfold is formed from an amine segment.

A feature of single crystals of most polymers is the dependence of the fold period (lamella thickness) on the temperature of crystallization. However, for a given solvent, nylon single crystals show an essentially constant lamella thickness independent of the crystallization temperature. It has been proposed (18) that this feature results from restrictions imposed by the hydrogen bonds. Whereas polyethylene with its short repeat distance (2.55 A) can fold and maintain crystallographic registry with little difficulty, the long repeat distance in nylon-66 (17.2 A) plus the necessity of forming hydrogen bonds limit the length of a molecular segment between folds to integral (or half-integral) multiples of the repeat distance. The growth mechanism postulated for a ribbon single crystal of nylon-66 is illustrated in Figure 8-13. The molecule folds on an amine unit with 3½ chemical repeat units traversing the lamella thickness. The direction of growth and thus the long dimension of the ribbon are parallel to the unit cell a-axis. The plane containing the folds at the surface of the crystal is parallel to the (001) plane of the unit cell. Thus, the linear molecular segments within the lamella are tilted with respect to the surface at an angle calculated to be 42°

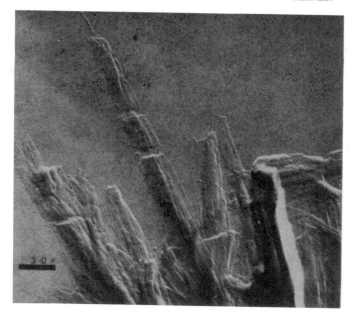

Fig. 8-12.   Nylon-66 single crystals precipitated on slowly cooling to room temperature (17).

with $46°$ being measured experimentally (15). The stabilizing influence of the hydrogen bonds in this model may explain why the thickness of a single crystal of nylon-66 (59 A) is substantially less than that of polyethylene (ca. 100 A). Thickening of the nylon crystals by annealing requires that sufficient thermal energy be added to break hydrogen bonds. In order to reform hydrogen bonds in a regular manner, the thickness must increase in multiples of the unit cell repeat unit. A tendency for doubling of the thickness of a lamella on annealing has been observed and explained (18, 19) as rearrangements of chainfolds by displacements perpendicular to the lamella surface as illustrated in Figure 8-14.

Similar growth mechanisims have been proposed for the growth of single crystals of nylons-610 and -612 (18) and probably are characteristic of the even-even series. A somewhat different chain-folding mechanism, not yet detailed, is needed for the even and the odd series such as nylon-6 (20, 21, 22), nylon-7 (23), and nylon-8 (19) where the molecular segments are normal to the surface of the lamella. The fold surface in nylon-6 appears to be parallel to the (120) plane (23).

Melt Crystallized Nylon

The infrared technique developed for identification of chainfolds (17) indicates that chainfolding in melt crystallized samples of nylon-66 is of the same type as

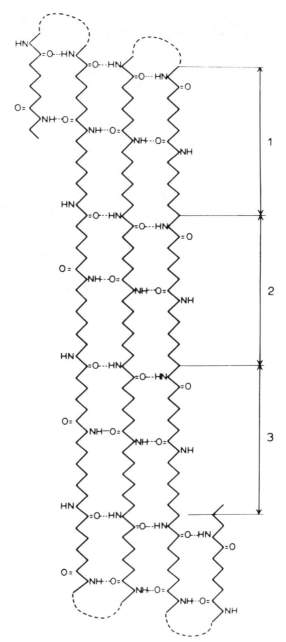

Fig. 8-13. Suggested chain-folded structure of nylon-66 with straight stems comprising 3 1/2 repeat units and the folds containing the diamine. The plane of folding contains the hydrogen bonds (18).

290

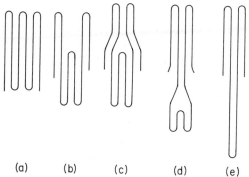

(a)   (b)   (c)   (d)   (e)

Fig. 8-14.   Proposed scheme for fold length increase leading to preferential doubling of the fold length: (a) to (e) progressive stages from single to doubled fold length (18).

in single crystals from solution. Samples which are initially quenched to low crystallinity show no regular folding; regular folds are formed on annealing. Additional infrared studies (24) show that the regularity of chainfolding is related to the type of spherulite growth and is discussed in the following section.

SPHERULITES

Occurrence

Spherulitic structure is characteristic of crystallization in an environment free from stress on the melt including both mechanical stress from agitation of the melt and thermal stress from strong temperature gradients. Thus in the carefully controlled environment of slow crystallization on a microscope slide, most crystalline polymers form spherulites. On the other hand, in commercial processing such as extrusion and molding, considerable stresses of both types are present and many crystalline polymers show a wide variety of nonspherulitic oriented structures (25). The nylons differ from most other crystalline polymers in that under the usual conditions of processing, the resulting structure is much more spherulitic. The reason is, we think, that the melt tends to supercool to a much greater degree. By the time crystallization begins, the stresses have been reduced greatly from the initial values and the environment approaches a quiescent condition. In contrast, a polymer such as polyoxymethylene crystallizes almost immediately, even during filling of a mold, resulting in a substantial fraction of highly oriented, nonspherulitic material.

The formation of a spherulite begins at a nucleus, either a foreign particle or a small aggregate of crystallized molecules, and proceeds in all directions until the supply of melt is depleted, generally from the intersection of boundaries of adjacent spherulites.

Birefringence

The characterization of spherulites in nylons is based on the nature of their birefringence. Birefringence results when the index of refraction is different when light passes through a crystal in different directions. It is characterized by a maximum and a minimum index which are mutually perpendicular. These directions are denoted the major optic axis and minor optic axis, respectively. A crystal of nylon is inherently birefringent by virtue of the anisotropy resulting from the alignment of the hydrogen bonds.

Birefringence is detected by means of crossed polarizers in the light path of an optical microscope. The polarizer is placed between the light source and the sample and the analyzer between the sample and the eyepiece. Without going into complicated theory, the observed optical effects will be explained in terms of rotation of plane polarized light. No rotation of the plane of light from the polarizer will occur if either the major or minor axis is parallel or perpendicular to the plane of the polarizer. Therefore, if the analyzer is crossed $90°$ with respect to the polarizer, no light will be transmitted if either the major or minor axis is parallel or perpendicular to the polarizer. Light will pass through the analyzer only if the principal axes are at angles other than $0°$ or $90°$ to the polarizer.

If a spherulite grows with a certain axis of the crystals parallel to its radius, as is the usual case, the spherulite itself will be birefringent. For example, consider the case in which the major axis of each polymer crystal is parallel to the radius of the spherulite. Under crossed polarizers there will be a diameter of the spherulite parallel to the plane of the polarizer; hence, for this diameter the major axes of the polymer crystals are parallel to the plane of the polarizer. Another diameter, normal to the first, has the major axes of the polymer crystals perpendicular to the plane of the polarizer and parallel to the plane of the analyzer. No light will be passed by the analyzer along these two diameters; however, light will be transmitted for the intermediate diameters. This is the idealized situation. In practice there will be roughly equal light and dark sectors yielding the familiar "Maltese cross" appearance. By the same arguments presented above, a similar Maltese cross effect will result if the radius of the spherulite is parallel to the direction of the minor axes of the crystals. It is important therefore to note that, in theory, birefringent spherulites will appear exactly the same when observed under crossed polarizers if the radial direction is parallel to either the major axis or the minor axis of the crystals.

In order to determine if the radius of a spherulite corresponds to a major or minor axis of crystals, an additional device such as a "first-order red wave plate" must be inserted between the sample and the analyzer. This wave plate is a crystal of quartz so cut that a beam of polarized white light is broken into its spectral components with different colors being rotated different amounts with respect to the optic axis of the quartz slice. In use, the wave plate is oriented so that with no sample between the crossed polarizers, only the red portion of the spectrum is rotated 90° and thus passed by the analyzer. Therefore, the black background becomes red. If the radius of a spherulite is parallel to the *major* axes of its component crystals, the arms of the Maltese cross become red; the quadrants parallel to the optic axis of the wave plate become blue and the alternate quadrants yellow. If the radius of the spherulite corresponds to the *minor* axes of the crystals, the quadrants parallel to the axis of the wave plate are yellow and the alternate quadrants blue.

If a spherulite grows with its radius parallel to the major axis of its component crystals, it is called a "positively birefringent spherulite" or simply a "positive spherulite." If the spherulite radius is parallel to a minor axis, it is termed a "negatively birefringent spherulite" or a "negative spherulite." As described above, these two types appear the same under crossed polarizers; a device which breaks the polarized light into its spectral components is necessary to observe this distinction. Figure 8-15 shows an optical micrograph of a compression molded specimen of nylon-66 with crossed polarizers. The photograph on the jacket is of the same specimen. Use of a first-order wave plate reveals that the spherulites are positively birefringent.

Types of Birefringent Structures

While the two types of spherulites described above are the simplest to identify, four different types of spherulites have been characterized in nylons from optical microscopic studies although not all types occur in all nylons. These are: positively birefringent, negatively birefringent, zero birefringent, and spherulitic aggregates. These aggregates are highly birefringent but have confusing optical properties. In addition to these well-characterized spherulitic growths, other less well-defined structures are observed in melt crystallized specimens. These include transcrystalline "spherulite brushes," grainy textured material, and "amorphous" structure. The most extensive studies of spherulites have been made on nylon-66 (24, 26, 27, 28, 29, 31), which shows all of these types and illustrates the nature of spherulitic structure found in nearly all nylons.

The type of spherulite formed by crystallization from the melt depends primarily on the temperature of crystallization. Once formed, annealing below the melting point and even through the Brill transition (see Chapter 9) will not change the sign of the birefringence, although its magnitude may vary somewhat. All types of spherulites are crystalline; differences in the birefringence are attributed to differences in the morphology, that is, the type and orientation of the spherulites and lamellae. Positive spherulites (that is, positively birefringent spherulites) are the most common type found in the nylons and the only type observed in nylon-6.

Positive spherulites of nylon-66 form exclusively when the melt is cooled below 250°C. Positive spherulites also grow at higher temperatures but in combination with other forms. Generally a straight extinction cross is observed

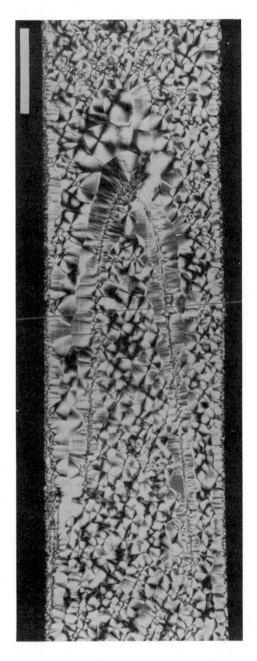

Fig. 8-15. Optical micrograph of compression molded nylon-66. Note transcrystalline spherulitic brushes at surface and row structures in the interior. Spherulites are positively birefringent. Bar scale is 0.01 in. (0.25 mm).

with crossed polarizers but at crystallization temperatures upward toward 250°C (28) or with a gold substrate (27), the extinction can show zig-zags in the crosses giving the appearance of a ringed structure similar to that shown in Figure 8-16. The ring spacing increases with crystallization temperature and parallels an increasing coarseness in the spherulitic texture (28). Experimental data from infrared (17, 14), x-ray diffraction (27), and electron microscopy (28), suggest that positive spherulites are composed of fine fibrils (folded-chain ribbons) growing along the crystallographic *a*-axis as in Figure 8-13. Thus the *a*-axis is parallel to the spherulite radius. The infrared spectrum (24) indicates fewer folds or less regular folding than in single crystal preparations; this is attributed to extensive branching of the fibrils. The ringed appearance of some spherulites may be due to a cooperative twisting of ribbons along the growth axis.

Negative spherulites of nylon-66 grow between 250°C and 270°C and are always accompanied by spherulitic aggregates. When held between 250 and 265°C, positive spherulites melt entirely and recrystallize to form negatively birefringent spherulites (26). Nucleation of these negative spherulites results from polymer aggregates or crystals and not from foreign particles. Negative spherulites grow faster than positive spherulites at a given temperature and have higher optical melting points (27). X-ray diffraction studies (27) show that negative spherulites are highly crystalline but have very little preferred orientation of the unit cells, although some evidence is found (28) for the crystallographic *b*-axis to be parallel to the spherulite radius. Infrared data (24) indicate very well-developed chain folding comparable with that in single crystals. Negative spherulites seem to be composed of lamella type crystal units rather than the branched fibrils found in positive spherulites; the lamella planes are roughly parallel to the spherulite radius. Ringed negative spherulites are not observed in nylon-66 but are the only type observed in nylon-11 (32) as shown in Figure 8-16.

Nonbirefringent spherulites grow in two temperature ranges which are at the limits of the growth temperatures within which negative spherulites can form. Because they are optically isotropic, they appear dark under crossed polarizers. One type of nonbirefringent spherulite of very fine texture crystallizes near 250°C. Since this is the temperature boundary below which positive spherulites are formed and above which negative spherulites develop, it has been suggested (30) that this type of non-birefringent spherulite results from a mixture of the growth patterns of positive and negative types. Nonbirefringent spherulites of larger diameter grow near the optical melting point at 265°C. No evidence of any preferred orientation of the unit cells has been found in nonbirefringent spherulites. They grow linearly with respect to time and do not appear to be an anomaly resulting from disruption of previously formed spherulites.

Spherulitic aggregates grow simultaneously with negative spherulites and at a faster rate of growth. They are strongly birefringent but have complex optical

Fig. 8-16.   Negatively birefringent, ringed spherulites of nylon-11 (32).

properties. Their appearance in the optical microscope varies with the thickness of the film in which they are grown but generally is similar to the positive spherulitic aggregate as shown in Figure 8-17. X-ray data from thin film specimens (27) indicate that the crystallographic b-axis is parallel to the radius and the (002) plane tends to be parallel to the surface. In thicker films, the orientation is of a similar nature but varies from place to place within an aggregate, even along a single radius. This variation of orientation over short distances is responsible for the complex optical properties.

   Transcrystalline "spherulitic brushes" and "row structures" are less well-defined structures. The characterization of the four general types of spherulites described above is based on studies of material grown under carefully controlled laboratory conditions, generally in the form of thin films. However, with less uniformity of thermal environment as is found in most commercial processing operations, structures in addition to well-defined spherulites are found although the spherulitic texture nearly always predominates. One of these nonspherulitic structures is a highly birefringent surface layer such as observed in Figure 8-15. This is typical of the "skin" found in extruded filaments and compression molded specimens. Similar structures are sometimes found in the interior of molded specimens as in Figure 8-15. These structures result from a localization of nuclei that greatly distort spherulitic growth. X-ray measurements of the transcrystalline skin in an extruded filament (29) show that this layer consists of

Fig. 8-17. Nylon-1010 spherulitic aggregates formed at 193°C following heating of the polymer at 203°C for 1/4 hr. 180X (28).

a closely spaced array of radial sections of spherulites resulting from numerous nucleation sites at the surface. Geometrical considerations prevent spherically symmetrical development and restrict growth to brushlike shapes. These are generally positively birefrigent like the spherulites adjoining them. A cross section of a typical commercial nylon-610 filament is shown in Figure 8-18. Transcrystalline row structures may result from the lining up of nuclei within the interior of a specimen. This morphology probably results from a combination of the flow pattern, supercooled melt and the presence of solid impurities. Such "row structures" are often seen in rod stock of large diameter.

Granular, nonspherulitic structures are frequently observed and seem to be less physically stable than spherulites. This structure appears as a grainy background and is not easily characterized optically. A study of the growth of negative spherulites (27) revealed that their growth was terminated by the formation of this granular structure thereby limiting the size of the spherulites. In contrast, the positively birefringent spherical aggregates grew at the expense of the granular material. Figure 8-19 shows a cross section of an injection molded bar containing an example of granular structure.

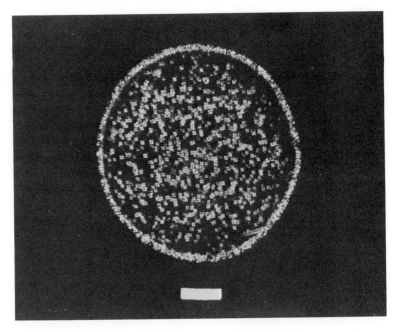

Fig. 8-18.   Cross section of oriented filament of nylon-610 showing spherulites embedded in an amorphous matrix with transcrystalline structure and small spherulites at surface.

"Amorphous" structures are formed if the melt is quenched to a low temperature, say -20°C. Although this material appears amorphous in the optical microscope, x-ray data show it to be crystalline but of a much lower degree than in positive spherulites. This is attributed to the lower thermal mobility of the molecules during the process of crystallization from the supercooled melt. Infrared studies (17, 24) show evidence of crystallinity but no regular chain folding in nonspherultic quenched films. However, annealing at temperatures above 180°C results in formation of regular folds although no change in optical appearance occurs. Heating in the range of 258° to 260°C, which is above the softening temperature, gives evidence for substantial rearrangement of fold conformations, possibly associated with the formation of negative spherulites. A similar type of nonbirefringent, nonspherulitic structure is seen on the surface of injection moldings as in Figure 8-19. The melt is effectively quenched on a cold mold wall. Since this initial deposit of melt hinders removal of heat from the interior mass of polymer, the remainder of the molding crystallizes principally in the form of positive spherulites.

The above discussion outlines the general types of spherulitic morphology in nylon-66 and the nature of their birefringence. Comprehensive studies are

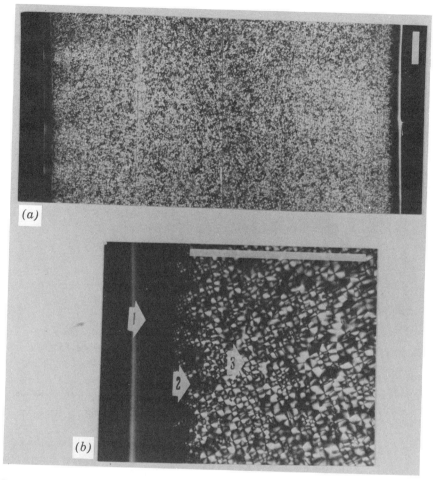

Fig. 8-19. (*a*) Cross section of injection molded bar of nylon-66. (*b*) Enlargement of surface region shows, 2: "amorphous skin", 2: granular structure and, 3: spherulites. Bar scale is 0.01 in. (0.25mm).

available in the literature on even-even (28), even-odd (32), odd-even (33), even (32), and odd nylons (32).

EFFECT ON PHYSICAL PROPERTIES

The variation in properties obtained from a given nylon resin depends primarily on the complex interrelationship between crystallinity and morphology. A series of experiments (34) with careful control of molding conditions and environment with nylon-66 has shown how mechanical properties depend on spherulite size in samples of equivalent crystallinity. Property differences associated with spherulite size are most pronounced in dry samples at low (23°C) temperatures and decrease or disappear with increase in water content or temperature. Dry samples with small spherulites show higher flexural modulus, higher upper yield stress, and lower ultimate elongation than do samples with large spherulites. Figure 8-20 shows the relationship between yield point and spherulite diameter for dry samples of approximately the same crystallinity (45 to 52%). As the spherulite size decreases, the yield point increases. Figure 8-21 shows the interrelationship between yield point and crystallinity for dry samples with and without visible spherulites. The yield points of spherulitic films are substantially higher than those without visible spherulites independent of the crystallinity. Use of reprocessed resin and purposeful addition of a nucleating agent to achieve a given spherulite size had a similar effect on the physical properties of dry polymer.

ORIENTATION

The concept of orientation like that of structure can be applied to several dimensional levels. Thus crystallinity is influenced by the orientation of

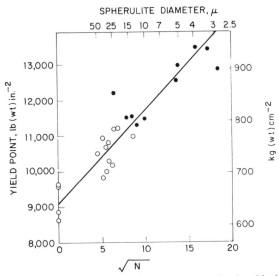

Fig. 8-20. Effect of spherulite size on the yield point of nylon-66: (o) compression-molded films; (●) injection-molded bars. N = number of spherulite boundaries per millimeter, (34).

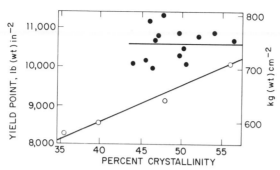

Fig. 8-21.   The effect of visible spherulites on the yield point of nylon-66 versus percent crystallinity relationship: (o) films without visible spherulities; (●) films with 30 to 65 μ spherulites (34).

hydrogen bonds with respect to the chain molecules. Spherulites and row structures result from various orientations of lamellae. However, the levels of orientation of most importance are those which explain the great improvement in physical properties which result from deformation of a nylon polymer. Representative of important commercial examples are fibers and filaments which constitute one type of orientation and strap which represents another type.

Filamentary or Uniaxial Orientation

Oriented fibers and filaments represent an area in which improvement in properties by controlled deformation has been developed to a high art. The tensile strength (Figure 8-22) and modulus increase greatly as the crystals are

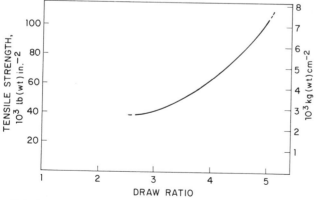

Fig. 8-22.   Relationship of tensile strength and orientation in nylon-66 (35).

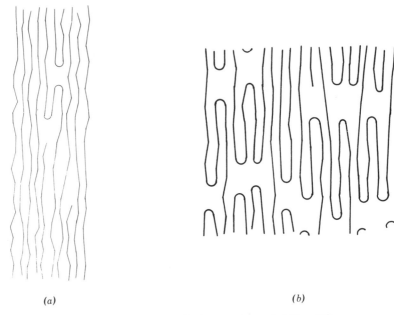

(a)   (b)

Fig. 8-23.   Model of (a) drawn and (b) drawn and annealed fiber (36).

oriented so that the chains become parallel to the fiber axis. This relationship between tensile strength and draw ratio is easily understood in terms of the classical two-phase model of morphology. X-ray diffraction measurements show that, as the draw ratio is increased, more chains become aligned parallel to the fiber axis and bring more covalent bonds into position for load bearing. However, the sensitivity of physical properties of nylon fibers to the spinning conditions and post treatments clearly show that more complex types of morphology are involved. It has been proposed (36) that initial drawing produces a structure in which the chains are highly extended and essentially parallel to the axis with relatively few chain folds. The degree of uniformity of the hydrogen bond directions is low, resulting in low crystallinity. Subsequent heat treatment results in breaking of hydrogen bonds. Enough mobility is thus provided to allow the chains to fold into small folded-chain aggregates similar to lamellae but with a high proportion of chains connecting one fold-chain unit to another. These concepts are illustrated in Figure 8-23. An alternate model (37) for the initial fiber consists of a highly folded arrangement of chains with much interpenetration of folded units along the chain axis as illustrated in Figure 8-24. This structure on annealing becomes regularized into folded-chain lamellae joined by tie molecules. Clearly much more work is needed to clarify the interrelationships between drawing, annealing, structure, and properties of fibers and filaments.

Fig. 8-24. Alternative model of drawn fiber (37).

Multiaxial Orientation

Strapping of nylon has a different type of orientation from that in fibers. Fibers are characterized as "uniaxial" because one axis, the chain axis of the crystal, is parallel to the fiber axis and the other crystal axes have cylindrical symmetry. It is possible by means of proper control of deformation processes to further refine the structure into "multiaxial" orientation. An example is nylon-66 strap (38).

By rolling a nearly rectangular billet in the solid state (at elevated temperatures), a type of multiaxial orientation termed "uniplanar-axial" can be created. This type of orientation is illustrated schematically in Figure 8-25. The chain axes are parallel both to the surface and the roll direction (strap length). This orientation of the covalent bonds of the chain axes in the longitudinal direction imparts desirable high tensile strength and modulus. In addition, the (010) hydrogen bond planes of the unit cells are parallel to the surface. The planar orientation combined with the axial orientation causes the hydrogen bonds to be oriented in the transverse direction. This provides a desirable strength and resistance to splitting in this direction without a sacrifice in tensile

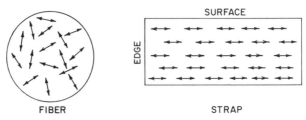

Fig. 8-25. Schematic illustration showing difference in orientation of hydrogen-bond directions (arrows) in highly oriented fiber and strap viewed in cross section. Chains are perpendicular to page.

strength resulting from orientation of the covalent bonds. In fact, this planar orientation also improves tensile strength in the longitudinal direction. The sensitivity of the tensile strength to the molecular orientation in strap is shown in Figure 8-26. The "degree of orientation" is expressed in terms of the width at half-maximum of x-ray diffraction peaks as defined in Ref. 38. Orientation increases as the width narrows. As in the case of fibers, the features of morphology associated with the molecular orientation are not well understood. Most likely folded-chain lamellae of the original billet are converted into some sort of substantially extended arrangement of molecules.

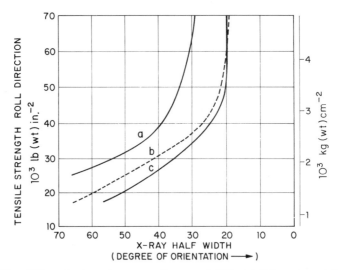

Fig. 8-26. Effect of orientation on tensile strength (38): (*a*) Orientation of (010) planes about roll direction. (*b*) Orientation of chains: roll direction versus transverse direction. (*c*) Orientation of chains: roll direction versus thickness direction.

REFERENCES

1. Kinoshita, Y., *Makromol. Chem.* **33**, 1 (1959).
2. Bunn, C.W., and E. V. Garner, *Proc. Rov. Soc. Lond.* **189A**, 39 (1947).
3. Keller, A., and A. Maradudin, *J. Phys. Chem. Solids* **2**, 301 (1957).
4. Holmes, D. R., C. W. Bunn, and D. J. Smith, *J. Polym. Sci.* **17**, 159 (1955).
5. Miyake, A., *J. Polym. Sci.* **44**, 223 (1960).
6. Trifan, D. S., and J. F. Terenzi, *J. Polym. Sci.* **28**, 443 (1958)
7. Geil, P. H., *Polymer Single Crystals,* Interscience, New York, 1963.
8. Statton, W. O., *J. Polym Sci.* **18C**, 33 (1967).
9. Campbell, G. A., *J. Polym Sci.* **7B**, 629 (1969).
10. Starkweather, H. W., Jr., and R. E. Moynihan, *J. Polym. Sci.* **22**, 363 (1956).
11. Starkweather, H. W., Jr., G. E. Moore, J. E. Hansen, T. M. Roder, and R. E. Brooks, *J. Polym. Sci.* **21**, 189 (1956).
12. Burnett, B. B., and W. F. McDevit, *J. Appl. Phys.* **28**, 1101 (1957).
13. McLaren, J. V., *Polymer* **4**, 175 (1963).
14. Badami, D. V., and P. H. Harris, *J. Polym. Sci.* **41**, 540 (1959).
15. Holland, V. F., *Makromol. Chem.* **71**, 204 (1959).
16. Sakaoku, K., H. Iwashima, M. Miyanoki and K. Nagumo, *Rept. Progr. Polym. Phys. (Japan)* **8**, 87 (1965).
17. Koenig, J. L., and M. C. Agboatwalla, *J. Macromol. Sci.* **B2**, 391 (1968).
18. Dreyfuss, P., and A. Keller, *J. Macromol. Sci.* **B4**, 811 (1970).
19. Dreyfuss, P., and A. Keller, *J. Polym. Sci.* **B8**, 253 (1970).
20. Geil, P. H., *J. Polym. Sci.* **44**, 449 (1960).
21. Nagai, E., and M. Ogawa, *J. Polym. Sci.* **3B**, 295 (1965).
22. Ogawa, M., T. Ota, D. Yoshizaki, and E. Nagai, *J. Polym. Sci.* **1B**, 57 (1963).
23. Holland, V. F., Paper presented at Elec. Micro. Soc. Am. Meeting, Milwaukee, August, 1960.
24. Cannon, C. G., and P. H. Harris, *J. Macromol. Sci.* **B3**, 357 (1969).
25. Clark, E. S., and C. A. Garber, *Intern. J. Polymeric Mater.* **1**, 31 (1971).
26. Khoury, F., *J. Polym. Sci.* **33**, 389 (1958).
27. Mann, J., and L. Roldan-Gonzalez, *J. Polym. Sci.* **60**, 1 (1962).
28. Magill, J. H., *J. Polym. Sci., Part A-2*, **4**, 243 (1966).
29. Barriault, R. J., and L. F. Gronholz, *J. Polym. Sci.* **18**, 393 (1955).
30. Cannon, C. G., F. C. Chappel and J. I. Tidmarsh, *J. Textile Inst.* **54**, T210 (1963).
31. Lindegren, C. R., *J. Polym. Sci.* **50**, 181 (1961).
32. Magill, J. H., *J. Polym. Sci., Part A-2*, **7**, 123 (1969).
33. Magill, J. H., *J. Polym. Sci.* **3A**, 1195 (1965).
34. Starkweather, H. W., Jr., and R. E. Brooks, *J. Appl. Polym. Sci.* **1**, 236 (1959).
35. Statton, W. D., *J. Polym. Sci.* **20C**, 117 (1967).
36. Dismore, P. F., and W. O. Statton, *J. Polym. Sci.* **13C**, 133 (1966).
37. Beresford, D. R., and H. Bevan, *Polymer* **5**, 247 (1964).
38. Dunnington, G. B., and R. T. Fields (to E. I. du Pont de Nemours and Co.), U.S. Patent 3,354,023 (Nov. 21, 1967).

CHAPTER 9

Transitions and Relaxations in Nylons

HOWARD W. STARKWEATHER, JR.

MELTING

Techniques for Measuring the Melting Point

From a thermodynamic viewpoint, the melting point is the temperature at which the crystal and the melt are in equilibrium. For theoretical considerations, one wishes to know the temperature at which the last vestige of crystallinity disappears. This can be obtained from the disappearance of birefringence in a thin sample as in the ASTM hot stage method (D 2117-64) or from the disappearance of the crystalline x-ray diffraction pattern. These values also correspond to the minimum temperature for melt fabrication.

Values based on differential thermal analysis are frequently reported (1). It should be understood that in a material having a fairly broad melting range, the temperature of a DTA melting peak may not be the thermodynamic melting point. The upper limit of the DTA peak, where the curve returns to a linear base line may give a good indication of the melting point, but it is sometimes hard to determine.

The results of a series of measurements of the melting points of four polyamides made in several laboratories using a number of techniques are summarized in Table 9-1 (2). In the Fisher-Johns method, a specimen is surrounded by a few drops of silicone oil, and a cover glass is placed over it. When the specimen melts, it no longer supports the cover glass, and the meniscus of the silicone oil begins to move. The temperature of this mechanical softening is a few degrees below the true melting point, but it is quite reproducible. For this reason, the Fisher-Johns method is included in the ASTM specifications for nylon plastics (D 789-66). Capillary methods depend on observing the disappearance of sharp edges. They are governed by the same factors as the Fisher-Johns method and are most suitable for materials which melt sharply to give liquids of relatively low viscosity.

Melting points obtained with the Kofler hot stage show more variety among laboratories. The higher values correspond fairly closely to the x-ray melting

Table 9-1. Comparison of the melting points of nylon resins determined by several techniques

Method	Melting point ($^{\circ}$C)			
	Nylon-11	Nylon-6	Nylon-610	Nylon-66
Fisher-Johns	191 (188-194)	220 (218-226)	219 (216-223)	260 (256-263)
Capillary	190 (188-193)	219 (214-223)	217 (212-221)	259 (254-266)
Kofler Hot Stage	191 (185-196)	221 (214-230)	220 (212-230)	261 (250-268)
X-Ray	192	226	227	267
DTA, start	188	220	221	261
peak	192	224	224	264
end	194	228	226	269

points and the upper ends of the DTA melting peaks.

None of the methods permits one to distinguish between nylon-6 and nylon-610 without reference to other properties, such as the density.

Effect of Chemical Structure

Reported values for the melting points of a large number of polyamides have been collected by Miller (3). When different values are given for a particular polymer, greater weight should be given to the higher values. In a homologous series of linear aliphatic polyamides of regularly repeating structure, the melting point increases as the concentration of amide groups is increased. Some of these data are shown in Figure 9-1. For a given concentration of amide groups,

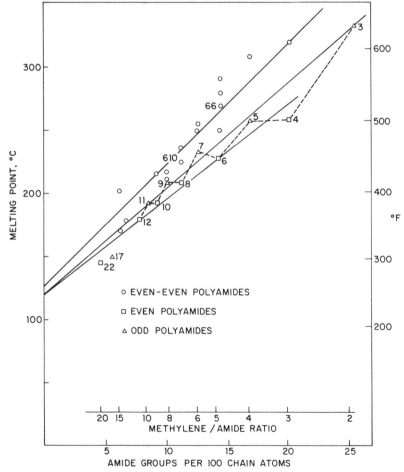

Fig. 9-1.   Melting point versus concentration of amide groups.

even-even polyamides, such as nylons-66 and -610, generally have the highest melting points. The melting points are significantly lower if the diamine and/or diacid from which the polyamide is derived has an odd number of carbon atoms. Among polyamides derived from ω-aminocarboxylic acids or the corresponding lactams, those having an even number of atoms (e.g. nylon-7), have higher melting points than those having an odd number of atoms (e.g., nylon-6). Extrapolation to zero amide groups gives a melting point considerably lower than that of linear polyethylene. This means that the methylene sequences in polyamides do not enjoy optimum packing.

The melting point is reduced by methyl branches or N-alkylation and increased by the incorporation of *para*-phenylene linkages. *Meta*-phenylene linkages, however, lower the melting point (3). These effects are illustrated in Table 9-2.

Theoretically, the melting points of copolymers in which only one kind of unit can crystallize vary according to the formula (4),

$$\frac{1}{T_m} - \frac{1}{T_m{}^0} = -\frac{R}{\Delta H_m} \ln X$$

where T_m and $T_m{}^0$ are the melting points in degrees Kelvin of the copolymer

Table 9-2. Effect on structural features on the melting points of nylon resin[a]

Nylon	Melting Point ($^\circ$C)
Branching and N-alkylation	
6	228
6-Methyl-6	185
11	194
2-Methyl-11	130
N-Methyl-11	80
66	269
N-Methyl-66	145
N, N'-Dimethyl-66	75
Para-and *meta*-phenylene linkages	
66	269
6T	371
6I	220
MXD6	246
610	227
PXD10	300

[a]Abbreviations are from Table 1-3, p. 7.

and homopolymer, ΔH_m is the heat of fusion, and X is the probability that a given crystallizable unit is followed by another one. When cocrystallization is possible, as for example when some of the adipic acid units in nylons-66 are replaced by terephthalic acid, the depression of the melting point is smaller(5). Many examples have been found for the isomorphous substitution of a p-phenylene or cyclohexyl linkage for a tetramethylene sequence (6,8).

Heat and Entropy of Fusion

The heat of fusion of a polymer can be determined by calorimetry, from the melting points of copolymers, from the effect of a diluent on the melting point, or from the dependence of the melting point on the molecular weight (4,9). The entropy of fusion is given by $\Delta H_m/T_m$. Unfortunately, many of the reported values from the various methods are in very poor agreement (9,10). The heat of fusion for nylons-66 and -6 is probably about 45 cal/g (9) for the crystalline fraction which is normally 40 to 55% by weight as calculated from the density (11). Thus, the measured heat of fusion is about 18 to 25 cal/g. The fact that the melting points of polyamides are higher than those of polyethylene and the corresponding polyesters is due not to high heats of fusion but to low entropies of fusion. Calorimetric studies (9) and infrared spectroscopy (12, 13) indicate that a large fraction of amide groups are hydrogen bonded above the melting point. The relatively low melt viscosity of nylon indicates that these groups are able to change bonding partners rather freely, but only a few of these changes need to be occurring at a particular moment. The consequence is that nylon melts are relatively well ordered and have relatively low entropies. Part of the entropy of fusion is due to the increase in volume. The remainder is attributable to rotation about the primary bonds in the polymer chain. In nylon-66, this rotational portion of the entropy of fusion is equivalent to that for 7.8 CH_2 groups in polyethylene (14). This can be understood if the amide groups are rigid (9) and rotation occurs only about bonds connecting two CH_2 groups, a conclusion which is supported by infrared spectroscopy (13).

Freezing

Polyamides, like other polymers, are subject to supercooling. In practical situations, the freezing points are frequently about $30°$ lower than the melting points (15). The following solidification temperatures have been reported (16): nylon-6, 170 to 190°C; nylon-66, 215 to 240°C; nylon-610, 175 to 195°C. In laboratory experiments, crystallization has been carried out much closer to the melting point, for example, at 205 to 215°C in the case of nylon-6 (17).

The Brill Temperature

In an x-ray diffraction pattern obtained at room temperature from a well-annealed sample of nylon-66, there are two strong peaks, at about $20°$ and $24°$ in 2θ when copper $K\alpha$ radiation is used with a nickel filter. As the temperature is increased, these peaks approach each other and become one at about $175°C$ (18). This point, which has been called the Brill temperature, varies among the even-even polyamides (19). The crystal structure at equilibrium is triclinic below and pseudohexagonal above this temperature. Nylon-66 which has been quenched quickly from the melt to room temperature has a single strong diffraction peak (14) and a structure which is distinct from the equilibrium structures. To form the triclinic structure, the polymer must be annealed above the Brill temperature. Thereafter, the changes are reversible (Figure 9-2). This suggests that the crystallographic changes are connected with the onset of molecular motion and some freedom for the amide groups to change hydrogen bonding partners. This conclusion is supported by studies of nuclear magnetic resonance (19,20).

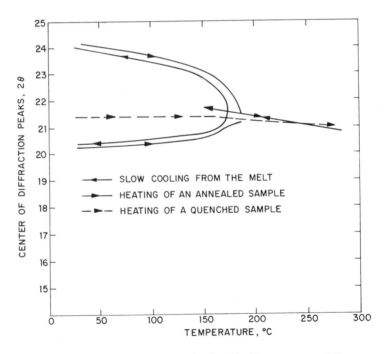

Fig. 9-2.   Changes in the diffraction pattern of nylon-66 with temperature (14).

RELAXATIONS

Experimental Techniques

Viscoelastic relaxations correspond to the onset of various types of internal motion with increasing temperature. They occur more rapidly as the temperature is increased and are observed at higher temperatures in high frequency experiments. These motions are detected through their interactions with mechanical, electrical, or magnetic fields. The techniques have been reviewed by McCrum et al. (21).

Mechanical experiments include creep, in which the stress or load is constant and the deformation is studied as a function of time, and stress relaxation, in which the strain is constant, and the diminution of stress with time is followed. In many respects, dynamic or sinusoidal mechanical experiments are among the most useful for studying relaxations. The torsion pendulum has been widely employed for this purpose. In this experiment, the square of the frequency is proportional to the torsion or shear modulus, G, and the internal friction or mechanical loss is related to the rate at which the amplitude of the oscillations decreases. The logarithmic decrement, Δ, is the natural logarithm of the ratio of the amplitudes of two successive oscillations. With this technique, the frequency is usually near one cycle per second.

Dielectric measurements are closely related to dynamic mechanical techniques. It is easier to vary the frequency of the measurement, but the method depends on the presence of electrical dipoles in the sample. When the material passes through a relaxation by increasing the temperature or decreasing the frequency and the modulus decreases, the dielectric constant increases, and the mechanical or electrical loss has a peak.

The internal motions can also be studied by means of nuclear magnetic resonance which can sometimes provide additional information about their mechanisms (19, 20).

The α, β, and γ Relaxations

The dependence of the logarithmic decrement on temperature obtained with a torsion pendulum on a sample of nylon-66 containing about 0.2% moisture is shown in Figure 9-3 (22). There are three peaks, labeled α, β and γ at 50, -80, and -140°C. The β peak has a secondary maximum at -60°C. In a dry sample (0.1% water) that had been slowly cooled there was no β peak, and the γ peak occurred at about -125°C (Figure 9-4). However, a peak was present at -55°C in a dry, quenched sample. This peak was removed by annealing at a temperature above the α peak.

It is seen from Figure 9-5 that when the water content of a slow-cooled specimen is increased, a peak appears at -70°C and shifts gradually to lower

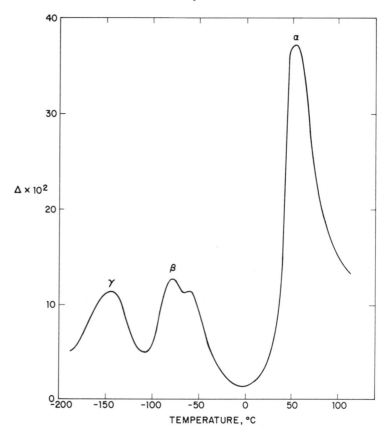

Fig. 9-3.   Internal friction of slightly wet nylon-66.

temperatures. Thus, there are two kinds of β peaks. One is associated with the presence of water. The other is related to a structural characteristic which is present in quenched specimens but not in slow-cooled or annealed specimens.

The γ relaxation is thought to be related to similar processes in other polymers that have short sequences of methylene groups (23). However, there must be some involvement of the amide groups since the relaxation affects the dielectric properties (21) and is somewhat modified by the presence of water (22). There is evidence that this relaxation is related to the presence of chain folds (24).

The α relaxation is believed to involve the motion of longer chain segments in the amorphous portions of the polymer. On the basis of dielectric data on cross-linked samples, Boyd concluded that these segments contain about 15 amide groups (25).

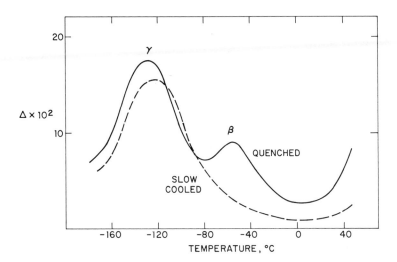

Fig. 9-4. Effect of thermal history on the low-temperature internal friction of dry nylon-66 (22).

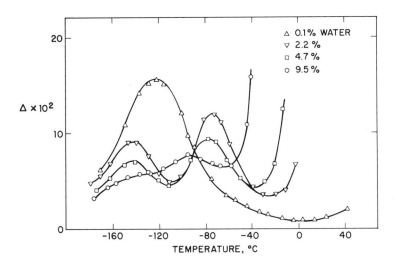

Fig. 9-5. Effect of water content on the low-temperature internal friction of nylon-66 (22).

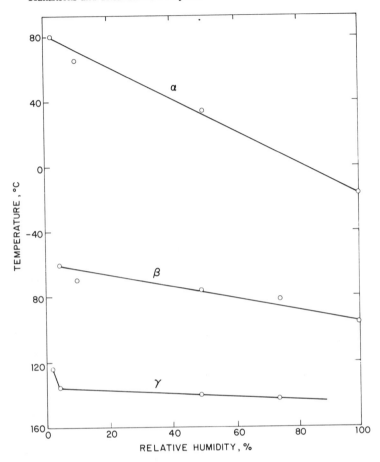

Fig. 9-6.   Dependence on relative humidity of the relaxation temperatures in nylon-66.

Effect of Water

The α, β, and γ relaxations all shift to lower temperatures as the water content is increased (Figure 9-6) (22). The effect is greatest for the α relaxation, from 80°C for an almost dry sample of nylon-66 to -15°C for a sample saturated with water. The β relaxation shifts from -60 to -95°C over the range of humidity at which it is observed. With the addition of a small amount of water to almost dry nylon-66, the temperature of the γ relaxation was reduced from -125 to -135°C, but there was only a small additional change at higher humidities.

It can be seen from Figure 9-5 that the height of the β peak first increases and then decreases as the water content is increased. This relaxation may be due to the onset of motion of water from one sorption site to another, a process which depends on the presence of both occupied and unoccupied sites. Thus, the amount of motion would be greatest at an intermediate water content.

Woodward et al. (26) studied the dynamic mechanical properties of nylon by means of resonant transverse vibrations in cylindrical rods. The frequencies varied from 100 to 2000 cycles per second. In a series of samples of nylon-66 containing from 0 to 6.4% water, the height of the γ peak decreased by half with increasing water content. Below the γ relaxation, the modulus was almost independent of the water content. Between the γ and β relaxations, the modulus increased with increasing water content. This is in contrast with the plasticizing effect of water at room temperature. Apparently, at low temperatures, water forms bonds between chain segments which are sufficiently stable to produce an increase in modulus.

Effect of Orientation

The viscoelastic properties of oriented crystalline polymers are highly aniso-tropic. In the case of nylon-66, this is seen most clearly in samples which have been oriented by uniaxial rolling (27). In this type of sample, the polymer chains are preferentially oriented in the direction of rolling (machine direction), and the hydrogen bonds are oriented in the transverse direction. The tensile modulus in these two directions was determined as a function of temperature at three levels of moisture.

At a given water content, the α relaxation occurs at the same temperature in both directions, but the decrease in modulus is greater in the machine direction. The modulus is higher in the machine direction at temperatures below the α relaxation and higher in the transverse direction above the relaxation. This behavior was interpreted in terms of crystals which are larger in the direction of the hydrogen bonds than in the direction of the polymer chains (27). Since the α relaxation is related to the amorphous portions, it is most prominent in the machine direction. The expansion due to the sorption of water is also greater in the machine direction than in the transverse direction (27).

At temperatures below the α relaxation, there is another kind of anisotropy associated with variations in the water content. In the machine direction, the tensile modulus at -45°C increases with increasing water content (27) just as it does in the unoriented nylon (26). This is shown in Figure 9-7. In this range, the modulus increases gradually with decreasing temperature to an extent which increases with increasing water content. In the transverse direction, the modulus at -45°C is almost independent of the water content. Between -45 and 0°C, the moduli of a dry sample and one conditioned to 50% relative humidity (R.H.) are much less dependent on temperature in the transverse direction (Figure 9-8) than in the machine direction. It has been suggested (27) that the β relaxation may occur in the machine direction and not in the transverse direction.

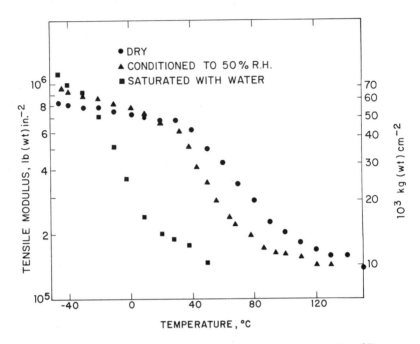

Fig. 9-7. Tensile modulus of oriented nylon-66 in the machine direction (27).

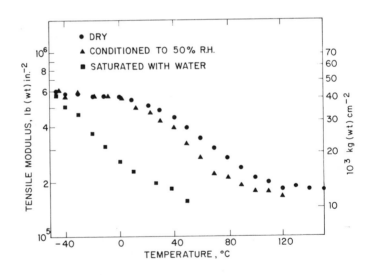

Fig. 9-8. Tensile modulus of oriented nylon-66 in the transverse direction (27).

318

SIGNIFICANCE OF RELAXATION PHENOMENA FOR PHYSICAL PROPERTIES

Water, Temperature, and Time as Complementary Variables

It is well known that increasing time and temperature have similar effects on viscoelastic properties (28). Reliable estimates of the long-term properties of a polymer at a given temperature can be obtained from the short-term properties at higher temperatures. With a hydrophilic material, like nylon, the water content is an additional complementary variable. Increasing the water content has about the same effect as an increase in temperature or the time scale of the experiment.

In a time-temperature superposition, a time-dependent mechanical property, such as creep, stress relaxation, or dynamic mechanical loss is determined at a series of temperatures. The data for the various temperatures are shifted along a log-time axis until they superimpose. Quistwater and Dunell (29,30) carried out a time-humidity superposition using filaments of nylon-66. The filaments were conditioned to various humidities, and forced longitudinal vibrations were applied at frequencies from 2 to 30 cycles per second. At 60°C, the maximum loss for the α relaxation occurred at 2% water, at 35°C, it was at 3.5% water, and at 9°C, it was at 6 to 8% water. It was concluded that the activation energy for the α relaxation is 60 to 80 kcal mole^{-1}.

Another study of time-humidity superpositions was reported by Onogi et al. (31). They carried out measurements of stress relaxation on films of nylon-6 which had been heat treated at 150 to 155°C. Stress relaxation experiments were conducted in tension on dry samples at temperatures from 25 to 77°C. It was found that log-log curves of apparent modulus, E, versus time could be superimposed by shifting along the log-time axis for data taken at temperatures higher than 50°C. The shift factors correspond to an activation energy of 110 kcal mole^{-1}.

In another series of experiments, stress relaxation measurements were made at 25°C over a wide range of relative humidities. At that temperature, relaxation was most rapid at 19 to 33% R.H., and the curves could be superimposed at those and higher humidities. For both time-temperature and time-humidity experiments, superposition was satisfactory only at conditions above the α relaxation.

Relaxation curves for three of the superpositions are shown in Figure 9-9. The upper curve is a time-temperature superposition for dry samples and is based on a reference temperature of 50°C. The other two curves are time-humidity superpositions for data taken at 50 and 60°C, respectively. The close correspondence of the upper two curves demonstrates the equivalence of temperature and humidity in their effect on stress relaxation. It was concluded that the two types of superpositions are equally reliable (31).

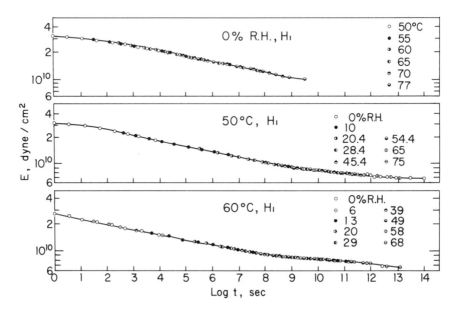

Fig. 9-9. Master relaxation curves for heat-treated nylon-6 film: (top) at 0% R.H., (middle) at 50°C, (bottom) at 60°C.

Tensile Properties

The yield point or upper yield stress is the first maximum in a plot of conventional tensile stress vs. strain. This property decreases with increasing temperature or water content. The yield point has been plotted against temperature at three levels of humidity in Figure 9-10 for nylons -66 (a), -6 (b), and -610 (c).

In nylon-66, the effect on the yield point on going from dryness to saturation with water is equivalent to an increase in temperature of about 80°C. Conditioning to 50% R.H. is equivalent to about half that change in temperature. The corresponding changes in temperature are somewhat less for nylons-6 and -610. At room temperature and above, the properties of nylons-66 and -610 are similar when saturated with water, and dry nylon-610 is similar to nylon-66 which has been conditioned to 50% R.H.

Once the yield point is known for nylon-66, -6, or -610, other features of the stress-strain curve can be predicted quite reliably (15). In Figure 9-11, the dependence on the yield point is shown for the lower yield stress, the ultimate strength, and the ultimate elongation. Data are included for the entire range of humidity and at temperatures from -40 to 100°C. The relationships among these properties are essentially the same regardless of whether the variations in yield

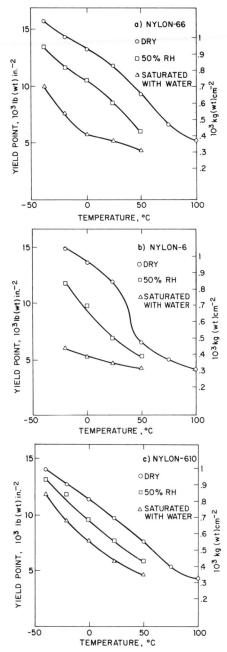

Fig. 9-10. Dependence of yield point at various relative humidities on temperature for nylons-66, -6, and -610.

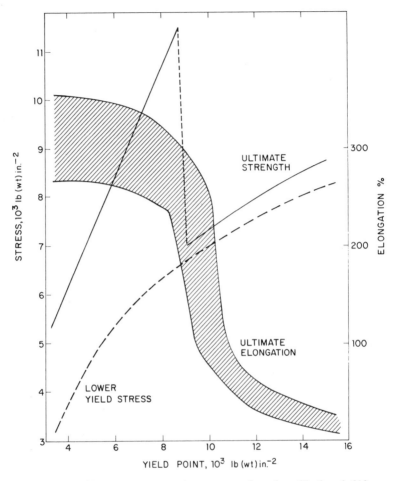

Fig. 9-11.   Relationships among stress-strain parameters for nylons-66, -6, and -610.

point are produced by changing the temperature, the humidity, or the polyamide, provided the molecular weights are similar. The ultimate strength and elongation are sharply higher when the yield point is less than 9000 lb (wt) in.$^{-2}$ (633 kg(wt)cm^{-2}). This corresponds to temperatures above the α relaxation.

Modulus

In Figure 9-12, the dependence of the flexural modulus on temperature is shown for injection molded samples of nylon-66 dry, conditioned to 50% R.H., and saturated with water (32). At each level of humidity, the data follow a sigmoidal

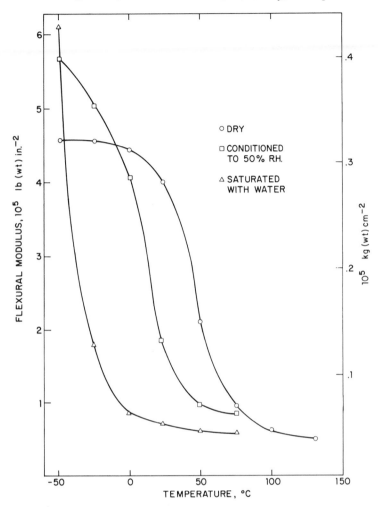

Fig. 9-12.   Flexural modulus of nylon-66.

pattern at a temperature corresponding to the α relaxation. The midpoints for the decrease in modulus are 50°C for the dry sample, just below room temperature at 50% R.H., and about -35°C for the sample saturated with water. Thus, the effect of the available range of water content on the flexural modulus is equivalent to a change in temperature of about 85°C.

At temperatures well above the α relaxation, the value of the flexural modulus is near 70,000 lb (wt) in.$^{-2}$ (4921 kg (wt) cm^{-2}) and only slightly dependent on temperature for all levels of moisture. At very low temperatures, the modulus increases with increasing humidity.

Toughness

The toughness or energy to break is very sensitive to the nature of the test, and it is important that comparisons be based on a well-defined procedure. The ultimate elongation and the elongation to yield are both higher at temperatures above the α relaxation. These increases in elongation generally outweigh any decrease in yield stress. Thus, the impact strength usually increases with increasing temperature or moisture content. In Table 9-3, these effects are illustrated for nylons-66 and -610 with data for the tensile impact strength (34).

Table 9-3. Effect of temperature and humidity on the toughness of nylon resins

Tensile impact strength $(\text{ft-lb (wt) in.}^{-2})$ (34)	Nylon-66	Nylon-610
Dry		
23°C	82[a]	79
−18°C	93	73
−40°C	85	68
Conditioned to 50% R.H.		
23°C	156	121
−18°C	113	73
−40°C	79	86

[a] $\times 2.14 = \text{cm-kg (wt) cm}^{-2}$.

REFERENCES

1. Ke, B., and A. W. Sisko, *J. Polym. Sci.* 50, 87 (1961).

2. Webber, A. C., private communication.

3. Miller, R. L., in *Polymer Handbook*, J. Brandrup and E. H. Immergut, Eds. Interscience Publishers, New York, 1966, Section III-1.

4. Flory, P. J., *Principles of Polymer Chemistry*, Cornell University Press, Ithaca, 1953, pp. 568-571.

5. Edgar, O. B., and R. H. Hill, *J. Polym. Sci.* 8, 1 (1952).

6. Allegra, G., and I. W. Bassi, *Adv. Polym. Sci.* 6, 549 (1969).

7. Levine, M., and S. C. Lemin, *J. Polym. Sci.* 49, 241 (1961).

8. Saotome, K., and H. Komoto, *J. Polym. Sci., Part A-1, Polym. Chem.* 4, 1475 (1966).

9. Dole M., and B. Wunderlich, *Die Makromol. Chem.* 34, 29 (1959).

10. Dole, M., *Adv. Polym. Sci.* 2, 221 (1960).

11. Starkweather, H. W., Jr., and R. E. Moynihan, *J. Polym. Sci.*, 22, 363 (1956).

12. Trifan, D. S., and J. F. Terenzi, *J. Polym. Sci.* 28, 443 (1958).

13. Cannon,C. G., *Spectrochim Scta.* **16**, 302 (1960).

14. Starkweather, H. W., Jr., J. F. Whitney, and Donald R. Johnson, *J. Polym. Sci., Part A* **1**, 715 (1963).

15. Starkweather, H. W., Jr., unpublished data.

16. Badische Anilin & Soda-Fabrik AG, "Technical Information Bulletins on Ultramid Resins", July, 1969.

17. Inone, M., *J. Polym. Sci.* **55**, 753 (1961).

18. Brill, R., *J. Prakt Chem.* [2], **161**, 49 (1942).

19. Slichter, W. P., *J. Polym. Sci.* **35**, 77 (1958).

20. Slichter, W. P., *J. Appl. Phys.* **26**, 1099 (1955).

21. McCrum, N. G., B. E. Read, and G. Williams, *Anelastic and Dielectric Effects in Polymer Solids,* J. Wiley & Sons, Inc., New York, 1967.

22. Trueman, T. L., unpublished data.

23. Willbourn, A. H., *Trans. Faraday Soc.* **54**, 717 (1958).

24. Bell, J. P., and T. Murayama, *J Polym. Sci., Part A-2,* **7**, 1059 (1969).

25. Boyd, R. H., *J. Chem. Phys.* **30**, 1276 (1959).

26. Woodward, A. E., J. M. Crissman, and J. A. Sauer, *J. Polym. Sci.* **44**, 23 (1960).

27. Starkweather, H. W., Jr., *J. Macromol. Sci.-Phys.* **B3(4)**, 727 (1969).

28. Tobolsky, A. V., *Properties and Structure of Polymers,* J. Wiley and Sons, Inc., New York, 1960.

29. Quistwater, J. M. R., and B. A. Dunell, *J. Polym. Sci.,* **28**, 309 (1958).

30. Quistwater, J. M. R., and B. A. Dunell, *J. Appl. Polym. Sci.* **1**, 267 (1959).

31. Onogi, S., K. Sasaguri, T. Adachi, and S. Ogihara, *J. Polym. Sci.* **58**, 1 (1962).

32. Kohan, M. I., private communication.

33. Plastics Department, E. I. du Pont de Nemours and Co., "Zytel® Nylon Resins, Design and Engineering Data," 1962.

34. Bragaw, C. G., *Mod. Plastics* **33**, No. 10, 199 (June, 1956).

CHAPTER 10

Properties of Molded Nylons

R. M. BONNER, M. I. KOHAN, E. M. LACEY,
P. N. RICHARDSON, T. M. RODER, AND
L. T. SHERWOOD, JR.

INTRODUCTION

The purpose of this chapter is to provide a general survey of the properties of the major types of injection molded nylons and their dependence on temperature, relative humidity, and other conditions of use. The properties of nylons whether molded, extruded, or otherwise shaped vary with processing conditions to the degree that the physical structure (Chapter 8) is altered. Poor polymer handling and fabricating procedures can introduce contamination or cause chemical changes that impair properties (Chapters 4, 5, and 6). Practically speaking, however, nylons molded in normal fashion in standard equipment exhibit typical values, and these are the values cited here. Some variation both above and below these values occurs routinely because exact duplication of thermal history from one molding situation to another is not to be expected.

Many properties change only a little with fabricating technique so that data obtained on molded specimens are often assumed to apply also to extruded objects unless highly oriented. Subsequent chapters will deal with the special properties of nylon products made by processes other than injection molding.

Attention centers on the homopolymers which make up the bulk of the nylon plastics business. Modifications such as copolymerization, plasticization, nucleation, reinforcement, and so forth significantly affect properties, and these effects will be noted where appropriate. However, this subject is dealt with more thoroughly in the next chapter. Also, because of the large number of modifications available (see Appendix), the precise properties of specific compositions in particular end-use situations must be obtained from the individual manufacturer.

In describing properties standard test methods such as those provided by the ASTM are used to the extent possible. However, essential as these tests are, they may not tell the designer all he needs to know. A careful analysis of end-use requirements will often pinpoint the critical properties and suggest nonstandard procedures to simulate such requirements. Examples of widely used but nonstandard tests are energy-to-break testing of actual production parts, immersion of stressed samples in zinc chloride solution to compare tendencies to craze under chemical attack, and tests to evaluate toughness under repeated impact.

The crystallinity of a nylon can be affected not only by processing technique but also by subsequent dry annealing or accelerated moisture conditioning (Chapter 17). Other secondary operations that follow the shaping process such as machining, assembly, and decorating can also alter the properties of nylons and are discussed in Chapter 17.

Nylons absorb more or less water depending on the type of nylon, the relative humidity, and the crystallinity of the part. The absorption of water can induce significant changes in modulus of elasticity, yield stress, toughness, and dimensions. These changes can be rationalized in terms of the effect of water on secondary crystallization (Chapter 8) and, in particular, on the α-transition temperature. The multiplicity of transitions and relaxation processes in nylons was considered in the preceding chapter. Particularly important is the α-transition which is often identified in the literature as the glass transition temperature (T_g). Because of its dependence on the method of measurement and its sensitivity to moisture, the reported α-transition values vary (Table 10-1). The α-transition is associated with motion in the noncrystalline part of the polymer. It is clear by comparing nylon-66 and nylon-6, which have the same concentration of amide groups, that a morphological parameter is involved. Nonetheless, trends are indicated: as the concentration of amide groups in linear, aliphatic, nylon homopolymers decreases, the α-transition temperature tends to decrease if dry, change but little at 50% R. H., and increase at 100% R. H.

It is often convenient with nylons to regard moisture content and temperature as complementary variables (Chapter 9). It should be noted that parts made from crystalline nylons retain their integrity and exhibit a measurable modulus almost to the melting point. This contrasts with the behavior of amorphous polymers such as poly(methyl methacrylate) and polycarbonate which lose all of their mechanical integrity above the T_g.

If a part has not reached equilibrium with the relative humidity in the environment, the amount and distribution of water will be determined by the temperature, the time of exposure, and the part thickness. Thus, a part of nylon-66 may absorb 2.5% water as a result of sufficiently long exposure to achieve equilibrium in a standard laboratory atmosphere of 50% R. H. and 23°C or by a relatively brief immersion in boiling water. In the latter case, the surface will be saturated, and the center essentially dry. Neither the properties nor the dimensions of such a part will be identical with one at equilibrium although this procedure may serve very well the practical purpose of achieving a desired level of toughness. A conditioning procedure frequently cited is Procedure A of ASTM D618, which involves 40 hr at 23°C (73°F) and 50% R. H. This will not

Table 10-1. α-Transitions for common nylon plastics (°C)

Nylon	CH$_2$/CONH	Dry Calculated From melt pt.[a]	Dry Calculated From addit. contr.[b]	Dry Reported From loss peak[c]	Dry Reported Other	Cond. to 50% R.H. From loss peak[c]	Cond. to 50% R.H. Other	Cond. to 100% R.H. From loss peak[c]	Cond. to 100% R.H. Other
66	5/1	82	67	80[e],78[i]	48[f]	35[e]	15[f]	-15[e]	-37[f]
6[d]	5/1	56	67	75[e],65[i]	41[f]	20[e]	3[f]	-22[e]	-32[f]
610	7/1	56	50	67[e],70[i]	42[g]	40[e]		10[e]	
612	8/1	52	44	60[h]	45[k]	40[h]	20[k]	20[h]	
11	10/1	36	35	53[i]	43[g]				
12	11/1	29	32	54[j]	42[g]			42[j]	

[a] Based on relation $T_g = (2/3) T_m$ in °K (Ref. 1).

[b] Based on additive contributions from CH$_2$ and CONH (Ref. 2).

[c] From temperature dependence of mechanical loss peak.

[d] Extracted nylon-6.

[e] Approximately 1 cps (Ref. 3).

[f] From inflection point of flexural modulus — temperature curve (Ref. 3).

[g] From break in DTA base line (Ref. 4).

[h] Approx. 1 cps (Ref. 5). The values at 50 and 100% R.H. are estimates based on incomplete data and reported moisture contents.

[i] 0.3 cps (Ref. 6).

[j] Approx. 1 cps (Ref. 7).

[k] From inflection point of flexural modulus — temperature curve (Ref. 8).

achieve equilibration for nylons except in the case of thin films or monofilament. The accelerated conditioning techniques of Chapter 17 are often preferred for the testing of typical moldings.

MECHANICAL PROPERTIES

The properties of nylon or of any plastic as measured in tension, compression, flexure, and shear are basic to mechanical behavior. Properly understood, they provide much of the information needed by the designer. It is essential, however, to keep in mind all the variables that can affect the behavior of a material. The moisture content of a nylon, as noted above, is an important factor in understanding its performance. There are also material, environmental, and stress factors as summarized in Table 10-2.

Table 10-2. Factors affecting mechanical behavior of nylons

Material factors
        Type of nylon
        Molecular weight ($\overline{M}_n$)
        Presence of modifiers (see Table 11-1)
        Moisture content
        Physical structure (crystallinity, morphology, orientation)
Environmental factors
        Temperature
        Relative Humidity
        Solvents or other chemicals
Stress factors
        Nature of stress (tension, compression, flexure, shear)
        Rate of loading
        Duration of stress

The material and environmental factors outlined in Table 10-2 are considered to the degree appropriate for the property under discussion and as permitted by the available information. It is important to realize that the mechanical properties – whether in tension, compression, flexure, or shear – are time-dependent. The steady application of low loads leading to deformation over long periods of time is dealt with in a separate section. The rapid application of a high load is considered under the heading of tensile properties and again under impact properties because of the nature of the available data.

Tensile Properties

The most widely quoted mechanical property is tensile strength (ASTM D638), but it is a single value of varying significance that can correspond to either yield or fracture and is a poor substitute for the actual stress-strain curve. The ASTM procedure recommends a rate of 0.2 in. (5.1 mm) per minute for rigid plastics in general but suggests 2.0 in. (51 mm) per minute for nylon because of the excessive time required to achieve the 30 to 300% elongation characteristic of most nylon compositions. A schematic curve of load versus extension for dry nylon homopolymers below the transition temperature is given in Figure 10-1. This is drawn to show the kind of curve that is actually observed on the recording chart of standard test equipment. The Hookean or essentially linear region, yield point, neck-down region, draw point, and break point are identified as tensile parameters, and corresponding stages in the physical appearance of the test specimens are depicted in Figure 10-2 for ASTM D638 Type I bars.

    Load-extension curves for dry, as-molded nylon-66 at 73°F (23°C) and at separation rates of 0.2 to 2.0 in. (5.1 to 51 mm) per minute are shown in Figure

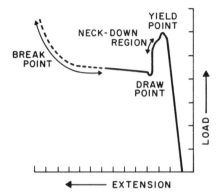

Fig. 10-1.   Schematic curve of load versus extension for dry nylon homopolymers as normally observed on the test machine chart.

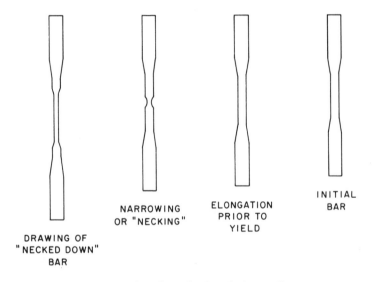

Fig. 10-2.   Appearance of test bars of nylons during tensile testing.

10-3. Because chart travel was kept equal to the testing speed, the extensions are on the same scale and are directly comparable except for ultimate extension at fracture, which was not determined. A distinct increase in load but little change in elongation at the yield point and a decrease in both load and elongation at the draw point are apparent with increasing rate of separation. The heat generated during drawing and the time available for its dissipation contribute to these changes.

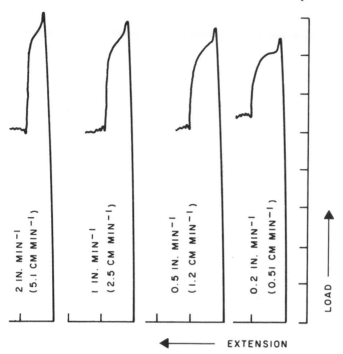

Fig. 10-3.   Effect of testing speed on the shape of the load versus extension curve for dry nylon-66 at 23°C (73°F).

Load-extension curves at 2.0 in. min^{-1} are shown for dry nylons-6, -610, -612, -11, and -12 in Figure 10-4. Nylons-66 and -6, with only five methylene groups per amide linkage and the greatest potential for hydrogen bonding between chains, have the highest yield and draw stresses. Nylons-11 and -12, with methylene/amide ratios of 10/1 and 11/1, have the lowest stresses. Other factors such as crystallinity and morphology have to be invoked to account for differences between nylon-66 and -6 and between nylons-11 and -12.

It is customary to calculate the stress corresponding to yield, necking, or draw by dividing the observed load by the initial cross-sectional area. Determination of percent elongation is also based on the original gauge length, but the material actually undergoing most of the extension is that in the neck-down region which may be only a small fraction of the original gauge length. Thus, the reported elongation can be much lower than the elongation actually experienced by the stretched material. For the same reason one cannot apply the simple multipliers provided in ASTM D638 to calculate values of true stress and strain unless properly applied to each point along the test specimen.

Figure 10-5 shows load-extension curves for nylon-66 at 23°C (77°F) dry and

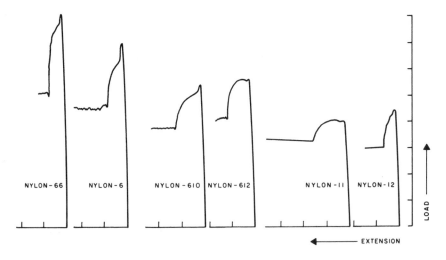

Fig. 10-4. Load versus extension curves at 2 in. min^{-1} (5.1 cm min^{-1}) and 23°C (73°F) for dry nylons-66, -6, -610, -612, -11, and -12.

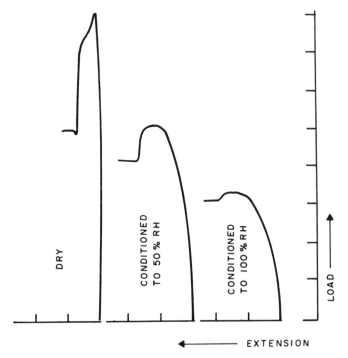

Fig. 10-5. Effect of moisture content on load versus extension curves for nylon-66 at 2 in. min^{-1} (5.1 cm min^{-1}) and 23°C (73°F).

at equilibrium with 50 and 100% R.H. The plasticizing effect of water is obvious and consistent with the lowering of T_g. The dry curve reflects the fact that the test temperature is well below T_g; the 100% R. H. curve, that the temperature is well above T_g. The 50% R. H. curve corresponds to an intermediate condition at which the test temperature and T_g are similar. As expected, increasing temperature (Figure 10-6) has an effect analagous to that of increasing moisture content.

Figure 10-7 shows the trend in tensile behavior of nylon-66 at high strain rates such as apply in impact (9). Stress at the yield point increases but elongation at yield shows little change and depends primarily on the moisture content and temperature of the material. This is consistent with the behavior at ordinary testing rates of 0.2 to 2 in. min^{-1} as described above (Figure 10-3). Similarly, ultimate elongation at the breakpoint is primarily controlled by whether or not the specimen is above or below the glass transition temperature. Above T_g elongation is relatively high; below, relatively low. In either case, elongation is essentially independent of the rate of jaw separation. However, in the transition region (e.g., nylon-66 at 50% R. H. tested at 23°C) there does

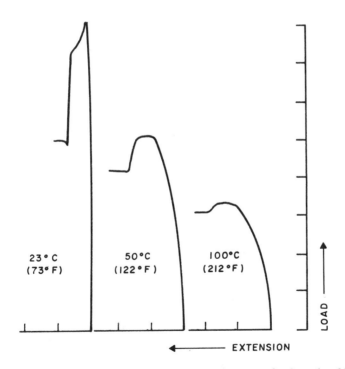

Fig. 10-6.   Effect of temperature on load versus extension curves for dry nylon-66 at 2 in. min^{-1} (5.1 cm min^{-1}) testing speed.

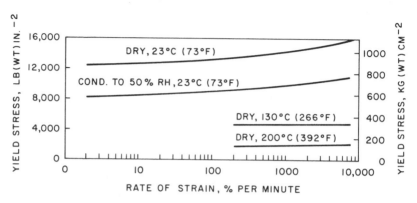

Fig. 10-7. Yield stress of nylon-66 versus rate of strain, dry at 23, 130, and 200°C (73, 266, and 392° F) and conditioned to 50% R.H. at 23°C (adapted from Ref. 9).

appear to be a dependence on rate. Comparable data at high strain rates are generally not available on other nylons, but there is information on nylon-6 that suggests behavior similar to that of nylon-66 (9).

Emphasis has been on the load-extension or stress-strain curve which provides the kind of information good design requires. Nonetheless, point values are used as guidelines and are more readily available. In Chapter 9 (p. 322, Figure 9-11) it was also pointed out that there is a predictable relationship among the stress-strain parameters for nylons-66, -6, and -610. Typical data are summarized for the commercial nylon plastics in Table 10-3. The glass-reinforced compositions, the nucleated 66, and the terpolymer illustrate the effects of modification, which is the subject of Chapter 11.

High elongation at fracture is often cited as desirable, but its merit depends on the intended use. In applications requiring toughness it is a helpful quality index because the work-to-break, the area under the stress-strain curve, normally increases as the elongation increases. In other applications, however, failure may be due to permanent deformation that occurs when a low strain (1 to 2%) is exceeded. High elongation at either fracture or yield is typically associated with a lowered modulus and stress to yield that indicate reduced resistance to permanent deformation.

Flexural Properties

The elastic modulus of nylon can be determined in tension, compression, or flexure, but the modulus in flexure is the most commonly encountered value because of the greater likelihood of a bending stress in use. Tensile modulus is often provided also, and the two values may not coincide. This is because the flexural test emphasizes the behavior of the surface, which is less crystalline than

Table 10-3. Tensile properties (ASTM D-638) of commercial nylons[a]

Nylon	66		Extracted 6		610		612		11	
	Dry	50% R.H.	Dry	50% R.H.	Dry	50% R.H.	Dry	50% R.H.	Dry	50% R.H.
Tensile stress[b], lb(wt)in.$^{-2}$, at:										
−40°F (−40°C)	15,700	14,900	17,400	—	12,000	12,100	13,600	13,500	9,900	
73°F (23°C)	12,000	11,200	11,800	10,000	8,500	7,100	8,800	8,800	8,500	7,800
170°F (77°C)	9,000	5,900	9,900	8,500	5,300	—	5,900	—	6,100	
Yield stress, lb (wt)in.$^{-2}$, at:										
−40°F (−40°C)	15,700	14,900	17,400	—	12,000	12,100	13,600	13,500		
73°F (23°C)	12,000	8,500	11,800	6,400	8,500	7,100	8,800	7,400		
170°F (77°C)	6,500	5,900	5,900	4,000	4,200	—	4,300	—		
Elongation at break, %, at:										
−40°F (−40°C)	20	20	5	—	20	30	15	30	37	
73°F (23°C)	60	300	200	300	20	220	150	340	120	330
170°F (77°C)	340	350	310	325	300	—	—	—	400	
Elongation at yield, %, at:										
−40°F (−40°C)	4	—	—	—	10	13	8	14		
73°F (23°C)	5	25	—	—	10	30	7	40		
170°F (77°C)	30	30	—	—	30	—	30	—		

Table 10-3. Tensile properties (ASTM D-638) of commercial nylons[a] (continued)

Nylon	12 Dry	12 50% R.H.	66 + 33% glass fiber Dry	66 + 33% glass fiber 50% R.H.	612 + 43% glass fiber Dry	612 + 43% glass fiber 50% R.H.	Nucleated 66 Dry	Nucleated 66 50% R.H.	66/610/6 Dry	66/610/6 50% R.H.
Tensile stress[b] lb(wt)in.$^{-2}$, at:										
−40°F (−40°C)									13,000	7,500
73°F (23°C)	8,000	7,600	27,000	18,000	28,000	24,000	15,300	9,100	7,400	12,700
170°F (77°C)									5,000	4,400
Yield stress, lb(wt)in.$^{-2}$, at:										
−40°F (−40°C)	11,500								—	12,700
73°F (23°C)	7,500	5,900					15,300	9,100	6,000	2,700
170°F (77°C)	3,500									
Elongation at break, %, at:										
−40°F (−40°C)									100	20
73°F (23°C)	250	250	3.0	3.6	4.0	5.0	15	175	300	370
170°F (77°C)									400	400
Elongation at yield, %, at:										
−40°F (−40°C)	10	20								
73°F (23°C)										
170°F (77°C)										

[a] Data compiled from trade literature.
[b] Highest stress whether at yield, fracture, or elsewhere in the stress-strain curve; lb(wt)in.$^{-2}$ × 0.0703 = kg(wt)cm^{-2}.

the core, and so leads to a lower modulus. A higher rate of strain is sometimes used in the tensile test and also contributes to a higher value.

Flexural modulus data for commercial nylons are given in Table 10-4. The effect of temperature on nylon-66, dry and conditioned to 50 and 100% R.H., is shown in Chapter 9 (Figure 9-12). The modulus decreases rapidly in the T_g region, which depends on the moisture content. Because of the crystallinity, however, the modulus changes little above T_g and is still about 25,000 lb (wt)in.$^{-2}$ (1760 kg(wt)cm^{-2}) at 450°F (232°C) (8). The effect of water on the modulus at several temperatures is illustrated for extracted nylon-6 in Figure 10-8.

The dry modulus increases, the 50% R.H. modulus is roughly similar, and the 100% R.H. modulus decreases with increasing amide group concentration (Figure 10-9). This is consistent with the concept that the amide group increases strength and rigidity via intermolecular hydrogen bonding but also promotes plasticization by water. A similar effect is shown in Figure 10-9 for glass-reinforced resins. The higher value for unmodified nylon-66 relative to -6 is associated with a somewhat lower degree of order in nylon-6 that is manifest also in its higher water absorption.

Other additives can affect the modulus by acting as nucleants if present in small amount or by acting as fillers, reinforcing agents, or plasticizers if present

Table 10-4. Flexural modulus (ASTM D790) of commercial nylons.[a]
10^3 lb(wt)in.$^{-2}$ ($\times$ 70.3 = kg(wt)cm^{-2})

Nylon		Dry			50% R.H.			100% R.H.		
	°F	-40	73	170	-40	73	170	-40	73	170
	°C	-40	23	77	-40	23	77	-40	23	77
66		470	410	100	500	175	82	--	70	--
Extracted 6		435	395	75	--	140	50	--	60	35
610		325	285	70	365	160	--	--	100	--
612		340	290	60	400	180	55	--	120	--
11		192	170	28	--	150	--	--	138	--
Plasticized 6		--	170	75	--	100	--	--	--	--
Nucleated, extracted 6		500	470	85	--	185	--	--	85	--
66 + 33% glass fiber		--	1,300	730	--	900	--	--	600	450

[a] Data from trade literature.

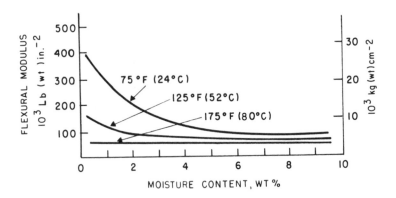

Fig. 10-8. Flexural modulus versus moisture content at 75, 125, and 175° F (24, 52, and 80° C) for extracted nylon-6 (adapted from Ref. 10).

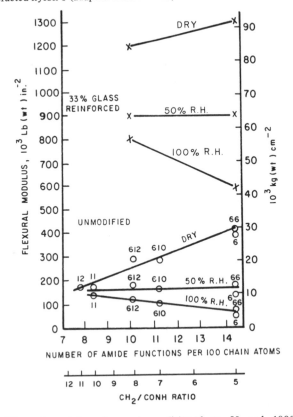

Fig. 10-9. Flexural modulus, dry and conditioned to 50 and 100% R.H., versus amide-group concentration.

in significant concentration. An example of a nucleated composition and of a plasticized one is provided in Table 10-4.

The flexural strength of most nylons is not routinely reported because fracture does not normally occur in standard tests. The stress corresponding to yield is sometimes identified with flexural strength although not always so stated. The rate of change of yield stress in flexure with moisture content and temperature for extracted nylon-6 is similar to that observed for the modulus (compare Figures 10-10 and 10-9).

Fracture in flexure does occur with reinforced, low elongation compositions, and the flexural strength of several dry, glass-reinforced compositions is shown in Table 10-5.

Compressive Properties

The stress-strain behavior of dry nylon-6 in compression at three temperatures is shown in Figure 10-11. Similar data for nylon-66, dry and conditioned to 50% R.H., are given in Chapter 15 (Figure 15-1).

In the ASTM D-695 compressive test unmodified nylons do not break, and compressive strength is meaningless. However, glass-reinforced nylons do break, and the trade literature provides compressive strength data for these materials (Table 10-6).

Because unreinforced nylons are not broken, several methods have been used to obtain information from compressive stress-strain curves such as those in Figures 10-11 and 15-1. A compressive modulus is sometimes calculated (Table

Table 10-5. Flexural strength (ASTM D790) of dry, glass-fiber reinforced nylons

Nylon	% Glass	Flexural strength[a]	
		$(lb(wt)in.^{-2})$	$(kg(wt)cm^{-2})$
66	13	24,000	1,700
66	33	38,000	2,700
6	14	22,500	1,600
6	30	35,000	2,500
610	20	26,000	1,800
610	40	35,000	2,500
612	33	37,000	2,600
612	43	43,000	3,000
12	30	26,000	1,800

[a] Data from trade literature.

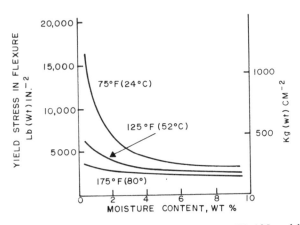

Fig. 10-10. Yield stress in flexure versus moisture content at 75, 125, and 175° F (24, 52, and 80° C) for extracted nylon-6 (adapted from Ref. 10).

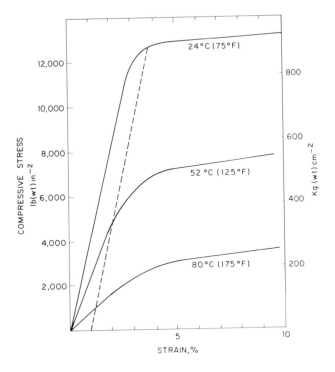

Fig. 10-11. Compressive stress versus strain for dry, extracted nylon-6 at 75, 125, and 175° F (24, 52, and 80° C) (adapted from Ref. 10). The dotted line indicates the 1% offset strain line for determining the 1% offset yield stress at 75° F (24° C).

Table 10-6. Compressive strength (ASTM D695) of dry, glass reinforced nylons[a]

Nylon	lb(wt)in.$^{-2}$	kg(wt)cm^{-2}
66 + 13% Glass fiber	25,000	1,750
66 + 33% Glass fiber	35,000	2,450
6 + 30% Glass fiber	19,000	1,340
610 + 30% Glass fiber	18,000	1,270
612 + 33% Glass fiber	23,000	1,620
11 + 30% Glass fiber	12,800	900

[a] Data from trade literature.

10-7). Very similar data, compressive stress at 1% deformation, are given in Table 10-8. If the stress-strain curve is linear to 1% deformation the stress values in Table 10-8 divided by 0.01 would be identical to the modulus.

The nonlinear nature of the stress-strain curve, shown in Figures 10-11 and 15-1, has resulted in the kind of data presented in Table 10-9. These numbers are known as compressive, 1% offset, yield stress values with the 1% offset an

Table 10-7. Compressive modulus (ASTM D695) of dry nylons[a]

Nylon	lb(wt)in.$^{-2}$	kg(wt)cm^{-2}
66	410,000	28,800
6	350,000	24,600
610	300,000	21,000
11	185,000	13,000

[a] Data from trade literature.

Table 10-8. Compressive stress at 1% deformation (ASTM D695) of dry nylons[a]

Nylon	lb(wt)in.$^{-2}$	kg(wt)cm^{-2}
66	4,900	340
610	3,000	210
612	2,400	170
66/610/6	800	56

[a] Data from trade literature.

Table 10-9. Compressive, 1% offset, yield stress (ASTM D695) of dry nylons[a]

Nylon	$lb(wt)in.^{-2}$	$kg(wt)cm^{-2}$
66	13,200	930
6	12,000	850
610	10,600	750
66 + 35% Glass fiber	14,000	980
6 + 30% Glass fiber	12,400	870

[a]Data from trade literature.

arbitrary choice. Figure 10-11 shows how the values are obtained. A line is drawn from the base line (zero stress) and 1% strain parallel to the initial straight line portion of the stress-strain curve. The stress at the intersection of the measured curve and the 1% offset line is the 1% offset yield stress.

The effect of rate of strain on the compressive stress-strain behavior of nylon-66 has been studied (11). Moisture content is uncertain, but it is clear that the stress corresponding to a given deformation increases with increasing strain rate.

Shear Properties

The shear strength of dry material at $23°C$ ($73°F$) is normally the only shear property reported for molded nylons (Table 10-10). It varies from 70% to almost 100% of the dry tensile stress (cf. Table 10-3), depending on the nylon. Modification produces the same effect already noted for the other mechanical

Table 10-10. Shear strength (ASTM D732) of dry nylons[a]

Nylon	$lb(wt)in.^{-2}$	$kg(wt)cm^{-2}$
66	9,600	680
6	8,500	600
610	8,400	590
612	8,600	610
66/610/6	5,700	400
66 + 13% Glass fiber	11,000	780
66 + 33% Glass fiber	12,500	880
6 + 30% Glass fiber	10,300	730
612 + 33% Glass fiber	11,000	780

[a]Data from trade literature.

properties, for example, reduction by copolymerization and enhancement by reinforcement. Shear strength, like flexural and compressive strength, is of more significance in low elongation, reinforced compositions, and the temperature dependence of shear strength for nylons-66 and -612 containing 33% glass fiber has been reported (Figure 10-12).

Shear modulus data have been obtained in dynamic mechanical testing where evidence for transitional phenomena, particularly from mechanical loss versus temperature data, has been the main goal (see Chapter 9, p. 313).

Impact Properties

Through the years nylon has been recognized as a tough plastic and has been used in many applications requiring impact resistance (see Chapter 20). At the same time it has been recognized that nylons are notch sensitive. Fillets and generous radii are used in the design of parts to insure absence of any notch effect. Impact strength increases rapidly as the notch radius becomes greater than 0.02 in. (2.5 mm), as shown in Figure 10-13 (8).

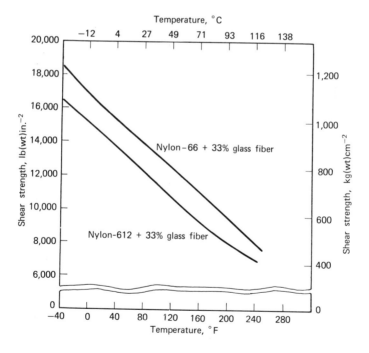

Fig. 10-12.   Shear strength versus temperature for dry nylons containing glass fibers (8).

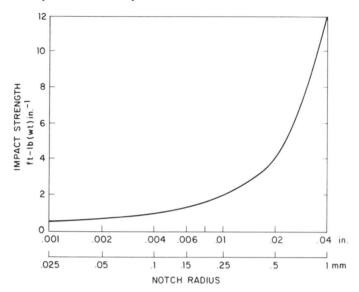

Fig. 10-13.   Impact strength at $23°C$ $(73°F)$ of nylon-66 conditioned to 50% R.H. versus notch radius (adapted from Ref. 8).

In the United States and a number of other countries, the common notched impact test is the Izod impact test (ASTM D256, Method A). The specimens are notched with a $45°$ "V" notch that has an apex with a curvature of 0.25 mm (0.01 in.) and are tested clamped as a cantilever beam. Results have traditionally been expressed in ft-lb (wt) in.$^{-1}$ of notch. The current procedure suggests use of the metric Newton-meters per meter (N-m m^{-1}) or, preferably, Joules per meter (J m^{-1}). Table 10-11 and Figure 10-14 (13) give standard Izod data for nylons. Increasing molecular weight or moisture content, glass fiber reinforcement, copolymerization, and plasticization increase impact strength. Moisture, copolymerization, and plasticization lower T_g so that the test temperature exceeds T_g; at temperatures low enough to be below the T_g of the modified composition, the impact strength may be lower than that of the unmodified resin.

The other common notched impact test is the Charpy impact test used in West Germany (DIN53453, 1965) and England (BS2782; 1965, 306E). The specimen, 120 x 15 x 10 mm, is supported at both ends and is notched with a "U" notch 2 mm deep with the radius of the corners not to exceed 0.2 mm. The results are expressed in cm-kg(wt)cm^{-2} of remaining cross section after notching. Typical values for nylons in this test are given in Table 10-12.

Table 10-11. Izod impact strength (ASTM D256)

Nylon	ft-lb(wt) in.$^{-1}$				
	Dry		Conditioned		
	$-40°C$ ($-40°F$)	$23°C$ ($73°F$)	$23°C$ ($73°F$)	% Water	Ref.
66 ($\overline{M}_n$ 18,000)	0.6	1.0	2.1	2.5	8
66 ($\overline{M}_n$ 34,000)		1.3	2.5	2.5	8
6 (extracted)	0.3	0.8	5.0	3.0	12
610	0.5	0.9	1.6	1.5	13
612	0.9	1.0	1.4	1.3	8
66/610/6		not broken			8
6 copolymer		not broken			14
66 plasticized		not broken			8
66 + 33% glass fiber		2.1			8
6 + 33% glass fiber		2.3			14
612 + 33% glass fiber		2.6			8

$$\text{ft-lb(wt) in.}^{-1} \times 5.45 = \text{cm-kg(wt) cm}^{-1}$$
$$\text{ft-lb(wt) in.}^{-1} \times 53.3 = \text{N-m m}^{-1} = \text{J m}^{-1}$$

Table 10-12. Notched impact strength (DIN 53453)

Nylon	cm-kg(wt) cm^{-2}		
	Dry $20°C$ ($68°F$)	Conditioned 4 Months $20°C$ and 65% R.H.	Ref.
66	2-3	12-20	15
6 (extracted)	3-6	Not broken	15
610	4-10	13-15	15
12	8		17
66 + 35% glass fiber	13	14	15
6 + 35% glass fiber	14	18	15

$$\text{cm-kg(wt) cm}^{-2} \times 0.466 = \text{ft-lb(wt) in.}^{-2}$$

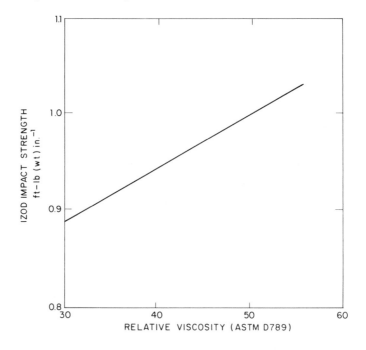

Fig. 10-14.   Izod impact strength of dry nylon-66 at 23°C (73°F) versus relative viscosity (adapted from Ref. 13).

Tensile impact strength as determined by ASTM D1822 measures the toughness of a specimen without a notch. For many years the specimen used in tensile impact tests varied, and the reader is cautioned when comparing data from different sources to be sure the same specimens were used. ASTM D1822 specifies the two shapes shown in Figure 10-15. With Type S (short) the extension is comparatively low, whereas with Type L (long) the extension is comparatively high. The Type S specimen gives greater reproducibility but less differentation among materials. Data for nylons are given in Table 10-13 for short and long specimens both dry and conditioned.

In contrast to the Izod impact test the addition of glass decreases the tensile impact strength. The dependence of the tensile impact strength of nylon-66 on relative viscosity is shown in Figure 10-16 (13).

A modified tensile impact type test has been used to subject specimens to a series of low-energy blows (17). The resistance to repeated impact is more important than resistance to single impact in many applications such as striker plates in automobiles and appliances, ladies' shoe heels, and gear teeth. In the

Table 10-13. Tensile impact strength (ASTM D1822)

| | ft-lb(wt)in.$^{-2}$ | | | | | |
| | Dry | | Conditioned | | | |
Nylon	Short	Long	Short	Long	% Water	Ref.
66 ($\overline{M}_n$ 18,000)	75	240	110	700	2.5	8
66 ($\overline{M}_n$ 34,000)		255		Not	2.5	8
6 (extracted)	70-120			broken		15
610	85-95					15
612	73	291	104	450	1.3	8
66 + 30% glass fiber	50					15
66 + 30% glass fiber		80-85				16
6 + 30% glass fiber	43					15
6 + 30% glass fiber		70-90				16
610 + 30% glass fiber		100				16

$$\text{ft-lb(wt)in.}^{-2} \times 2.14 = \text{cm-kg(wt)cm}^{-2}$$

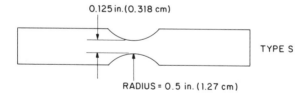

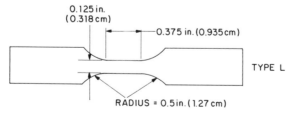

Fig. 10-15. Tensile impact specimens (ASTM D1822). Thickness = 0.0625-0.125 in. (0.1588-0.6350 cm).

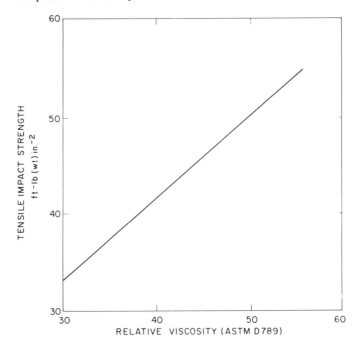

Fig. 10-16.   Tensile impact strength of type S specimens of nylon-66 conditioned 96 hr at 23°C (73°F) and 50% R.H. versus relative viscosity (adapted from Ref. 13).

cited test two criteria of failure were used to evaluate over a range of rates of loading nylon-66 conditioned to 50% R.H. and five non-nylons including polycarbonate and die-cast metals. The criteria were the number of blows to failure and to a 20% reduction in cross-sectional area. The nylon-66 was distinctly superior to all other materials tested by either criterion.

A pendulum impact machine used for Izod and tensile impact tests can be applied to commercial moldings by designing suitable jigs to hold the actual part. Such a device has proved helpful as a production quality control tool to select molding conditions (18).

In the brittleness temperature test (ASTM D746) standard specimens are immersed in a nonreactive, nonsolvent, heat transfer medium and impacted with a striking arm traveling at 6.0 to 7.0 ft sec^{-1}. The brittleness temperature is the temperature at which 50% of the specimens break. The typical values cited in Table 10-14 provide evidence of a considerable degree of toughness at low temperatures. It is seen that increasing the molecular weight and the hydrocarbon nature of the nylon lowers the brittleness temperature and increasing the water content raises the brittleness temperature.

Table 10-14. Brittleness temperature (ASTM D746)

Nylon	Dry		Conditioned			
	(°F)	(°C)	(°F)	(°C)	% Water	Ref.
66 ($\overline{M}_n$ 18,000)	−112	−80	−85	−65	2.5	8
66 ($\overline{M}_n$ 34,000)	−148	−100	−121	−85	2.5	8
610	−130	−90	−80	−62	1.5	13
612	−195	−126	−165	−110	1.3	8

Hardness

The hardness of nylons is generally reported in the trade literature in terms of Rockwell values (ASTM D785); the higher the number, the greater the hardness. Rockwell hardness is an indentation hardness and should not be considered a measure of the abrasion or wear resistance of nylon, which are discussed later in this chapter. Indentation hardness has been correlated with yield stress of various plastic materials including nylon-66 (19).

The Rockwell hardness scales, which indicate indenter diameter and load, are identified by a letter. Nylons are usually reported on the R-scale, although some data on the overlapping and more severe M-scale are available. There is no correlation between scales, and it is apparent from the data (Table 10-15) that no reliable conversion can be made even within the family of nylon resins. As expected, nucleation causes an increase in hardness. Plasticizers such as monomer in unextracted 6 or water decrease hardness, and glass reinforcement increases hardness.

MECHANICAL DURABILITY

Creep

According to ASTM E6 creep is the time-dependent part of strain resulting from force, but the word creep is frequently used in a more general sense. It is concerned with the total long-term deformation from stress whether constant, variable, continuous, or cyclic. Creep information is essential to design in applications involving long-term loads. It is also of considerable interest in theoretical studies of the fundamentals of material behavior. The theoretical aspects of creep are beyond the scope of this chapter and are available in other publications (20, 21).

Table 10-15. Rockwell hardness (ASTM D785) of nylons[a]

Nylon	Conditioning treatment		
	Dry	Equilibrium, 50% R.H.	Unspecified[b]
66	R118-120, M79	R108-109, M59-60	
Nucleated 66	R123, M91		
6	R116-119	R97-101, M49	
Nucleated 6	R121	R109	
Unextracted 6	R100-107		
612	R114		
610	R110-111		
11			R107-108
12			R106-108, M31
66 + 30-33% glass fiber	R121-123, M96-101		
6 + 30% glass fiber	R121, M92		

[a] Data from trade literature.

[b] Conditioning normally has only a small effect on the hardness of nylons that absorb little moisture.

Nylon creep data for design requirements are becoming more readily available (8,21,22,23). An example provided in Figure 10-17 shows the creep behavior in tension of dry nylon-66 at 20°C (68°F) at stresses ranging from 1000 to 6000 lb(wt)in^{-2} (70-420 kg(wt)cm^{-2}). Total strain is plotted against log time. The time of onset of a rapid rate of change of strain with time is seen to decrease by a factor of about 100 as the stress is increased by a factor of 6. For design purposes strains in the range of 1% and below are of most interest.

Figures 10-18 and 10-19 present the same information in four different ways and illustrate the methods used to show creep data. All concern the creep in flexure at 23°C (73°F) of nylon-66 conditioned to 50% R.H. Figure 10-18a is of the same kind as Figure 10-17, a plot of percent strain versus log time at various levels of stress. Figure 10-18b, constructed from Figure 10-18a, is a plot of isometric stress versus log time at three levels of strain. This can be useful to the designer when he has established a permissible strain level and wants to know how long a stress can be sustained before reaching that level of strain. Figure 10-19a, also constructed from Figure 10-18a, is a plot of isochronous stress versus strain at four times. It is useful when the duration of load is fixed and the relationship between stress and strain for this specific period of time provides

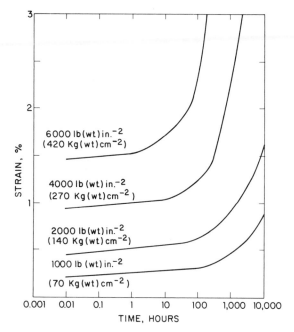

Fig. 10-17.   Creep in tension of dry nylon-66 at 20°C (68°F) (adapted from Ref. 21).

the basis for design. Figure 10-19*b* is a plot of creep (apparent) modulus versus log time. The creep modulus is a secant modulus as defined by ASTM D638 or D790 and is the ratio of stress to strain at a particular time. Creep data in the form of Figure 10-19*b* are convenient because engineering formulas often involve modulus rather than strain.

Much of the creep data shown were measured as creep in flexure of beam shaped test bars. The usual simple beam equations are applicable. Stress, S, is calculated as the maximum stress in the outer "fiber." It is related to load, P, at the midpoint, the length between supports, L, and the width, b, and thickness, d, of the beam:

$$S = \frac{3PL}{2bd^2}$$

Strain, r, is calculated as the maximum strain in the outer "fiber" and is related to the deflection, D, at the midpoint:

$$r = \frac{6Dd}{L^2}$$

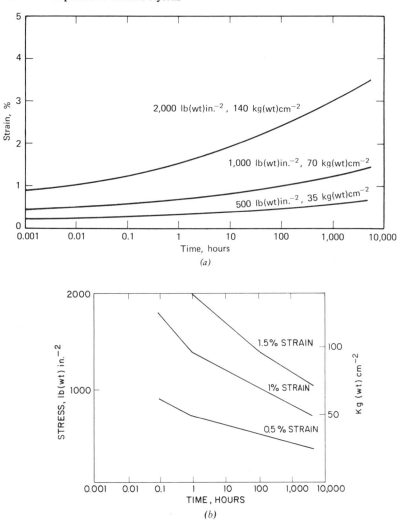

Fig. 10-18. Creep in flexure of conditioned nylon-66 at 23°C (73°F) and 50% R.H.: (a) strain and (b) stress versus log time (adapted from Ref. 8).

In recent years creep data on commercial products have been included in *Modern Plastics Encyclopedia* (23). The 1971-1972 edition gave data on 35 nylon resins including glass-reinforced products. These data are presented in tabular form listing creep (apparent) moduli at several time periods for practical levels of temperature and stress.

Figure 10-20 illustrates for nylon-66 the increase in creep caused by increasing relative humidity or temperature. The creep at 50% R.H. of nylon-612

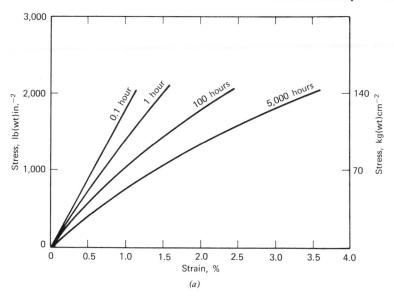

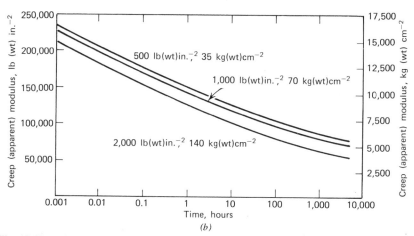

Fig. 10-19. Creep in flexure of conditioned nylon-66 at 23°C (73°F) and 50% R.H.: (a) stress versus strain and (b) creep (apparent) modulus versus log time (adapted from ref. 8).

(8) is less than that of nylon-66 (8) and nylon-6 (22) (Figure 10-21); this trend correlates with moisture absorption. Figures 10-22 and 10-23 show that nucleating agents decrease creep in nylon-6 (21) and nylon-66 (8). The large diminution in creep achieved by adding glass fiber is shown for nylon-66 (8) in Figure 10-24.

The dimensional recovery upon the removal of a load producing creep is of interest in design; however, data concerning recovery are not as available. Figures

10-25 and 10-26 (8) give an indication of the behavior of nylon-66 at 23°C (73°F) and 50% R.H. after short and long periods of loading. Recovery is more rapid after short loading times, but a measurable deflection remains even after a lengthy recovery period.

The related subject of long-term strength, creep rupture, is important to the design of pipe and pressure vessels and is discussed in Chapter 12, p. 449, and Chapter 18, p. 604.

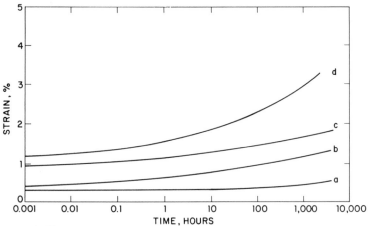

Fig. 10-20.   Effect of humidity and temperature on creep at 1000 lb (wt) in.$^{-2}$ (70 kg(wt)cm^{-2}) stress of nylon-66: (a) dry, 20°C (68°F) in tension (21); (b) 50% R.H., 23°C (73°F) in flexure (8); (c) 50% R.H., 60°C (140°F) in flexure (8); (d) dry, 125°C (257°F) in flexure (8).

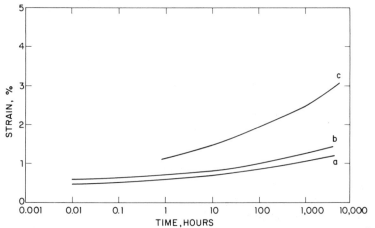

Fig. 10-21.   Effect of type of nylon on creep in flexure at 1000 lb(wt) in.$^{-2}$ (70 kg(wt)cm^{-2}) stress at 23°C (73°F) and 50% R.H.: (a) nylon-612 (8); (b) nylon-66 (8); (c) nylon-6 (22).

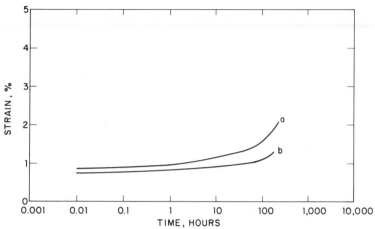

Fig. 10-22. Effect of nucleation on creep in tension at 2900 lb (wt) in.$^{-2}$ (203 kg (wt) cm^{-2}) stress at 20°C (68°F) of dry nylon-6: (*a*) standard grade and (*b*) nucleated grade (adapted from Ref. 21).

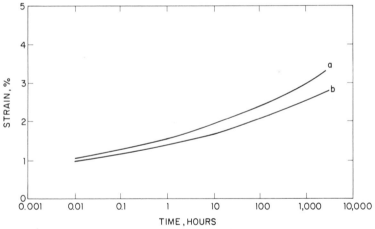

Fig. 10-23. Effect of nucleation on creep in flexure at 2000 lb (wt) in.$^{-2}$ (140 kg (wt) cm^{-2}) stress at 23°C (73°F) for nylon-66 conditioned at 50% R.H.: (*a*) standard grade and (*b*) nucleated grade (adapted from Ref. 8).

Deformation Under Load

The deformation under load test (ASTM D621) measures the resistance of a plastic to dimensional change from 10 sec to 24 hr after applying a compressive load at 73, 122, or 158°F (23, 50, or 70°C). It may be described as a short-term compressive creep test measured on cubes with 0.5 in. (12.7 mm) sides. The merit of this test is open to question particularly with the increased availability of creep data. Typical values for nylons are given in Table 10-16.

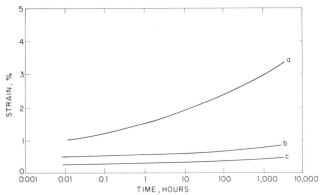

Fig. 10-24. Effect of the addition of glass fiber on the creep in flexure at 2000 lb (wt) in.$^{-2}$ (140 kg (wt) cm^{-2}) stress at 23°C (73°F) for nylon-66 conditioned to 50% R.H.: (*a*) unmodified, (*b*) 13% glass fiber, and (*c*) 33% glass fiber (adapted from Ref. 8).

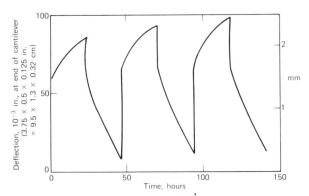

Fig. 10-25. Creep and recovery of nylon-66 subjected to a cyclic stress of 1000 lb (wt) in.$^{-2}$ (70 kg(wt) cm^{-2}) at 23°C (73°F) and 50% R.H. measured as the deflection at the end of a cantilever test bar (3.75 × 0.5 × 0.125 in. = 9.5 × 0.127 × 0.32 cm) (8).

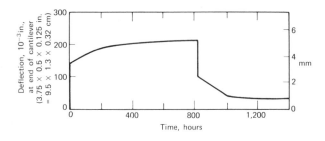

Fig. 10-26. Creep and recovery of nylon-66 subjected to a cyclic stress of 1500 lb(wt)in.$^{-2}$ (105 kg(wt)cm^{-2}) at 23°C (73°F) and 50% R.H. measured as the deflection at the end of a cantilever test bar (3.75 × 0.5 × 0.125 in. = 9.5 × 0.127 × 0.32 cm) (8).

358

Table 10-16. Deformation under load (ASTM D621) of dry nylons (24)

Nylon	Percent[a] 2000 lb(wt)in.$^{-2}$ (140 kg(wt)cm^{-2})	4000 lb(wt)in.$^{-2}$ (280 kg(wt)cm^{-2})
66	1.4	1.5
6	1.6	1.8
610	4.2	- -
612	1.6	- -
66/610/6	20	- -
66 + 30% glass fiber		0.8
6 + 30% glass fiber		.9
610 + 30% glass fiber		.9
612 + 30% glass fiber		1.0

[a] Deformation at 122° F (50° C) after 24 hr.

Fatigue Resistance

Nylons have excellent fatigue resistance. Applications as diverse as gears and brush filaments depend on this ability of nylon plastics to withstand cyclic or vibrational stresses.

A common method of obtaining fatigue data is by means of a Sonntag-Universal machine. This device may be used to test a suitable specimen in tension, compression, or flexure and is characterized by provision for a constant maximum stress. Alternate tension and compression comprise a common mode of test. The "Wohler method" implies a constant maximum stress and includes all tests run on the Sonntag machine. To be meaningful, data obtained by the "Wohler method" should indicate the nature of the stress. In the Krouse technique (Method A in an earlier version of ASTM D671) a constant maximum deflection in flexure is maintained, and the initially applied stress changes if the modulus of the specimen varies during test. Because of problems in the Krouse device, ASTM D671 is now a flexural fatigue test that involves only constant maximum force, and a new constant deflection method is under development.

All fatigue tests involve rapid cycling (e.g., 1800 cycles/min) so the question of heating (22) and change in modulus is pertinent. For this reason the specimen is sometimes cooled during testing. Thus, fatigue data should specify the technique used, the nature of the stress, the cycle, and any provision for cooling as well as the usual details with regard to the condition of the nylon and the nature of the environment.

To estimate fatigue endurance, stress is plotted against the number of cycles to failure (S-N curve), and the maximum stress corresponding to 10^6 or 10^7 cycles without failure is sometimes cited as the endurance value or "limit." This may not be a true limit which implies no reduction in stress level with further increase in the number of cycles.

The response of nylon-66 to a cyclic flexural stress is shown in Figure 10-27 (8). The fatigue endurance for the dry resin is greater than that of the resin conditioned to 50% R.H. Axial fatigue data for nylon-66 with alternate tension and compression are provided in Figure 10-28. Slightly higher stresses to failure can be sustained in this test than in the flexural fatigue test. As expected from the effect of water, the fatigue endurance level of nylon-66 decreases with increasing temperature (Figure 10-29). The important effect of cooling on performance is clearly demonstrated. Nylon-66 and nylon-612 are comparable

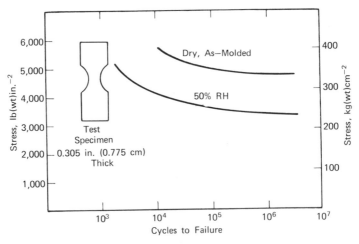

Fig. 10-27. Sonntag flexural fatigue of nylon-66 at 23°C (73°F) constant maximum stress, and 1800 cycles per minute (8).

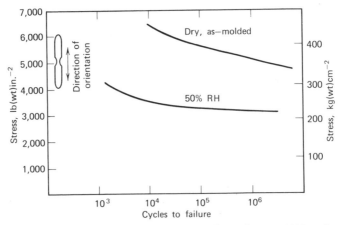

Fig. 10-28. Sonntag axial fatigue of nylon-66 at 23°C (73°F) and 1800 cycles per minute with alternate tension and compression (8).

when tested at 50% R.H. in alternate axial tension and compression (Figure 10-30).

Only limited data are available in which the fatigue behavior of nylon resins has been determined in atmospheres other than air. As indicated in Figure 10-31 gasoline vapors have little effect on the fatigue endurance level of nylon-66.

Glass reinforced nylons have higher fatigue strength (26,27,28) than the unmodified resins and are superior in this regard to other glass fortified compositions (26, 27). Short (0.125 in. = 0.32 cm) fibers yield products with better fatigue performance than long (0.5 in. = 1.3 cm) fibers (27).

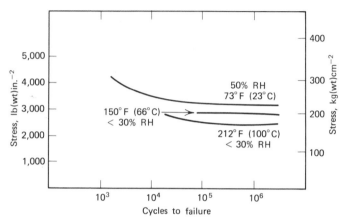

Fig. 10-29. Effect of temperature on Sonntag axial fatigue of nylon-66 at 1800 cycles per minute with alternate tension and compression (8).

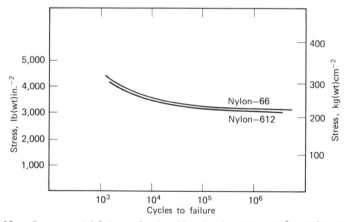

Fig. 10-30. Sonntag axial fatigue of nylon-66 and nylon-612 at 23°C (73°F), 50% R.H., and 1800 cycles per minute with alternate tension and compression (adapted from Ref. 8).

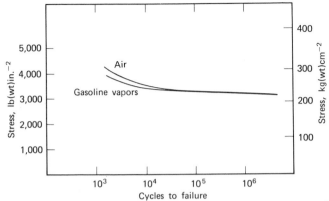

Fig. 10-31.   Effect of gasoline vapors on Sonntag axial fatigue of nylon-66 at $23°C$ ($73°F$) and 1800 cycles per minute with alternate tension and compression (8).

Abrasion Resistance, Friction, and Wear

Much of the early success of nylon as a plastic was due to its abrasion resistance and its ability to be used as an unlubricated bearing (see Chap. 20, pp. 626-629). Nylon composition compounded especially for low friction and wear are described in Chapter 11 (p. 437).

By tradition in the United States the abrasion resistance of nylons is measured in a nonstandard Tabor abrasion test adapted from ASTM D1044. Results are expressed as milligrams lost in 1000 cycles with conditions specified as a CS-17 wheel with a 1000 g load. Data for a number of nylons are given in Table 10-17. Copolymerization and the addition of glass reduce the abrasion resistance according to this test. Abrasion resistance of polymers has been extensively reviewed in Ref. 29.

Table 10-17. Tabor abrasion of nylons[a]

Nylon	mg loss per 1000 cycles[b]
66 ($\overline{M}_n$ 18,000)	6-8
66 ($\overline{M}_n$ 34,000)	3-5
6	5
610	5-6
612	5-7
6 copolymer	16
66/610/6	9-14
66 + 13% glass fiber	12
66 + 33% glass fiber	14
610 + 30% glass fiber	18

[a] Data from trade literature.
[b] CS-17 wheel, 1000-g load.

Data on friction and wear generally are not published in trade literature probably because reproducibility of frictional measurements is poor. A general discussion of friction and plastics is beyond the scope of this chapter and may be found in Ref. 20. Table 10-18 gives data on the static and dynamic coefficient of friction for nylon-66 (8) against itself, acetal homopolymer, and steel.

Figure 10-32 shows the effect of normal pressure on the coefficient of friction for nylons-66, -610, and -6 (22). The references caution against using these values of coefficients for situations other than the specific apparatus used for testing.

The best information concerning friction and wear of nylon comes from experience with its use as a bearing. Chapter 18, p. 602, and Ref. 8, 22, 30, 31, and 32 give suggestions concerning the design and use of nylon in bearings. Chapter 18, p. 600, and Ref. 8 and 33 give design information for nylon in gears. Details concerning the predictions of wear in bearings are covered in Ref. 8, 30, 34, and 35.

Table 10-18. Coefficient of friction at 23°C (73°F) of unlubricated nylon-66 containing 2.5% water (8)

Other surface	Coefficient of friction[a]	
	Static	Dynamic
Nylon-66	0.36−0.46	0.11−0.19
Acetal homopolymer	0.13−0.20	0.08−0.11
Steel	0.31−0.74	0.17−0.43

[a]Determined by the thrust washer method with a pressure of 20 lb(wt)in.$^{-2}$ (1.4 kg(wt)cm^{-2}) and at a surface speed of 95 ft min^{-1} (29 m min^{-1}).

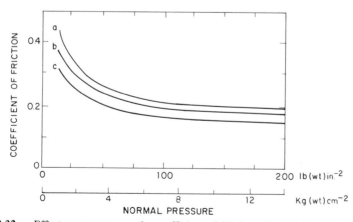

Fig. 10-32. Effect on pressure on the coefficient of friction of unlubricated nylons: (a) nylon-610, (b) nylon-66, and (c) nylon-6 (22).

WATER ABSORPTION, DENSITY, AND DIMENSIONAL STABILITY

Water Absorption

The important effect of water on the properties of nylons has been noted many times. The treatment of water and temperature as complementary variables was discussed in Chapter 9. Data on water absorption are provided in Chapter 5 in connection with material handling requirements for injection molding and in Chapter 17 in the context of conditioning to a desired moisture level. The specific information available in these chapters is indicated in Table 10-19. The

Table 10-19. Summary of data on water absorption by nylons

Nylon	Information	Location of data
6 66 610 612 11 12	Water absorption after 24 hr (ASTM D-570), equilibration at 50% R.H., and saturation.	Table 5-1 (p. 157)
6	Effect of geographical location on moisture content.	Table 5-2 (p. 158)
66	Rate of moisture absorption of pellets at 23°C and 50, 75, and 100% R.H.	Figure 5-1 (p. 159)
66	Rate of moisture absorption of regrind at 23°C and 50, 75, and 100% R.H.	Figure 5-2 (p. 160)
6 66	Rate of moisture absorption at 20°C, 100% R.H. at thicknesses of 2 and 5 mm.	Figure 17-1a and 17-1b (p. 558)
610	Rate of moisture absorption at 20°C, 100% R.H. at thicknesses of 1, 2, and 5 mm.	Figure 17-1c (p. 558)
6 66	Relationship between part thickness and time in water at 20-100°C to absorb 2.5 or 7.0% water.	Figure 17-2a and 17-2b (p. 559)
610	Relationship between part thickness and time in water at 20-100°C to absorb 1.2 or 3.0% water.	Figure 17-2c (p. 559)
66	Relationship between part thickness and time in potassium acetate solution at 121°C to absorb 2.5% water.	Figure 17-3 (p. 561)

expansion due to water absorption is considered in the following section on dimensional stability.

The effect of crystallinity on the water absorption of nylons-66 and -610 is illustrated in Figure 10-33. Lowering the amide-group concentration and the relative humidity diminishes the effect of a change in crystallinity. There is little effect of temperature over the normal range of 20 to 100°C on the amount of water absorbed for nylons-6, -66, -610, and -612, but a major effect is reported for nylon-11 at 100% R.H. (from 1.9 at 20°C to 2.9 at 100°C). Presumably, a similar phenomenon applies to nylon-12 although no verification has as yet been published. The dependence of equilibrium moisture content on relative humidity

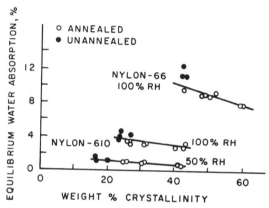

Fig. 10-33.   Water absorption versus crystallinity of nylon films (adapted from Ref. 36 using crystallinity correction in Ref. 39).

for typical moldings of commercial nylons is shown in Figure 10-34. The expected relation to amide group concentration is clear and is depicted graphically in Figure 10-35.

The water absorption of modified compositions obviously depends on the nature of the modification. Plasticized resin such as unextracted nylon-6 will have the same moisture content as extracted polymer but will show lower weight gain by virtue of loss of monomer. Even unmodified resins may contain about one percent of extractable material so that precise data will require a moisture analysis as well as weight gain information. On the other hand, resins containing glass or other inert additives absorb water in proportion to the fraction of nylon present.

Density

The measurement of specific gravity and density is discussed in Chapter 3, p. 87. The variation of density with temperature for commercial nylon plastics is

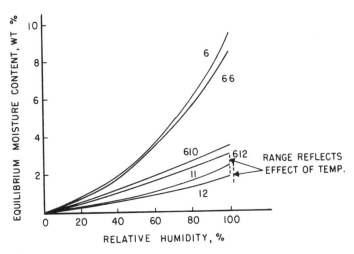

Fig. 10-34. Dependence of moisture content on relative humidity for nylons (data from trade literature).

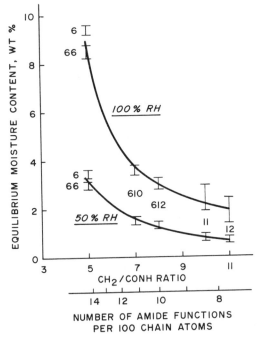

Fig. 10-35. Dependence of moisture content of aliphatic nylons on amide-group concentration.

366

illustrated in Figure 10-36. Curves such as these are instructive in that they provide an estimate of the rate of change. Each curve should be a band, however, to reflect the variation of density with crystallinity. The melt densities will also vary because of differences in pressure as noted earlier in Table 4-2 (Chapter 4, p. 124).

The estimation of crystallinity was discussed in Chapter 8 where the utility of the concept in accounting for changes in properties was recognized although a simple two-phase system consisting of pure amorphous and crystalline regions does not in fact exist. Density has proved to be a convenient index of "percent crystallinity," and linear relationships such as those shown in Figure 10-37 have been reported. Crystallinities over about 60% have never been achieved from the melt, and extension of the line to 100% is based on extrapolations of infrared

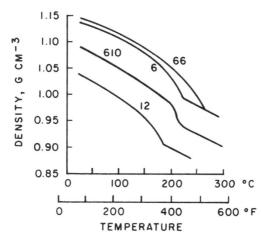

Fig. 10-36.   Density versus temperature for commercial nylons (adapted from Refs. 15 and 37).

data and theoretical values calculated from unit cell dimensions. The literature does not agree on the precise position of these lines. For example, nylons-6 and -66 are sometimes shown to have similar but non-identical density-crystallinity relationships with a cross-over reflecting a higher amorphous and lower unit cell density for nylon-6. There is, however, much variation in the crystallographic data (40), and the distinction between 6 and 66 is within experimental error and justifies use of a single line for the sake of simplicity. The typical range for commercial products, shown in Figure 10-37, is based on the specific gravity values given in Table 3-1 (Chapter 3, p. 88).

The density of nylons from omega-aminoacids decreases fairly regularly with decreasing amide-group concentration (Figure 10-38). The density of diamine-

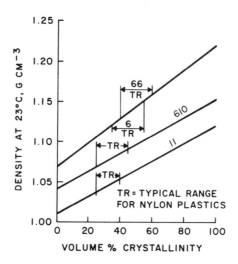

Fig. 10-37.   Density versus crystallinity for commercial nylons (data from Refs. 38, 39, 40 and Table 3-1)

diacid derived polymers shows a saw tooth dependence on amide-group concentration (Figure 10-39).

The effect of moisture on density is small. For nylon-66 and nylon-610 the absorption of water has been shown to either increase or decrease density because of the variation in the partial specific volume of water in the polymer (43). For example, the partial specific volume of water in nylon-66 increases from 0.46 cm^3 g^{-1} at 0% R.H. to 1.0 at 100% R.H. At equilibrium in 50% R.H. where a value of 0.73 cm^3 g^{-1} can be assumed and 2.5% water is absorbed, the calculated density, based on 1.140 for the dry polymer, is 1.145 g cm^{-3}. This also assumes no change in crystallinity of the nylon occurs during equilibration to 50% R.H.

Dimensional Stability

Dry nylon parts made by any means are subject to dimensional changes caused by water absorption, crystallization, stress relief, and thermal expansion or contraction. Nylon articles are, however, found in applications where dimensional stability is important because the major factors, water absorption and stress relief, often offset each other.

In most real situations the humidity of the environment is cyclical, and the moisture content of nylon parts will fluctuate around an average value. Once this average value is approached the subsequent dimensional change depends on how much the humidity varies and the time available at ambient temperature for the nylon part to respond. Experience has shown that under normal conditions the

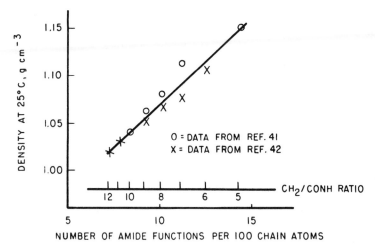

Fig. 10-38.   Density versus amide-group concentration for nylons from omega-aminoacids.

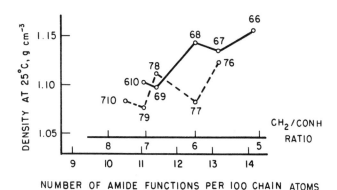

Fig. 10-39.   Density versus amide-group concentration for nylons from hexamethylenediamine (– – – – –) and heptamethylenediamine (––––) (adapted from Ref. 41).

change due to cyclical variations in humidity is small, usually less than 0.2% in any one direction (8).

The equilibrium moisture content of a nylon depends on amide-group concentration, crystallinity, relative humidity, and sometimes on temperature as noted in the above discussion on water absorption. The expansion effect of the moisture can also vary because the specific volume of water in nylon may depend on humidity as mentioned in the preceding section on density. Thus, an exact prediction of the dimensional change resulting from exposure to a specified relative humidity involves knowledge not only of the kind of nylon and part geometry but also of thermal history and absorption isotherms. Conservative, that is, high, first approximations can be made, however, if it is assumed that the volume change equals the moisture absorption typical of commercial moldings (Figure 10-34), the part is stress-free, and the change is isotropic. The maximum linear change in dimensions is, accordingly, estimated to be one-third of the moisture content. Calculated and reported dimensional changes are shown in Figure 10-40 for nylon-66 and in Table 10-20 for nylons-6, -610, and -12.

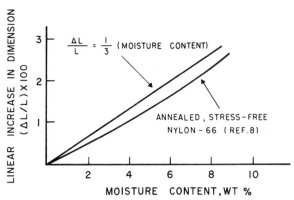

Fig. 10-40.   Estimated and actual linear change in dimension of annealed, stress-free nylon-66 with moisture content.

Table 10-20. Calculated and reported dimensional changes of nylons-6, -610, and -12 due to absorption of water

	Nylon-6 (Extracted)[a]	Nylon-610[a]	Nylon-12[b]
At 20°C, 65% R.H.:			
Reported % water absorbed	3.5-4.0	1.8-2.0	- -
Reported % max. linear change	1.1-1.3	0.5-0.6	- -
Calc'd linear change = (1/3)			
(% water)	1.2-1.3	0.6-0.7	- -
At 20°C, 100% R.H.:			
Reported % water absorbed	8.5-10.0	3.2-3.8	1.4
Reported % max. linear change	2.8-3.0	0.9-1.2	0.2
Calc'd linear change = (1/3)			
(% water)	2.8-3.3	1.1-1.3	0.5

[a] Absorption and expansion data from Ref. 15.

[b] Expansion data from Ref. 45.

Stresses frozen into nylon parts are discussed in Chapter 17 in relation to annealing and conditioning. Relief of stress does take place under typical end-use conditions although more slowly. The net effect typical of combined stress relief and water absorption is shown for an injection molded plaque of nylon-66 in Figure 10-41. This is the situation that usually occurs and accounts for the small overall change. The linear change is seen to be below 0.5% for relative humidities up to 90%. Crystallization also contributes to the shrinkage but is normally a minor factor.

Mold shrinkage constitutes a design factor often considered in the present context because it is convenient to relate part dimensions to cavity size rather than dry, as-molded dimensions. In general, mold shrinkage increases with increasing crystallinity so that high mold temperatures and thick parts cause high shrinkage. On the other hand, increasing injection pressure tends to reduce shrinkage. Dimensional changes that are based on cavity size and represent the combined effects of mold shrinkage, stress relief, and water absorption are shown for a nylon-66 and a nylon-612 composition in Figure 10-42.

Low mold shrinkage reflects quenching in the mold that causes higher frozen-in stress which leads to higher shrinkage when relieved. Annealed parts, therefore, have similar dimensions despite differences in mold temperatures, and unannealed parts in use will eventually approach comparable dimensions. This situation is illustrated for extracted nylon-6 in Figure 10-43 which shows the effect of thickness on shrinkage of a molded bar. Gate size, pressure, and

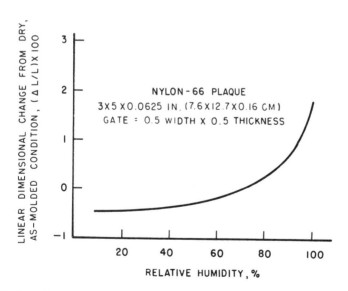

Fig. 10-41.   Combined effect of stress relief and water absorption on linear dimensional change of a nylon-66 plaque (8).

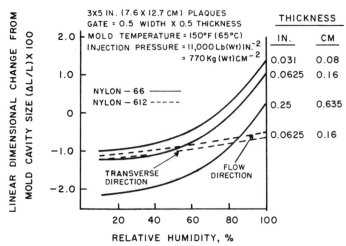

Fig. 10-42. Dimensional change of unannealed plaques of nylon-66 and -612 based on cavity size (adapted from Ref. 8).

temperatures were adjusted to vary mold shrinkage, but the total shrinkage after annealing is the same (44).

The coefficient of linear thermal expansion for nylons varies from about 6 to 12×10^{-5} $(\Delta L/L)$ $^\circ C^{-1}$ or 3 to $7 \times 10^{-5}\,^\circ F^{-1}$ depending on the nylon, its moisture content, and the temperature (see section on thermal properties). Thus, a $100^\circ C$ $(180^\circ F)$ change in temperature produces a change in linear dimension of 0.6 to 1.2% or about 0.01 in./in. (0.01 cm/cm).

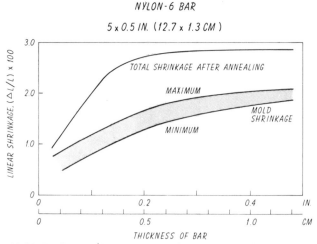

Fig. 10-43. Mold shrinkage and annealing shrinkage versus thickness for a nylon-6 bar (adapted from Ref. 44).

Modified nylons can exhibit markedly different dimensional behavior than the parent resins. For example, nucleation and glass reinforcement can affect the anisotropy such that both as-molded and in-use dimensions will differ from those of unmodified resin. The performance of a nylon-66 containing 33% glass fiber is illustrated in Figure 10-44. The dimensional change in the flow direction is much less than that in the transverse direction, but in either direction the stability is better than that of the parent resin (compare Figure 10-42). The water absorption of this composition is only two-thirds of the unmodified resin; the coefficient of thermal expansion, about one-third. All of these factors contribute to the superior dimensional behavior of glass-reinforced resins.

The above information is intended to provide an understanding of the factors affecting dimensions of nylon parts and to give estimates of the amount of change to be expected. For precise work the manufacturer's literature sometimes suffices, but careful testing of the actual part is often necessary.

ENVIRONMENTAL RESISTANCE

Resistance to Chemicals and Other Substances

The trade literature commonly evaluates the resistance of proprietary nylon plastics to a wide variety of chemicals, food preparations, and the like in terms such as excellent (resistant), good (resistant under some conditions), and unsatisfactory (not resistant). This is often adequate even though information on the effects of concentration, temperature, and time and data on absorption,

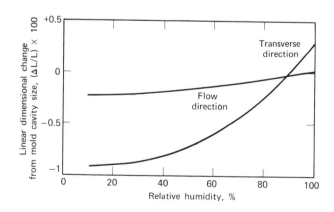

Fig. 10-44.   Dimensional change of a plaque of glass reinforced nylon-66 based on cavity size (adapted from Ref. 8).

dimensional changes, and alteration of mechanical properties are limited. Additonal insights into the kinds of materials that might affect performance and should therefore concern the designer are available. Examples of helpful generalizations with regard to percent crystallinity, amide group concentration, copolymerization, temperature, and stress as well as the type of solvent, acid, base, and so on are provided in the sections on solvent attack and hydrolysis in Chapter 2 (pp. 68-71) and in the discussion of stress cracking in Chapter 17 (p. 555).

The best assurance of performance comes from the many successful applications of nylons not only as moldings but also as film, filament, and fiber. For example, automotive parts of nylon-66 were examined after extended service and exposure to gasoline, greases, oils, and salts (46). Unmodified resin showed excellent retention of properties when used as chassis and body components, and heat stabilized formulations performed equally well in hot engine and transmission applications.

The absorption of water by nylons is reviewed in an earlier part of this chapter, and the effect of water is noted in discussing the individual properties. This is an obvious consideration when examining aqueous solutions, but it can also be a factor if the nylon is exposed to large quantities of organic solvents or substances that may contain relatively small amounts of water.

By way of example, the amount of various solvents absorbed by several nylons is shown in Table 10-21, and the effect of crystallinity is illustrated in Table 10-22.

Resistance to Elevated Temperatures in Air

The thermal oxidation of nylons from a chemical point of view was reviewed in Chapter 2. Here we will consider the changes in physical properties resulting from thermal oxidation. It is this resistance to air oven aging of molded specimens and not thermal decomposition above the melting point that is commonly meant by "heat stability" in the trade literature. The efficacy of stabilizers will be noted, but a discussion of stabilizer systems is reserved for Chapter 11, p. 426.

The retention of tensile strength of standard and stabilized grades of nylon-66 (8) and nylon-6 (10) is compared in Figures 10-45 and 10-46. Stabilization is obviously very effective. There is less difference between the standard and stabilized grades of 33% glass-fiber-reinforced nylon-66 (8) (Fig. 10-47) because glass alone causes a large improvement in durability. Additional data on glass-reinforced nylon are given in Figure 11-9 of Chapter 11.

Figures 10-48 and 10-49 show the effect of stabilization on the retention of impact strength by nylon-66 (8) and nylon-6 (10). The impact strength of standard resin decreases more rapidly than tensile strength, but a many fold improvement by stabilization is again observed.

Table 10-21. Absorption of various solvents by commercial nylons

Nylon	% Wt gain[a]			
	6	66	610	11
Water	11	10	4	1.8
Methyl alcohol	19	14	16	9.5
Ethyl alcohol	17	12	13	10.5
n-Propyl alcohol	18	12	17	--
n-Butyl alcohol	16	9	17	--
Ethylene glycol	13	10	4	--
Glycerol	3	2	2	--
Benzyl alcohol	55	38	40	--
Phenylethyl alcohol	31	9	29	--
Acetaldehyde	16	14	16	--
Benzaldehyde	26	10	35	--
Methylene chloride	20	16	24	19
Chloroform	34	27	40	33
Trichlorethylene	5	4	20	--
Perchlorethylene	2	2	3	11
Carbon tetrachloride	4	2	3	4.5
Acetone	4	2	5	4.5
Methyl acetate	3	2	5	5.5
Aliphatic hydrocarbons	1	1	1	1
Benzene	1	1	4	7.5
Toluene	1	1	3	6.8
Cyclohexane	1	1	1	1

[a] Low crystallinity specimens equilibrated at room temperature. Adapted from Ref. 47.

In recent years much air oven aging of plastics has been done to determine an Underwriters Laboratories "temperature index" (48). Mechanical and electrical tests are carried out on specimens aged at four elevated temperatures, and the time for a 50% reduction in the property at each temperature is determined. The logarithm of the time to 50% property reduction is plotted against the reciprocal of the absolute temperature (Arrhenius curves). The temperature in °C corresponding to the extrapolated time of 40,000 hours is the UL temperature index. A 1971 listing of indices (49) includes 85 nylon compositions. The index depends on the property measured. We have already seen that impact strength decreases more rapidly than tensile strength. Electrical properties change least. Thus an unstabilized nylon-66 is assigned a value of 75°C without qualification, 85° if resistance to impact is not essential, and 105° if only electrical properties are essential. Heat stabilized nylon-66 has a value of 105°C whether impact strength, tensile strength, or electrical property retention is the criterion of

Table 10-22. Effect of crystallinity on solvent absorption

	% Wt gain[a]					
	6		66		610	
	b	c	b	c	b	c
Water	11	9	10	7.5	4	3
Methanol	19	3	14	9	16	9
Ethanol	17	3	12	3	13	8
Benzene	1	0	1	0	1	1
Toluene	1	0	1	1	3	1
Acetone	4	3	2	1	5	1
Benzyl alcohol	55	11	38	3	40	11
Methylene chloride	20	8	16	10	24	13
Chloroform	34	19	27	5	40	21

[a] Equilibrated at room temperature. Adapted from Ref. 47.

[b] Low crystallinity.

[c] High crystallinity.

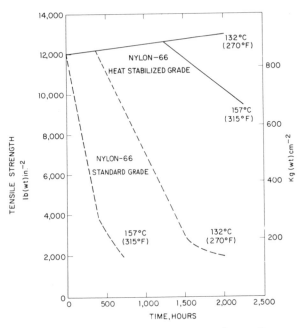

Fig. 10-45. Effect of temperature and stabilizers on the tensile strength of nylon-66 exposed in a circulating air oven (adapted from Ref. 8).

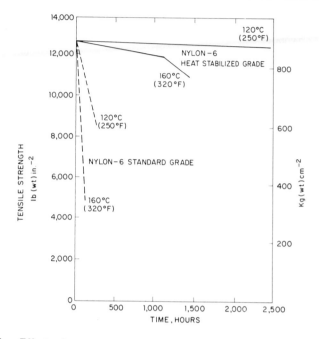

Fig. 10-46. Effect of temperature and stabilizers on the tensile strength of nylon-6 exposed to air oxidation (adapted from Ref. 10).

performance. The index also changes with thickness. For example, where resistance to impact is not required, a heat stabilized nylon-6 is rated at 95°C if the thickness is 0.05 in. (1.3 mm) and 100°C if 0.125 in. (3.2 mm). Better performance of thicker parts is consistent with the observation that oxidation occurs largely at the surface (see Chapter 2, p. 72).

Resistance to Hot Water

For many years nylons have been used in washing machines and dishwashers where resistance to change by water is required. Other applications require stability in the presence of steam. It is well recognized, however, that nylons are subject to some degradation from hot water, and the chemistry of the changes that occur is discussed in Chapter 2 (p. 70). Commercially this is acknowledged by the availability of hydrolysis resistant grades of nylons with improved performance in hot water uses (8, 10, 13). It has been clearly shown that changes which take place may not be simply related to hot water but rather to the combination of hot water and air (50). Thus, the oxygen content of water is

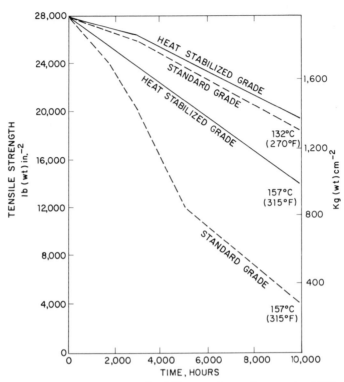

Fig. 10-47. Effect of temperature and stabilizers on the tensile strength of 33% glass fiber reinforced nylon-66 exposed in a circulating air oven (adapted from Ref. 8).

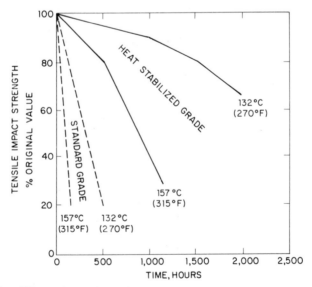

Fig. 10-48. Effect of temperature and stabilizers on the tensile impact strength of nylon-66 exposed in a circulating air oven (adapted from Ref. 8).

378

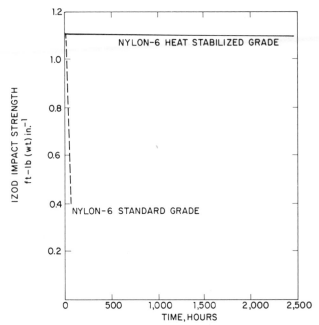

Fig. 10-49.   Effect of stabilizers on the Izod impact strength of nylon-6 exposed to air at 120°C (250°F) (adapted from Ref. 10).

important to the useful life of a nylon. For example, stagnant hot water in which oxygen is used up and is not replaced causes fewer changes to nylon than flowing hot water in which oxygen is continuously available. As would be expected the changes are also temperature dependent. Figure 10-50 shows the change in solution viscosity of hydrolysis stabilized nylon-66 that takes place upon exposure to water at 71°C (160°F) and 120°C (250°F) for periods up to 1200 hr (13). Standard and stabilized grades of nylon-66 are compared at 140°C (284°F) in Figure 10-51, which is a plot of tensile strength versus exposure times in water up to 100 hr (13). The standard resin shows an appreciable loss of strength after 24 hr.

Similar date for flowing water tests at 77°C (170°F) for standard and stabi-lized nylon-66 are given in Figure 10-52 (8). The tensile strength of the standard product is reduced to 1000 lb(wt)in.$^{-2}$ (70 kg(wt)cm^{-2}) after 2000 hr.

Figure 10-53 shows the results of stagnant boiling water tests on nylon-66 containing 33% glass fiber (8). Stabilization of the nylon again gives improved performance.

A stabilized nylon-66 when tested in water at 71°C (160°F) and at 120°C (250°F) for 1000 hr showed essentially no change in tensile strength but about a 50% decrease in tensile elongation (13). Similar results were obtained on

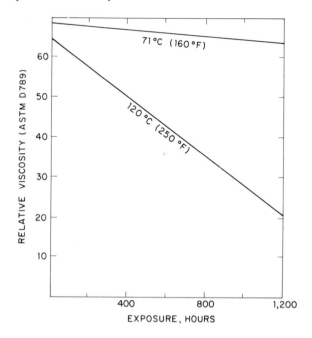

Fig. 10-50.   Effect of exposure to hot water on the relative viscosity of hydrolysis stabilized nylon-66 (adapted from Ref. 13).

exposure of a stabilized nylon-66 to steam at 120°C (250°F) (8). These results are similar to those of aging in air at elevated temperatures or of weathering in that elongation or impact strength decreases before tensile strength is affected. Service lives for standard and stabilized nylon-66 based upon a 50% decrease in impact strength for four temperatures are given in Table 10-23 (8). It is based upon stagnant water tests, and a 30 and 50% reduction in service time is suggested for use in flowing water.

Weather Resistance

Nylon, like most plastic materials, is degraded by ultraviolet light and has relatively poor weatherability unless a stabilized grade of resin is used (51, 52). Weathering causes embrittlement, color change, and loss of surface gloss. The chemical changes taking place were discussed in Chapter 2, p. 73.

The effect of outdoor weathering on the tensile strength and elongation of standard and stabilized nylons is shown in Figures 10-54 to 10-57. Data on 0.125 in. (3.2 mm) thick tensile test bars of nylon-66 exposed in Florida, Arizona, and Delaware are reported in Figures 10-54, 10-55, and 10-56 (8). In

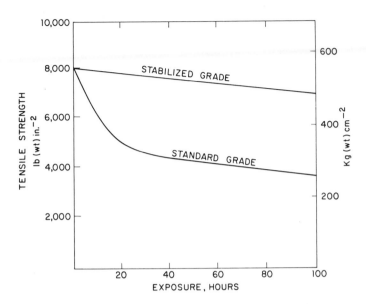

Fig. 10-51. Effect of exposure to water at 140°C (284°F) on the tensile strength of standard and hydrolysis stabilized nylon-66 (adapted from Ref. 13).

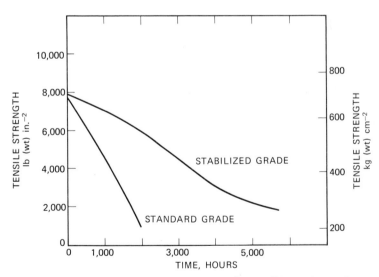

Fig. 10-52. Effect of exposure to flowing water at 77°C (170°F) on the tensile strength of standard and hydrolysis stabilized nylon-66 (8).

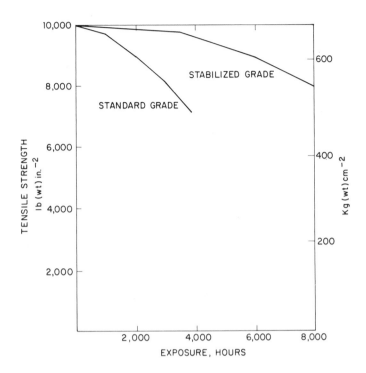

Fig. 10-53. Effect of exposure to stagnant boiling water on the tensile strength of standard and hydrolysis stabilized nylon-66 containing 33% glass fiber (adapted from Ref.8).

Table 10-23. Service life of standard and stabilized nylon-66 in stagnant hot water (8)

Temperature		Service life[a] (hr)	
°C	°F	Standard Grade	Stabilized Grade
100	212	1,500	5,000
93	200	2,000	6,500
82	180	3,000	10,000
71	160	8,000	25,000

[a] Based upon a 25 to 50% decrease in impact resistance and elongation.

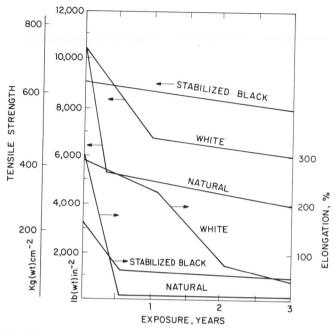

Fig. 10-54. Effect of weathering in Florida on the tensile strength and elongation of natural, white, and black stabilized nylon-66 (adapted from Ref. 8). Tensile bars were exposed at a 45° angle and were facing south. They were tested as received and had moisture contents ranging from 2 to 3%.

each case the composition containing sufficient carbon black to be recommended as a weatherable grade shows the least change in tensile strength with exposure. The behavior of a white pigmented nylon-66 is intermediate between that of the natural resin and the one containing carbon black. Figure 10-57 shows the effect of weathering in Florida on nylon-6 monofilament having a thickness of 0.015 in. (0.37 mm) (10). A nylon-6 containing only enough carbon black to color it does not give the weatherability of one containing sufficient carbon black to be classified as stabilized. The weathering behavior of nylons-6, -66, and -610 is the same based on exposure studies in Central Europe and Australia (15). For Central Europe, embrittlement times of 3, 5, and 10 years are given for resins containing 0, 0.5, and 2% carbon black; for Australia, times of 0.1 to 1, 3, and 10 years are reported.

Accelerated weathering data obtained with an X-W Weather-Ometer® for nylon-66 are given in Figure 10-58. The same general behavior is seen as in outdoor weathering – the resin containing carbon black shows the least change and the natural resin the most change.

The Florida and X-W Weather-Ometer® weathering behavior of glass-rein-

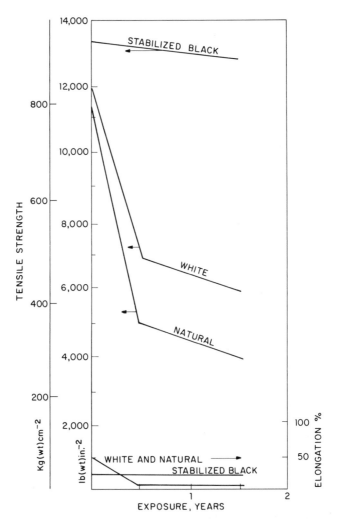

Fig. 10-55.   Effect of weathering in Arizona on the tensile strength and elongation of natural, white, and black stabilized nylon-66 (adapted from Ref. 8). Tensile bars were exposed at a 45° angle, were facing south, and were initially exposed in a dry, as-molded condition. They were tested as received without conditioning.

forced nylon-66 (8) and -610 (51) is shown in Figure 10-59. The addition of glass fiber both increases the tensile strength of the composition and improves its weatherability.

Radiation Resistance

Nylon is intermediate among plastic materials in resistance to radiation. Thus, on exposure to 6 megarads of gamma radiation from a cobalt 60 source, nylons-66

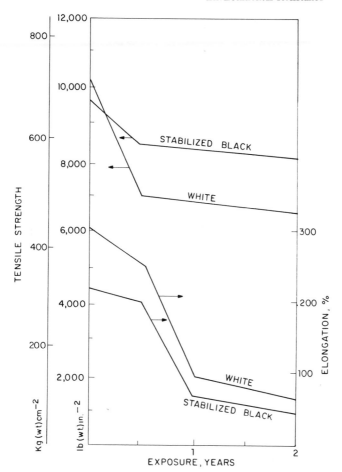

Fig. 10-56.   Effect of weathering in Delaware on the tensile strength and elongation of white and black stabilized nylon-66 (adpated from Ref. 8). Tensile bars were exposed at a 45° angle, were facing south, and were conditioned at 2.5% water before exposure. They were tested as received without conditioning.

and -11 were superior to polyethylene and poly(vinylidene chloride) but inferior to polystyrene and certain polyesters based on the amount of gas evolution (53). Similarly, in the neutron-gamma radiation flux of an Oak Ridge National Laboratory pile reactor, nylons-66 and -610 ranked ahead of acrylics and cellulosics and behind polystyrene and mineral filled phenolics (54). Aromatic nylons such as MXD-6 are more resistant than their aliphatic analogs.

Both cross-linking and chain scission occur and to a degree that depends on

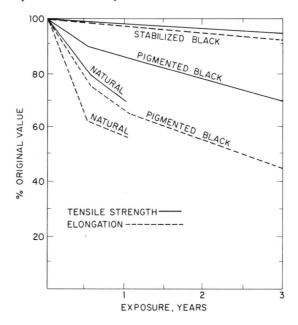

Fig. 10-57.   Effect of weathering in Florida on the tensile strength and elongation of natural, black pigmented, and black stabilized nylon-6 (adapted from Ref. 10). Specimens of monofilament 0.015 in. (0.38 mm) thick were exposed and tested.

the time and temperature of irradiation, the crystallinity and moisture content of the nylon, and post-irradiation treatment. Nylon-66 and nylon-610 exposed to pile radiation (54) show little initial change in tensile strength and modulus, but the modulus rapidly increases with increased time of exposure. Specific gravity and water absorption are also little affected initially but eventually show an increase at an exposure which is about the same and similar to that at which the modulus begins to change rapidly. Elongation and impact strength, however, decrease at low dosage levels.

A detailed discussion of the radiation chemistry of polyamides is available in Ref. 55.

The immediate consequences of irradiation are the breaking of bonds and the formation of free radicals, as in photo-oxidation (Chapter 2). The hydrogens adjacent to the amide nitrogens are most easily removed, and hydrogen is the chief gaseous product. Carbon monoxide, carbon dioxide, and methane are also produced.

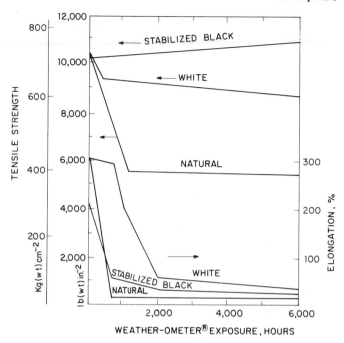

Fig. 10-58.   Effect of exposure in an X-W Weather-Ometer® on the tensile strength and elongation of natural, white, and black stabilized nylon-66 (adapted from Ref. 8). This was carried out with 0.125 in. (3.2 mm) tensile bars which were conditioned to 50% R.H. at 23°C (73°F) before testing.

THERMAL PROPERTIES

Melting point determinations and their significance are covered in Chapters 3 and 9.

The deflection temperature is the temperature at which a 0.01 in. (0.25 mm) deflection takes place when a bar is heated under one of two loads (ASTM D648). Values for dry nylons are given in Table 10-24. Significant errors can result with nylons unless moisture content is controlled and the bar is annealed for relief of molded-in stresses. Adding glass fibers produced a large increase in the deflection temperature (see also Chapter 11, p. 431). In general, the deflection temperature follows the melting point of the resin.

The coefficient of linear thermal expansion (ASTM D696) for dry nylons becomes larger as the hydrocarbon content of the nylon increases (Table 10-25). Adding glass markedly reduces the expansion coefficient as expected because of

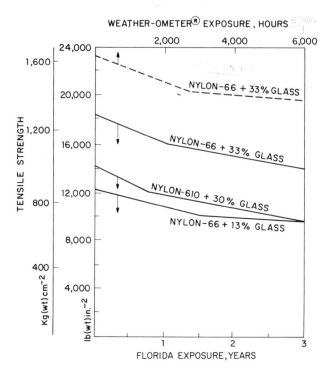

Fig. 10-59.   Effect of exposure in Florida or in an X-W Weather-Ometer® on the tensile strength of glass-reinforced nylon-66 (8) and nylon-610 (51). The nylon-66 was exposed as tensile bars which were at a 45° angle and facing south. They were conditioned to 50% R.H. at 23°C (73°F) before testing. The tensile bars from the Weather-Ometer® were tested without conditioning. It is estimated this corresponds to conditioning to 40% R.H. at 23°C (73°F). No conditions for the nylon-610 were given.

the lower coefficient for glass. Water has the opposite effect; saturation of nylon-66, for example, raises the value from 4.5 to 6.5 (in./in.) °F⁻¹. The coefficient would be expected to respond to temperature as it does to water, and this is shown to be true in Table 10-26.

The specific heat of dry nylons at room temperature is 0.3 to 0.4 (Table 10-27). Predictably, it increases with increasing temperature and is largest in the vicinity of the melting region (Table 10-28). The heat capacity of nylons-6 and -7 has been reported for a wide range of low temperatures, 70 to 360°K (-203 to 87°C) (58). At -203°C (-333°F) the specific heat of extracted nylon-6 is 0.105.

Values of thermal conductivities for nylons at room temperature are given in Table 10-29. The thermal conductivity of nylon-66 was measured from 0 to

Table 10-24. Deflection temperature (ASTM D648) of dry nylons[a]

Nylon	66 lb(wt)in.$^{-2}$ (4.6 kg(wt)cm^{-2})		264 lb(wt)in.$^{-2}$ (18.5 kg(wt)cm^{-2})	
	($^{\circ}$F)	($^{\circ}$C)	($^{\circ}$F)	($^{\circ}$C)
66 ($\overline{M}_n$ 18,000)	470	243	220	104
66 ($\overline{M}_n$ 34,000)	470	243	220	104
6 (extracted)	365	185	152	67
610	315	157	150	66
612	330	166	180	82
11	300	149	130	54
12	284	140	124	51
66 + 30% glass fiber			480	249
6 + 30% glass fiber			420	216
612 + 30% glass fiber			410	210

[a] Data from trade literature.

Table 10-25. Coefficient of linear thermal expansion (ASTM D696) for dry nylons[a]

Nylon	$(\Delta L/L)$ ($^{\circ}$F^{-1})[b]	$(\Delta L/L)$ ($^{\circ}$C^{-1})[b]
66	4.5×10^{-5}	8.1×10^{-5}
6	4.6×10^{-5}	8.3×10^{-5}
610	5.0×10^{-5}	9.0×10^{-5}
612	5.0×10^{-5}	9.0×10^{-5}
11	5.5×10^{-5}	9.9×10^{-5}
12	5.5×10^{-5}	9.9×10^{-5}
66 + 33% glass fiber	1.3×10^{-5}	2.3×10^{-5}
6 + 30% glass fiber	2.1×10^{-5}	3.8×10^{-5}
612 + 33% glass fiber	1.3×10^{-5}	2.3×10^{-5}

[a] Data from trade literature.
[b] $\Delta L/L$ = change in length divided by original length, for example, in./in. or cm/cm.

389

Table 10-26. Effect of temperature on the coefficient of linear thermal expansion of dry nylon-66 and nylon-612 (8)

Temperature		Nylon-66		Nylon-612	
($^{\circ}$F)	($^{\circ}$C)	(ΔL/L) ($^{\circ}$F^{-1})	(ΔL/L) ($^{\circ}$C^{-1})	(ΔL/L) ($^{\circ}$F^{-1})	(ΔL/L) ($^{\circ}$C^{-1})
-40	-40	3.5×10^{-5}	6.3×10^{-5}	4.0×10^{-5}	7.2×10^{-5}
32	0	4.0×10^{-5}	7.2×10^{-5}	4.5×10^{-5}	8.1×10^{-5}
73	23	4.5×10^{-5}	8.1×10^{-5}	5.0×10^{-5}	9.0×10^{-5}
170	77	5.0×10^{-5}	9.0×10^{-5}	6.0×10^{-5}	10.8×10^{-5}

Table 10-27. Specific heat of dry nylons at room temperature[a]

Nylon	Specific heat[b]
66	0.35
6	0.4
610	0.4
612	0.4
11	0.3
12	0.3
66 + 35% glass fiber	0.33
6 + 35% glass fiber	0.34
610 + 35% glass fiber	0.35

[a] Data from trade literature.

[b] Specific heat is dimensionless. It is the heat capacity (cal g^{-1} $^{\circ}$C^{-1}) relative to that of water at 15°C.

100°C (32 to 212°F) and was found to increase slightly, from 0.21 to 0.22 kcal hr^{-1} m^{-2} ($^{\circ}$C/m)$^{-1}$, over this range (59). The thermal conductivity of a melt of nylon-6 also increases only slightly, from 0.11 to 0.12 kcal hr^{-1} m^{-2} ($^{\circ}$C/m)$^{-1}$, when the temperature is raised from 220 to 280°C and, similarly, that of nylon-610 from 0.094 to 0.104 kcal hr^{-1} m^{-2} ($^{\circ}$C/m)$^{-1}$ over the interval 190 to 260°C (60). For heat transfer calculations involving melting and freezing of nylon-66 the equations given by Heyman (61) are particularly useful (see Chapter 5, p. 188).

Table 10-28. Change of specific heat with temperature for dry nylons

Temperature		Specific heat		
°C	°F	Nylon-6 (57)	Nylon-66 (56)	Nylon-610 (56)
0	32	0.33	0.31	0.33
40	104	0.40	0.37	0.42
80	176	0.47	0.47	0.51
120	248	0.55	0.55	0.53
160	320	0.64	0.64	0.59
200	392		0.65	0.69
220	428		0.66	1.95
240	464		0.80	0.74
260	500		2.75	0.63
280	536		0.75	0.64

Table 10-29. Thermal conductivity of dry nylons[a]

Nylon	Thermal conductivity	
	$(Btu\ hr^{-1}\ ft^{-2}\ (°F/in.)^{-1})$	$(Kcal\ hr^{-1}\ m^{-2}\ (°C/m)^{-1})$
66	1.7	.22
6	1.7	.22
610	1.5	.19
612	1.5	.19
11	1.5	.19
12	1.5	.19
66 + 30% glass fiber	1.5	.19
6 + 30% glass fiber	1.7	.22
610 + 30% glass fiber	1.5	.19
12 + 30% glass fiber	1.1	.14

[a] Data from trade literature.

The effect of pressure and temperature on the specific volume of solid nylon-66 was calculated by Griskey and Shou (63). They also have calculated enthalpy and entropy data for nylon-66. Similar enthalpy and entropy data for nylon-610 were reported by Griskey, Din, and Gellner (64). The effect of temperature and pressure on the melt densities of several nylons is given in Table 4-2 of Chapter 4.

ELECTRICAL PROPERTIES

One of the large applications for nylon resins is in the electrical field. Injection molded parts include a variety of connectors, coil forms, terminals, switch parts, harnesses, and many other devices. These are used in low-frequency moderate-voltage applications that do not demand exceptional dielectric properties but do require an unusual combination of the electrical, mechanical, thermal, and chemical properties characteristic of nylons. Nylons are adequate dielectric materials like most plastic resins. They are, however, clearly superior to the traditional dielectrics that they have replaced, such as paper and textile related products, natural resinous products, and wood-flour-filled phenolic resin, which is especially poor in arc resistance. Differences do exist among nylons and are due mainly to different amounts of absorbed moisture (Table 10-30).

The dielectric constants of nylons at any moisture level decrease with increasing frequency (Figure 10-60) because of the reduced ability of absorbed water molecules to respond to higher frequencies. Dry, as-molded nylons have approximately the same dielectric constants and show only a small dependence on frequency at room temperature. As expected, the difference between nylons and the dependence on frequency increases with relative humidity and in proportion to the moisture content. It has been shown that the dielectric constants of nylons show large increases with temperature. At 212°F (100°C), 50% R.H., and 100 Hz, values near 20 are reported for both nylon-66 and nylon-612.

The dissipation factor of nylons varies up and down over the frequency range, passing through a number of inflections that are related to dielectric relaxation. These relaxations were discussed in Chapter 9, p. 313, and they have been the subject of detailed studies with nylons-66 and -610 (65). The effect of temperature, moisture content, and frequency on the dielectric properties of nylon-66 has been studied particularly extensively (67). Emphasis was again on definition and elucidation of transitional phenomena. It was proposed that about 15 chain segments are involved in cooperative motion in the observed transitions by noting the density of side chains, generated by radiation, necessary to eliminate the transition. Table 10-30 shows that the differences among common nylons in the dry condition are not large at any frequency in the range of 60 to 10^6 Hz. The dissipation factor is significantly larger when the nylon is saturated with moisture. Increasing temperature appears to cause a smaller increase than moisture.

Nylon	Condition	Volume resistivity (ASTM D257) (ohm-cm)	Dielectric constant (ASTM D150)			Dissipation factor (ASTM D150)		
			(60-100 Hz)	(10^3 Hz)	(10^6 Hz)	(60-100 Hz)	(10^3 Hz)	(10^6 Hz)
66	Dry	10^{15}	4.0	3.9	3.6	.02	.03	.02
	50% R.H.	10^{13}	6.0	5.1	4.1	.04-.18	.04-.16	.05-.09
	100% R.H.	10^9	31	29	18	.50	.23	.28
6	Dry	10^{15}	3.8	3.7	3.4	.01	.02	.03
	50% R.H.	10^{13}	13	8.3	4.2	.18	.20	.12
	100% R.H.	10^8	--	--	25	--	--	--
610	Dry	10^{15}	4.0	3.9	3.6	.04	.04	.03
	50% R.H.	10^{13}	7.4	6.1	3.7	.04	.04	--
	100% R.H.	10^{11}	15	12	4.0	.03	--	.1
612	Dry	10^{15}	4.0	4.0	3.5	.02	.02	.02
	50% R.H.	15^{13}	6.0	5.3	4.0	.04	.04	.03
	100% R.H.	10^{11}	14	--	--	--	--	--
11	Dry	10^{15}	--	--	--	--	--	--
	50% R.H.	10^{14}	--	--	--	--	--	--
	100% R.H.	10^{13}	--	--	--	.03	--	.06
12	Dry	10^{15}	4.2	3.8	3.1	.04	.05	.03
66 + 33% glass fiber	Dry	10^{15}	--	4.5	3.7	--	.02	.02
	100% R.H.	10^9	--	25	11	--	--	--

[a] Data from trade literature.

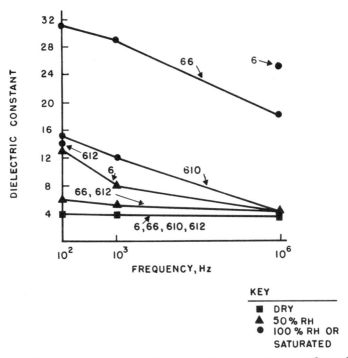

Fig. 10-60. Dielectric constant versus frequency of various nylons at 73°F (23°C) and different humidities.

Dry nylons have similar volume resistivities of 10^{14} to 10^{15} ohm-cm. Absorption of moisture decreases resistivity in approximate proportion to the amount of water absorbed (Table 10-30 and Ref. 66). Volume resistivity is not frequency dependent, but it does decrease with increasing temperature. Values of 10^9 to 10^{10} ohm-cm have been reported for dry nylons-66, -610, and -612 at 212°F (100°C) (8).

The dielectric strength of nylons is frequently reported in the trade literature although these data are of little or no value for design purposes. The dielectric strength of nylon is generally several hundred volts per mil (0.125 mm) of thickness. Even in very thin coil forms of 10 mils thickness, this would provide a dielectric strength of several thousand volts, probably at least a tenfold safety factor for common usage. Another factor that makes comparison of trade literature dielectric strength values difficult is its inverse dependence on thickness, which is frequently not reported. When the thickness is reported, it is often 0.1 or 0.125 in. (2.5 – 3.2 mm), and unreported thicknesses are probably in this range also. Nylons-66, -6, -610, and -11 are reported to have dielectric strengths (ASTM D149, step-by-step test) of 270 to 425 V/mil (10,800 to

17,000 V/mm) at dry conditions, but inconsistencies in the data preclude making meaningful comparisons among these resins. Short-time test values are somewhat higher. In addition to the dependence on thickness, dielectric strength decreases moderately with increasing moisture content and more significantly with increasing temperature. Dielectric strength is also dependent on frequency, but all available data are obtained using frequencies of 60 to 100 Hz which fall within the range of commercial power frequencies.

Nylons have excellent arc resistance. After about 100 sec in the ASTM D495 test they melt rather than carbonize and arc. The arc resistance of nylons is not affected by moisture content or temperature up to about 212°F (100°C), which includes the useful temperature range.

The additives in modified nylons can change significantly the electrical properties. Polar materials in some heat stabilizer formulations or fire retardant compositions increase dielectric constant and dissipation factor and decrease resistivity. The addition of carbon black decreases arc resistance. Glass reinforcement, however, has little effect on the electrical properties of nylons. (Table 10-30).

OPTICAL PROPERTIES

Most nylons are nearly opaque in thicknesses over about 0.1 in. (2.5 mm) and transparent below about 0.02 in. (0.5 mm). Intermediate thicknesses, which include many commercial articles, are translucent. The degree of light transmission can be high and is responsible for widespread application of nylons as lenses for dome lights in automobiles. The translucency is often adequate for viewing material in a nylon container. The haze (forward scattering fraction deviating by more than 2.5° from the incident beam – ASTM D1003) of nylon-66 with a thickness of 0.05 mm (0.002 in.) has been reported to be 26.5% (21).

The opacity is due to light scattering at the boundaries of spherulites which are crystalline aggregates (Chapter 8). Increasing crystallinity and the number of spherulites via use of nucleating agents causes a further reduction in light transmission. Increasing crystallinity via annealing of fabricated objects has no effect because the sperulitic structure has been determined by the initial thermal history. Stretching as in forming (Chapter 15) can destroy the spherulitic texture and increase transparency. Particulate additives in the form of fillers, pigments, or reinforcing fibers decrease light transmission.

Reducing the rate of crystallization and the spherulitic nature of the nylon product, however it is shaped, enhances transparency. This can be accomplished by quenching the melt during fabrication, by modifying the polymer via copolymerization or plasticization, or by use of a nylon with a large unit cell or a stiff chain. Thus, the larger unit cells of nylon-11 and -12 make for slower crystallization and greater light transmission than occurs with nylon-66 or -6. The nylon from the mixture of 2,2,4- and 2,4,4-trimethylhexamethylene-diamine and terephthalic acid (TMDT) is sufficiently disordered and of sufficient chain stiffness to inhibit crystallization completely, and products from TMDT are amorphous and transparent in all thicknesses.

In general, the refractive index decreases with increasing wave length of illumination and increasing temperature; see Table 10-31. Reported values for nylon vary, and conditions are not always specified. It is presumed that the usual sodium D-line (5893 Å) illumination and 20°C apply.

Using atomic refractivities the refractive index for nylon-66 was calculated to be 1.544 for a density of 1.14 g cm^{-3} (71). The same calculation applied to nylons-610, -6, and -11 yields, in order, 1.535, 1.538, and 1.530 for densities of 1.08, 1.13, and 1.04 g cm^{-3}. For a given nylon a change in density of 0.01 causes a corresponding change of about 0.006 in the refractive index.

Absorption in the infrared region is discussed in Chapter 3. Light transmission in the ultraviolet, visible, and near infrared regions is illustrated for nylon-66 in Figures 10-61 and 10-62.

ECOLOGICAL AND TOXICOLOGICAL CHARACTERISTICS

Ecology

The stability of nylon plastics contributes to their value in both industrial and consumer applications. As was shown in Chapter 1 (Table 1-1) much of the nylon is used as an engineering material in the automotive, appliance, and

Table 10-31. Refractive index (68, 69, 70, 72)

Nylon	Iso[a]	‖[b]	⊥[c]	Δ[d]	Molded (undrawn)
66	1.54	1.580-1.582	1.519-1.520	0.060-0.063	1.532
610	1.52	1.57	1.52	0.05	1.532 (25°C)
6	1.54	1.575-1.58	1.52-1.525	0.05-0.06	1.530
11	1.52	1.55	1.51	0.04	

[a] Isotropic = $(n_{\|} + n_{\perp})/3$.

[b] Parallel to the optic axis, generally the direction of orientation.

[c] Perpendicular to the optic axis.

[d] Birefringence = $n_{\|} - n_{\perp}$.

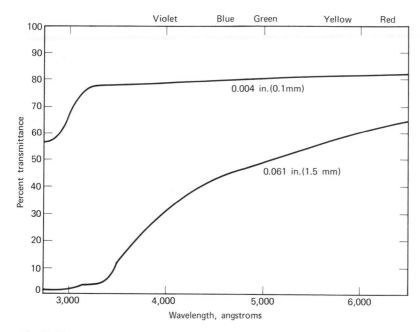

Fig. 10-61. Ultraviolet and visible light transmittance of nylon-66 (8).

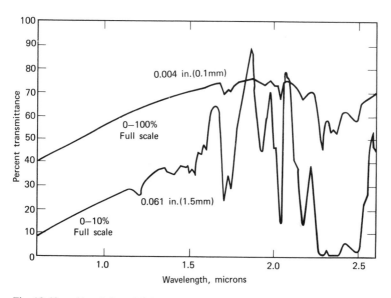

Fig. 10-62. Near infrared light transmittance of nylon-66 (8).

electrical equipment industries where little is discarded into the solid waste stream. Also, nylons are not widely used in applications such as industrial packaging which do rapidly become solid waste matter. Further, as noted in Chapters 5 and 6, scrap and trim from fabrication processes are commonly mixed with virgin polymer and recycled.

Consumer applications of nylon plastics include bristles for various types of brushes and bags for oven cooking of food. These account for less than 20% of the nylon plastic produced and, with strapping, are the only normally disposable uses of nylon. The polymer is not segregated at refuse collection points for recycling because of the extremely small amounts present in solid waste. The total quantity of nylon plastic manufactured each year in the United States represents about 0.01% of the refuse (garbage plus rubbish) collected annually.

Earlier discussions have pointed out that nylons degrade only slowly in sunlight. Organic acids, aldehydes, carbon monoxide, and carbon dioxide are formed and are assimilated with other natural decomposition products. Nylons are very slow to decompose on burial in the soil because they are highly resistant to both bacterial and fungal attack. For this reason there is no information available on the products of biodegradability (73). Nylons are unaffected by fresh or salt water temperatures up to 120°F (49°C) for periods of at least fifteen years. In sanitary landfills, the stability of nylon is an excellent characteristic because it means good compaction and no leaching of by-product materials into the soil or contamination of subsurface water. It remains as an inert fill.

The products of thermal degradation from nylons depend upon the polymer structure, the temperature, and the environment in which the heating occurs (Chapter 2, pp. 46 and 71). At a temperature of 305°C in air the slow degradation of nylon-66 and -6 gives rise to carbon dioxide, water, ammonia, and minor amounts of amines. About half the nitrogen is eliminated as ammonia (50). Under normal incineration conditions (800°C with air) polyamides produce mainly carbon dioxide and carbon monoxide with small amounts of ammonia and hydrogen cyanide. These same gaseous products are obtained from burning proteinaceous materials such as wool, silk, hair, or leather. Under these conditions in air at 800°C, nylon-66 yields lower levels of both ammonia and hydrogen cyanide than does silk (74). A benefit of incineration is that nylon provides a fuel source to assist in the burning of other less combustible materials present in refuse.

In summary, most nylon plastics do not enter the waste stream and constitute a negligibly small fraction of refuse. Where modern methods of incineration and landfill are practiced nylons present no more of a problem than other materials and are easily handled. They produce less hazard in terms of either air or water pollution than many other products in daily use. Thus, the effect of nylon plastics on the environment is not now and is not expected to become significant.

Toxicology

Most nylons present no toxicological problems because they are insoluble in body fluids and are biologically inert. The areas of caution that do exist are due to additives or low-molecular-weight fragments remaining after polymerization. Unmodified nylons-66, -610, -6, and -11, as well as certain copolymers, have been cleared, subject to carefully defined limits of extractables present, for food contact applications by the Food and Drug Administration (75). Nylon-6 must be treated to remove almost all of the approximately 10% of water-extractable material, largely monomer (see Chapter 2, p. 19), left after polymerization and must be subsequently processed under conditions minimizing reversion to the equilibrium monomer content. The monomer ε-caprolactam has been reported to be a convulsant poison in sufficient doses although its hydrolysis product, ε-aminocaproic acid, is not (76).

The situation with respect to dermatological hazards is similar. The polymers are inert, but additives and, in the case of nylon-6, the monomer can create problems. One manufacturer's literature reports that about 1% of the subjects tested developed an allergic skin reaction after prolonged contact with unextracted nylon-6 film (77).

Although nylon has been widely used as a suture material, it has had limited applicability in implants because of its reported susceptibility to deterioration and/or fragmentation over a prolonged period of time (78).

FLAMMABILITY

Increasing emphasis on product safety and ecology has prompted several studies on the flammability of plastics (79-84). We are in a period of considerable test development, and it is beyond the scope of this chapter to cover the details of test methods and their relative merits. Recent issues of *Modern Plastics Encyclopedia* (85) have included charts that provide flammability results from the seven kinds of test described in Ref. 83. In the 1971 issue 95 entries for nylons are listed under Underwriters' Laboratories Inc. flammability ratings.

Unmodified nylons are rated as self-extinguishing according to ASTM D635 and SE-2 (self-extinguishing class 2) according to UL. Increasing the hydrocarbon nature of the polymer decreases fire resistance, and nylon-12 is described as borderline between slow burning and self-extinguishing in the ASTM test.

Addition of glass fibers, plasticizers, and other modifiers often changes the UL rating from SE-2 to SB (slow burning). However, some compositions with special additives, including some containing glass or other modifiers, are identified as SE-O (self-extinguishing class O). Fire resistance improves from SE-2 to SE-O. Fire retardant additives are described in Chapter 11, p. 427.

Nylon-66 is reported to have an oxygen index (ASTM 2863) of 28 to 29%

(80) which indicates it is difficult to burn compared to most plastic materials. Nylons also have low values of smoke density in the National Bureau of Standards smoke generation test (80). Heats of combustion for nylons are discussed in Chapter 2, p. 71.

PROPERTIES OF NEW NYLONS

A very large number of polyamides have been synthesized since Carothers first made nylons-66 and -610, but few have been evaluated as plastic materials. Also, failure to specify test conditions, particularly the moisture content of the polymer, limits the utility of available data. Tables 10-32 through 10-37 provide information on nylons that are only recently commercial or have been discussed as commercial possibilities. These include 7, 69, 13, 1313, TMDT (polymer from trimethylhexamethylenediamine and terephthalic acid; see Table 1-4), MXD6 (from *meta*-xylylenediamine and adipic acid), and HPXD8 (from *trans*-hexahydro-*para*–xylylendiamine and suberic acid). These tables illustrate the variation in reported data that preclude unambiguous comparisons.

Nylon-7 (Table 10-32) would be expected to be very similar to nylon-6; -69 (Table 10-33) to -610; -13 and -1313 (Table 10-34), to -11 or -12. Differences of some value in specific applications can also be expected but cannot be generalized and will certainly not all accrue to one particular polymer. Cost is a major problem in commercializing these newer nylons.

TMDT (Table 10-35) was first introduced in Europe and recently in the United States. It is a stiff, transparent, amorphous material with low mold shrinkage and good retention of properties in water up to 80°C. Abrasion resistance and solvent resistance are, however, reduced below that of other nylon plastics (86). The xylylenediamines are potentially available at low cost, and

Table 10-32. Properties of nylon-7

Property	Units	Value[a]
Melting point	°C (°F)	223(433)
Specific gravity	- -	1.13
Tensile strength at break	$kg(wt)cm^{-2}$	590
Elongation at break	%	100-200
Water absorption, 1-hr boil	%	1.65
Volume resistivity	ohm-cm	2×10^{14}
Dielectric constant	- -	4.4

[a] Data from Ref. 87. Test conditions were not specified. $kg(wt)cm^{-2} \times 14.2 = lb(wt)in.^{-2}$

Table 10-33. Properties of nylon-69

Property	Units	Value[a]
Melting point	°C (°F)	215(419)
Specific gravity	- -	1.08
Tensile strength	lb(wt)in.$^{-2}$	7,480
Elongation at break	%	155
Notched izod impact	ft-lb(wt)in.$^{-1}$	2.8
Deflection temperature at 64 lb(wt)in.$^{-2}$	°C (°F)	154(309)
Flexural modulus	lb(wt)in.$^{-2}$	193,000
Volume resistivity, 50% R.H.	ohm-cm	0.2×10^{14}
Power factor, 50 cps, 50% R.H.	- -	0.019
10^3 cps, 50% R.H.	- -	0.105
10^6 cps, 50% R.H.	- -	0.048
Dielectric constant, 50 cps, 50% R.H.	- -	5.85
10^3 cps, 50% R.H.	- -	4.86
10^6 cps, 50% R.H.	- -	3.34

[a] Data from Ref. 88. Test conditions were not specified except for the electrical measurements which may be in error because of exposure to a salt solution. lb(wt)in.$^{-2}$ × 0.0703 = kg(wt)cm^{-2}.

MXD6 and HPXD8 (Tables 10-36 and 10-37) represent attempts to capitalize on this situation. MXD6 can be either transparent and amorphous or opaque and crystalline depending on molding conditions and subsequent annealing. Inclusion of a ring in the polymer chain raises T_g because of added rigidity, but it also increases melt viscosity, which can impair melt fabricability.

REFERENCES

1. Beaman, R. G., *J. Polym. Sci.* 9, 470 (1953).
2. Weyland, H. G., P. J. Hoftyzer, and D. W. Van Krevelen, *Polymer* 11, 79 (1970).
3. Kohan, M. I., unpublished information.
4. Gordon, G. A., *J. Polym. Sci., Part A-2* 9, 1693 (1971).
5. Illers, K.-H., *Makrom, Chem.* 38, 168 (1960).
6. Thomas, A. M., *Nature* 179, 862 (1957).
7. Rhode-Liebenau, U., *Kunststoffe* 55, 302 (1965).
8. Anon., Plastics Department, E. I du Pont de Nemours and Co., Tech. Bulletin, "Zytel® Nylon Resins Design Handbook," 1972.
9. Anon., PlasTech Equipment Corp., private communication.
10. Anon., Plastics Division, Allied Chemical Co., "Plaskon® Nylon Engineering Data."
11. Hsiao, C. C., *ASTM Proceedings* 56, 1425 (1956).

Table 10-34. Properties of nylon-13 and nylon-1313[a]

Property	Units	Nylon-13	Nylon-1313
Melting point	°C (°F)	180(356)	174(345)
Specific gravity	- -	1.01	1.01
Tensile strength	lb(wt)in.$^{-2}$	5,550	5,700
Yield stress	lb(wt)in.$^{-2}$	4,780	4,660
Elongation at break	%	130	130
Tensile modulus	lb(wt)in.$^{-2}$	113,000	114,000
Hardness	Rockwell M	34	40
Linear coefficient of thermal expansion	$(\Delta L \times 10^{-4}/L)\,°C^{-1}$	1.18	1.03
Deflection temperature at 264 lb(wt)in.$^{-2}$	°C (°F)	51(124)	53(128)
Water absorption, 23°C	%	1.04	0.75
Abrasion resistance	mg/100 cycles	184	213
Coefficient of friction			
Static	- -	0.40	0.39
Kinetic	- -	0.30	0.25
Volume resistivity, dry	ohm-cm	- -	8.6×10^{14}
50% R.H.	ohm-cm	16×10^{14}	8.5×10^{14}
100% R.H.	ohm-cm	15×10^{14}	1.1×10^{14}
Power factor, 50 cps, dry	- -	- -	0.050
50% R.H.	- -	0.015	0.039
100% R.H.	- -	0.019	0.066
Dielectric constant, dry	- -	- -	3.05
50% R.H.	- -	3.95	3.10
100% R.H.	- -	4.52	3.15

[a] Data from Ref. 89. The moisture content and temperature of test specimens were not given except as shown; they are most likely low (dry, as-molded or short-term storage at 50% R.H. and 23°C). lb(wt)in.$^{-2}$ × 0.0703 = kg(wt)cm^{-2}.

12. Anon., Gulf Nylon 401 Tech. Data Sheet.
13. Anon., Hydrocarbons and Polymers Division, Monsanto, Tech. Bulletin, "Nylon Resins."
14. Anon., Plastics Division, Allied Chemical Co., "Plaskon® Nylon Resins and Molding Compounds Composite Data Sheet."
15. Anon., Badische Anilin- und Soda-Fabrik AG, Tech. Information Data Sheets, "Ultramid® A, B, and S Resins," 1969.
16. Theberge, J. E., and N. T. Hall, *Mod. Plast.* **46** (7), 114 (July, 1969).
17. Heater, J. R., and E. M. Lacey, *Mod. Plast.* **41** (5), (May, 1964).
18. Heyman, E., *SPE J.* **24** (5), 49 (May, 1968).
19. Baer, E., R. E. Maier, and R. N. Peterson, *SPE Tech. Pap.* **7**, 17-1 (1961).

Table 10-35. Properties of poly(trimethylhexamethylene terephthalamide)

Property	Test	Units	Value[a]	Comment
Density	DIN 53479	g cm^{-3}	1.12	Test same as ASTM D792
Flex strength	DIN 53452	kg(wt)cm^{-2}	1250	Test same as ASTM D790
Impact strength, 20°C	DIN 53453	kg(wt)cm cm^{-2}	No break	Differs from ASTM tests
Impact strength, −50°C	DIN 53453	kg(wt)cm cm^{-2}	> 60	Differs from ASTM tests
Notched impact, 20°C	DIN 53453	kg(wt)cm cm^{-2}	10-15	Differs from ASTM tests
Notched impact, −50°C	DIN 53453	kg(wt)cm cm^{-2}	3-5	Differs from ASTM tests
Yield stress	DIN 53455	kg(wt)cm^{-2}	850	Similar but not identical to ASTM tests
Yield elongation	DIN 53455	%	9.5	Similar but not identical to ASTM tests
Ultimate stress	DIN 53455	kg(wt)cm^{-2}	600	Similar but not identical to ASTM tests
Ultimate elongation	DIN 53455	%	70	Similar but not identical to ASTM tests
Elastic modulus	- -	kg(wt)cm^{-2}	29,000	Same as ASTM D790
Hardness	DIN 53456	kg(wt)cm^{-2}	1400	Differs from ASTM tests
Abrasion resistance	Tabor	mg/100 cycles	25	Nonstandard test
Water absorption	DIN 53472	mg	40	Differs from ASTM tests
Water absorption, 20°C, 65% R.H.		%	~ 3	
Deflection temp.,	DIN 53461			
18.5 kg(wt)cm^{-2}		°C	130	Same as
4.6 kg(wt)cm^{-2}		°C	140	ASTM D648
Specific heat	- -	cal g^{-1}	0.35	
Thermal conductivity	- -	kcal m^{-1} hr^{-1} °C^{-1}	0.18	
Coeff. thermal expansion	- -	°C^{-1}	~ 6 × 10^{-5}	
Flammability	ASTM D635	- -	Nonburning	
Vol. resistivity, dry	DIN 53482	ohm-cm	> 5 × 10^{14}	Same as ASTM D257
Dielec. const., dry, 10^3 cps	DIN 53483	- -	3.5	Same as ASTM D150
Dissipation factor, dry, 10^3 cps		- -	0.028	Same as ASTM D150
Refractive index, n_D^{20}			1.566	

[a] Data from Ref. 86. Unless otherwise indicated the properties are assumed to be those of dry, as molded material at 20°C. kg(wt)cm^{-2} × 14.2 = lb(wt)in.$^{-2}$.

403

Table 10-36. Properties of poly(*meta*-xylylene adipamide)

Property	Units	Value[a]
Melting point	°C (°F)	243(469)
T_g	°C (°F)	68(154)
Specific gravity	--	1.20-1.22
Deflection temperature	(ASTM D648)	
at 66 lb(wt)in.$^{-2}$	°C (°F)	75 to over 200
		(167 to over 392)
at 264 lb(wt)in.$^{-2}$	°C (°F)	70 to 200
		(158 to 392)
Flexural modulus	lb(wt)in.$^{-2}$	550,000
Stress at yield	lb(wt)in.$^{-2}$	15,900
Stress at break	lb(wt)in.$^{-2}$	6,100
Elongation at break	%	37

[a] Data from Refs. 3 and 90. The properties can be assumed to be for dry, as-molded material except where ranges are shown. Some variation in properties with molding conditions is to be expected because the product can be either transparent and amorphous or opaque and crystalline, but the large variation in deflection temperature also reflects discrepancies between the two sources of data.

Table 10-37. Properties of poly(*trans*-hexahydro-*para*-xylylene suberamide)

Property	Test	Units	Value[a]
Melting point	--	°C (°F)	295(563)
T_g	--	°C (°F)	86(187)
Density	ASTM D1505	g cm^{-3}	1.12
Hardness	ASTM D785	Rockwell L	110
Yield stress	ASTM D638	lb(wt)in.$^{-2}$	12,000
Elongation at break	ASTM D638	%	30
Tensile modulus	ASTM D638	lb(wt)in.$^{-2}$	360,000
Notched Izod impact strength	ASTM D256	ft-lb(wt)in.$^{-1}$	0.7
Deflection temperature			
at 66 lb(wt)in.$^{-2}$	ASTM D648	°C (°F)	130(266)
at 264 lb(wt)in.$^{-2}$		°C (°F)	90(194)
Deformation under load at 50°C (122°F)	ASTM D621	%	0.3
Water absorption, 24-hr boil	--	%	3.5
Stiffness, as molded	ASTM D747	lb(wt)in.$^{-2}$	350,000
Stiffness, after 24-hr boil			320,000

[a] Dry, as molded unless indicated otherwise. Data from Ref. 25. lb(wt)in.$^{-2}$ × 0.0703 = kg(wt)cm^{-2}.

20. Nielsen, L. E., *Mechanical Properties of Polymers*, Reinhold Publishing Corp., New York, 1962, Chap. 3.

21. Ogorkiewicz, R. M., Ed., *Engineering Properties of Thermoplastics*, Wiley-Interscience, a division of John Wiley and Sons, Ltd., London, 1970, Chaps. 4 and 9.

22. Hachmann, H., Z. Wirt. *Fertigung* 61 (4), 165 (April, 1965).

23. Anon., *Mod. Plast. Encycl.* 48 (10A), 604 (1971).

24. Anon., *Mater. Eng.* 76 (1), 50 (July, 1972).

25. Watson, M. T., and G M. Armstrong, *SPE Tech. Pap.* 11, X-3 (1965).

26. Theberge, J. E., *Mod. Plast.* 45 (10), 155 (June, 1968).

27. Dally, J. W., and D. H. Carillo, *Polym. Eng. Sci.* 9, 434 (1969).

28. Williams, J. C. L., D. W. Wood, I. F. Bodycot, and B. N. Epstein, Soc. Plast. Ind. Tech. Conf. No. 23, Reinf. Plast. Div., Sec. 2-C (1968).

29. Aver, E. E., in *Mechanical Properties of Polymers*, N. B. Bikales, Ed.. Wiley-Interscience, a division of John Wiley and Sons, Inc., New York, 1971.

30. Ricour, B., and R. A. Scherer, *SPE J.* 28, 41 (1972).

31. Erhard, G., and E. Strickle, *Kunststoffe* 62 (1), 2 (Jan., 1972); *ibid.* (4), 232 (Apr., 1972); *ibid.* 282 (May, 1972).

32. Hachmann, H., and E. Strickle, *Konstruktion* 16 (4), 121 (April, 1964).

33. Hachmann, H., and E. Strickle, *Konstruktion* 18 (3), 81 (Mar., 1966).

34. Lewis, R. B., ASME Paper No. 63-WA-325 (1963).

35. Steijn, R. P., *Metals Eng. Quarterly*, May, 1967, p. 9.

36. Starkweather, H. W., Jr., G. E. Moore, J. E Hansen, T. M. Roder, and R. E. Brooks, *J. Polym. Sci.* 21, 189 (1956).

37. Weiske, C. D., *SPE Tech. Pap.* 13, 676 (1967).

38. Muller, A., and R. Pfluger, *Plastics (London)* 24, 350 (1959).

39. Starkweather, H. W., Jr., and R. E. Moynihan, *J. Polym. Sci.* 22, 363 (1956).

40. Miller, R. L., in *Polymer Handbook*, J. Brandrup and E. H. Immergut, Eds., Interscience Publishers, a division of John Wiley and Sons, New York, 1966, p. III-16.

41. Kinoshita, Y., *Makromol. Chemie* 33, 1 (1959).

42. Champetier, G., and J. P. Pied, *Makromol. Chemie* 44, 64 (1961).

43. Starkweather, H.W., Jr., *J. Appl. Polym. Sci.* 2, 129 (1959).

44. Anon., Plastics Division, Allied Chemical Co., Tech. Data Report 58-8.

45. Griehl, W., and D. Ruestem, *Ind. Eng. Chem.* 62 (3), 16 (1970).

46. Cash, F. M., and H. H. Goodman, Talks presented to Soc. Automotive Engrs, Mid-Year Meeting, Detroit, Michigan, May, 1970.

47. Weiske, C. D., *Kunststoffe* 54, 626 (1964).

48. Reymers, H., *Mod. Plast.* 47 (9), 78 (Sept., 1970).

49. Anon., *Mod. Plast. Encycl.* 48 (10A), 593 (1971)

50. Harding, G. W., and B. J. MacNulty, in *Thermal Degradation of Polymers*, S.C.I. Monograph No. 13, London, 1961, pp. 392-412.

51. Rugger, G. R., in *Environmental Effects on Polymeric Materials*, D. V. Rosato and R. T. Schwartz, Eds., Interscience Publishers, a division of John Wiley and Sons, New York, 1968, Chap. 4.

52. Winslow, F. H., W. Matreyek, and A. M. Trozzolo, *SPE J.* 28, 19 (1972).

53. Krashansky, V. J., B. G. Achhammer, and M. S. Parker, *SPE Trans.* **1**, 133 (1961).
54. Sisman, O., and C. D. Bopp, U. S. Atomic Energy Commission, ORNL-928, "Physical Properties of Irradiated Plastics," June 29, 1951.
55. Zimmerman, J., in *Radiation Chemistry of Macromolecules,* M. Dole, Ed., Academic Press, New York, 1972, Vol. 2.
56. Wilhoit, R. C., and M. Dole, *J. Phys. Chem.* **57**, 14 (1953).
57. Marx, P., C. W. Smith, A. E. Worthington, and M. Dole, *J. Phys. Chem.* **59**, 1015 (1955).
58. Kolesov, V. P., I. E. Paukov, and S. M. Skurator, *Zh. Fiz. Khim,* 36 770 (1962).
59. Kline, D. E., *J. Polym. Sci.* **50**, 441 (1961).
60. Fuller, T. R., and A. L. Fricke, *J. Appl. Polym. Sci.* **15**, 1729 (1971).
61. Heyman, E., *SPE J.* **23**, 37 (1967).
62. Haug, W. A., and R. G. Griskey, *J. Appl. Polym. Sci.* **10**, 1475 (1966).
63. Griskey, R. G., and J. K. P. Shou, *Mod. Plast.* **45** (10), 139 (June 1968).
64. Griskey, R. G., M. W. Din, and G. A. Gellner, *Mod. Plast.* **44**, 129 (Nov. 1966).
65. Curtis, A. J., *J. Res. Nat. Bur. Stand.* **65A**, 185 (1961).
66. Muller, A., and R. Rfluger, *Kunststoffe* **50**, 203 (1960).
67. Boyd, R. H., *J. Chem. Phys.* **30**, 1276 (1959).
68. Sweeny, W., and J. Zimmerman, in *Encyclopedia of Polymer Science and Technology,* John Wiley and Sons, New York, 1969, Vol. 10, pp. 59 and 568.
69. Roff, W. J., and J. R. Scott, *Handbook of Common Polymers,* CRS Press, a Division of the Chemical Rubber Co., Cleveland, 1971, p. 215.
70. Korshak, V. V., and T. M. Frunze, *Synthetic Heterochain Polyamides,* Daniel Davey and Co., Inc., New York, 1964, pp. 443 and 455.
71. Nose, S., *J. Polym. Sci.* **2B**, 1127 (1964).
72. Brauer, G. M., and E. Horowitz, in *Analytical Chemistry of Polymers,* G. M. Kline, Ed., Interscience Publishers, a Division of John Wiley and Sons, New York, 1962, Part III, p. 49.
73. Rodriguez, F., *Chem. Technol.* 1971, 409.
74. Anon., Underwriters Lab. Bull. Res. No. 53, July, 1963.
75. Anon., U. S. Code of Federal Regulations, Title 21, Part 121.2502.
76. Goldblatt, M. W., M. E. Farquharson, G. Beneett, and B. M. Askew, *Brit. J. Ind. Medicine* **11**, 1 (1954).
77. Anon., Plastics Department, E. I. duPont deNemours and Co., "General Guide to Products and Properties of Zytel® Nylon Resins," July, 1971.
78. Wesolowski, S. A., and J. D. McMahon, *SPE J.* **24**, 43 (June 1968).
79. Lyons, J. W., *The Chemistry and Uses of Fire Retardants,* Wiley-Interscience, a Division of John Wiley and Sons, Inc., New York, 1970, p. 412.
80. Hilade, C. J., *Flammability Handbook for Plastics,* Technomic Publishing Co., Stamford, Conn., 1969.
81. Einhorn, I. N., in *Reviews in Polymer Technology,* I. Skeist, Ed., Marcel-Dekker Inc., New York, 1972.
82. Warren, P. C., *SPE J.* 27 (2), 17 (Feb., 1971).
83. Reymers, H., *Mod. Plast.* **47** (10), 92 (Oct., 1970).

84. Blair, J.A., and R. B. Akin, *Constr. Specifier* **24** (7), 39 (July, 1971).
85. Anon., *Mod. Plast. Encycl.* **48** (10A), 572 (1971).
86. Doffin, H., W. Pungs, and R. Gabler, *Kunststoffe* **56**, 542 (1966).
87. Korshak, V. V., and T. M. Frunze, *Synthetic Hetero-chain Polyamides,* Daniel Davey and Co., Inc., New York, 1964, p. 460.
88. Anon., Emery Industries, Org. Chemicals Div., Tech. Bulletin 450, 1967.
89. Perkins, R. B., J. J. Roden, III, A. C. Tanguary, and I. A. Wolff, *Mod. Plast.* **46** (5), 136 (May, 1969).
90. Anon., Oronite Chem. Co., Tech. Bulletin, Dec., 1957.

CHAPTER 11

Modified Nylons

M.I. KOHAN

INTRODUCTION

The facility with which the structure of a nylon can be altered to meet new property requirements has been cited as a factor in the steady growth of the nylon plastics industry. Diamine, diacid, amino acid, and lactam intermediates that are available at reasonable cost can be combined almost at will to provide a variety of homopolymers and copolymers. Alternatively, starting with a given nylon, a number of materials can be added to yield, either via simple blending or chemical interaction, another variety of compositions having desired processing behavior, improved stability, or other specific properties. Table 11-1 is indicative of the many routes to modified nylons.

Modifications can be combined to yield still other useful compositions such as glass-reinforced nylons containing graphite for lubricity or thermally stabilized, plasticized copolymers. On the other hand, many modifying agents affect more than one property, and care must be taken to be sure that such effects do not impair the ability of the nylon composition to meet cost goals or to serve potential applications. For example, nucleating agents are added to nylons to accelerate crystallization and reduce cycle times in injection molding, but they normally also cause some loss in toughness. Plasticizers lower modulus, but they normally retard the rate of crystallization, which may complicate processing, and they may restrict the end-use temperature range. In some processes the broadened melting range of the plasticized composition may be more important than the retarded rate of crystallization and so simplify processing.

In brief, nylon technology affords many opportunities for fitting properties to needs, but a detailed knowledge of processing methods, end-use requirements, and appropriate evaluation techniques is essential to assure the designer of the new nylon that the modified composition and the applications are well matched in terms of cost and performance.

For the most part, commercially available nylon modifications are proprietary with the polymer supplier. The purpose of this chapter is not to judge specific compositions in specific applications but to provide some insight into the range of properties attainable in nylon plastics and the many techniques for producing desired changes.

Compounding occurs also in the plant of the processor, particularly when the additives involve low concentrations of processing aids or colorants. Comments on the preparation of modified nylons are to be found in Chapter 2, pp. 29 to 31, and, to the extent extrusion compounding is involved, in Chapter 6.

Processing technique (Chapters 5 and 6) can have an important effect on the mechanical properties of nylons. This is especially true if the products obtained involve different levels of orientation. For example, the tensile strength of monofilament can be as much as six times that of an injection molded bar.

Forming (Chapter 15) and post-fabrication processes such as annealing, conditioning, assembly, or decorating (Chapter 17) can also affect properties. Cast nylon-6 (Chapter 13) made via polymerization in the mold differs in some respects from melt processed nylon-6. These alternative ways of shaping or treating a nylon product are, as indicated, discussed elsewhere. Our focus here is on the kinds of compositional changes listed in Table 11-1.

Table 11-1. Some routes to modified nylon plastics

Polymer modification
   Physical form
   Molecular weight
   Copolymerization
   Chemical reaction
   Blend or graft other polymers

Colorants and processing modifiers
   Pigments and dyes
   External lubricants
   Mold release agents
   Nucleating agents
   Viscosity thickeners
   Blowing agents

Chemical property modifiers
   Antioxidant systems to retard thermal oxidation and/or reduce development
     of color
   Stabilizers to inhibit weathering
   Stabilizers to inhibit hydrolysis
   Fire retardants

Physical property modifiers
   Plasticizers
   Fillers and reinforcing agents
   Lubricity aids
   Antistatic agents

POLYMER MODIFICATION

Physical Form

The preferred physical form of a polymer differs with the fabricating technique — granules that are nominally cylindrical or rectangular with sides, lengths, or diameters in the range, 0.06 to 0.12 in. (1.5 to 3.0 mm), for injection molding or extrusion, and powders with diameters of 10 to 100 μ (0.01 to 0.1 mm) for fluidizid bed coating or rotational molding. In the latter case powder diameter is less critical, and larger sizes have also been used. The melt synthesis of nylons (Chapter 2) permits cutting the polymerizer extrudate to the granular form required by nylon fabricators. Powders involve the added cost of precipitation from solution or low temperature grinding (Chapter 14, p. 480). Anionic polymerization of lactams can be carried out in the presence of suitable solvents to yield powders directly (1), but the process is not yet commercial.

Stable aqueous dispersions and gels have been prepared using microcrystalline nylons with diameters of 50 to 100 Å (0.005 to 0.01 μ) (2). Preparation involves partial hydrolysis with a corresponding loss in molecular weight. Potential utility appears to be in the ability of these aqueous systems to produce continuous, ultrathin coatings.

Molecular Weight

Molecular weight is controlled by use of monofunctional additives, by adjustment of the ratio of diamine to diacid, or by variation in the water content (Chapter 2, pp. 37 to 40). The highest molecular weights require extended polymerization cycles or solid-phase polymerization of prepolymers, both of which mean added cost. In general, toughness increases rapidly with molecular weight up to a point and then changes only slowly, and all commercial nylon plastics exceed this critical level. The main purpose of varying molecular weight is to adjust the melt viscosity to the level preferred for different processes (see Chapters 4 and 6): over about 40,000 poise (0.6 lb(wt) sec in.$^{-2}$) for making blown film, blow molding, or free extrusion of rod and profiles; 20 to 40,000 poise (0.3 to 0.6 lb(wt) sec in.$^{-2}$) for extrusion of tubing; 10 to 20,000 poise (0.15 to 0.3 lb(wt) sec in.$^{-2}$) for making flat film; and less than 10,000 poise (0.15 lb(wt) sec in.$^{-2}$) for extrusion coating, monofilament extrusion, and injection molding. The viscosities cited are only approximate, depend on the specific product (size, wall thickness) as well as processing equipment, and are to some degree adjustable by changing processing temperature. Almost all commercial nylon plastics have number-average molecular weights in the range of 11,000 to 40,000.

Chemical Constitution

An obvious and important example of variation in nylon structure is provided by the variety of homopolymers. This is excluded from Table 11-1 because discussion of the different homopolymers pervades this book and is not peculiar to this chapter. The number of homopolymers will certainly continue to grow, however, and to extend the range of properties attainable via nylon technology.

Nylon Copolymers

Copolymerization is a familiar route to lower melting point, lower modulus, increased solubility, and increased transparency (see Chapter 16). Theory predicts that the melting point should decrease in a characteristic fashion depending on the heat of fusion of the major component and the mole fraction of monomeric impurities (Chapter 9, p. 309), but theory is rarely applicable if the major component falls below about 70%. Some typical melting-point-composition curves based on weight fraction and covering the full range of components show minimum melting compositions that melt at temperatures well below that of either homopolymer and at a weight fraction rich in the monomers that yield the lower melting homopolymer (Figure 11-1). The rigidity and solvent resistance of nylon-66/6 decrease as the minimum melting point is approached (Figure 11-2; Ref. 3). The melting point of a polymer is constant after a relatively low level of molecular weight is reached, and for nylon-66/6 copolymers this molecular weight increases as the minimum melting composition

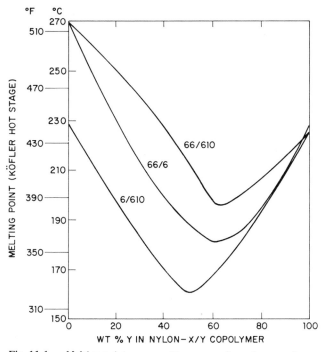

Fig. 11-1.   Melting point–composition curves for nylon copolymers.

is approached (4). The molecular weight corresponding to fiber forming capability shows a similar dependence on nylon-66/6 composition and is highest at the minimum melting composition (4). The percent crystallinity and density of nylon-66/6 copolymers follow a pattern of change similar to the melting-point-composition curves (Figure 11-3). On the other hand, water absorption depends more on the hydrocarbon to amide ratio and shows a more nearly regular change with composition (5, 12).

An x-ray study of nylon copolymers (5) showed the predominance of a single type of pattern rather than mixed crystals regardless of composition. The way in which the crystal structure varied with composition depended on the components, that is, the transition from dimensions approximating one homopolymer to those characteristic of the coingredient was sharp or broad depending on the components (Figure 11-4). The rate of change in modulus with composition appeared to relate to the rate of change in crystal structure. As expected, the rate of crystallization decreases as the amount of coingredient increases (6, 7).

In some nylon copolymers isomorphism occurs, and the melting-point-composition curve does not show a minimum. Nylon-A6/AT shows isomorphism if A is a linear aliphatic diamine with six to twelve carbon atoms (T = terephthalic acid) but not if A is *meta*-xylylene diamine (8). Isomorphism is found also if the methylene group in a polyamide is replaced by an ether or thioether group (9).

Our discussion to this point has assumed that the copolymer components form crystalline homopolymers, and this is usually the case with the commercially available nylon intermediates. Nylon-6PIDA is not a commercial entity

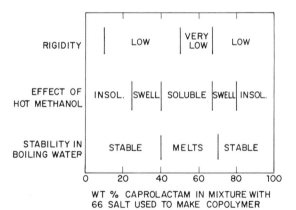

Fig. 11-2.   Rigidity, solubility, and hot water stability of nylon-66/6 copolymers (3).

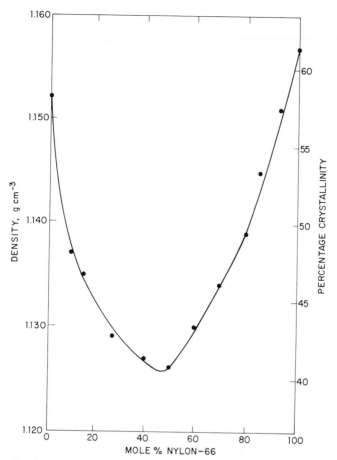

Fig. 11-3. Densities and percentage crystallinities for nylon-66/6 copolymers (7).

but provides an example of an amorphous homopolymer. PIDA is "phenyl-indane dicarboxylic acid":

PIDA

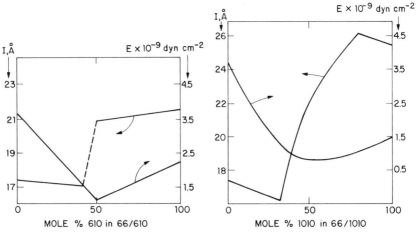

Fig. 11-4. X-ray repeat distance (I) along chain axis and Young's modulus (E) for nylon-66/610 and nylon-66/1010 (5).

Nylon-66/6PIDA becomes amorphous at 30 mole % or more PIDA (10). The α-transition increases with increasing % PIDA in spite of the decreased melting point because of the increased chain stiffness (10).

Multi-ingredient nylons involve more monomers than do copolymers, and such nylons can lead to still further reductions in crystallinity. Changes in properties not attainable in copolymers can result. The ternary diagram (Figure 11-5) for nylon-66/610/6 reveals regions of complex shape corresponding to specified levels of stiffness, solubility, and softening temperature (11). It is clear that only a terpolymer has a softening point below 160°C (320°F). The extension of the softening point region beyond the limits of stiffness and solubility emphasizes the fact that different properties vary in their response to changes in copolymer composition. The effect of changes in composition on crystallinity, frequency of amide groups, and intrinsic chain rigidity all have to be considered.

Nitrogen-Substituted Nylons

Elimination of the hydrogen atom on the amide nitrogen destroys hydrogen bonding and brings about lower stiffness, lower melting point, and increased solubility (13-16). The degree of change depends on the degree and distribution of nitrogen substitution (14, 15). For example, the melting point of a 40% N-methylated nylon-66 is 210°C if the polymer is made from a 60/40 mixture of hexamethylenediamine and N,N'-dimethylhexamethylenediamine and 185°C if made from a 20/80 mixture of hexamethylenediamine and N-methylhexa-methylenediamine (14). This is reasonable because the properties reflect the availability of a crystallizable entity (nylon-66) which is greater in the 60/40

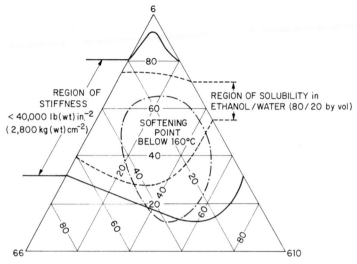

Fig. 11-5.   Stiffness, softening point, and solubility of nylon-66/610/6 terpolymers (11).

mixture. Even a polymer consisting exclusively of monoalkylated diamine (50% N-substitution) would be disordered by the head-to-head, head-to-tail arrangement of amine segments:

$$-NH(CH_2)_n\underset{\underset{R}{|}}{N}C(CH_2)_mC\underset{\underset{R}{|}}{N}(CH_2)_nNHC(CH_2)_mC\underset{\underset{R}{|}}{N}(CH_2)_nNHC(CH_2)_mCNH(CH_2)_n\underset{\underset{R}{|}}{N}C(CH_2)_mC-$$

with the carbonyl groups shown and the labels:

head-to-head   tail-to-head   tail-to-tail

The size of the alkyl substituent appears to have only a minor effect on melting point, but it decreases water absorption in accord with the increased hydrocarbon character of the polymer (14, 15). The use of a diacid with a sufficiently long methylene chain can lead to a crystallizable nylon even if fully alkylated (17): N,N'-dimethyl6 and hexadecanedioic acid give an amorphous polymer, but N,N'-dimethyl6 and octadecanedioic acid yield a crystallizable nylon. A tendency towards block polymer formation when using mixtures of primary and secondary amines was attributed to the difference in reactivity (17).

N-substitution can be brought about by reaction with the polymer as well as by polymerization of N-substituted diamines. The commercially available series of alkoxyalkylated nylon-66 (Chapter 2, p. 31) provides an example. These are soluble in the lower alcohols and are used to provide protective coatings and, when mixed with epoxides, high-strength adhesives (18). Cross-linking to reduce solubility and increase temperature resistance can be effected by heating with acid (18, 19):

$$
\begin{array}{ccccc}
| & & | & \overset{+}{H} & | & & | \\
NCH_2OCH_3 & + & NH & \longrightarrow & N-CH_2-N & + & CH_3OH \\
| & & | & & | & & | \\
CO & & CO & & CO & & CO \\
| & & | & & | & & |
\end{array}
$$

Reaction of nylon with ethylene oxide leads to a product with poly(ethylene oxide) side chains attached to the amide nitrogen atom (20, 21):

$$
\begin{array}{c}
-NCO- \\
| \\
(CH_2CH_2O)_nH
\end{array}
$$

Melting point and solubility are less affected than in the case of the N-alkyl or N-alkoxyalkyl nylons. However, the stiffness of nylon-66 is lowered by an order of magnitude, and the α-transition is lowered to below -40°C. Tensile strength is reduced and elongation is increased. Cross-linking of the ethylene oxide treated nylon occurs on heating in the presence of acid (20).

Polymer Combinations

Polymer combinations include blends of different nylons, blends of nylons and other polymers, block copolymers, and graft copolymers.

Extrusion compounding of different nylon homopolymers leads to products that are in part block copolymers and in part physical blends. Reaction between the homopolymers would result in a random copolymer if the residence time were long enough (Chapter 2, p. 46). Differential thermal analysis of nylon-6/nylon-11 blends indicates that the melting point of the higher-melting polymer is depressed slightly, but that of the lower-melting nylon is unaffected (22). Freezing points follow a similar pattern except for an increase in the freezing point of nylon-11 in the blend containing 10% nylon-11 (Figures 11-6(a) and 11-6(b)). Crystallinities of the constituent polymers were estimated by comparing the area under the DTA melting peaks for the blends with the areas for the pure homopolymers (Figure 11-6(c)). Densities of the constituent polymers were then calculated and used to predict the densities of the blends (Figure 11-6(d)). Close correspondence of the observed and calculated densities of the blends was obtained.

Incompatible hydrocarbon polymers such as polyethylene can be melt blended with nylons to yield compositions that have improved permeability (Table 11-2) and are processible into film, filament, and bottles (23).

Improvements in moisture absorption, impact strength, flexibility, molding characteristics, and structure uniformity are claimed for blends made from nylons and ethylene/alkyl acrylate ester copolymers (28). Mixtures of a low-melting nylon terpolymer and an ethylene/unsaturated carboxylic acid copolymer have been milled together to yield 0.125-in. (0.318-cm) sheets with only slight haze (29).

Other polymers can be grafted onto nylon by using the nylon chain as the initiator for growth of the polymer. The ethylene oxide treated nylon described in the preceding section provides an example in which the N-H bond of the amide serves as initiator. Graft copolymers are also made by exposing nylons to high energy radiation to generate free radicals and then treating with vinyl

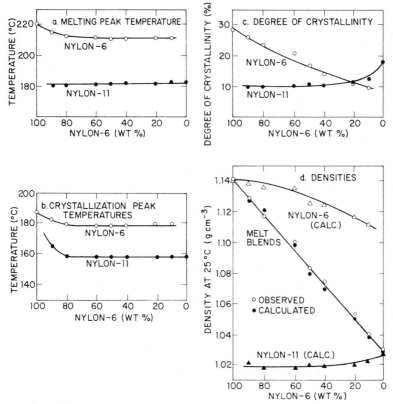

Fig. 11-6.   Properties of nylon-6/nylon-11 blends (22): (a) melting peak temperatures; (b) crystallization peak temperatures; (c) degree of crystallinity; (d) densities.

Table 11-2. Permeability of polyethylene/nylon blends (23)

Blend, wt %		Loss/yr[a], wt %		
Polyethylene (density 0.945)	Nylon-6	n-Heptane	Methyl Salicylate	Methyl Alcohol
100	0	2.18	6.7	3
80	20	10.4	0.3	1.6
20	80	7.6	0.5	- -
0	100	- -	- -	ca. 400

[a]Filled and sealed 4-oz bottles, 30 to 40 mil wall thickness, 102°F, 50% R.H., extrapolated from measurements for up to 33 days.

monomers such as vinyl acetate, acrylonitrile, vinyl fluoride, and acrylates (24). It is also possible to soak the nylon in the vinyl compound as a liquid or in solution and irradiate in a single step. The site of attachment of the vinyl polymer to the nylon is the carbon atom adjacent to the amide nitrogen. These polyvinyl grafts lead to improved dyeability, static resistance, light durability, and resistance to soiling, property improvements that are especially advantageous in fibers.

Graft copolymers can be obtained by the joining of premade polymers. One example is the reaction of novolacs (non-cross-linked phenol-formaldehyde resins) with nylons-11, -610, -66, and -6 (25). Incorporation of 20-25 wt % of the novolac produced only minor to moderate changes in melting point and solubility. Another example involves mixing the melts of a nylon and a hydrocarbon polymer together with a small amount of free radical forming catalyst (26). Properties of nylon-6/low density polyethylene graft copolymers are illustrated in Table 11-3. Important changes occur in water absorption, flow properties, and tensile strength without significant change in melting point.

It is also possible to achieve a graft copolymer by polymerizing an AB-type monomer such as caprolactam in the presence of an ethylene copolymer containing acid groups (27). A product is obtained that is partly nylon-6 homopolymer and partly a graft copolymer. Improved impact strength at room temperature and at -40°C is claimed. The polymerization of nylons including AABB types has been carried out in the presence of 1 to 30 wt % of compatible phenol-formaldehyde resins that do not become infusible within 1 to 3 hr at 180 to 270°C. Polymers of improved stiffness result (30).

COLORANTS AND PROCESSING MODIFIERS

Colorants and processing modifiers are to be found in a great many commercial nylon plastics. They are more likely to be used in in-plant compounding,

Table 11-3. Properties of nylon-6/polyethylene graft copolymers (26)

Nylon-6/PE (wt ratio)	Density (g cm^{-3})	Melting range (°C)	% H$_2$O (1 hr in boil. H$_2$O)	Melt index (g/5 min at 235°C)	Tensile strength	
					(kg(wt)cm^{-2})	(lb(wt)in.$^{-2}$)
90/10	1.09	221-226	3.7	8.9	425	6050
70/30	1.02	219-225	1.7	6.2	250	3560
50/50	1.01	217-226	1.3	2.3	210	2990
30/70	0.96	212-225	0.9	1.2	175	2490
10/90	0.93	177-200	0.6	0.5	160	2280

however, than are the additives dicussed elsewhere in this chapter. They are both most often used in small quantities, that is, 1 wt % or less. Our concern here is with the nature of the additives and their effect on properties. Processing technique is described in Chapter 2, p. 30, and Chapter 6, p. 212. The dyeing and decorating of nylon articles that have already been processed into the desired shape are discussed in Chapter 17.

Colorants

Colorants include organic dyes, normally soluble in the polymer, and pigments, which are generally insoluble and may be either organic or inorganic. An opaque pigment is advantageous in leading to color depth at low concentration, in providing superior light fastness, and in avoiding color change if excessive pigment is used. On the other hand, transparent dyes or very finely divided pigments are necessary for light transmission in translucent or transparent articles.

A colorant for nylons must be (1) dispersible to avoid nonuniform color in the product, (2) stable to the nylon melt temperature, and (3) stable to reduction by the nylon (31). Some colorants are not satisfactory because they are too volatile or decompose at the necessary temperatures which are often in excess of 260°C (500°F). For example, hydrated iron oxide loses water and changes from yellow to red. The organic azo, vat, and quinacridone colorants are reduced to yellow or colorless forms by molten nylon. The inorganic chromates and molybdates are sufficiently effective oxidants to also be unsuitable for nylons (31). A wide variety of colorants do, however, fulfill the requirements of dispersability and stability and so make a broad line of colors possible in nylon plastics (Table 11-4).

Sample formulations that provide distinctive opaque colors in 80-mil (2.0-mm) thickness and are stable to the temperatures encountered with the high melting nylon-66 are given in Table 11-5. Sample translucent formulations are indicated in Table 11-6. Space does not permit discussion of blends of colorants

Table 11-4. Colorants widely used in nylon plastics. (Based on listing in 1969-1970 edition of *Modern Plastic Encyclopedia*, pp. 1010-1012.)

Class	Generic Name	Color	Color Index-2nd Ed.	
Dye	Nigrosine	Wide color range	- -	
	Induline	Wide color range	- -	
Organic pigment	Anthraquinone	Red	- -	
	Phthalocyanine	Blue	Pig. Blue	15, 74160
	Phthalocyanine	Green	Pig. Gr	7, 74260
	Carbon black	Black	Pig. Blk	7, 77266
Inorganic pigment	Titanium dioxide	White	Pig. White	6, 77891
	Zinc sulfide	White	Pig. White	7, 77975
	Lithopone	White	Pig. White	5, 77115
	Zinc oxide	White	Pig. White	4, 77947
	Cadmium sulfoselenide	Maroon, red, orange	Pig. Red	108, 77196
	Cadmium mercury	Maroon, red, orange	Pig. Red	113, 77201
	Ultramarine	Red	- -	
	Ultramarine	Pink	Pig. Vio	15, 77007
	Ultramarine	Violet	Pig. Vio	15, 77007
	Chrome-tin	Pink	- -	
	Cadmium sulfide	Yellow	Pig. Yell	37, 77199
	Titanium pigment	Light yellow	Pig. Yell	53, 77788
	Cobalt aluminate	Blue	- -	
	Chrome cobalt aluminate	Turquoise	- -	
	Ultramarine	Blue	Pig. Blue	29, 77007
	Ultramarine	Green	- -	
	Ceramic black	Black	- -	
Special	Metallic oxides	Browns	- -	
	Aluminum plastic grades	Silver	- -	
	Bronze plastic grades	Red gold to yellow gold	- -	

Table 11-5. Sample opaque color formulations for nylon plastics (32)

Color	Approx. wt %	Colorant	Colorant supplier
White	0.5	Ti-Pure® R-960	Pigments Dept., Du Pont
Yellow	0.5	Cadmium Yellow 1470	Harshaw Chemical Co.
Orange	0.5	Cadmium Orange 1510	Harshaw Chemical Co.
Red	0.5	Cadmium Red 1540	Harshaw Chemical Co.
Blue	0.7	Ultramarine Blue UB9055	Pigments Div., Chemtron Corp.
Turquoise	0.5	Ferro Turquoise V7610	Ferro Corp.
Green	0.25	Ultramarine Blue UB9055	Pigments Div., Chemtron Corp.
	0.25	Cadmium Yellow 1470	Harshaw Chemical Co.
Black	0.5	Induline Base 5G	American Cyanamid

Table 11-6. Sample translucent color formulations for nylon plastics (32)

Color	Approx. wt %	Colorant	Colorant supplier
Yellow	0.2	Irgacet® Yellow 2GL	Geigy Corp.
Blue	0.1	Monastral® Blue G BT-383D	Pigments Dept., Du Pont
Green	0.1	Monastral® Green G GT-751D	Pigments Dept., Du Pont
Pink	0.1	Sulfo Rhodamine B	GAF Corp.

or the relative efficiencies of specific colorants. The effect of combining a rutile titanium dioxide with cobalt, ultramarine, cadmium sulfide, cadmium sulfoselenide, or assorted brown colorants has been studied as a means of defining nylon colorants in CIE space (31).

In general, the incorporation of colorants has little or no effect on the properties of nylon plastics, but it is advisable to confirm adequate performance of the color modified composition if the intended end-use requires close quality control. Light fastness may be a factor particularly in transparent or translucent formulations.

Processing Modifiers

Processing modifiers include additives to facilitate the flow of the granular nylon, to improve mold release, to accelerate crystallization, to alter melt viscosity, to avoid color formation particularly in rework, and to provide foamed structures.

Coating nylon granules with a lubricant can be advantageous in injection molding because of increased bulk density, decreased tendency to bridge in hoppers, decreased pressure drop particularly in ram machines and therefore lower injection pressures or better filling of mold cavities, and reduced or more uniform screw retraction times in reciprocating screw machines. The use of 0.005 to 0.10 wt % of ethylenebisdiamides, $RCONHCH_2CH_2NHCOR$, of aliphatic monocarboxylic acids with 12 to 20 carbon atoms (33) has been described. Improvement in bulk density and thermal stability is claimed if 0.005 to 0.20 wt % of ethylenebispelargonamide (the acid has nine carbon atoms) is used (34). Metallic stearates are also effective. There is little or no effect on the properties of the product unless the lubricant is a nucleating agent also.

Certain surface coated or melt incorporated additives facilitate ejection of molded parts and lead to shorter cycle times. Nylon-6 coated with 0.1 to 0.5% of a purified form of sodium stearate exhibits improved mold release without loss in clarity (35). Also described as mold release agents, particularly for nylon-6, are 0.25 to 2.0% behenic acid, the aliphatic monocarboxylic acid with 22 carbon atoms (36); 0.01 to 2.0% aliphatic monoamines with at least 14 carbon atoms (37); and 0.02 to 1.5% of a variety of fatty compounds including aliphatic monoalcohols of 12 to 36 carbon atoms and esters of these alcohols with fatty acids with 13 to 35 carbon atoms (38). The use of 0.05 to 0.20% zinc stearate as well as nucleating agent was found to assist mold release in copolymers containing at least 75 wt % nylon-66 (39).

Crystallization occurs via the development of nuclei which then serve as sites of crystal growth (Chapter 8). Crystalline aggregates of microscopic size known as spherulites result. Addition of nucleating agents accelerates crystallization and leads to smaller and more uniform spherulites. Within limits, crystallinity and spherulitic content are independently variable depending on thermal history in processing and subsequent annealing (40). Finely divided materials that are solid in the vicinity of the melting point of the nylon, do not agglomerate, and provide polar surfaces capable of adsorbing amide groups, can serve as nucleating agents. These include 0.005 to 5% of sodium phenylphosphinate, sodium isobutylphosphinate, magnesium oxide, mercuric bromide, mercuric chloride, cadmium acetate, lead acetate, and phenolphthalein as well as silver halides and very fine silicas and aluminas (40). Also cited as nucleating agents are 0.0001 to 0.1% of graphite, molybdenum disulfide, lithium fluoride, and talcum (41).

Increased modulus, yield stress, and hardness but decreased elongation result. The difference between several commercial nucleated and non-nucleated nylons has been reviewed (42, 43).

The nucleating tendency of a wide variety of phosphinates useful as antioxidants for nylons has been defined (44). The combination of 0.05 to 0.5% calcite and 0.5 to 2.0% of a fatty alcohol is claimed to accelerate crystallization without substantially increasing total crystallization (45). Note that some of the same materials are cited as effective for lubrication, mold release, and nucleation, and the use of 0.1 to 0.5% of the salts of saturated or unsaturated fatty acids for combined mold release and nucleation has been claimed (46).

Additives that affect the melt viscosity of nylons have been noted in Chapter 4.

It is occasionally desirable to use a nylon with exceptional color stability during processing, particularly if processing involves a large fraction of work. Antioxidants in low concentration (below 5 wt %) are indicated, but care in selection is necessary because not all antioxidants prevent discoloration. Stabilization of nylons is the basic issue in question, and this is discussed in the following section on chemical modifiers.

Foamed nylon is not an important commercial factor at this time, although a variety of foaming techniques for nylons have been described (48, 49). Open-cell, foamed nylon that is saturated with lubricating oil makes excellent bearings and bushings. Difficulty in achieving low densities has been ascribed to low melt viscosity and high melting points for most nylon plastics compared to the decomposition temperature of most blowing agents. Oxamic acid and its esters (50) and the combination of hydroxyalkyldiamines as cross-linking agents and methyl or ethyl esters as blowing agents (50, 51) have been used to expand nylon at 250 to 280°C to yield products with densities of about 0.2 g cm^{-3} or 12 lb ft^{-3}.

A number of foaming systems have been applied to the promoted anionic polymerization of lactams, which involves higher melt viscosities and lower temperatures than most other commercial nylon plastics. All of the familiar methods of inducing foaming are used: for example, a volatile solvent plus an agent to facilitate bubble formation (52), decomposable compounds such as formates and formamides (53), and injection of an inert gas (54, 55). Product densities vary with the nylon and the foaming technique, but values as low as 0.05 g cm^{-3} or 3 lb ft^{-3} have been reported (52).

CHEMICAL MODIFIERS

Some additives improve the resistance of nylons to chemical attack, that is, thermal oxidation, photooxidation, hydrolysis, and combustion. Evaluation of the relative performance of these additive systems is handicapped because of scant attention in the scientific literature and the abundance of patents with an almost equal number of testing techniques. Patents claim improved resistance to thermal oxidation or to photooxidation or both. The improvement claimed may be retention of color or selected, often nonstandard, mechanical properties or retention of both properties and color. Prediction of long-term (i.e., years) performance almost always rests on extrapolation of the results of accelerated tests. The variety of ASTM, Underwriters Laboratory, and industry tests for flame resistance attests to the problem of quantifying resistance to combustion. It should be recognized that the utility of any given additive system or combination of systems depends on the specific performance criteria and service conditions of the intended application, and failure of a candidate composition sometimes reflects inadequate analysis of end-use requirements rather than lack of a suitably modified nylon.

Stabilizers

Additives that inhibit thermal and photooxidation in nylons were reviewed in 1964 (56), and several patents for stabilizer systems in nylons were listed in a 1960 publication (57). Modern Plastics Encyclopedia contains a list of antioxidants that includes alkylated phenols, polyphenolic compounds, aromatic amines, and organic phosphites, and specific members of each of these organic additives are identified as useful in nylons.

Incorporation of organic antioxidants means improved resistance to thermal oxidation, sometimes referred to in the nylon literature simply as heat stability. However, some antioxidants are also said to increase resistance to light (for example, derivatives of hypophosphorous acid, $H_2P(=O)$ OH, Ref. 47, and phenyl halophosphines, Ref. 58); others, to both light and hydrolysis (for example, phosphite esters, Ref. 59). Probably many of the antioxidants provide a measure of protection against both thermal and photo oxidation because of similarity in the free radical mechanisms involved (Chapter 2, pp. 72 and 73).

Aromatic amines and phenols tend to discolor nylons. Synergistic combinations for resistance to thermal oxidation are reported, for example, phosphorus acid derivatives plus 2-mercaptobenzimidazole for nylons in general (60) or N,N'-di(β-naphthyl)paraphenylenediamine plus mercaptobenzothiazole for nylon-6 (61). Other organic stabilizers reported to be effective against thermal and photooxidation include stable radicals such as nitroxides and hydrazyls (62) and mixtures of UV absorbers and antioxidants for low-melting nylons (63).

Many stabilizer systems for nylons include inorganic compounds. Copper derivatives in combination with assorted halides are indicated to provide excellent resistance to thermal oxidation (64-66). Manganese derivatives with or without tungsten or molybdenum compounds to minimize discoloration are cited as effective against photooxidation (67, 68). Metal halides plus a phosphorous compound with or without manganese or cobalt compounds improve resistance to thermal oxidation (69, 70).

Carbon black remains the additive of choice where long-term, outdoor durability is required.

In long-term exposure of nylons of low crystallinity, the contribution of a slow increment in crystallinity to changes in properties should not be overlooked (57).

Fire Retardants

The year 1970 saw a rapid increase in the number of commercially available nylon compositions, with and without reinforcing agents, said to have superior fire retardancy. The technology is complex, the literature is meager, and there exist a number of tests that include specimens of assorted sizes. At this writing no generalizations as to retardant systems for nylons are possible. It is broadly true that those nylons that have high amide-group concentrations to some degree have intrinsically better resistance to combustion than those nylons with low amide-group concentrations.

Fire retardant compositions described in patents include: (1) 5 to 20 wt % of an organic halide plus 3 to 15% of an oxide of tin, lead, copper, iron, zinc, or antimony (71), (2) 5 to 30% of a halide of zinc, cadmium, lead, or rare earth metal or 3 to 15% of one of these halides plus 1 to 5% of copper oxide or one of several metals (72), and (3) 0.5 to 25% of a melamine compound (73). The effect of these systems, which involve relatively large concentrations of additives, on the processing and solid-state properties of nylons have not yet been detailed, but the effect of up to 30% cadmium chloride on the tensile properties of nylon-66 (72) and of 5% melamine on the notched Izod impact strength of nylon-6 (73) was small.

PROPERTY MODIFIERS

Property modifiers are those materials added to nylons to extend the useful range of physical properties; they may be plasticizers on one end of the scale or high modulus, reinforcing agents on the other. Such modifiers are normally present in relatively large amounts, that is, from 5 to 20 wt %. Other additives are employed to improve specific properties such as lubricity or the tendency to develop a static charge, and concentrations vary considerably.

Plasticizers

Plasticizers are normally high-molecular-weight polar compounds of low volatility that are capable of withstanding processing temperatures and interrupt hydrogen bonding between the nylon chains by association with the amide function (Chapter 2, page 68). These include a variety of monomeric carbonamides and sulfonamides (74, 80), phenolic compounds (75), cyclic ketones (75), mixtures of phenols and esters (76), sulfonated esters or amides (77), N-alkylolarylsulfonamides (78), selected aliphatic diols (79, 81), and phosphite esters of alcohols such as 9-phenylnonanol-1 (82). Unextracted nylon-6 is one kind of plasticized nylon; as noted earlier (Chapter 2, page 20), nylon-6 is for most purposes extracted to remove monomer and oligomers that comprise about 10 wt % of the as-made polymer.

The choice of plasticizer depends on one or more of a number of variables such as cost, compatibility with the specific nylon homopolymer or copolymer, solvent sensitivity and anticipated solvent exposure, processing behavior, or low-temperature performance. Water may be regarded as a non-processible plasticizer, and plasticizers lower the α-transition of nylon in the same manner as water (Chapter 9, Figures 9-6 and 9-12). This means lower modulus and tensile strength and higher ultimate elongation and impact strength at least until the temperature is lowered below the α-transition of the particular composition. Plasticizers lower the melting point by an amount related to the mole fraction of plasticizer.

Fillers and Reinforcing Agents

Fillers and reinforcing agents are commonly differentiated on the basis of whether the additive and the polymer matrix do or do not interact. This sharp dividing line is contested (83), but it is true that the use of a suitable agent to bond the additive more effectively to the polymer results in important changes in the properties of the composite.

Inert inorganic fillers such as talc, kiesrlguhr, asbestos, or mica can be used in nylons (75). The presence of 0.5 to 20% kaolin in nylon-6 is reported to lead to smaller spherulites (nucleation effect), but the kaolin was found to be located primarily at interspherulitic boundaries (84). A 17-mμ silica produced a larger increment in modulus and melt viscosity than a larger size-200mμ silica in both extracted and unextracted nylon-6 (85). The properties of a silica-filled, extracted composition are compared with an unfilled, extracted polymer in Table 11-7.

Reinforcements for plastics include a variety of fibrous and nonfibrous materials of which glass fiber is the most important in nylons, although asbestos fibers and glass spheres are also used. Asbestos fibrils in thermoplastics other than nylon have been studied (86), and the properties of these compositions depend on the amount and alignment of the fibrils in much the same manner as the properties of glass fiber containing nylons depend on the glass fibers as discussed below. The propterties of a commercial, asbestos-reinforced nylon are shown in Table 11-8.

Table 11-7. Silica-filled, extracted nylon-6 (85)

	No filler	10 wt % 17 mμ SiO$_2$
Density, g cm^{-3}	1.129	1.191
Rockwell hardness	R94	R106
Izod impact strength, ft-lb(wt)in.$^{-1}$	2.9	1.5
Tensile strength, lb(wt)in.$^{-2}$	8870	8070
Elongation, %	253	140
Flexural modulus, lb(wt)in.$^{-2}$		
23°C (73°F)	130,000	173,000
50°C (122°F)	95,000	120,000
100°C (212°F)	67,000	87,000
150°C (302°F)	52,000	80,000
Mandrel bend, breaks/bends	0/15	0/20
Fatigue endurance limit, lb(wt)in.$^{-2}$	2500	3000
Creep rate, mils in.$^{-1}$ hr^{-1} at 100 hr[a]		
(stress not given)	1.2	0.44
Melt viscosity, 270°C, poise at shear stress		
of 78.4 lb(wt)in.$^{-2}$ [a]	300	7200

[a] Unextracted polymer.

Notes: Reference does not say, but data suggest mechanical properties were measured on specimens conditioned to 50% R.H. lb(wt)in.$^{-2}$ × 0.0703 = kg(wt)cm^{-2}; ft-lb(wt)in.$^{-1}$ × 5.45 = cm-kg(wt)cm^{-1}.

Table 11-8. Properties of a commercial asbestos-reinforced nylon (87)

Tensile strength, lb(wt)in.$^{-2}$	12,000
Elongation, %	3.0
Modulus of elasticity, lb(wt)in.$^{-2}$	700,000
Flexural strength, lb(wt)in.$^{-2}$	17,000
Flexural modulus, lb(wt)in.$^{-2}$	700,000
Izod impact strength, ft-lb(wt)in.$^{-1}$	0.6
Deflection temperature, °F @ 264 lb(wt)in.$^{-2}$	380
Deflection temperature, °F @ 66 lb(wt)in.$^{-2}$	460
DUL @ 4000 lb(wt)in.$^{-2}$ and 122°F, %	0.9
Coefficient of linear thermal expansion, °F^{-1} × 10^5	2.2
Rockwell hardness	M80-90
Water absorption, %	0.9
Specific gravity	1.3
Linear mold shrinkage, in./in.	
1/8-in. average section	.006
1/4-in. average section	.010
1/2-in. average section	.014

Notes: Ref. does not say, but data suggest measurements made on dry specimens.
lb(wt)in.$^{-2}$ × 0.0703 = kg(wt)cm^{-2}; ft-lb(wt)in.$^{-1}$ × 5.45 = cm-kg(wt)cm^{-1}.

Glass Reinforcement

Glass-reinforced nylons have generated considerable interest because of their excellent combination of strength, rigidity, and chemical resistance. A number of commercial grades involving variations in both the glass content and the type of nylon are available. These offer a degree of choice in level of reinforcement, processing behavior, and chemical resistance. Compositions which include lubricity aids or fire retardants are also commercial.

The Glass Fiber and Theory of Reinforcement. The glass fiber commonly used is an E-glass, which can vary somewhat in composition (88, 89). Typical properties of the glass fiber are: a specific gravity of 2.54, a virgin tensile strength of 500,000 lb (wt) in.$^{-2}$ (35,000 kg (wt) cm^{-2}), a modulus of 10,500,000 lb (wt) in.$^{-2}$ (740,000 kg (wt) cm^{-2}), and a coefficient of linear thermal expansion of 2.8 $\times$ 10^{-6} °F^{-1} (5.0 $\times$ 10^{-6} °C^{-1}). Normal handling reduces the tensile strength of commercial fibers to about 250,000 lb (wt) in.$^{-2}$ (17,500 kg (wt) cm^{-2}).

Fiber aspect ratio (length/diameter) is a critical factor in reinforcement unless a minimum value is exceeded: L/D must be greater than the ratio of fiber tensile strength (T_f) to twice the strength of the bond $(T_{f\text{-}p})$ between the fiber and the matrix. The critical length, L', is therefore defined: $L' = D\, T_f/2T_{f\text{-}p}$. Assuming the nylon to fiber bond equals the shear strength of nylon, about 9000 lb (wt) in.$^{-2}$, L' for commercial fiber (K-glass diameter = 0.525 mils = 0.0133 mm; G-glass = 0.375 mils = 0.0095 mm) is about 8 mils (0.2 mm) (90).

A simple law of mixtures applies to a continuous filament comprising a core of fiber surrounded by a tightly adhering annulus of polymer:

$$T_c = T_f V_f + T_p V_p$$

$$E_c = E_f V_f + E_p V_p$$

where T = tensile strength, E = modulus, V = volume fraction, c = composite, f = fiber, and p = polymer. Although useful as guidelines to first approximations, these equations do not accurately describe reinforced thermoplastics not only because the fibers are discontinuous but also because the fibers are not precisely aligned and are not perfectly adherent to the polymer matrix (91, 92). Modified equations for discontinuous fibers in carefully oriented composites have been suggested. An example is: $T_c = T_f V_f(1 - L'/2L) + T_p V_p$ where L is the actual fiber length and exceeds L', but best use of this equation requires knowledge of the fiber length distribution (91). Tensile strengths approaching 80% of theory have been reported for dry nylon-66 containing 33% glass fiber (90).

Glass fiber is coated with a finish to hold the strands together, avoid breakage, and improve the fiber-to-polymer bond. The choice of finish or binder depends on the polymer (88, 93, 94). The importance of the binder and the effect of fiber length are both illustrated in Figure 11-7.

Properties of Glass Fiber Reinforced Nylons.

Nylons modified with 6 to 60 wt % glass fiber have been offered commercially. Stiffness increases with increasing glass content for all nylons, but tensile and notched impact strengths tend to dip at low glass levels and to level off at high glass levels (Figure 11-8, also Refs. 96 and 97). The deflection temperature (ASTM D648, 264 lb(wt)in.$^{-2}$) approximates the melting point of the nylon at about 15% glass (96, 98). Most commercial compositions contain 20 to 40% glass fiber. No optimum concentration can be defined because this depends on the application and a combination of property, processing, and cost requirements.

The properties of nylons containing 30 to 33% glass fiber are summarized in Table 11-9, and the effect of the glass on each property is estimated. High stiffness, high strength in tension, flexure, compression, or shear, improved notched impact strength, much higher deflection temperature, reduced water absorption, and reduced coefficient of linear thermal expansion are to be noted. Data on tensile impact strength are meager, but results on nylon-66 suggest that

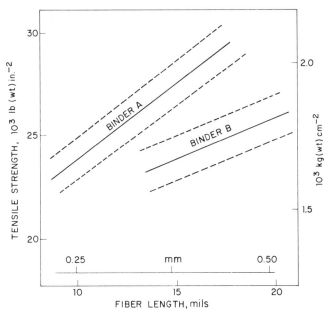

Fig. 11-7.   Effect of fiber length and binder on tensile strength of dry, as-molded, glass-reinforced nylon (90).

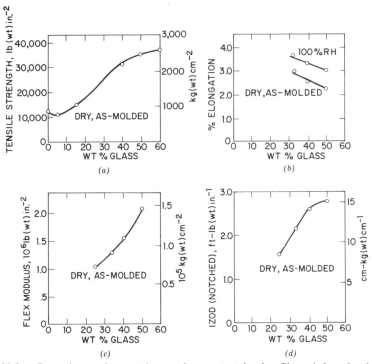

Fig. 11-8. Dependence of properties on glass content in glass-fiber-reinforced nylon-66 (90). (a) Tensile strength; (b) elongation; (c) flex modulus; (d) Izod impact strength.

it is slightly lower for the glass-reinforced resin than for the unmodified resin (50 versus 60 ft lb (wt) in.$^{-2}$ for 33 versus 0% glass, Ref. 90). Low creep and excellent fatigue resistance have been reported for glass-reinforced nylon-66 and nylon-610 (90, 98-100).

Absorption of moisture reduces rigidity by a factor that depends on the amount of glass, the binder, and the type of nylon. However, a nylon-66 with 33% glass fiber has a higher modulus at 100% relative humidity than the unmodified nylon when dry (600,000 versus 400,000 lb (wt) in.$^{-2}$, Ref. 90). Resistance to hydrolysis, weathering, and oxidation can be improved significantly by incorporation of glass fiber (Figure 11-9).

Some commercial compositions contain "long" glass fibers. Higher notched impact strength but lower tensile impact strength, lower-work-to-break over a range of strain rates, and inferior fatigue performance have been reported for nylon-6 and nylon-66 with "long" as opposed to the more commonly used "short" glass fibers (100,101).

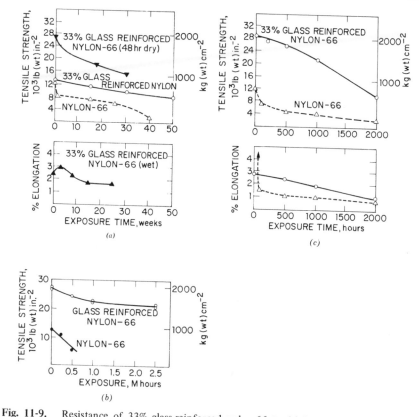

Fig. 11-9. Resistance of 33% glass-reinforced nylon-66 to (a) hydrolysis at 100°C, (b) Weather-Ometer® exposure, and (c) air-oven aging at 200°C (90).

Properties are affected by molding conditions largely because of the effect of mechanical working on fiber length (102). Loss in mechanical properties can be minimized by limiting the working of the melt. After four successive moldings of 100% rework (that is, fifth molding compared with first), a nylon-66 containing 33% glass fiber showed the following changes (102):

20% decrease in tensile strength
15% decrease in flex modulus
40% decrease in Izod impact
45% decrease in tensile impact
5% decrease in shear strength
1×10^{-3} in./in. increase in mold shrinkage of a 5-in. bar.

The average fiber length decreased from 31.4 mils in the virgin resin to 21.7 after the first molding and 13.6 after the fifth (0.80 to 0.55 to 0.35 mm). Raising the

Table 11-9. Dry, as-molded properties of glass fiber reinforced nylons[a]

Property	ASTM	Units	Nylon-66 30-33% glass
Specific gravity	D792	- -	1.37-1.38
Melting point	D789	$^\circ C/^\circ F$	257/495
Mold shrinkage, 1/8-in. (0.32-cm) thick, flow dir.	- -	$(\Delta L/L) \times 10^3$	3.5-5
Water absorption, $73^\circ F$ ($23^\circ C$)/ 24 hr	D570	%	1.0
Water absorption, saturation	D570	%	5.7
Tensile Strength, $73^\circ F$ ($23^\circ C$)	D638	lb(wt)in.$^{-2}$	26,000-28,000
Tensile elongation, $73^\circ F$ ($23^\circ C$)	D638	%	2-4
Flexural modulus, $73^\circ F$ ($23^\circ C$)	D790	lb(wt)in.$^{-2}$	1,300,000
Flexural strength, $73^\circ F$ ($23^\circ C$)	D790	lb(wt)in.$^{-2}$	38,000
Shear strength, $73^\circ F$ ($23^\circ C$)	D732	lb(wt)in.$^{-2}$	12,500
Compressive strength, $73^\circ F$ ($23^\circ C$)	D695	lb(wt)in.$^{-2}$	35,500
Izod impact strength, $73^\circ F$ ($23^\circ C$), 1/8-in. (0.32-cm) bar	D256	ft-lb(wt)in.$^{-1}$	1.7-2.1
Izod impact strength, $73^\circ F$ ($23^\circ C$), 1/4-in. (0.635-cm) bar	D256	ft-lb(wt)in.$^{-1}$	- -
Rockwell hardness	D785	M-scale	101
Deformation under load, 4000 lb(wt)in.$^{-2}$, $122^\circ F$ ($50^\circ C$)	D621	%	0.8
Deflection temp., 264 lb(wt)in.$^{-2}$	D648	$^\circ C/^\circ F$	249/480
Coeff. of linear thermal expansion	D696	$\Delta L/L \times 10^5 \ ^\circ C/^\circ F^{-1}$	2.3-2.9/1.3-1.6
Thermal conductivity	- -	btu hr^{-1} ft^{-2} ($^\circ$F/in.)$^{-1}$	1.5
	- -	kcal hr^{-1} m^{-2} ($^\circ$C/m)$^{-1}$	0.19
Specific heat	- -	- -	0.3-0.4
Dielectric strength, step by step	D149	V mil^{-1}	400-440
		kV mm^{-1}	16-17
Volume resistivity	D257	ohm-cm	5×10^{14}-5×10^{15}
Surface resistivity	D257	ohm	3×10^{14}
Dielectric constant, 10^3 cycles	D150	- -	4.0-4.5
Dissipation factor, 10^3 cycles	D150	- -	0.02

[a] Data from trade literature and Ref. 95.

Note. lb(wt)in.$^{-2} \times 0.0703$ = kg(wt)cm^{-2}; ft-lb(wt)in.$^{-1} \times 5.45$ = cm-kg(wt)cm^{-1}.

Table 11-9. (continued)

Nylon-6 30% glass	Nylon-610 30% glass	Nylon-612 33% glass	Nylon-11 30% glass	Nylon-12 30% glass	Approx. effect of glass on property[b]
1.37	1.31	1.32	1.25	1.23	1.2
216/420	216/420	213/415	185/365	175/347	1.0
3.5	3.5	3	--	1.5	0.25
1.2	0.2	0.2	0.12	--	0.7
6.5	2.1	2.0	1.4	1.0	0.7
23,500	21,000	24,000	14,000	17,400	2-2.5
3-4	3-4	4.5	3-4	3-4	0.05
1,200,000	1,100,000	1,200,000	875,000	1,000,000	3-4
35,000	32,000	--	20,000	22,000	-- } Yields
12,000	11,000	11,000	--	--	-- } if
23,000	20,000	23,000	13,000	--	-- } unmodified
--	--	--	--	--	} 1.5-2.5
2.3	2.4	2.6	2.2	3.0	
92-100	93	(R118)	--	--	> 1
0.9	0.9	--	--	--	< 1
216/420	216/420	210/410	171/340	174/345	$+ (100\text{-}170°\text{C})/(200\text{-}300°\text{F})$
3.1-3.8/1.7-2.1	2.7/1.5	2.3/1.3	5.4/3	13.5/7.5	0.3-0.4
1.6-3.3	1.6-3.5	--	--	3.9	} 1
0.20-0.41	0.20-0.43	..	--	0.48	
0.35	0.34	--	--	--	0.75
435	440	490	--	--	} ≥1
17	17	19	--	--	
5×10^{14}	8×10^{15}	10^{15}	--	--	1
9×10^{13}	10^{15}	10^{15}	--	--	$10\text{-}10^3$
4.3	3.8	3.7	--	--	1.0-1.3
0.03	0.015	0.024	--	--	1

[b] Property value for reinforced resin divided by value for unmodified resin. This varies somewhat with the nylon and proprietary composition.

screw speed or back pressure and lowering the rear zone temperature increased fiber breakage. Other studies have recommended a nozzle with a large opening, a screw with a long transition zone, a compression ratio of 2/1, use of a cone-shaped pin at the juncture of the sprue and runners to reduce set-up time, a sharp corner to flat surface type of shut-off ring, and an increasing temperature profile from rear to front (103). The corrosiveness of compositions that contain glass fibers has been said to be controlling in machine wear (104), but the weight of the evidence is that abrasion is controlling and can be made negligible via suitable design (105).

Glass Spheres

The solid glass spheres used in plastics are made from A-glass rather than E-glass from which the fibers are made (see above). A-glass has a 16% lower virgin tensile strength than E-glass, a 4.5% lower modulus, an almost double coefficient of thermal expansion, and a 3% lower density (106). The spheres are generally less than 325 mesh with an average diameter of 30 μ (0.0012 in.).

As with the fibers, efficient bonding to the polymer matrix is essential to good performance. Sphericity means a significantly reduced effect on melt viscosity and allows higher loadings. The directionality of the properties of the molded object are similarly reduced. The large property changes in tensile strength and modulus observed in fiber reinforcement where alignment in the flow direction is a factor are therefore not achieved, but the compressive strength is high (Table 11-10). A loss in Izod impact strength and no improvement in deflection temperature also characterize the nylon-66 containing 40 wt % glass spheres.

Table 11-10. Properties of a nylon-66 containing 40 wt % glass spheres (107,108)

Specific gravity	1.43
Tensile strength, lb(wt)in.$^{-2}$	14,200
Ultimate elongation, %	2.5
Flexural modulus, lb(wt)in.$^{-2}$	731,000
Compressive strength, lb(wt)in.$^{-2}$	36,500
Izod impact strength, ft-lb(wt)in.$^{-1}$	0.6
Rockwell hardness	R117
Deformation under load, 4000 lb(wt)in.$^{-2}$, 122°F (50°C)	0.8
Deflection temp., 264 lb(wt)in.$^{-2}$, °C/°F	74/165
Coefficient of linear thermal expansion, $\Delta L/L \times 10^5$ °C^{-1}/°F^{-1}	5.2/2.9

Note: Reference does not say, but data suggest measurements made on dry specimens.
lb(wt)in.$^{-2}$ × 0.0703 = kg(wt)cm^{-2}; ft-lb(wt)in.$^{-1}$ × 5.45 = cm-kg(wt)cm^{-1}.

Lubricity Aids

Lubricity aids for nylons include molybdenum disulfide, polytetra-fluoroethylene, and graphite. Nylon-66, -610, or -6 with 10 to 90 wt % of MoS_2 have been described as filled compositions that have unusual lubricating qualities, good dispersion, and the capacity of being cold pressed and sintered (109). Machined parts containing 0.25 to 4.0% MoS_2 are said to have equivalent or better frictional and wear properties than unmodified parts and about half of the coefficient of linear thermal expansion (110). Dry bearings of MoS_2-modified nylon-66 can be run at up to 50% higher PV (pressure $\times$ velocity) ratings than those of unmodified resin (111). The wear rate against steel is reduced by about 50%. Dry modulus is increased by 45%, but both Izod and tensile impact strengths are decreased by up to 35% (111).

Nylons containing polytetrafluoroethylene or graphite are commercially available, but information on the properties of compositions that do not also contain glass reinforcement is scant. The improvement in wear resistance increases with increasing amounts of graphite at least up to 10% based on Russian work (11). Graphite lowers the coefficient of friction of nylon-11 about as much as MoS_2 (0.18 to 0.11) and yields a composition with inferior impact resistance but better resistance to compression than the MoS_2 modification (113).

Other Modifiers

Antistatic agents are principally of concern in the fiber industry. In plastics they tend to be hygroscopic liquids of low volatility which can migrate to the surface and form a continuous layer of higher conductance (114). The development of a static charge is a factor in most nylons only if the polymer is dry, and it is not normally a problem in the nylon plastics industry. The absorption of water at ambient conditions frequently provides the requisite surface conductivity in nylon articles. Use of antistats in nylon plastics is rare.

Nylons are considered to be generally resistant to fungus attack (115), but nylon-11 is indicated to be commercially available with added fungicides (116).

REFERENCES

1. Sahler, W.A. (to Dr. Plate GmbH), U.S., Patent 3,484,415 (Dec. 16, 1969); Grechele, G.B., and G.F. Martinis, *J. Appl. Polym. Sci.* 9, 2939 (1965).

2. Battista, O.A. (to FMC Corp.), U.S. Patents 3,299,011 (Jan. 17, 1967) and 3,536,647 (Oct. 27, 1970); *Indus. Res.* 11 (3), 66 (Mar., 1969).

3. Stastny, F., *Kunststoffe* 40, 273 (1950).

4. Batzer, H., and A. Moschle, *Makromol. Chemie* 22, 195 (1957).

5. Baker, W.O., and C.S. Fuller, *J. Am. Chem. Soc.* 64, 2399 (1942).

6. Hybart, F.J., and B. Pepper, *J. Appl. Polym. Sci.* 13, 2643 (1969).

7. Harvey, E.D., and F.J. Hybart, *J. Appl. Polym. Sci.* 14, 2133 (1970).

8. Yu, A.J., and R.D. Evans, *J. Polym Sci.* 42, 249 (1960).

9. Saotome, K., and H. Komoto, *J. Polym. Sci., Part A-1*, 4, 1475 (1966).

10. Ridgway, J.S., *J. Polym. Sci., Part A-1*, 7, 2195 (1969).

11. Catlin W.E., E.P. Czerwin, and R.H. Wiley, *J. Polym. Sci.* 2, 412 (1947).

12. Muller, A., and R. Pfluger, *Kunststoffe* 50, 203 (1960).

13. Baker, W.O., and C.S. Fuller, *J. Am. Chem. Soc.* 65, 1120 (1943).

14. Biggs, B.S., C.J. Frosch, and R.H. Erickson, *Ind. Eng. Chem.* 38, 1016 (1946).

15. Wittbecker, E.L., R.C. Houtz, and W.W. Watkins, *Ind. Eng. Chem.* 40, 875 (1948).

16. Lewis, J.R., and R.J.W. Reynolds, *Chem. Ind.* 1951, 958.

17. Saotome, K., and H. Komoto, *J. Polym. Sci., Part A-1*, 5, 107 (1967).

18. Anon., Belding Chemical Industries Tech. Bull. VIII-A.

19. Cairns, T.L., H.W. Gray, A.K. Schneider, and R.S. Schreiber, *J. Am. Chem. Soc.* 71, 655 (1949).

20. Haas, H.C., S.G. Cohen, A.C. Oglesby, and E.R. Karlin, *J. Polym. Sci.* 40, 427 (1955).

21. Haas, H.C., and S.G. Cohen (to Polaroid Corp.), U.S. Patent 2,835,653 (May 20, 1958).

22. Inoue, M., *J. Polym. Sci., Part A-1*, 1, 3427 (1963).

23. Mesrobian, R.B., and C.J. Ammondson (to Continental Can Co.), U.S. Patent 3,093, 255 (June 11, 1963).

24. Magat, E.E., and D. Tanner (to E.I. du Pont de Nemours and Co.), U.S. Patents 3,412,175 (Nov. 19, 1968), 3,412,176 (Nov. 19, 1968), and 3,413,378 (Nov. 26, 1968).

25. Ravve, A., and C.W. Fitko, *J. Polym. Sci., Part A-1*, 4, 2533 (1966).

26. Craubner, H., and G. Illing (to H. Roemmler G.m.b.H.), U.S. Patent 3,261,885 (July 19, 1966).

27. Anon. (to Allied Chemical Corp.), Brit. Patent 1,205,424 (Sep. 16, 1970); Kray, R.J., and R.J. Bellet (to Allied Chemical Corp.), U.S. Patent 3,388, 186 (June 11, 1968).

28. Anspon, H.D., and H.E. Robb (to Gulf Oil Corp.), U.S. Patent 3,472,916 (Oct. 14, 1969).

29. Murdock, J.D., N. Nelan, and G.H. Segall (to Canadian Industries Ltd.), U.S. Patent 3,236,914 (Feb. 22, 1966).

30. Vaala, G.T. (to E.I. du Pont de Nemours and Co.), U.S. Patent 2,378,667 (June 19, 1945).

31. Webber, T.G., Plastics Department, E.I. du Pont de Nemours and Co., Paper presented at the SPE RETEC, Palisades Section, June 10, 1966.

32. Anon., Plastics Department, E.I. du Pont de Nemours and Co., Unpublished Technical Note, "Dry Coloring of Zytel® Nylon Resins," 1970.

33. Cocci, A.J. (to E.I. du Pont de Nemours and Co.), U.S. Patent 2,948,698 (Aug. 9, 1960).

34. Silverman, B. (to Monsanto Co.), U.S. Patent 3,475,365 (Oct. 28, 1969).

35. Lowery J.H., and E.C. Schule (to Allied Chemical Corp.), U.S. Patent 2,856,373 (Oct. 14, 1958).

36. Voigt, J.L. (to N.V. Onderzoekingsinstituut), U.S. Patent 2,982,744 (May 2, 1961).

37. Kessler, C.F., and H.R. Spreeuwers (to N.V. Onderzoekingsinstituut), Ger. Patent 1,127,076 (April 5, 1961).

38. Voigt, J.L. (to N.V. Onderzoekingsinstituut), U.S. Patent 3,008,908 (Nov. 14, 1961).

39. Hansen, J.E. (to E.I. du Pont de Nemours and Co.), U.S. Patent 3,009,900 (Nov. 21, 1961).

40. Brooks, R.E., J.F. Cogdell, Jr., and C.K. Rosenbaum (to E.I. du Pont de Nemours and Co.), U.S. Patent 3,080,345 (Mar. 5, 1963).

41. Anon. (to Farbenfabriken Bayer AG), Brit. Patent 851,300 (Oct. 12, 1960).

42. Schneiner, L.L., *Plast. Tech.* **13** (4), 37 (April, 1967).

43. Toelcke, G.A., M.N. Riddell, G.N. Bellucci, R.J. Welgos, and J.L. O'Toole, *Mod. Plast.* **45** (16), 117 (Dec., 1968).

44. Ben, V.R. (to E.I. du Pont de Nemours and Co.), U.S. Patent 2,981,715 (Apr. 25, 1961).

45. Robb, H.E., G.D. Newman, Jr., and D.L. McCullough (to Gulf Oil Corp.), U.S. Patent 3,400,087 (Sep. 3, 1968).

46. Anon. (to Farbenfabriken Bayer AG), Brit. Patent 889,403 (Feb. 14, 1962).

47. Gray, H.W. (to E.I. du Pont de Nemours and Co.), U.S. Patent 2,510,777 (June 6, 1950).

48. Ferrigno, T.H. *Rigid Plastic Foams,* 2nd ed., Reinhold Publishing Corp., New York, 1967, p. 362.

49. Benning, C.J., *Plastic Foams,* Wiley-Interscience, New York, 1969, Vol. 1, p. 608.

50. Becke, F., and K. Wick (to Badische Anilin- und Soda-Fabrik), Can. Patent 600,453 (June 21, 1960).

51. Becke, F. and K. Wick (to Badische Anilin- und Soda-Fabrik), U.S. Patent 3060,135 (Oct. 23, 1962).

52. Beyerlein, F., and H. Wilhelm (to Badische Anilin- und Soda-Fabrik AG), Ger. Patent 1,177,340 (Sep. 3, 1964).

53. Hyde, T.G. (to E.I. du Pont de Nemours and Co.), U.S. Patent 3,449,269 (June 10, 1969).

54. Fisher, C.F. (to E.I. du Pont de Nemours and Co.), Fr. Patent 1,326,136 (March 25, 1963).

55. Kohan, M.I. (to E.I. du Pont de Nemours and Co.), Fr. Patent 1,330,984 (May 20, 1963).

56. Neiman, M.B., Ed., *Aging and Stabilization of Polymers,* English Edition, Consultants Bureau, New York, 1965 (Russian Edition, 1964), pp. 249-251 and 261-265.

57. Epstein, M.M., and C.W. Hamilton, *Mod. Plast.* 37 (11), 142 (July, 1960).

58. Thompson, L.M. (to Monsanto Co.), U.S. Patent 3,180,849 (Apr. 27, 1965).

59. Rothrock, D.A., Jr., and R.F. Conyne (to Rohm and Haas Co.), U.S. Patent 2,493,597 (Jan. 3, 1950).

60. Stamatoff, G.S. (to E.I. du Pont de Nemours and Co.), U.S. Patent 2,630,421 (Mar. 3, 1953).

61. Tazewell, J.H. (to The Firestone Tire and Rubber Co.), U.S. Patent 3,477,986 (Nov. 11, 1969).

62. McQueen, D.M. (to E.I. du Pont de Nemours and Co.), U.S. Patent 2,619,479 (Nov. 25, 1952).

63. Stokes, K.B. (to General Mills), U.S. Patent 3,454,412 (July 8, 1969).

64. Stamatoff, G.S. (to E.I. du Pont de Nemours and Co.), U.S. Patent 2,705,227 (Mar. 29, 1955).

65. Wilhelm, H., G. Mueller, R. Pflueger, and H. Doerfel (to Badische Anilin- und Soda-Fabrik), U.S. Patent 3,458,474 (July 29, 1969).

66. Hermann, K.H., and H. Rudolph (to Farbenfabriken Bayer AG), U.S. Patent 3,477,986 (Nov. 11, 1969).

67. White, T.R. (to British Nylon Spinners), U.S. Patent 2,984,647 (May 16, 1961).

68. Edgar, O.B. (to Imperial Chemical Industries), U.S. Patent 3,458,470 (July 29, 1969).

69. Stamatoff, G.S. (to E.I. du Pont de Nemours and Co.), U.S. Patent 2,640,044 (May 26, 1953).

70. Costain, W., and H.J. Palmer (to Imperial Chemical Industries and British Nylon Spinners), U.S. Patent 3,160,597 (Dec. 8, 1964).

71. Busse, W.F. (to E.I. du Pont de Nemours and Co.), U.S. Patent 3,418,267 (Dec. 24, 1968).

72. Busse, W.F. (to E.I. du Pont de Nemours and Co.), U.S. Patent 3,468,843 (Sept. 23, 1969).

73. Michael, D., and W. Kosiol (to Farbenfabriken Bayer AG), Brit. Patent 1,204,835 (Sept. 9, 1970).

74. Coffman, D.D. (to E.I. du Pont de Nemours and Co.), U.S. Patent 2,214,405 (Sept. 10, 1940).

75. Gordon, W.E. (to E.I. du Pont de Nemours and Co.), U.S. Patent 2,309,729 (Feb. 2, 1943).

76. Vaala, G.T. (to E.I. du Pont de Nemours and Co.), U.S. Patent 2,456,344 (Dec. 14, 1948).

77. Walker, I.F. (to E.I. du Pont de Nemours and Co.), U.S. Patent 2,473,924 (June 21, 1949).

78. Sido, G.R. (to Monsanto Chemical Co.), U.S. Patent 2,499,932 (Mar. 7, 1950).

79. Hurwitz, M.J. (to E.I. du Pont de Nemours and Co.), U.S. Patent 2,615,002 (Oct. 21, 1952).

80. Dazzi, J. (to Monsanto Chemical Co.), U.S. Patent 2,757,156 (July 31, 1956).

81. Metz, E.A. (to Nypel of Delaware), U.S. Patent 3,475,368 (Oct. 28, 1969).

82. Illing, G. (to H. Rommler GmbH), Brit. Patent 1,019,348 (Feb. 2, 1966).

83. Seymour, R.B., *Mod. Plast. Encycl.* 46 (10A), 372 (Oct., 1969).

84. Solomko, V.P., I.A. Uskov, T.A. Molokoyedova, and S.S. Pelishenko, *Polym. Science (USSR)* 6, 2435 (1964).

85. Symons, N.K.J. (to E.I. du Pont de Nemours and Co.), U.S. Patent 2,874,139 (Feb. 17, 1959).

86. Noga, E.A., and R.T. Woodhams, *SPE J.* 26 (9), 23 (Sept., 1970).

87. Anon, Fiberfil Division, Dart Industries Inc., Tech. Bulletin "Nylode" Asbestos Reinforced Nylon, Dec. 1969.

88. Mettes, D.G., in *Handbook of Fiberglass and Advanced Plastics Composites*, G. Lubin, Ed., Van Nostrand Reinhold Co., New York, 1969, Chapter 7.

89. Paulus, H.J., *Mod. Plast. Encycl.* 46 (10A), 392 (Oct., 1969).

90. Williams, J.C.L., D.W. Wood, I.F. Bodycot, and B.N. Epstein, Soc. Plast. Ind. Tech. Conf. No. 23, Reinf. Plast. Div., Sec. 2-C (1968).

91. Lees, J.K., *Polym. Eng. Sci.* **8**, 195 (1968).

92. Lees, J.K., *Polym. Eng. Sci.* **8**, 186 (1968).

93. Hartlein, R.C., Soc. Plast. Ind. Tech. Conf. No. 25, Reinf. Plast. Div., Sec. 16-B (1970).

94. Sterman, S., and J.G. Marsden (to Union Carbide Corp.), Can. Patent 823,150 (Sep. 16, 1969).

95. Jones, R.F., SPE Regional Tech. Conf., Hartford, Conn., Oct., 1969, p. 1.

96. Anon., Liquid Nitrogen Processing Corp., Tech. Bulletin, "LNP Glass-fortified Nylon Thermocomp® Series RF, QF, PF," No. 203-769.

97. Reichelt, W., *SPE Tech. Pap.* **13**, 919 (1967).

98. Anon., Plastics Department, E.I. du Pont de Nemours and Co., Tech. Bulletin, "Du Pont Glass-Reinforced Zytel® Nylon Resins," Sep., 1969.

99. Theberge, J.E., *Mod. Plast.* **45** (10), 155 (June, 1968).

100. Dally, J.W. and D.H. Carillo, *Polym. Eng. Sci.* **9**, 434 (1969).

101. Theberge, J.E., and N.T. Hall, *Mod Plast.* **46** (7), 114 (July, 1969).

102. Filbert, W.C., Jr., *SPE J.* **25** (1), 65 (Jan., 1969).

103. Lachowecki, W., *Plast. Des. Process* **9** (7), 28 (July, 1969).

104. Olmsted, B.A., *SPE J.* **26** (2), 42 (Feb., 1970).

105. Butler, R.A., *Brit. Plast.* **43** (6), 139 (June, 1970).

106. Strauch, O.R., SPE Regional Tech. Conf., Hartford, Conn., Oct., 1969, p. 14.

107. Strauch, O.R., *SPE J* **25** (9), 38 (Sep., 1969).

108. Wotitzky, H.J., SPE Regional Tech. Conf., Hartford, Conn., Oct. 1969, p. 24.

109. Stott, L.L. (to The Polymer Corp.), U.S. Patent 2,849,415 (Aug. 26, 1958).

110. Stott, L.L. (to The Polymer Corp.), U.S. Patent 2,855,377 (Oct. 7, 1958).

111. Powers, T.E., *Mod. Plast.* **37** (10), 148 (June, 1960).

112. Vlasova, K.N., M.A. Rudyk, L.A. Nosova, A.N. Pichugin, and G.P. Ivanova, *Sov. Plast.* **1965** (4), 35.

113. Anon., Aquitaine-Organico, "Rilsan" Tech. Data Sheets Nos. 2 and 5, 1969.

114. Valko, E.I., and G.C. Tesoro, in *Encyclopedia of Polymer Science and Technology*, Interscience Publishers, a division of John Wiley and Sons, New York, 1965, Vol. 2, pp. 204-229.

115. Scullin, J.P., M.D. Dudarevitch, and A.I. Lowell, in *Encyclopedia of Polymer Science and Technology*, Interscience Publishers, a division of John Wiley and Sons, New York, 1965, Vol. 2, pp. 379-401.

116. Anon., Aquitaine-Organico, Tech. Bulletin, "Rilsan® Nylon 11:Nomenclature and Description of Principle Grades." Sep., 1968.

CHAPTER 12

Properties of Extruded Nylons

THOMAS M. RODER

INTRODUCTION

Nylon compositions used for extrusion are often identical to those used in injection molding. Most extruded products, however, have one or more properties that are characteristic of the configuration and are peculiarly important to a specific application. For example, burst strength is used to rate tubing, bend recovery is important in monofilament, and optical properties are significant in film. Some of these properties can be found in tabulations such as the chart on films published annually in the *Modern Plastics Encyclopedia*, but many pertain to a specific form and can only be found in the manufacturer's literature.

The thickness of many nylon extruded products is often less or greater than that of most injection moldings. For example, film is commonly 0.002 in. (0.05 mm) thick or less, and wire coating is usually less than 0.010 in. (0.25 mm) in thickness. Monofilament and strapping are at most in the low end, 0.020 to 0.040 in. (0.5 to 1 mm), of the injection molding thickness range. On the other hand, extruded stock shapes of nylon are often an inch or more in thickness and are beyond the practical limit of injection molding. These unusual thicknesses associated with extruded products influence cooling rate and have a significant effect on crystalline structure and on the properties of nylons. Because of the greater difficulty in control of frozen-in stress in thick, extruded shapes, annealing is more commonly practiced on extruded articles. Annealing procedures are considered in Chapter 17, and the physical structure of nylons is discussed in Chapter 8.

Products such as tubing and sheeting exhibit properties that may differ from those of injection moldings because of the orientation associated with the extrusion process. For example, tensile strength in the machine direction may be considerably greater than in the transverse direction. This effect is magnified in thin sections like film unless precautionary steps are taken to minimize orientation. Some films such as poly(ethylene terephthalate) are biaxially oriented to achieve goal properties, but this has not been the case with nylons to date. For other nylon products such as monofilament and strapping, orientation is necessary to obtain the desired properties.

There are also some special considerations inherent in the testing and sampling of extruded products. Unlike injection molding, the extrusion process does not always lend itself to convenient and well-controlled methods of sample preparation. Samples for many tests must be cut or machined, and the nature of the cut edge can influence test results. Specimen thickness, which is always subject to some variation in extrusion, will affect test results, and in film the variation can be significant. Thin sections can also yield widely divergent results from sample to sample because of rapid absorption and desorption of moisture. This of course occurs with all nylon shapes but is usually confined to the surface and is not a factor in the time involved in testing. In fact, the problem of keeping film dry is responsible for the practice of reporting data only for samples equilibrated to 50 or 100% R.H.

FILM AND COATINGS

The choice of a nylon as a film or coating is usually based on properties that are relatively independent of thickness. High melting point, abrasion resistance, and chemical properties most often are the reasons for selecting a nylon.

Film

A nylon film is used in food packaging and other applications both by itself and in composite structures with one or more other resins such as polyethylene, ionomer resin, and poly(vinylidene chloride). The composites can be made by laminating nylon film to another film, by coextrusion, by coating a substrate with nylon, or by coating nylon with another resin. The properties of the nylon portion of these composite structures are not significantly different from those of nylon film. The mechanical properties of the composite usually lie between those of the individual films if adhesion between the layers is adequate to prevent delamination. On the other hand, resistance to permeation by gases and liquids is the sum of the resistances of the individual layers. This enhancement of barrier properties is an important reason for using composite structures.

The properties of several types of nylon film, compared in Table 12-1, are taken from the trade literature and represent typical values or ranges. Wide variations can result from changing extrusion conditions or, especially, from changing processing equipment. Flat film extruded on to a chill roll is cooled or quenched rapidly and is therefore less crystalline than the more slowly cooled blown film. The latter tends to be less transparent and to have lower elongation, higher yield strength, and higher specific gravity than chill roll film. Unfortunately, the principal source of information is trade literature, and processing technique is highly proprietary.

In some cases data are given in the machine direction (MD) as well as in the transverse direction (TD). Orientation in the machine direction causes a higher yield strength and lower elongation. Processing conditions are adjusted to minimize this difference and achieve balanced properties in the two directions. Serious lack of balance can interfere with some applications of nylon film.

Permeability of gases and liquids for several nylon resins is compared in Table 12-2. As the data indicate, these nylons are excellent barriers to many common gases and organic liquids. In general, nylons have high permeability to polar substances such as water and methanol. The polyethylene or ionomer resin used with nylon in composite structures provides resistance to such polar substances.

A number of nylon resins that are suitable for film applications may be used in food packaging and other food contact applications. These include 66, 610, 6, 11, and certain copolymers of 66 with 6 or 610, but the nylons most used are 66 and 6. The FDA regulation (1) limits the allowable level of extractables in specific solvents. It is advisable to consult the resin supplier to ascertain which specific compositions meet the requirements of this regulation.

Nylon films can be heat sealed successfully by impulse or radio frequency methods and also by hot bar techniques under carefully controlled conditions. For the high melting nylons such as 66, heat stabilized formulations are necessary to avoid brittle seals.

Table 12-1. Properties of nylon film

Property[a]	Unit[b]	Method	Nylon-66	Nylon-6[c]	Nylon-11	Nylon-12
Specific gravity			1.14	1.13	1.03	1.01
Tensile strength	lb(wt)in.$^{-2}$	ASTM D882	MD 11,000	MD 9,000-12,000 TD 10,000-13,000	11,000	MD 8,790-12,300 TD 6,540-7,800
Yield strength	lb(wt)in.$^{-2}$	ASTM D882	5,000	MD 3,900 TD 3,600		MD 5,950-5,980 TD 5,940-5,960
Elongation	%	ASTM D882	250	MD 350-400 TD 400-500	250-400	MD 150-245 TD 290-330
Tensile modulus	lb(wt)in.$^{-2}$	ASTM D882	175,000	MD 90,000-100,000 TD 105,000-125,000		MD 124,000-144,000 TD 117,000-127,000
Tear strength	g mil^{-1}	ASTM D1922	MD 25-50 TD 30-150	MD 50-90 TD 50-70	400-500	MD 450 TD 500
Haze	%	ASTM D1003		1.5-4.5		
Gloss		ASTM D2457 (20°)		70-100		

[a] 73°F (23°C), 50% R.H.

[b] lb(wt)in.$^{-2}$ × 0.0703 = kg(wt)cm^{-2} ; g mil^{-1} = g (0.025 mm)$^{-1}$.

[c] Extracted.

Sources: Trade literature and *Modern Plastics Encyclopedia.*

Table 12-2. Permeability of nylon films

Permeant	Unit[a]	% R.H.	Nylon 66	610[b]	6[c]	12
Oxygen	cm³ mil (100 in.²)⁻¹ day⁻¹ atm⁻¹	0	--	--	2.6	--
		50	2	4.3	6	--
		100	12	7	--	52-92
Carbon dioxide	cm³ mil (100 in.²)⁻¹ day⁻¹ atm⁻¹	Unspecified	--	--	--	--
		0	--	--	9.7	--
		50	10	--	25	--
Nitrogen	cm³ mil (100 in.²)⁻¹ day⁻¹ atm⁻¹	Unspecified	--	--	--	156-336
		0	--	--	0.9	--
		50	0.7	--	2(2)	--
Water vapor	g mil (100 in.²)⁻¹ day⁻¹	Unspecified	--	--	--	12.8-18
		50	0.9	0.9	0.6	--
		100	20	14	19-20 at 100°F (38°C)	--
Gasoline	g mil (100 in.²)⁻¹ day⁻¹	Unspecified	--	--	--	0.7
		50	0.2	--	--	--
Motor oil at 125°F (52°C)	g mil (100 in.²)⁻¹ day⁻¹	50	0.1	--	0.00	--
Toluene	g mil (100 in.²)⁻¹ day⁻¹	50	2	--	--	--
Trichloroethylene	g mil (100 in.²)⁻¹ day⁻¹	50	--	--	0.05	--

[a] The unit of permeability corresponds to quantity lost (volume for a gas, weight for a liquid) × thickness per unit area × time (× pressure difference if the permeant is a gas). Many combinations of units are found in the literature. Conversions of the above units to others commonly encountered follow:

cm^3 mil (100 in.²)⁻¹ day⁻¹ atm⁻¹ × 0.394 = cm^3 mm m⁻² day⁻¹ atm⁻¹
cm^3 mil (100 in.²)⁻¹ day⁻¹ atm⁻¹ × 6 × 10⁻¹³ = cm^3 cm cm⁻² sec⁻¹ (cm of mercury)⁻¹
cm^3 mil (100 in.²)⁻¹ day⁻¹ atm⁻¹ × 6 × 10⁻³ = barrers
g mil (100 in.²)⁻¹ day⁻¹ × 0.394 = g mm m⁻² day⁻¹.

[b] The permeability of nylon-612 has not been reported but is most likely comparable to that of nylon-610.

[c] Extracted.

447

Wire Jacketing

Nylons are used as a jacket on wire and cable because they have mechanical strength, abrasion resistance, chemical resistance, and the ability to endure elevated temperatures. In most electrical applications nylon acts as a protective coating over a primary insulation such as poly(vinyl chloride) although its electrical properties are adequate without other insulation in low voltage applications. Nylon-610 and -612 have somewhat better electrical properties than nylon-6, -66, and copolymers that absorb more water although nylon-6, 66/6 copolymers, 610, and 612 are commonly employed in electrical applications. The electrical properties of various nylons are discussed in Chapter 10.

When nylon is used over a poly(vinyl chloride) primary insulation the latter resin must be chosen for compatibility with nylon as well as for other properties. Some polyester plasticizers in poly(vinyl chloride) will cause embrittlement and failure of the combined coating. Plasticizers such as the high boiling phthalates do not exhibit this harmful interaction. In fact, the nylon outer jacket reduces plasticizer loss and prolongs useful life of the combined coating. Primary insulating materials such as polyethylene, polyester film, or fluorocarbon resins have no compatibility problem with nylons (3).

Tensile properties have been reported for nylon coatings stripped from wire with a primary insulation of poly(vinyl chloride). The data shown in Table 12-3 were measured on 5-mil (0.13-mm) jackets that had been conditioned at 50% R.H. and room temperature for several days or dried at 175°F (80°C) under vacuum. After aging five months at 50% R.H., nylon-610 exhibited improved tensile properties probably due to slow crystallization.

A mandrel wrap test for nylon wire jackets has been developed to show differences related to cracking tendency in service. No correlation was found

Table 12-3. Tensile data for nylon wire jackets (4)

Nylon	Condition	Yield[a] strength $(lb(wt)in.^{-2})$	Ultimate[a] strength $(lb(wt)in.^{-2})$	Ultimate elongation (%)
610	Dried	6500-6700	7200-7400	290-300
610	50% R.H. unaged	4000-4200	7800-9700	320-390
610	50% R.H., aged 5 months	6100-6300	10,800	340
6[b]	Dried	- -	6500-10,900	200-330
612	Dried	6500-6900	8000-9100	370-400

[a] $lb(wt)in.^{-2} \times 0.0703 = kg(wt)cm^{-2}$.

[b] Extracted.

between cracking and tensile properties or other standard toughness measurements. In the mandrel wrap test at least ten turns of jacketed wire is wrapped on a mandrel of 1 to 2.5 times the wire diameter. The ends are tied to prevent unwrapping prior to transfer of the coil from the mandrel to an oven at 265 to 300°F (130 to 148°C). After 15 min the coil is placed in a desiccator to cool for 30 min and is then straightened. Tough nylon jackets do not crack as they are unwound from the coil even after the severe drying step described above. This procedure has shown that tougher jackets of nylon-610 result from modifications of extrusion conditions that increase the temperature of the water quench bath or of the wire (5).

SHAPES

Tubing

Nylon tubing is used both for conveying fluids and mechanical applications such as the lining of flexible cable. The abrasion resistance, strength, and other mechanical properties of injection moldings, which were discussed in Chapter 10, are important in the mechanical applications of tubing. In fluid conveying applications, one property characteristic of extruded nylon tubing, burst strength, is of special importance. This property is most conveniently reported in terms of hoop strength at failure, so that tubing of varying dimensions can be compared. The hoop strength is related to burst pressure and tubing dimensions according to the following formula:

$$\text{hoop stress} = \frac{\text{pressure} \times \text{outside diameter}}{2 \times \text{wall thickness}}$$

Table 12-4 compares several different nylons at a variety of conditions. Hoop strength is a tensile measurement and should show some similarity to conventional tensile strength values if the rate of loading approaches that of standard tensile testing procedures. The values in the tabulation are generally lower than the resin tensile strength indicating that the rate of loading or pressuring was significantly less. Nylon tubing is often chosen instead of metal tubing because of its flexibility and ease of installation. Flexural moduli of tubing samples are also reported in Table 12-4.

Contact with a variety of chemicals is implicit in fluid conveying applications of tubing. Nylons are highly resistant to both attack and permeation by a wide variety of nonpolar materials, especially gasoline, oils, other hydrocarbons, and fluorocarbon refrigerants and propellants. Certain active, polar substances, such as hydrochloric acid or zinc chloride solutions, can cause stress cracking of nylons as discussed in Chapter 17. Studies of tubing show that nylon-6 is highly

Table 12-4. Properties of nylon tubing (6,7)

Nylon	Test conditions	Hoop strength[a] (lb(wt)in.$^{-2}$)	Flexural modulus[b] (lb(wt)in.$^{-2}$)
66	75°F (24°C), 2.35% moisture (average), oil inside	8,800	--
66	75°F, dry	7,500	300,000
66	75°F, 50% R.H.	4,500	--
66	75°F, moisture saturated	3,100	--
610 (black)	75°F, 0.9% moisture (average)	7,500	--
610 (black)	75°F, dry	6,500	203,000
610 (black)	75°F, 50% R.H.	5,800	53,000
610 (black)	75°F, moisture saturated	5,000	--
6 (plasticized)	75°F, dry	6,000	110,000
6 (plasticized)	75°F, 50% R.H.	2,600	40,000
6 (plasticized)	75°F, moisture saturated	2,100	--

[a] Hoop strength is based on short term tests to failure.

[b] lb(wt)in.$^{-2}$ × 0.0703 = kg(wt)cm^{-2}.

susceptible to zinc chloride stress cracking, 66 is less susceptible, and 610, 612, and 11 are much more resistant.

The resistance to permeation by fluorocarbons is important in the use of nylons as the inner core resin in reinforced hose constructions that convey refrigerant in automotive air conditioning systems. These hoses must contain the refrigerant, sometimes at elevated temperatures, for a period of years. Permeation of fluorocarbon refrigerant through a proprietary hose with a nylon inner core of 0.025-in. (0.62-mm) thickness is compared in Table 12-5 with nitrile rubber hose. Both are used in automotive air conditioning systems although the wall thickness of the latter must be made 5 to 6 times greater than that of the nylon liner in order to contain the refrigerant.

Stock Shapes and Profiles

Extruded rod, slab, and other stock shapes represent a major use of nylon in extrusion. These stock shapes are subsequently machined or stamped into a great variety of nylon parts similar to those commonly made by injection molding. Machining from stock shapes can be less expensive than injection molding in

Table 12-5. Comparison of permeation constants for hoses of nylon and nitrile rubber (8)

Hose type	Freon® refrigerant	Permeation constant, $K \times 10^5$ g in. day^{-1} [lb(wt)]$^{-1}$		
		140°F (60°C)	180°F (82°C)	220°F (104°C)
Nylon[a]	R-12	0.012	0.0223	0.0225
Nylon[a]	R-22	0.545	0.63	0.77
Nitrile rubber	R-12	1.02	1.94	2.53
Nitrile rubber	R-22	U	U	U

Note: U = Unmeasurably large.

[a] Plasticized 66/6 copolymer.

low volume applications. Small flat parts that can be stamped from nylon sheet are sometimes less expensive than molded parts even in high volume because of the cost and difficulty of running complex multicavity molds with high rework ratios. Other economic factors in the choice between machining and molding are discussed in Chapter 19.

Because stock shapes and injection moldings are both low in orientation, manufacturers of stock shapes usually report only standard property information for molded test bars. However, the thicker stock shapes can differ significantly from standard test bars because slow cooling of the melt results in a high level of crystallinity. Tensile yield strength can therefore be higher, elongation lower, and wear resistance improved. These differences in tensile properties have sometimes been observed in test bars machined from thick injection moldings (see Chapter 10). Most other properties such as flexural modulus, hardness, permeability, and absorption of solvents are also affected, but the changes do not significantly alter end-use application from that of injection moldings.

ORIENTED FORMS

Monofilament

The most familiar and important use of nylon monofilaments is in brush bristles. Monofilaments, however, are also used for fishing line, sewing thread, clothing stiffeners, and a variety of woven screens for filters. In some applications such as brush bristles and clothing stiffeners, high flexural modulus is needed. Others such as fishing line and sewing thread require flexibility but with retention of tensile strength. The rigidity requirement for these varied applications is achieved by varying thickness and by selection of the proper nylon.

Table 12-6 presents modulus data at several conditions of temperature and humidity for the different types of nylon monofilaments. Tensile strength and elongation are also shown. These properties are typical of some commercial products, but they are dependent on the degree of orientation. Tensile strengths in excess of 100,000 lb (wt)in.$^{-2}$ [7000 kg(wt)cm^{-2}] have been developed by orientation of the larger nylon monofilaments.

Several other properties of monofilaments are important in brush applications. Bend recovery of the monofilament, which is related to the resistance to matting of the brush bristles, is measured in a mandrel wrap test (9). Ten or more loops of monofilament conditioned to 73°F (23°C) and 50% R.H. are wrapped around a 3/32-in. (0.24-cm) mandrel under the tension of a weight in grams equal to one-half the square of the filament diameter in mils (1 mil = 0.025 mm). After the filament has been on the mandrel for 4 min, it is cut off and placed in water at 73°F to relax for 1 hr. Bend recovery is defined as the difference between the original number of loops and the number remaining expressed as a percent of the original number. In a standard 0.012-in. (0.3mm) caliper, bend recoveries are 92% for nylons-610 and -612 and 90% for nylon-66. The bend recovery of nylons-610 or -612 is equivalent to that of hog bristle and superior to other common natural and synthetic brush filaments. Bend recovery varies with temperature and humidity as well as time under load, relaxation time, strain rate, and total strain.

Table 12-6. Properties of nylon monofilaments

Property	Conditions	Units[a]	Nylon-66	Nylon-610	Nylon-612	Nylon-6[b]
Tensile modulus	73°F (23°C) 50% R.H.	lb(wt)in.$^{-2}$	650,000	530,000	580,000	280,000
	73°F (23°C) 100% R.H.	lb(wt)in.$^{-2}$	230,000	333,000	420,000	140,000
	120°F (59°C) 50% R.H.	lb(wt)in.$^{-2}$	360,000	310,000	350,000	- -
Tensile strength	73°F (23°C) 50% R.H.	lb(wt)in.$^{-2}$	62,000	56,000	58,000	- -
Elongation	73°F (23°C) 50% R.H.	%	35	35	26	- -

[a] lb(wt)in.$^{-2}$ × 0.0703 = kg(wt)cm^{-2}.

[b] Unextracted.

Sources: Trade literature.

In industrial brushes that are often subjected to severe service, fatigue resistance rather than bend recovery usually determines life. Fatigue leads to complete failure after many cycles of flexing due to breaking or splitting of the monofilaments. This property is measured using tufts of monofilament that are rotated against an impact bar (10). The tufts are held in four chucks on a motor-driven rotating head, and the chucks are individually gear-driven so that successive impacts occur at eight points around the periphery of each tuft. The standard test is conducted at 73°F (23°C), 50% R.H. , and 1000 impacts per minute. The maximum strain is 4%, and testing time is kept proportional to the reciprocal of monofilament diameter to attain equivalent fatigue values for various sizes. A 0.012-in. (0.3-mm) monofilament is tested for 107 min with 0.156-in. (3.9-mm) distance between the chuck face and the impact bar. The fatigue resistance or flex life is the percentage of the original number of monofilaments remaining unharmed after the standard test. Monofilament of nylons-610 and -612 exhibit fatigue resistance values of 98%; nylon-66, 90%. The fatigue resistance of nylons -610 or -612 is equivalent or superior to other natural and synthetic bristles.

Monofilaments of nylon are also superior to both natural and other synthetic bristles in abrasion resistance, as might be expected from the performance of nylons in other fabricated forms. Abrasion resistance of 0.012-in. (0.3-mm) monofilaments is measured by rotating a twisted-in-wire brush against sandpaper and measuring monofilament weight loss after a standard test period (11). The abrasion rates of nylons-610 and -612 are 40 mg/hr and of nylon-66, 46 mg/hr. The comparable value for hog bristle is 160 mg/hr.

Strapping

Nylon strapping is used in competition with steel in commercial packaging and shipping. This application requires that the strapping be oriented to maximize tensile strength. The tensile properties of commercial nylon-66 strapping are shown in Table 12-7. Tensile strength depends on the level of orientation and is many times higher than the tensile strength of test bars or other relatively unoriented nylon-66 samples. The data in Table 12-7 also show the excellent retention of tensile properties at both the high and low temperatures that may be encountered in shipping and storage.

In addition to tensile strength the time dependent properties related to creep are important in strapping applications. The ability of nylon to recover from initial elongation allows it to remain tight on a package that has compacted and is an advantage over both steel and other plastics with less recovery. This property is illustrated in Figure 12-1. Nylon-66 recovers to less than 1% elongation when the load is removed. This difference between the original length and the recovered length is the permanent set or creep. The time-dependent behavior of nylon-66 under stress is shown in a different way in Figure 12-2. Retention of tension under constant strain or elongation is plotted versus time.

Table 12-7 (12). Properties of nylon-66 strapping

Property	Unit[a]	Test Method	0°F (−18°C)	73°F (23°C)	120°F (49°C)
Tensile strength	lb(wt)in.^{2}	ASTM D638	70,000	65,000	55,000
Elongation to break	%	ASTM D638	14	18	20
Energy to break	in.-lb(wt)in.$^{-2}$	ASTM D638	5,000	6,000	6,000

[a] lb(wt)in.$^{-2}$ × 0.0703 = kg(wt)cm^{-2} ; in.-lb(wt)in.$^{-2}$ × 0.179 = cm-kg(wt)cm^{-2}.

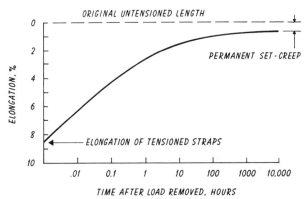

Fig. 12-1. Recovery of nylon-66 strapping from a tension of 35,000 lb (wt) in.$^{-2}$ (2500 kg (wt) cm^{-2}) applied for one week.

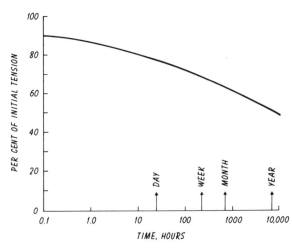

Fig. 12-2. Relaxation of nylon-66 strapping from an initial tension of 25,000 lb (wt) in.$^{-2}$ (1800 kg (wt) cm^{-2}) (6.7% strain).

454

Approximately one-half of the original tension is retained in one year. The data in both figures were obtained using 0.02 × 0.5 in. (0.5 × 12.7 mm) strapping.

REFERENCES

1. Code of Federal Regulations, 21 CFR 121.2502.
2. Hughes, R.L., and D.G. Simpson, *Plast. Tech.* 5 (11), 41 (Nov., 1959).
3. Anon., "Wire and Cable Coaters' Handbook," E.I. du Pont de Nemours & Co., 1968, p. 108.
4. Bonner, R.M., E.I. du Pont de Nemours & Co., Plastics Department, Technical Bulletin, TR-110, April 2, 1963.
5. Bonner, R.M., E.W. Kjellmark, Jr., R.E. Shaw, "Tougher Wire Jackets of Nylon", presented at the Twelfth Annual Symposium of the U.S. Army Signal Corps Research and Development Laboratories, Asbury Park, N.J., Dec. 4-6, 1963.
6. Richardson, P.N., E.I. du Pont de Nemours & Co., Plastics Department, Technical Bulletin TR-93, 1962.
7. Anon., Polymer Corp., Technical Bulletin NyT-4 (Revised), June, 1968.
8. Goldman, R. F., and D.D. Rudy, E.I. du Pont de Nemours & Co., Organic Chemicals Department, "Freon" Products Division, Technical Bulletin RT-51, Feb., 1971.
9. Anon., E.I. du Pont de Nemours & Co., Plastics Department, Technical Bulletin No. 5.
10. Anon., E.I. du Pont de Nemours & Co., Plastics Department, Technical Bulletin No. 6.
11. Anon., E.I. du Pont de Nemours & Co., Plastics Department, Technical Bulletin No. 3.
12. Guthrie, B.R., E.I. du Pont de Nemours & Co., Plastics Department, Technical Bulletin, "Comparative Performance of Dymetrol® Nylon Strapping."

Monomer Casting

GEORGE CARLYON

INTRODUCTION

The nylon monomer casting process provides the designer great latitude in areas of application that may be either too difficult or too expensive by extrusion or molding methods. This process, although not generally economical for small thin parts weighing less than one pound, is very practical in a range up to several hundred pounds. Low volume production may be justified because very simple, inexpensive molds can be utilized to achieve an almost unlimited range of shapes. Undercuts and quite complex parts can be made by use of molds that can be disassembled to allow removal of the casting.

Cast nylon-6 combines the machinability, structural rigidity, and toughness of metal with the lightness, resiliency, lubricity, and chemical and abrasion resistance of plastics. Advantages over extruded or molded nylon-6 result from its higher molecular weight and crystallinity. These include higher modulus, higher deflection temperature, improved solvent resistance, lower moisture absorption, and better dimensional stability.

CHEMISTRY OF MONOMER CASTINGS

Pure, dry caprolactam does not polymerize (1). Nylon-6 polymer and fiber are made in the presence of water (Chapter 2, p. 18). Anhydrous, acid catalyzed polymerization takes place, but it is a poor reaction with low conversion (2, 9, 12). Anhydrous, base catalyzed polymerization at a high temperature (250°C, 482°F) can yield a highly viscous product in 10 min (3). Addition of an imide,

$$\overset{O}{\underset{\|}{C}}-N-\overset{O}{\underset{\|}{C}}-$$

$-C-N-C-$, was first shown to greatly accelerate the base catalyzed polymerization of a lactam in the case of nylon-4 (4). A similar result was subsequently demonstrated with caprolactam (5, 6). It is this effort on cocatalyzed, anionic polymerization of lactams that has fathered the current technique for monomer casting of nylons. Caprolactam is the predominant commercial monomer because of the combination of cost and polymer properties it affords.

Mechanism

The steps involved in the generally accepted mechanism (5, 6) for cocatalyzed, anionic polymerization of caprolactam are:

Lactam anion formation by addition of strong base

$$\left(\begin{array}{c} \overset{O}{\underset{\|}{C}}-NH \\ (CH_2)_5 \end{array} \right) + MB \rightleftharpoons \left(\begin{array}{c} \overset{O}{\underset{\|}{C}}-N^{\ominus} \\ (CH_2)_5 \end{array} \right) M^+ + HB$$

"catalyst"

Propagation by addition of lactam anion to imide

$$
\left(\begin{array}{c} O \\ \| \\ C-N \\ (CH_2)_5 \end{array}\right)^{-} + \left(\begin{array}{c} O \\ \| \\ C-N \\ (CH_2)_5 \end{array}\right)\!\!-\!\begin{array}{c} O \\ \| \\ CR \end{array} \rightleftharpoons \left[\left(\begin{array}{c} O \\ \| \\ C-N \\ (CH_2)_5 \end{array}\right)\!\!-\!\!\left(\begin{array}{c} O^{\ominus} \\ \| \\ C-N \\ (CH_2)_5 \end{array}\right)\!\!-\!\begin{array}{c} O \\ \| \\ CR \end{array}\right]
$$

"cocatalyst"

$$
\Updownarrow
$$

$$
\left(\begin{array}{c} O \\ \| \\ C-N \\ (CH_2)_5 \end{array}\right)\!\!-\!\!\begin{array}{c} O \\ \| \\ C(CH_2)_5 \end{array}\!\!-\!\!\begin{array}{c} O \\ \ominus \| \\ N-CR \end{array}
$$

(I)

Regeneration of lactam anion

$$
(I) + \left(\begin{array}{c} O \\ \| \\ C-NH \\ (CH_2)_5 \end{array}\right) \rightleftharpoons \left(\begin{array}{c} O \\ \| \\ C-N \\ (CH_2)_5 \end{array}\right)\!\!-\!\begin{array}{c} O \\ \| \\ C(CH_2)_5 \end{array}\!\!-\!NHCR + \left(\begin{array}{c} O \\ \| \\ C-N \\ (CH_2)_5 \end{array}\right)^{\ominus}
$$

In the absence of added imide, a relatively slow and rate-controlling initiation step is necessary in which imide is generated by addition of lactam anion to a lactam molecule. It is elimination of the initiation step which allows the cocatalyzed, anionic polymerization of caprolactam to occur at temperatures 100°C and more below those required for simple anionic polymerization and to occur without an induction period (5). On the other hand, there is no polymerization at all if the cocatalyst, but not the anion, is present.

A variety of strong bases and their salts can be used as catalysts and include alkali metals and their hydrides, hydroxides, alkoxides, carbonates. and amides (7,9). Cocatalysts include not only imides and lactams with a variety of electron attracting groups attached to the nitrogen atom but also reagents capable of rapid combination with the lactam to form the required cocatalyst in place (8, 9). Examples are N-acetylcaprolactam, acetic anhydride, carbon dioxide, phenyl isocyanate, cyanuric chloride, phenyl N-phenylbenzimidoether, and many, many others.

Rate, Monomer Content, and Thermal Effects

The rate constant at $220°C$ for the cocatalyzed, anionic polymerization of caprolactam has been estimated to be 10^7 times greater than the rate constant for polymerization in the presence of water (9). There is very little accurate rate information available although many catalyst-cocatalyst systems have been compared on the basis of the time required to achieve reasonable conversion to polymer (8). Several systems can achieve 95% conversion in minutes at an initial temperature of $150°C$.

The low-molecular-weight, extractable fraction is usually lower in cast nylon-6 than in hydrolytic polymer although the same ratio of polymer to monomer at equilibrium applies (7) because the maximum temperature in the cast process is normally much lower and can even be below the melting point of the polymer (see Figure 2-5).

The kinetics and equilibria for a variety of lactam polymerizations have been reviewed (9, 10, 11).

Lactam polymerization can be used to make thick objects because the maximum temperature rise due to polymerization is only about $50°C$ $(90°F)$ (9) compared to the $186°C$ $(335°F)$ rise for polymerization of methyl methacrylate. The maximum temperature rise for crystallization of nylon-6 is about $56°C$ $(101°F)$ and is, therefore, greater than that for polymerization.

The relationship between temperature, time, conversion, and crystallization in monomer casting of caprolactam has been discussed at some length (12), but uncertainties in heat capacities, heat transfer, and heats and rates of reaction make the relationships highly empirical. Reported increases in temperature of about $80°C$ indicate contribution from crystallization as well as polymerization.

Side Reactions

Any compound capable of interaction with the lactam anion or the imide affects the rate of polymerization. Water, alcohols, and other materials containing active hydrogen atoms will diminish the content of lactam anion more or less depending on the position of the following equilibrium:

$$\begin{pmatrix} \overset{\displaystyle O}{\overset{\|}{C}} - N \\ (CH_2)_5 \end{pmatrix}^{\ominus} + ROH \rightleftharpoons \begin{pmatrix} \overset{\displaystyle O}{\overset{\|}{C}} - NH \\ (CH_2)_5 \end{pmatrix} + RO^{\ominus}$$

The RO $^{\ominus}$ may result in loss of imide but with regeneration of lactam anion:

$$
\begin{matrix} O & & O \\ \| & | & \| \\ -C-N-C- \end{matrix} \quad + \quad RO^{\ominus} \quad \longrightarrow \quad \begin{matrix} O \\ \| \\ -C-OR \end{matrix} \quad + \quad \begin{matrix} & O \\ \ominus & \| \\ -N-C- \end{matrix}
$$

Compounds such as amines may have hydrogen atoms insufficiently reactive to neutralize the anion but capable of rapid reaction with the imide:

$$
\begin{matrix} O & O \\ \| & \| \\ -C-N-C- \end{matrix} \quad + \quad RNH_2 \quad \longrightarrow \quad \begin{matrix} O \\ \| \\ -C-NHR \end{matrix} \quad + \quad \begin{matrix} O \\ \| \\ -NH-C- \end{matrix}
$$

Catalyst and cocatalyst concentrations are typically of the order of one mole percent of the lactam or less; clearly, even small amounts of active hydrogen impurities can have profound effect on polymerization.

The activity of the anion-imide system for polymerization of caprolactam decreases with time. Wichterle suggested that imide could disappear via a self-condensation reaction, and the condensation product could cyclize to yield a compound with an active hydrogen atom that would cause loss of anion (13).

Reduction products such as hexamethyleneimine and 6-aminohexanol may appear if the anion is prepared above 100°C with sodium metal (7).

Molecular Weight

Highly viscous products, corresponding to higher molecular weights and broader distributions than result from hydrolytic polymerization, are typically obtained in anionic polymerization unless suitable measures are taken. If the concentration of imide exceeds that of the anion, the number average molecular weight depends on the reciprocal concentration of imide in accord with the mechanism which requires one imide end per chain. Molecular weight can therefore be limited by increasing the amount of imide. Self-condensation of imide destroys this relationship if the anion concentration is too high. Simple amides can be added to provide additional end groups via transamidation involving the amide anion:

$$
\begin{pmatrix} O & & O \\ \| & & \| \\ C - N & - & C-polymer \\ & | & \\ & (CH_2)_5 & \end{pmatrix} + \begin{matrix} O \\ \| \ominus \\ RC-NR' \end{matrix} \;\rightleftarrows\; \begin{matrix} O & R' & O \\ \| & | & \| \\ RC-N & - & C-(CH_2)_5 \end{matrix} \begin{matrix} & O \\ \ominus & \| \\ -N-C-polymer \end{matrix}
$$

Similar reactions involving chain amide ions cause branching.

Continued heating of the initial product leads to a redistribution which approaches the most probable. The rate of redistribution can be retarded by treatment of the initial product with water or acid to remove the anion. The literature on molecular weight effects in anionic polymerization of lactams is meager and has been best summarized by Wichterle (7).

MANUFACTURING PROCESS

The nylon casting process involves four steps — melting the monomer, adding the catalyst and activator, mixing the melts, and casting.

Caprolactam must be converted from flake to liquid form by melting under very closely monitored conditions of temperature and atmospheric humidity. The average normal moisture content of flake caprolactam is approximately 0.015%. Because it is very hygroscopic, exposure to high humidity will result in rapid moisture absorption. As explained above, excess moisture will cause decomposition of the catalyst and incomplete polymerization. When the optimum melt temperature has been established constant control must be maintained for satisfactory polymerization.

Two equally divided solutions of lactam melt are prepared, one to include the proper type and amount of catalyst, and the other the activator or cocatalyst. The solutions are then thoroughly mixed before pouring into suitable molds. The temperature versus time relationship for a typical casting of nylon-6 is noted in Figure 13-1.

Equipment

For prototypes, samples, or limited production requirements, a very simple "bench top" operation is adequate. The equipment may consist of a scale, electric hot plates, two stainless-steel pails with covers, thermometers, and stirring rods. The pails should be capable of containing 2.5 times the size of the total melt required. Equal amounts of lactam are added to each container and heated to the proper temperature. The catalyst is added to one melt and the activator or cocatalyst to the other. Under ideal conditions this would be accomplished with constant stirring under an inert gas atmosphere. When this is not available the containers must be kept covered as much as possible to avoid moisture absorption. When the proper melt conditions have been established the solutions may be mixed by dumping from one container to the other several times and pouring into a properly prepared mold with minimum delay.

Where production requirements justify the investment, a suitably sized bulk handling system is preferable (Figure 13-2). The caprolactam is melted in the

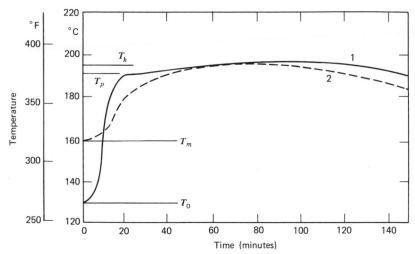

Fig. 13-1.   Temperature versus time in a typical cast polymerization of nylon-6
1 – Temperature at center of large casting; 2 – temperature of mold; T_o – initial
temperature of polymerization (melt); T_p – final temperature of polymerization;
T_k – maximum temperature; T_m – initial temperature of mold.

larger vessel, held at the proper temperature through thermostatic controls, and
transferred into the mixing vessels as required. At reduced temperatures the
melting vessel may also serve as a storage unit to hold production requirements
for several days. It is best, however, to use the complete contents of the mixing
vessels prior to shutdown of the system. All vessels are stainless steel with an
inert gas system and are jacketed to accommodate either steam or hot oil heating
units. The mixing vessels include means of constantly stirring the melt. All
piping, valves, pumps, controls, and any components coming into contact with
the melt are stainless steel. The system is designed for either gravity or
mechanical transfer of the melt from unit to unit and into the molds.

Molds

There are many mold materials that may be utilized for sample or low volume
parts. The material must be capable of containing a low-viscosity liquid at an
approximate temperature of 200°C and be designed to allow for a part
shrinkage of 13 to 15%. The dependence of specific gravity on temperature for
monomer and polymer is shown in Figure 13-3. The mold must be cleaned and
maintained at a temperature of 160 to 200°C prior to use. In general, plain hot
rolled 15 gauge sheet metal is the most practical because of its ease of
fabrication, heat transfer characteristics, durability, and low cost (Figure 13-4).
For medium-volume production (1000 and over) cast aluminum molds may be

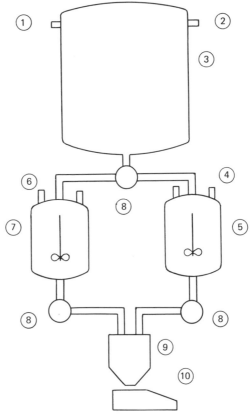

Fig. 13-2. Schematic of bulk handling system for monomer casting of nylon-6. 1 – Caprolactam inlet; 2 – N$_2$ inlet; 3 – heated storage vessel; 4 – activator inlet; 5 – mix vessel with stirrer; 6 – catalyst inlet; 7 – mix vessel with stirrer; 8 – feed pumps; 9 – mixing device; 10 – mold.

considered. Aluminum castings offer the advantage of relatively light weight, may be cast in more complicated contours, and may be cored to provide the constant, accurate heat control essential to this operation (Figures 13-5). Permanent tooling for cast nylon parts may be fabricated from metal weldments or machined from steel castings. They may be quite similar to molds used for injection molding but will not be as massive as those subject to high molding pressures (Figure 13-6).

Regardless of the materials used, proper temperature must be maintained during pouring to insure a sound casting. Because of variation in heat transfer and the mass of the pour, exact temperatures are unpredictable but should fall in the range of 150 to 180°C.

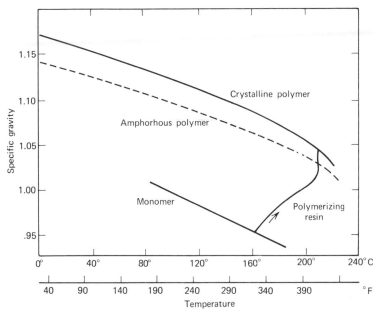

Fig. 13-3.   Specific gravity of caprolactam monomer and polymer versus temperature.

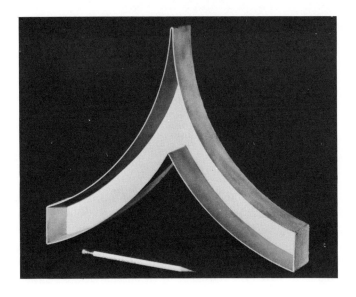

Fig. 13-4.   Example of sheet metal mold.

Fig. 13-5.   Example of cast aluminum mold.

Fig. 13-6.   Example of mold machined from steel castings.

Fillers and Colors

Additives such as pigments and various insoluble fillers present a processing problem because of the difficulty in achieving a homogeneous dispersion. If the density of the additive is more or less than the melt, pigment may be concentrated in the lower or upper segment of the casting with undesirable results. Many commercial dyes or colorants cannot be used because they are soluble in the melt, have little resistance to chemical attack, are unstable at the operating temperatures involved, or have adverse effects on the polymerization cycle. All additives must first be very thoroughly dried.

Annealing

The casting process may introduce stresses that must be relieved before machining or placing the part in use. In some instances this problem may be avoided by removing the heat of polymerization very slowly by cooling in an oven or packing the part in an insulating material for up to 24 hr. Where this approach is impractical other methods may be considered. Air annealing is performed in circulating ovens operated at 140 to 180°C. The best results, however, for parts with heavy cross sections, are accomplished with an oil bath utilizing highly refined mineral oil. The submersion time ranges from 0.5 to 4 hr depending upon configuration and part thickness.

Safety

Any hot melt, of course, can cause severe burns. Safety glasses and protective clothing should always be used when preparing or pouring the melt. When handling the catalyst, gloves should be worn to avoid contact with the skin; catalyst exposed to water yields a strongly basic solution. If exposure does occur, the affected area should be immediately and thoroughly washed with water. Adequate ventilation must be provided to avoid inhalation of any fumes.

PROPERTIES OF CAST NYLON-6

Several of the properties of cast and molded nylon-6 are compared in Table 13-1. The higher molecular weight and crystallinity of the cast polymer results in higher tensile strength, modulus, and deflection temperature and better retention of properties with changing humidity because of reduced water absorption (Figure 13-7). The chemical resistance of nylons has been discussed in Chapters 2 and 10. The increased crystallinity of cast nylon-6 serves to improve the already excellent performance of nylons in resisting a wide variety of chemicals. Among other advantages claimed for cast nylon-6 are better wear and fatigue resistance and lower creep (14).

EXAMPLES OF APPLICATIONS

Examples of commercial application of monomer cast nylon-6 are given in the illustrations that follow.

Table 13-1. Properties of cast and molded nylon-6

	ASTM method	Units	Cast nylon-6	Molded nylon-6
Specific gravity	D792	--	1.15	1.13
Specific volume	D792	in.^{3} lb^{-1} (cm^3 g^{-1})	24.0 (0.87)	24.5 (0.88)
Mechanical				
Tensile strength	D638	lb(wt)in.$^{-2}$	11,000-14,000[a]	7,000-12,000[a]
Elongation	D638	%	10-60	20-300
Tensile modulus	D638	lb(wt)in.$^{-2}$	350,000-470,000	100,000-470,000
Compressive strength	D695	lb(wt)in.$^{-2}$	12,000	6,700-13,000
Flexural yield strength	D790	lb(wt)in.$^{-2}$	7,000-17,500	7,000-16,000
Izod impact strength	D256	ft-lb(wt)in.$^{-1}$	0.8-3.0[b]	0.8-No break[c]
Rockwell hardness	D785	--	R110-120	R100-110
Flexural modulus	D790	lb(wt)in.$^{-2}$	80,000-450,000	80,000-400,000
Thermal				
Thermal expansion	D696	($\Delta L/L$) × 10^5 °F^{-1} (°C^{-1})	5.0 (9.0)	4.6 (8.3)
Deflection temperature	D648			
66 lb(wt)in.$^{-2}$		°F (°C)	400-425 (204-218)	300-370 (149-188)
264 lb(wt)in.$^{-2}$		°F (°C)	200-425 (93-218)	140-180 (60-82)
Melting point	D789	°F (°C)	435 (224)	425 (218)
Electrical				
Dielectric strength, short time, 0.125 in. (3.2 mm) thick	D149	V mil^{-1} (kV mm^{-1})	300-400 (11.8-15.8)	440-510 (17.3-20.1)
Dielectric constant, 60-10^6 cycles	D150	--	3.7[d]	3.5-5.5[d]
Dissipation factor, 60-10^6 cycles	D150	--	0.02[d]	0.01-0.06[d]
Volume resistivity	D257	ohm-cm	10^{12}-10^{15}	10^{12}-10^{15}

[a] Ranges given in the table are due mostly to changes in moisture content.　[c] Bar cross section = 0.5 × 0.5 in. (12.7 × 12.7 mm).
[b] Bar cross section = 0.125 × 0.5 in. (3.2 × 12.7 mm).　[d] Dry only.

Note: lb(wt)in.$^{-2}$ × 0.0703 = kg(wt)cm^{-2}; ft-lb(wt)in.$^{-1}$ × 5.45 = cm-g(wt)cm^{-1}.

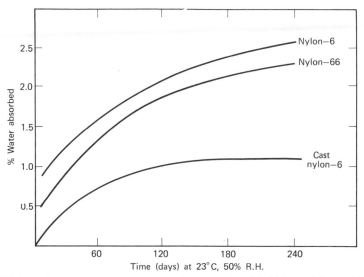

Fig. 13-7. Water absorption versus time for cast nylon-6 (0.12 in. or 3.2 mm thick).

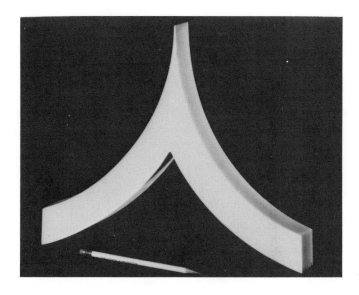

Fig. 13-8. Feed stock control wedge for 3-roll mill (made from tooling illustrated in Figure 13-4).

Fig. 13-9. Conveyor bucket (made from tooling illustrated in Figure 13-5).

Fig. 13-10. Cyclone separator head (made from tooling illustrated in Figure 13-6).

470

Fig. 13-11. Blank for machining large-diameter gears used on dryers in paper industry (cast in a sheet metal mold).

REFERENCES

1. Hermans, P.H., D. Heikens, and P. F. van Velden, *J. Polym. Sci.* **30**, 81 (1958).

2. Van der Want, G.M., and Ch. A. Kruissink, *J. Polym. Sci.* **35**, 119 (1959).

3. Joyce, R.M., and D.M. Ritter (to E.I. du Pont de Nemours and Co.), U.S. Patent 2,251,519 (Aug. 5, 1941); Hanford, W.E., and R.M. Joyce, *J. Polym. Sci.* **3**, 167 (1948).

4. Ney, W.O., Jr., and M. Crowther (to Arnold, Hoffman and Co.), U.S. Patent 2,739,959 (Mar. 27, 1956).

5. Sebenda, J., and J. Kralicek, *Coll. Czech. Chem. Comm.* **23**, 766 (1958).

6. Hall, H.K., Jr., *J.Am. Chem. Soc.* **80**, 6404 (1958).

Fig. 13-12.   Drive screw (22 in. or 0.56 m long) for moving containers in canning machine (machined from cast rod stock).

7. Wichterle,O., J.Sebenda, and J. Kralicek, *Fortschr. Hochpolym. Forsch* **2**, 578 (1961).
8. Stehlicek, J., J. Sebenda, and O. Wichterle, *Coll. Czech. Chem. Comm.* **29**, 1236 (1964).
9. Hall, H.K. Jr., publication in press.
10. Dainton, F.S., and K.J. Ivin., *Quart. Rev.* **12**, 61 (1958).
11. Frisch, K.C., and S.L. Reegen, Eds., *Ring-Opening Polymerization*, Marcel Dekker, New York, 1969.
12. Reimschuessel, H.K., in *Ring-Opening Polymerization,* K.C. Frisch and S.L. Reegen, Eds., Marcel Dekker, New York, 1969, Chap. 7.
13. Wichterle, O., *Makromol. Chem.* **35**, 174 (1960).
14. Baum, B., *SPE Tech. Pap.* **13**, 628 (1967).

CHAPTER 14 _____

Powdered Nylons

D. S. RICHART

INTRODUCTION

While the bulk of nylon plastics are processed by well-known methods such as extrusion and injection molding, a number of processes are in current use based on powdered nylons. The most important of these are fluidized bed coating, electrostatic coating, pressing and sintering, rotational molding, and fabric adhesive application.

The fluidized bed coating process has the most important commercial significance at the present time. However, electrostatic coating looms as a far greater potential outlet for powdered resins. Pressed and sintered parts based on powdered nylon will probably always be a low-volume specialty market. The use of powdered nylons in fabric adhesives is in a very early state of development and could reach a position of commercial consequence in years to come. Rotational molding of powdered plastics, especially polyolefins, is gaining increased acceptance; however, the future of nylon in this process is difficult to predict.

FLUIDIZED BED COATING

The use of powdered plastics in coating processes is of relatively recent origin. Reports of flame spraying or plasma spraying of polymeric materials first started to appear in the literature in 1949-1950 (1). In 1949, a British patent was issued to Schori Metallising Ltd. for a coating method based on rolling or covering heated metal parts in thermoplastic powders followed by postheating to fuse into a film (2).

Early work on the fluidized bed process was carried out by Erwin Gemmer in the laboratories of Knapsack-Griesheim in 1952. Gemmer's original work was aimed at developing uses for the industrial gases which KG produced (3). In 1955, the first basic patent on fluidized bed coating was issued to Gemmer (4) in Germany. This patent covered thermoplastic materials and gave examples based on polyethylene, wax, and bitumens in combination with fillers such as slate dust. The first British patent was issued in 1956 to Knapsack-Griesheim A.G. covering essentially the same subject matter as the German patent (5). A second patent was issued in Great Britain in 1961 covering thermosetting resins (6). In a paper presented at the 7th German Plastics Convention in 1957, Gemmer (7) described the new "Whirlsinter" process in more detail and discussed the application of coatings using both high and low pressure polyethylenes, acrylics, and nylons-6 and -11.

The first United States patent describing the basic process issued in the United States in 1958 (8) followed by many others disclosing process improvements. Rights to the fluidized bed coating process were acquired by The

Polymer Corporation, Reading, Pennsylvania, in a series of agreements starting in 1955. Up to the present, some forty United States patents have been issued to Knapsack-Griesheim A.G. (now combined in Hoechst) and The Polymer Corporation. The first commercial plant designed exclusively for fluidized bed coating was erected in California in 1959 by The Polymer Corporation. In the decade that followed, the process was adopted in many industries, and at present there are over 300 commercial installations in the United States and Canada utilizing fluidized bed coating.

The basic principles of fluidized bed coating are deceptively simple. A powdered plastic material is placed in a fluidizing container and maintained in a fluidized state by the passage of gas through a porous plate. In this condition the powder exhibits many characteristics of a fluid such as mobility, hydrostatic pressure and observable upper boundary layer. Individual particles are in random motion throughout the fluidized chamber. The object to be coated is heated above the melting point of the plastic material and immersed in the fluidized bed for several seconds during which time the plastic particles melt on the surface and form a coating layer. Depending on the heat capacity of the substrate a postheating operation may or may not be required in order to fuse the coating to a continuous layer. Figure 14-1 shows a schematic diagram of a fluidized bed indicating the basic elements and the behavior of plastic particles in contact with the heated metal surface. Following is a more detailed discussion of the various elements of fluidized bed coating.

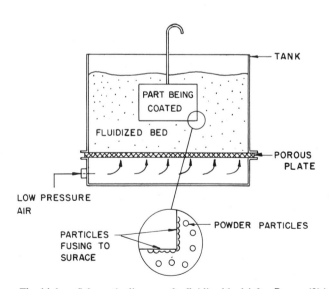

Fig. 14-1.   Schematic diagram of a fluidized bed (after Pascoe (9)).

The Fluidized Bed

The use of both fixed and fluidized beds in the chemical process industries has been practiced for many years. In a fixed bed, the particles are motionless and supported by each other and the walls of the container. In a fluidized bed, particle density is lower, the particles have motion relative to each other and the sides of the container, and the bed exhibits typical characteristics of mobility and hydrostatic pressure. In a smoothly fluidizing bed such as is desirable for coating, the upper surface is relatively smooth resembling a lightly boiling liquid. Some powders have a tendency to "channel," a condition where flow paths of the fluidizing gas establish themselves through the powder. In this case, the fluidized bed gives the appearance of "geysers" with the powder only partially fluidized and having very little relative motion. This is an undesirable state for coating for obvious reasons. Another defect occurring in fluidized beds is called "slugging," Here, a discreet bubble of the fluidizing gas rises through the powder, breaks the surface, and blows powder out of the fluidizing container. This defect is primarily a result of too high a length-to-diameter ratio of the fluidized bed and can be eliminated by increasing the diameter of the fluidized bed. The problem of channeling can be reduced by vibration of the fluidizing container or by mechanical agitation and circulation of the powder in the fluidized bed (10).

An excellent discussion of the terms relating to the fluidized state is given in an editorial article in the 1949 issue of *Industrial and Engineering Chemistry* (11). More detailed information is also given by Frantz (12) and Parent et al. (13).

The powdered coating material is placed in the fluidizing chamber, and the fluidizing gas, usually air, is introduced into the plenum chamber (Figure 14-1). Either compressed air or blowers can be used; in any case, however, the air must be dry and free from oil. To this end, filters and dryers are included in the fluidizing air system. When the air is introduced, pressure builds up in the plenum chamber, and the air is forced through the pores in the porous plate. During this period the pressure drop across the bed increases; there is no relative motion of the particles and the density remains constant. These conditions describe a fixed bed. As the flow rate of the air increases, a point is reached where the pressure drop peaks, drops slightly, and then increases to a steady level and remains stable. This high point in the pressure drop/gas flow plot (Figure 14-2) is the onset of fluidization; the bed starts to expand and particle mobility begins. The bed is unstable and cannot be maintained in this condition. A slight increase in flow rate beyond this point gives a smoothly fluidized bed which is suitable for coating. The pressure required for fluidization is very low: 0.33 lb(wt)in.$^{-2}$ ft^{-1} (0.077 kg(wt)cm^{-2} m^{-1}) of powder height; a gas flow rate of 260 to 460 ft^3 hr^{-1} ft^{-2} (80-140 m^3 hr^{-1} m^{-2}) of porous plate area is typical for most beds.

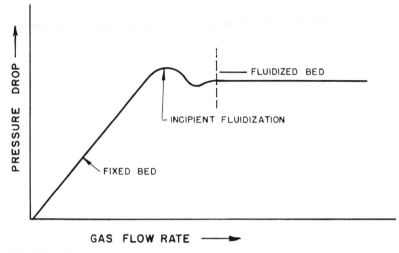

Fig. 14-2.   Pressure drop versus gas flow rate.

As would be expected, the fluidizing characteristics of a powder are related to particle size. For coating purposes, a fairly wide distribution in particle size can be tolerated. Table 14-1 shows the typical particle size distribution for a compounded nylon-11 either ground to a powder by cryogenic grinding techniques or precipitated from solution.

Table 14-1. Typical particle size range for fluidized bed coating powder (nylon)

Mesh	Microns	% Retained on indicated screen
70	210	0-5
100	149	15-30
140	105	45-60
200	74	65-80
325	44	80-95
Thru 325 mesh	<44	5-15

Nylon Coating Materials

In the years since fluidized bed coating originated, only a limited number of materials have been used commercially. These were described in 1962 by Richart (14) in some detail and have not changed materially since that time. Table 14-2 is a relative rating in various performance factors of the materials in current use.

Table 14-2. Comparison of fluidized bed coating materials

	Nylon-11 or nylon-12	Epoxy	Vinyl	Cellulosic	Polyethylene (low-density)	Chlorinated polyether
Chemical resistance						
Exterior durability	F	F	F	G	F	F
Water (salt or fresh)	G	G	E	VG	VG	E
Solvents						
Alcohols	E	E	E	E	E	E
Gasoline	E	E	E	G	VG	VG
Hydrocarbons	E	E	G	G	VG	E
Esters, Ketones	G	F	P	P	G	VG
Chlorinated	E	E	P	P	F	P
Salts	VG	E	E	VG	E	E
Ammonia	G	P	E	P	E	E
Alkalies	G	VG	E	F	VG	E
Mineral acids						
Dilute (10%)	F	E	E	G	E	E
Concentrated (> 30%)	P	G	G	P	VG	E

Oxidizing acids						
Dilute (10%)	P	G	E	P	VG	E
Concentrated (> 30%)	P	P	G	P	P	G
Organic acids:						
Acetic, formic, etc.	P	F	F	P	VG	E
Oleic, stearic, etc.	VG	E	E	F	VG	E
Mechanical and physical properties						
Abrasion resistance	E	VG	G	VG	F	VG
Flexibility	VG	F	E	G	VG	F
Impact resistance	VG	G	E	E	F	G
Max. service temp., °F/°C	180/82[a]	350+/177+	200+/93+	180/82	160/71	250/121
Dielectric strength	G	E	VG	VG	E	VG
Decorative properties						
Color range	E	E	E	E	F	P
Color retention	VG	F	VG	E	VG	G
Initial gloss	G	G	VG	E	F	G
Gloss retention	E	P	G	E	F	P

[a]Up to 300°F (149°C) in nonoxidizing environment

Key. E = excellent, VG = very good, G = good, F = fair, P = poor.

While a variety of polyamide resins have been evaluated over the years, nylon-11 has achieved the highest degree of acceptance. Nylon-66 cannot be utilized in the fluidized bed process because the extremely high temperatures required to melt this polymer result in degradation of the material before it can fuse and flow-out to a continuous film. Nylon-6 has been used in several commercial applications, but at the present time there is practically none being sold for fluidized bed coating. Low-melting copolymers can also be applied by fluidized bed coating techniques. At the present time, however, these materials have not achieved commercial significance.

While the workhorse of nylon coatings has been nylon-11, the more recently available nylon-12 is being seriously evaluated by many in the industry, and the first commercially coated parts in the United States were delivered to customers in 1970. The physical and chemical properties of nylons-11 and -12 are quite similar, and the ultimate choice as to which one to use will probably be decided by economic considerations rather than performance (15). In addition, both exhibit similar characteristics in pigmentation, formulation, grinding, and other powder manufacturing variables. Powders based on nylons-11 and -12 can be mixed in all proportions with no significant or noticeable effect, if they are of approximately the same molecular weight.

The physical properties of nylons-11 and -12 used in fluidized bed coating are given in Table 14-3.

There are two types of nylon coating powders generally available: the precipitated grades and the melt mixed grades. The precipitated grades are made using a solution/precipitation technique and are normally only available in limited stock colors such as black, white, and gray. In a modification of the precipitated powder process, coloring is achieved by the dry blending of small amounts of colors with intensive mixing on a high-speed mixer such as the Henschel or Papenmeier. This technique can lead to nonhomogeneous color because the pigments may not always be dispersed equally over the surface of the powder. In addition, fluidization may sometimes be adversely affected, and the electrostatic charging characteristics may be altered significantly.

In the melt mix process, the nylon is extruded with colorants, antioxidants, plasticizers, and other additives as may be required. The extruded ribbon is cut and then ground to a powder using cryogenic grinding techniques. Powders prepared by this method are somewhat more angular in nature than those prepared by precipitation; however, particle size of both products is essentially the same. Nylon coating powders prepared by the melt mixing process can be readily modified during the compounding operation. Addition of plasticizers allows the use of lower temperature preheats by decreasing the film modulus and improving melt flow. Ultraviolet stabilizers and opacifying pigments can be added to improve exterior durability. Also, special additives such as molybdenum disulfide can be added to alter frictional characteristics.

Table 14-3. Typical properties of nylon-11 and -12

	Nylon-11	Nylon-12
Tensile strength at yield	5700-6400 lb(wt)in.$^{-2}$	5700-6400 lb(wt)in.$^{-2}$
Elongation at yield	20%	20%
Ultimate tensile strength	6900-9200 lb(wt)in.$^{-2}$	7000-9250 lb(wt)in.$^{-2}$
Ultimate elongation	200-380%	200-380%
Flexural modulus	155,000-165,000 lb(wt)in.$^{-2}$	170,000-180,000 lb(wt)in.$^{-2}$
Impact strength – notched (Charpy)	1.4-4.4 ft-lb(wt)in.$^{-2}$ (3-9.5cm-kg(wt)cm^{-2})	2.3-6.1 ft-lb(wt)in.$^{-2}$ (5-13cm-kg(wt)cm^{-2})
Specific gravity	1.03-1.04	1.01-1.02
Melting point	377°F 186°C	349°F 176°C
Dielectric strength	508-635 V mil^{-1} (20-25k V mm^{-1})	635-760 V mil^{-1} (25-30k V mm^{-1})
Volume resistivity	10^{13} ohm-cm	10^{14} ohm-cm
Dielectric constant at		
60 Hz	3.7	3.4
10^6 Hz	3.5	3.0
Loss factor at		
60 Hz	0.10	0.08
10^6 Hz	0.07	0.06
Water absorption		
20°C (68°F) 65% R.H.	1.05%	0.95%
20°C (68°F) immersion	1.85	1.50
Coefficient of linear thermal expansion	1.1-1.5 × 10^{-4}/°C	1.04-1.20 × 10^{-4}/°C

Note. lb(wt)in.$^{-2}$ × 0.0703 = kg(wt)cm^{-2}.
Properties measured after conditioning at 20°C and 65% R.H. 14 days.

In fluidized bed coating, the part to be coated is normally heated 50 to 150°C (ca. 100 to 300°F) higher than the melting point of the coating material. For nylons-11 and -12 [melting point 186°C (367°F) and 176°C (349°F), respectively] preheat temperatures are generally in the range of ca. 240 to 360°C (464 to 680°F). Depending on the mass (heat capacity) of the article to be coated, a postheat may be required to completely fuse the coating. Typical coating conditions for several representative part configurations are given in Table 14-4.

Substrate Preparation, Priming, and Masking

While a variety of substrate materials including metals, glass, and even other plastics can be coated using the fluidized bed process, the vast majority of

Table 14-4. Coating conditions for nylons -11 and -12.

Part description	Preheat		Dip time (sec.)	Postheat		Coating thickness (mils/microns)
	Oven temp.	Time (min.)		Oven temp.	Time (sec.)	
Steel panel	650°F	5	10	500°F	30	16/412
3 × 4 × 1/16 in.	(343°C)			(260°C)		
(7.6 × 10.1 × 0.16 cm)						
Steel bar	500°F	20	10	500°F	0-15	13/336
3 × 4 × 1/2 in.	(260°C)			(260°C)		
(7.6 × 10.1 × 1.3 cm)						
Steel panel	500°F	20	10	500°F	15	14/360
3 × 4 × 1/4 in.	(260°C)			(260°C)		
(7.6 × 10.1 × 0.64 cm)						
Steel wire	575°F	5	10	500°F	1	7/180
0.12 in. (3 mm) diameter	(300°C)			(260°C)		
Welded steel wire	600°F	10	10	500°F	45	
fabrication	(316°C)			(260°C)		
Heavy wire 0.188 in.						11.5/296
(4.8 mm)						
Thin wire 0.092 in.						6/155
(2.4 mm)						

articles coated are metals. In general, any substrate which can withstand the preheat temperatures required — typically 120 to 340°C (248 to 644°F) — can be coated. For example, even wood has been coated; however, the coatings are unsatisfactory due to volatiles given off during the fusion of the coating.

In the case of metallic substrates, the article must be free from grease, dirt, corrosion, and mill scale. Ferrous metals must frequently be sandblasted to remove the mill scale and corrosion. If the surface is relatively free from corrosion, a simple alkaline wash followed by water rinse and drying or vapor degreasing is adequate surface preparation. If superior corrosion resistance is required, a zinc phospate treatment is recommended. While iron phosphate treatment gives an excellent surface for coating, little or no improvement in corrosion resistance is obtained over a well-prepared plain steel surface. Galvanized surfaces present special problems in fluidized bed coatings. The quality of galvanizing appears to vary quite substantially from one applicator to another and even by the same applicator. Various degrees of adhesion ranging from practically none to excellent can be obtained over galvanized surfaces. This is probably due to the fact that for fluidized bed coating, primers are applied at extremely low film thicknesses and do not completely cover the galvanized surface which tends to be porous.

Most of the thermoplastic coating materials such as nylons, butyrates, vinyls, and so on used in fluidized bed coating require the use of primers. These are usually applied as solvent solutions. If the application is primarily decorative and a high degree of adhesion is not required, adequate adhesion can be obtained with nylons and polyethylenes by the use of very high temperature (310 to 370°C, 590 to 698°F) preheats. At these high metal temperatures, the material probably is oxidized at the surface yielding some carbonyl groups and modest adhesion. However, adhesion is usually lost after mild exposure to a corrosive atmosphere; for example, 1 hr in boiling water. Some special problems occur in primers for fluidized bed coating that are not present in the case of typical paints. In the case of nylon, for example, preheat temperatures are frequently in the range of 260 to 340°C (500 to 644°F). Depending on the mass of the object, it may take from 3 to 10 min to get the article up to temperature. Exposure to the high temperatures for too long a period of time will cause the primer to cross-link excessively or degrade. This problem was investigated by Nagel (16), who found that "preheat resistance" of the primer could be greatly improved by using very thin films – in the neighborhood of 2 to 3 microns (ca. 0.1 mil).

In some cases it is necessary that certain parts of the article such as threads, bolt holes, flanges, and so on remain uncoated. To accomplish this these areas must be masked. The simplest method, especially where only a few parts are to be coated, is to use a pressure sensitive tape. The tape is applied to the areas which are to remain free from coating and removed as soon as the article comes out of the postheat oven and the coating is still molten. The pressure sensitive adhesive becomes oxidized during the preheat and coating operations and frequently leaves a residue which is difficult to clean from the metal surface.

Another method for preventing selected areas from being coated is to use cold contact masking. In this case, the preheated article is removed from the oven by holding the areas where no coating is desired by forceps or clamps coated with Telflon® or other heat resistant material. The article is coated in the fluidized bed and the clamp released. Because it is cold, the clamp is not coated. Initially, the masking obtained by this technique is quite good. However, as more objects are coated, the clamping devices become heated through contact with the metal and eventually become coated themselves. A variety of other methods of masking can also be used (17).

The Coating Process

Before immersing the article to be coated into the fluidized bed, it must first be preheated. In most commercial coating lines, this is accomplished using gas-fired air circulating ovens. Properly designed, this type of oven is the most economical and trouble free for general purpose use. In designing the preheat oven, thought must be given to the total mass of metal which must be heated so it has sufficient capacity. Circulation must also be carefully controlled so as to avoid

®Trademark of E. I. du Pont de Nemours & Co.

dead spots or hot spots in the oven. Ideally, the oven temperature should not vary more than ±5°F (±3°C) throughout its entire volume. Exhaust of the preheat oven to the atmosphere is normally not required, if sufficient drying time has been allowed for the primer which is normally applied from a solvent solution.

In designing a fluidized bed coating system it is important to know the required part temperature for coating rather than the actual oven temperature. The rate at which a part is heated in a circulating air oven is affected by the thermal conductivity, mass, and cross section of the part and the thermal capacity, air velocity (heat transfer), and temperature of the oven. If a high line speed is desirable, common practice is to maintain a higher oven temperature than the part temperature required but use a shorter preheat time. During most coating operations there is a "delay time" of from 5 to 15 sec between the time the part is removed from the preheat oven and dipped into the fluidized bed. This time must be taken into consideration in designing the coating system because the part is cooling rapidly during this period.

In articles having different cross-sectional areas such as fabricated wire goods it is sometimes difficult to maintain a uniform temperature throughout the part because of the "preheat delay" and the differential cooling rate of wires of varying diameter. This problem can be alleviated by a knowledge of the heating and cooling rates of wires of different diameter and proper choice of oven temperature, preheat time, and cooling time. Techniques for obtaining uniform coatings by controlling these variables are more fully described by Pettigrew (18).

Other methods of preheating can also be used such as induction, resistance, or infrared. These tend to be more complex and difficult methods of heating compared with circulating air ovens. In certain instances, however, these specialized methods of preheating are more effective. A good example is in continuous wire coating where, because of the very low cross-sectional area, the loss of heat during a delay time might cause the metal to cool below the fusion temperature of the coating. It is, therefore, necessary to reduce or eliminate the delay time. This can be accomplished by the use of an induction heater inside the fluidized chamber (19). There are a number of commercial installations in Germany that are based on the use of this technique and coat multiple wire strands which are subsequently fabricated into fencing.

The preheated part is then dipped or immersed in the fluidized bed. If manual dipping is used, most operators tend to give an up-and-down/sideways movement to the part during the coating operation. Care must be observed in coating parts having flat surfaces perpendicular to the air flow through the fluidized bed. If this condition occurs, the powder above the flat area loses fluidization, becomes more dense, and gives a heavier coating thickness on the top of the part. Symmetrical parts can be coated with a very high degree of uniformity if they are rotated about the vertical axis of symmetry. The effect of part motion and configuration is discussed in greater detail by Pettigrew (20).

As would be expected, the coating thickness obtained on a part is a function of the part temperature, the immersion time in the fluidized bed, the heat capacity of the part and the nature of the coating material. Figure 14-3 shows the coating thickness versus dip time for nylon at various preheat temperatures. These curves, of course, would also vary with the cross section (mass) of the substrate. Where a coating line has been set up to coat a specific article, it is frequently desirable to use an automatic dipping device. This can be accomplished by the use of timing mechanisms and air actuated pistons. Two types of mechanisms are normally employed; those which impart the "major motion"—the actual moving of the part into and out of the fluidized bed, and the "minor motion"—the motion imparted to the part while it is in the fluidized bed. It is not necessary to dip the article into the fluidized bed; the bed can also be raised to cover the object being coated (21).

If the article being coated is of insufficient mass to provide enough thermal energy to fuse out the coating, postheating is required. Much the same type of ovens are used for postheating as in preheating. However, some precautions in design must be taken that are not necessary with the preheat oven. Because of

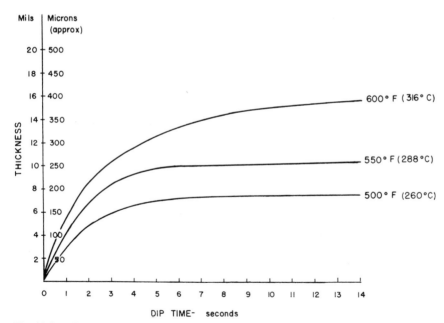

Fig. 14-3.   Coating thickness of nylon versus dip time on 1/8-in. (3.2-mm) steel substrate.

the high temperature involved, some degradation of the coating material occurs and fumes are generated. Sometimes the fumes have a deleterious effect on the coating itself and lead to wrinkling, roughness, or other surface imperfections. In addition, a possible explosion hazard exists if the concentration of volatile ingredients reaches too high a level. For these reasons it is well to vent the postheat oven to the atmosphere and to introduce in the neighborhood of 20 to 40% fresh air during recirculation. Postheating temperatures are frequently well below preheat temperatures since the metal is still hot and the coating material already fused with the exception of the surface layer. For nylon, postheat temperatures in the neighborhood of 230°C (446°F) are typically used for a period of 1 to 5 min depending upon the mass of the substrate.

After the coating is fused, it is common practice to cool the parts down rapidly so they can be handled. This is accomplished by water spraying or water immersion. The method of cooling can sometimes have a significant effect on the appearance of the coated article. In the case of nylon, for example, water quenching will give a more glossy surface than a similar part allowed to cool down to room temperature without forced cooling.

Safety

As with any finely divided material, normal precautions must be exercised in working with plastic coating powders during manufacture and application to prevent inhalation or the possibility of dust explosions. In powder manu-facturing plants, all equipment should be provided with totally enclosed motors and explosion proof switches.

A central vacuum cleaning system should also be provided with convenient outlets throughout the manufacturing area. Housekeeping rules must be rigorously enforced; periodic cleaning of all overhead lights, pipes, beams, ductwork, and the like should be carried out to prevent accumulation of powder.

Coating operators and manufacturing personnel should be provided with protective clothing and dust masks, as required. In coating areas, exhaust systems should be installed to remove the fine particles that normally accompany the coating operation. Most fluidized bed tanks are equipped with a peripheral exhaust and a few with dust collection systems which continuously feed powder back into the fluidized bed.

A comprehensive study on the nature of dust explosions and experimental work evaluating various commercial plastics were carried out by Jacobson et al (22) of the Bureau of Mines. They derived an "index of explosibility" which rates the relative explosibility of materials compared with Pittsburgh coal dust having a rating of 1.0. This index is empirical in nature but provides a rating which is consistent with laboratory experiments and practical experience. Table 14-5 lists the explosibility index for a number of materials used in powder coating methods.

It is apparent that while the index of explosibility is fundamentally a property of the base resin, it can be altered considerably by additives, such as fillers which will dilute the volumetric concentration of resin. On the other hand, the index of explosibility of material such as PVC is increased by the addition of plasticizers or other ingredients having a higher index of explosibility.

Summary – Advantages and Disadvantages

It can be seen that the various operations required in fluidized bed coatings are quite interrelated, and it is necessary, therefore, to consider the process as a complete system.

Obviously, no single coating method is the best for all applications and conditions. Following is a comparison of the advantages and disadvantages of fluidized bed coating with conventional coating systems.

Advantages

Since coating powders are 100% nonvolatile, essentially no by-products are evolved during the coating operation. With the increasing emphasis on reduction

Table 14-5. Explosibility index of various powders (22)

Material	Explosibility index
Epoxy resin (unmodified)	> 10
Epoxy (one part anhydride type 1% catalyst)	7.2
Cellulose acetate butyrate molding compound	8
Polyethylene	$4->10$
Fluorocarbon resins	< 0.1
Nylon-66 from filter	> 10
PVC resin (unmodified)	< 0.1
PVC – dioctyl phthalate mixture (67-33)	2.9

Relative explosion hazard rating	Index of explosibility
Weak	< 0.1
Moderate	0.1-1.0
Strong	1.0-10
Severe	> 10

of solvent emissions and pollution control, this factor is of major commercial significance. Fluidized bed coatings are normally applied at ten to twenty times the film thickness of conventional solution coatings and provide complete coverage of corners and sharp edges. Coated parts possess a much longer service life in corrosive enviroments and outdoor weathering and much better resistance to scratching, abrasion, and mechanical damage.

Disadvantages

Because of their greater film thickness, fluidized bed applied coatings are not economical for many applications. In addition, the powder reservoir required for coating adds to increased inventories and costs. This problem is compounded if very large parts are being coated or if a variety of colors is involved. It is frequently difficult to develop new applications unless a bed of powder is already on hand since so much powder is required to fill the bed in relation to the amount applied to the prototype part. It is even more difficult to develop a new application when a new coating material is required.

ELECTROSTATIC COATING

A relatively new coating technique is based on electrostatic principles, utilizes powdered nylon and other plastics, and is rapidly gaining commercial acceptance. It is called "electrostatic powder spray coating". In this method, a plastic powder is charged to a high potential and the charged particles are attracted to the object to be coated (usually metallic) which is maintained at ground potential. The object to be coated is usually at room temperature, and the plastic powder layer, which is held to the substrate by mechanical and electrostatic forces, is subsequently fused into a continuous film by heating.

The first commercial electrostatic spray units were introduced in Europe by SAMES, Grenoble, in 1962 and soon thereafter in the United States. For the next several years a reasonable level of experimental work followed, but few commercial installations resulted. Among the first installations were automatic systems with 8 or 16 guns for continuous coating of gas distribution piping. The acceptance of electrostatic powder coating in this application was largely fortuitous because experimental coating lines already built, based on powder flocking techniques, were not giving an acceptable product. Very few articles appeared in the literature during this period. The first general discussion of the process in the United States was given by Miller (23) at an SPE Regional Technical Conference, "Plastic Powders II," in 1967. At a similar conference held in 1965, "Powder Processing of Plastic Materials," the subject was mentioned, but no details were given (24).

The electrostatic powder systems can be considered a general extension of conventional electrostatic paint spraying technology which originated in the early 1940s. The basic principles are the same; however, in one case, the film forming ingredients are dissolved or dispersed in organic solvents while in the other they are dispersed in air. In the years since its introduction, electrostatic powder spraying techniques have been widely accepted, and interest is accelerating rapidly. Several of the largest automobile manufacturers are initiating projects to determine the feasibility of this system for the application of automotive topcoats. The reason for this interest is apparently the result of a real effort on the part of these manufacturers to control and reduce air pollution.

It is estimated that as of the end of 1970 there were some 200 commercial installations of electrostatic powder spray equipment in the United States, mostly hand guns, and perhaps another 50 or so units in research and experimental laboratories. The process has been accepted on a broader scale in Europe with an estimated 400 to 500 systems in operation, of which 35 to 50 are automated continuous lines. Also indicative of the greater European interest is the international symposium, "Powder Coatings 1970," held in London. Of the 23 papers presented, 7 dealt with electrostatic coating in some respect.

Theory

Under the influence of an electric field, a particle takes on the charge

$$Q = KE_0 a^2 \qquad (14\text{-}1)$$

where K = a constant dependent on the nature of the powder, primarily its dielectric constant; E_0 = the strength of the electric field in the charging zone; and a = average radius of the particle. The charged particles are attracted to the grounded substrate with a force F, according to the relationship

$$F = QE \qquad (14\text{-}2)$$

where E = the field intensity where the particle is located. Ideally, for optimum charging and attraction to the substrate, the equipment should produce a high electrical field, and the material should have a fine particle size. This is generally true. However, other factors are involved. After the particles are charged, they move mainly along the lines of the electric field but are modified in their flight due to the force of gravity, the influence of the air stream in which they are carried, the resistance of the air through which they are passing, and their own electrical repulsion. In most electrostatic spray devices, an attempt is made to remove the particles from the influence of the air stream in which they are carried to the charging nozzle so that the electrostatic force is the only one acting on the charged particles.

As the charged particles hit the substrate, some of the charge is lost by conduction and drains off to ground. Most plastic materials, however, are good insulators and do not lose their charge readily. Thus, after a layer of powder builds up on the substrate, it accumulates a charge so that further particles are repelled. On cold substrates, therefore, most materials have a limiting thickness which can be applied by electrostatic powder deposition. If the polarity of the field charging the particles were reversed at this point, further deposition of powder would continue to take place. However, a point would eventually be reached where the weight of the deposited powder layer would cause it to fall off the substrate due to gravitational forces. A thorough discussion of the various factors influencing the charging and deposition of plastic powders is given by Oesterle and Szasz (25).

Nylon Spray Powders

In general, the same materials (Table 14-2) used in fluidized bed coating are also used for electrostatic spray coating. Some European powder suppliers have indicated availability of experimental polyester, acrylic, and melamine-alkyd powders, but so far there appears to be little application or performance experience with these products. Nylons-11 and -12 are the only commercially used polyamides (Table 14-3) and are the types referred to throughout this discussion.

Most materials coat best when charged negatively. In electrostatic units where it is not possible to switch from positive to negative charging, negative charging is usually used. Different materials show optimum charging and coating characteristics at various charging levels. Nylon coating powders have optimum application characteristics when the charging voltage is in the range of 30 to 45 kV.

Optimum particle size for spraying is in the range of about 10 to 100 microns with a high percentage in the 30 to 50 micron diameter range. A powder of this particle distribution will be typically applied to give coatings 130 to 180 microns (ca. 5 to 7 mils) in thickness. If the particle size is extremely fine (for example, 100% <30 microns) continuous coatings of about 25 microns (1 mil) can be obtained.

Nylons are more sensitive towards moisture than most other thermoplastic materials. The powder may have a moisture content of less than 0.1% by weight to about 2% depending upon the conditions under which it was prepared and stored. This can have a significant effect on how the powder accepts the charge and performs during electrostatic application. A high moisture content will adversely affect the charging and dry, free-flow characteristics of the powder. This problem is further aggravated because of the fine particle size (high surface area) required for electrostatic coating powders. If damp powder is a problem, it can usually be remedied by placing the powder in a fluidized bed and fluidizing overnight. (As indicated earlier, the fluidized bed should have driers and filters.) If a fluidized bed is not available, the powder can be dried by placing it in shallow trays in a convection oven at $120°C$ ($248°F$) for several hours.

Because of the crystalline nature of nylon and the sharp transition from solid to melt phase, nylon powders show a tendency to crater during fusion. Cratering is a common phenomena occurring in paints and generally relates to surface tension effects. This is probably also the case with nylon as well because there seems to be a general correlation between the tendency to crater and the melt viscosity of the nylon used.

Another phenomenon related to the high-melting-point/low-melt-viscosity characteristics of nylon is called "edge pull." This is a condition occurring during fusion in which the coating "pulls away" and becomes thinner at sharp edges than at other portions of the article being coated. The tendency toward edge pull can be reduced by using resins of higher melt viscosity. However, this approach can only be followed to a certain degree; eventually, the melt viscosity becomes so high that the powder particles will not flow out uniformly and form a smooth, continuous film.

Equipment

The currently available electrostatic spray coating units are all based on the same principle but differ in design and certain of the major subsystem components. A schematic diagram of a typical electrostatic spray system is shown in Figure 14-4. Important details common to all units are as follows.

1. *Power source.* In order to charge the particles to a high potential, a high-voltage source is required. Most units incorporate a transformer for this purpose. For ideal operation and maximum versatility controls should be incorporated so that the voltage can be varied over a range of 0 to about 100 kV and the sign either positive or negative. The designs of various high-voltage sources suitable for electrostatic equipment are discussed by Lever (26).

2. *Powder transport.* Powder must be transported from the reservoir to the charging zone of the gun. This is accomplished by carrying the powder in an air stream. In some equipment the powder in the reservoir is kept in a fluidized state and carried to the charging zone using an air Venturi (Figure 14-4). The air feeding the Venturi is operated by a solenoid valve connected with the trigger mechanism of the hand gun. In other systems, the powder is not fluidized but mechanically agitated to keep it in a state of mobility. Problems can occur when the air flow is turned off due to settling and accumulation of the powder in the transport lines. When the gun is started again, the powder that has accumulated in the lines spurts out, and the spray must be diverted from coating in order to stabilize the flow of the air/powder dispersion. Once a stable spray has been

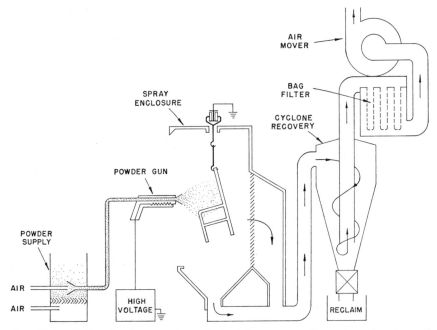

Fig. 14-4. Schematic diagram of an electrostatic spray coating system. (courtesy of Ransburg Electro-Coating Corp.)

established, coating can begin. Sharp bends in the powder transport lines and into the gun must be avoided. Most powders will form very hard deposits at sharp bends in the lines.

As in fluidized bed coating, positive means should be taken to remove oil and moisture from the air stream due to the deleterious effect of these materials on the dry, free-flow characteristics of the powder. In addition, air flow controls should be present for accurate regulation of the air pressure both in the fluidized bed reservoir (if used) and the powder transport lines.

3. *Nozzle dispersion.* A third important feature in electrostatic powder spray equipment is the design of the nozzle and means to control the powder velocity at the exit of the gun. Ideally, the charged powder should be completely separated from the air stream which carried it to the charging zone so that the only force acting on the powder will be that of electrostatic attraction to the substrate. This is difficult to accomplish. In one system, means are provided for a baffle or deflector combined with a nozzle which rotates, partially overcoming the forward motion of the particles by imparting a transverse motion. Another means is to have a movable baffle that interrupts the air stream to a greater or lesser degree.

The maximum powder output of currently available electrostatic equipment is in the range of 25 to 35 kg hr^{-1} (50 to 80 lb hr^{-1}). This must be taken into consideration in designing how many guns are required to coat various objects at desired production rates. In addition, the amount of overspray must also be factored in this calculation. For some objects, such as flat panels, overspray is very low—in the range of 5 to 10%. On the average, 20 to 40% overspray occurs; for wire work the amount of overspray can be as high as 60%.

The Substrate

Most of the information given in the discussion on fluidized bed coating regarding cleaning, jigging, masking, and priming is also relevant for electrostatic coating. A more detailed description of cleaning solutions and methods of surface preparation for different types of metals in electrostatic spray coating is given in the literature (27). As in fluidized bed coating, the article must be held, and the same general techniques to patch the scar are applicable. If cold parts are being coated, masking can be done at this point. A vacuum system with a small opening can be effectively used to remove powder from areas where no coating is desired. This technique is useful in certain instances; however, it is somewhat tedious and time-consuming. In any application where the coated part will be exposed to a corrosive environment or used outdoors, it is important that a primer is used. Most nylon primers used in fluidized bed coating can also be used for electrostatically applied coatings. Primers are applied from solution as relatively thin films, 2.5 to 12 microns (0.1 to 0.5 mils). While the problem of "preheat resistance" of primers, discussed with relation to fluidized bed coating, is generally not a problem in electrostatically applied coatings, the problem of adequate primer cure must be taken into consideration. Some nylon primers require a minimum heat treatment (cure) in order to develop optimum adhesion. Because of the relatively modest postheating conditions necessary to fuse the powder layer, especially on thin substrates, adequate cure of the primer is not always developed, and erratic adhesion may result. This problem can be resolved by proper choice of primer or by preheating the primed metals prior to coating.

In fluidized bed coating the object is heated above the melting point of the coating powder, but in electrostatic coating the substrate may be at room temperature, warm, or heated to a high temperature. On a cold substrate the charged powder builds up to a maximum thickness, after which no further powder will adhere to the substrate because the charged powder already present repels particles approaching the surface. By warming or mildly heating the substrate, a heavier layer can be obtained. In the case of a hot substrate, the powder melts upon contacting the surface, and the charge is lost almost immediately. This allows the coating to be built up in thickness to that approximating the general range capable with fluidized bed coating.

The Process

During the spraying operation, the volume (pressure) of air transporting the powder should be kept as low as possible. For current systems, this is in the range of 2 to $2.5m^3$ hr^{-1} (70 to 80 ft^3 hr^{-1}). Best electrostatic projection and a lower degree of overspray are obtained when the concentration of powder in the air stream is maximized but consistent with adequate transport of the powder through the lines.

Powder overspray can be considerable depending on the configuration of the substrate being coated. In any sizeable installation, therefore, a powder collection and recovery system is necessary. Most systems are designed with the coating area enclosed except for openings for the spray guns and the parts to enter and exit. Exhaust is provided by cyclones or bag after-filter (Figure 14-4). Since much of the overspray settles by gravity, the recovery system normally includes a sloping bottom collector feeding to the exhaust. In some designs expanded metal mesh or perforated metal panels are located behind the work being coated. These are grounded, and much of the overspray is deposited on them. These panels are located over powder collector trays or an exhaust port and are cleaned periodically by mechanical rapping or vibration.

Because the powder layer is held to the substrate by relatively weak mechanical and electrostatic forces, it is desirable to transfer the coated article to the fusion oven as soon as possible after coating. If undisturbed, the electrostatically applied powder coating layer will adhere to the substrate for several days. However, excessive mechanical shock must be avoided. As in fluidized bed coating, if batch ovens are used the temperature should be uniform throughout the volume of the oven to within $\pm10^{\circ}F$ ($\pm6^{\circ}C$). In a continuous coating line where a conveyor runs through tunnel ovens it is desirable to vary the temperature profile of the oven so that the temperature increases to the fusion temperature in the first third of the oven, holds that temperature long enough to fuse the powder, and decreases prior to the exit end. In order to properly fuse the powder layer into a continuous film, the substrate should remain in the oven long enough for it to reach the temperature of the oven. For general purpose work, an oven temperature of $220^{\circ}C$ ($428^{\circ}F$) is suitable for nylons-11 and -12. The time required to fuse the coating will depend on the thickness of the substrate, the heating capacity of the oven, and the temperature. Figure 14-5 shows the time required to fuse a nylon powder coating on substrates of varying thicknesses.

After the coating is fused, parts are frequently quenched in a bath of cold water so they can be handled quickly. Nylon coatings of higher gloss are obtained if the coating is water quenched immediately after removal from the

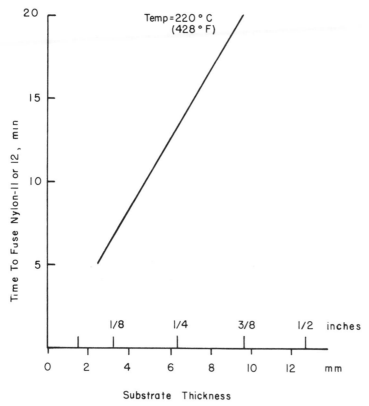

Fig. 14-5.   Time to fuse nylon versus substrate thickness.

fusion oven. If the part is allowed to air dry, the coating will have a "satin finish."

Safety

All of the comments made in the fluidized bed coating section with regard to this subject are applicable. Electrostatic spray grade powders are finer in particle size and represent an even greater hazard with regard to dust explosions than do fluidized bed coating powders.

Design of dust collection systems for use in electrostatic spraying systems must be carefully considered. Because the fine powders are collected and conveyed through the system in an air stream, significant electrostatic charges can accumulate as a result of triboelectric charging. The dust collectors should be grounded at all critical areas and use of insulating materials avoided.

All safety features on the electrostatic equipment should be regularly checked, and repair or modification of the high-voltage generators and cables should not be attempted by unskilled personnel.

Summary

In common with fluidized bed coating, the electrostatic coating powder coating method has the same advantages over conventional coating techniques. That is, systems are 100% nonvolatile—no solvents are given off. Material utilization is high, and with proper collection systems almost 100% of the material can be used. The system lends itself well to automation, and once set up and operating uniformly it can run consistently for long periods of time. Because of the low film thickness that can be applied (1 to 3 mils, ca. 25 to 75 microns), the low fusion temperatures possible (150 to 220°C, ca. 300 to 430°F), and the high utilization of material obtained, electrostatic powder coating approaches or exceeds the economics of conventional solution painting. It is expected that with the continued emphasis on pollution and the favorable economics of electrostatic powder spraying, this process could comprise as high as 50% of the industrial paint lines by 1980.

ELECTROSTATIC FLUID BED COATING

As the name indicates, electrostatic fluid bed coating is a method incorporating principles of both fluidized bed and electrostatic techniques. In operation, a relatively standard fluidized bed apparatus is used combined with a plurality of pinpoint electrostatic charging means. It is essential that the entire fluidized bed construction be of materials which are good electrical insulators. This poses some problems with the porous plate fluidizer; however, porous sintered polyethylene has been used satisfactorily for this purpose.

The depth of powder fluidizing is kept at a very low level—barely above the electrostatic charging electrodes. When the charging electrodes are energized, the particles become charged and are attracted toward any grounded object. The high degree of efficiency in charging powders by the electrostatic fluid bed is due to several factors. One is that the finest particles are in a higher concentration in the surface above the fluidizing bed than are coarse particles. Because of the smaller diameter, these finer particles are charged to a higher level and are, therefore, more attracted to the grounded object. As a result, over a period of time coarse particles tend to accumulate at the bottom of the bed near the porous plate. Another factor involved is the insulating nature of the fluidized bed chamber. After a period of continuous operation, the inside of the fluidized chamber becomes charged, thus acting to further repel the charged powder in the chamber. A schematic of an electrostatic bed is shown in Figure 14-6.

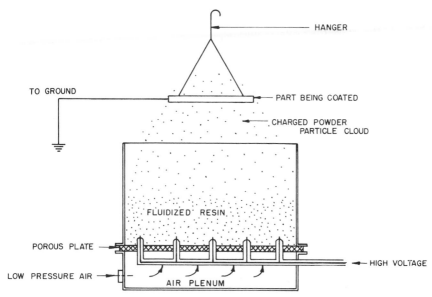

Fig. 14-6. Schematic diagram of an electrostatic fluidized bed. (courtesy of Ransberg Electro Coating Corporation.)

Most articles cannot be lowered into the electrostatic fluid bed since the danger of coming in contact with the charging electrodes exists and may result in a spark. Thus, articles to be coated must be moved across the top of the fluidized bed, preferably continuously. Electrostatic fluidized bed, therefore, is ideal for coating such items as continuous coil or sheet stock on one side and continuous lengths of essentially two-dimensional wire goods such as chicken wire and woven wire fencing. Other items with relatively small dimensions in the cross section (2 to 3 in., 5 to 7.5 cm) and of relatively open network such as expanded metal mesh and seat springs can also be coated by this process. Development of the electrostatic fluid bed is of relatively recent origin; there are few commercial installations at the present time. Because of the limitations with regard to part configuration, electrostatic fluidized bed coating techniques will probably never receive the widespread acceptance of either conventional fluidized bed coating or electrostatic spray coating.

Nylon powders charge very well and generally are quite satisfactory for use in the electrostatic bed coating method. Because charging is so effective, fresh metal must be exposed to the charged particles at a uniform, continuous rate. Otherwise, the powder builds up on the part and results in a variation of film thickness. Whereas spraying time for a nylon powder to achieve a coating thickness of 5 to 7 mils (130 to 180 microns) is in the order of 5 to 8 sec using electrostatic spray equipment, the exposure time in the electrostatic bed to achieve the same thickness is only 2 to 4 sec. Also, because the charging is so effective, care must be taken to make sure no extraneous surfaces are close by at ground potential. Otherwise, they soon become coated with a heavy powder layer. For this reason many electrostatic fluidized beds are completely enclosed with the exception of the inlet and outlet ports.

APPLICATIONS AND MARKETS FOR NYLON COATINGS

Nylon coatings are used in many applications where the corrosion resistance, weatherability, abrasion resistance, and decorative appearance of nylon are required. Major market areas are in the food handling and pharmaceutical industries where smooth, nonadherent coatings are applied to many articles to promote ease of cleaning and hygienic conditions. Nylon coatings can be used in contact with food and food products; they can be sterilized by autoclaving with no adverse effects. In architectural applications, items such as window frames, bus shelters, lighting standards, and hardware are coated with nylon. Where corrosion resistance is a problem, boat ladders, skin diving tanks, marine lamps, hardware, and other items are protected with nylon coatings. Chain sprockets and guides, ball bearing races, valve needles, ball joints, and many other articles are coated where the low coefficient of friction and excellent wear characteristics of nylon are important.

An application worth singular mention, because it illustrates the outstanding performance which can be obtained with nylon coatings, is the coating of spline shafts. A spline shaft transmits the torque from the transmission to the drive shaft in automotive vehicles. In addition, the spline shaft is a support member for the drive shaft and there is axial motion between the two. Under these severe conditions, on tests conducted on heavy duty trucks, the nylon coated shafts were in excellent condition after completion of a 230,000 mile (370,000 km) test whereas standard hardened steel shafts showed galling and siezing in the 5,000 to 50,000 mile (ca 8,000 to 80,000 km) range (28).

Literally thousands of articles are being coated with nylon powders by fluidized bed or electrostatic coating methods. Figure 14-7 shows a number of typical examples. Other applications have been discussed by Stott (29, 30), Toder et al. (31), Pascoe (32), Crater (33), Dixon (34), Gilbert (35), and in numerous trade journals (36-38).

Nylon coating powders are much more widely used and accepted in Europe than in the United States. It is estimated that somewhat over 4 million pounds (ca. 1815 tons) of nylon coating powders were sold in Europe in 1970 while the figure for the United States was certainly below 250,000 pounds (ca. 113.5 tons). The reason for this disparity is not well established; certainly, one of the factors holding back widespread application in the United States has been the high cost of the material. Table 14-6 shows the relative cost per thousand square feet per mil for various commercially available powders in the United States. It can be seen that nylon is still among the most expensive. As a result, nylon is used only in those applications where the quality demands it or the functional characteristics of good abrasion resistance and lubricity are required. Very little has been reported in the literature with regard to the market size for the various materials used in powder coating. However, Gemmer (39) recently presented the information shown in Table 14-7. These data indicate that while the total amound of material used in fluidized bed coating is far greater than for electrostatic spray coating, the rate of increase for electrostatic spray applied materials is far greater—in the range of 100% per year.

At the present time the bulk of electrostatic spray grade material is based on epoxy resin. This is true in both Europe and the United States. For the first time, however, nylon coatings can be applied by the use of electrostatic powder deposition in economical thicknesses to compete in market areas where it has not been possible to even consider nylon in the past. This has been, of course, due to the fact that nylon has not been available in the form of a paint or conventional coating material because of its limited solubility in most common solvents. There is a current trend in the United States today which could result in much greater acceptance of nylon coatings. The high cost of chrome and nickel plating chemicals are causing many platers to look for other methods to produce acceptable decorative finishes for such items as refrigerator shelves, shopping carts, automotive components, and other decorative wire goods.

There is a much higher level of activity in Europe in all phases of powder coating than in the United States. For example, it is estimated there are some 20 to 25 equipment suppliers and 30 to 35 producers of powder for electrostatic application in Europe while in the United States the respective numbers are approximately 6 and 10. However, the signs of increased activity in the United States are unmistakable. In the next several years, the numbers, especially with regard to powder suppliers, could change significantly.

In summary, while current markets for powdered nylons are relatively small, the ability to apply thin films combined with a greater acceptance of powder coating techniques in the industry could lead to significant new applications and increased volume for nylon coating powders.

FABRIC ADHESIVES

Until recently, the only method for joining fabrics has been by sewing. In the last ten years adhesive bonding of fabrics has been employed, and today fabric composites can be purchased in any dry goods store. A typical composite is a

Fig. 14-7.   Typical application for nylon coatings. (Courtesy of Aquitaine-Organico.)

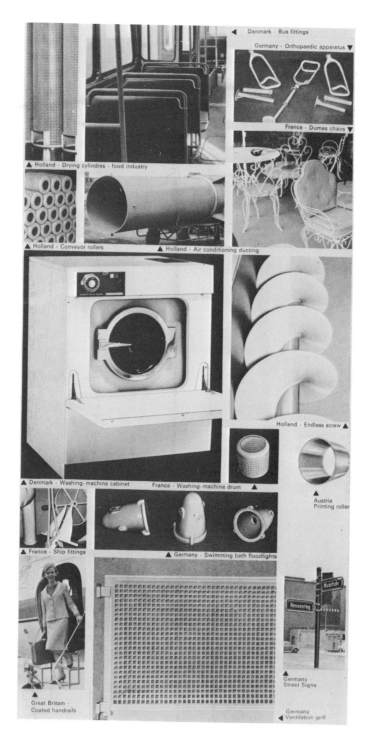

Denmark - Bus fittings

Germany - Orthopaedic apparatus ▼

France - Dumas chairs ▼

▲ Holland - Drying cylindres - food industry

▲ Holland - Conveyor rollers

▲ Holland - Air conditioning ducting

Holland - Endless screw ▲

▲ Austria
Printing roller

▲ Denmark - Washing-machine cabinet

France - Washing-machine drum ▲

▲ France - Ship fittings

▲ Germany - Swimming bath floodlights

Germany -
Street Signs

Great Britain -
Coated handrails

Germany
◄ Ventilation grill

Fig. 14-7.   Continued

Table 14-6. Comparative cost of various coating powders[a]

Material	$ per 1000 ft^2 per mil[b]
Cellulosics	7.50-8.20
Epoxies	7.90-10.20
Nylon	11.00-14.50
Polyethylene	1.50-3.00
Penton®	46.00
Vinyl – melt mix	6.00-7.50
Vinyl – dry blend	3.50-5.00

[a] Based on published price lists as of December, 1970. Assumes application by electrostatic powder spray and purchases of 5,000 lbs or more.

[b] 1000 ft^2 per mil = 93 m^2 per 0.025 mm.

Table 14-7. Estimate of world-wide sales of coating powders in metric tons (Ref. 39). (Thousand pounds)

	Fluidized bed coating	Electrostatic powder coating
1967		
Europe	4,000 (8,800)	200 (440)
World	10,000 (22,000)	600 (1,320)
1968		
Europe	6,450 (14,190)	500 (1,100)
World	15,000 (33,000)	1,600 (3,520)
1969		
Europe	7,400 (16,280)	1,200 (2,640)
World	17,000 (37,400)	3,400 (7,480)

finely knit tricot bonded to wool—the so-called bonded wool. The tricot serves as both a lining and stabilizer for the woolen fabric. This type of laminate is usually produced in relatively long lengths and sold by the yard. The adhesive is applied as a solution or latex, the two fabrics joined together under heat and pressure, and the adhesive cross-linked (40). Typical adhesives are thermosetting acrylics or polyurethanes. These possess good resistance to dry cleaning solvents after curing.

Another area where adhesive bonding is gaining acceptance is in joining interliner fabrics to the face fabric in garment construction. The interliner fabric is used to provide form stability, reinforcement, and "hand" to the face fabric; it is normally not visible in the finished garment. Typical examples are suit fronts, collars, waist bands, and coats. The traditional method involves holding the interliner to the face fabric, until the garmet is stitched, by basting threads which are then removed. In the bonding method, the adhesive is applied to long rolls of the interliner fabric, fused in place, and the bolts are sold to the garment manufacturer. During manufacture, the interliner fabric is cut to shape and joined to the face fabric using steam or electric presses. This method greatly reduces the degree of operator skill and time required to build the garment.

Nylon copolymers are gaining acceptance as adhesives in "fusible" interliner fabrics and will be considered in more detail in the following sections (see also Chapter 16).

Types of Adhesives

Since the interliner fabric must be cut and joined to the face fabric during garment manufacture, the adhesive must be fusible. Thus, the thermosetting types used to bond continuous laminates are generally not suitable. Instead, thermoplastic adhesives are used. The general properties of the adhesive required for fusible interliners are as follows:

1. Must fuse at temperatures of 220 to 248°F (104 to 120°C) for steam presses or 248 to 300°F (120 to 149°C) for electric presses at pressures of 2 to 6 lb(wt) in.$^{-2}$ (141 to 423 g(wt)cm^{-2}).

2. Must have good adhesion to both the interliner and face fabric.

3. Must not have too low a melt viscosity or the adhesive will "strike through" the face fabric.

4. Must not have too high a melting point or melt viscosity or it will not bond adequately.

5. Must have good resistance to dry cleaning solvents at both room and elevated temperatures.

6. In most cases, the adhesive must also have good resistance to hot water and detergents (laundering).

7. In some cases, the adhesive must not soften or delaminate when the fabric is heat treated to cure wash-and-wear treatments.

8. Must have good application characteristics (to the interliner fabric) by whatever means used.

Several resins are currently used as the adhesive in fusible interliners. Poly(vinyl acetate) has good adhesion but limited resistance to dry cleaning solvents and is used in applications such as women's coats, where the fabrics are bulky and the garment is dry cleaned only several times during its expected lifetime. Various grades of polyethylene, normally those in the 0.91 to 0.93 density range with a melt index of 10 to 40, are also used where the life expectancy of the garment is not too long and limited dry cleaning resistance is adequate.

Since 1964, plasticized PVC has been used commercially as the adhesive on interliner fabrics, and many millions of garments have been produced with this system. While PVC based adhesives are adequate for a wide variety of applications, there appears to be a definite swing towards polyamide-based adhesives as the preferred types.

Practically all of the nylon resins used for fabric adhesives are copolymer or terpolymer resins, based largely on the monomers used to make nylons-6 and -66. Typical properties of some commercially available fabric adhesives are given in Table 14-8. Various grades are available in differing particle size ranges depending on the method of application and at different melt indices depending on whether steam or electric presses will be used for bonding. The type of interliner fabric and the weight, resiliency, bulk, and tightness of weave of the face fabric are also factors in the choice of adhesive. In addition, the bonding temperature, pressure, degree of bond strength desired, and resistance to dry cleaning or laundering required in the laminate must also be considered.

While the bond strength of the laminate is dependent on many of the above factors, it is most directly related to the amount of adhesive applied. Normal ranges are from 0.5 to 0.8 oz yd^{-2} (16 to 26 g m^{-2}). Bond strength for a typical cotton interliner fabric bonded to various face fabrics after dry cleaning and

Table 14-8. Properties of commercial nylon fabric adhesives (41)

	Copolymer for steam pressing	Copolymer for electric pressing	Soluble terpolymer
Melting point	110-120°C	120-130°C	140-150°C
Relative viscosity			
(0.5% in m-cresol)	1.5	1.4	1.4
Water absorption			
65% R.H. at 20°C	2.5%	1.5%	--
Immersion at 20°C	6%	3-4%	--
Tensile strength at			
yield	3135 lb(wt)in.$^{-2}$	3135 lb(wt)in.$^{-2}$	2565 lb(wt)in.$^{-2}$
Tensile strength at			
break	4560 lb(wt)in.$^{-2}$	3705 lb(wt)in.$^{-2}$	4275 lb(wt)in.$^{-2}$
Elongation at yield	20%	20%	15%
Elongation at break	400%	330%	350%
Tensile modulus	54,150 (lb(wt)in.$^{-2}$	51,300 lb(wt)in.$^{-2}$	74,100 lb(wt)in.$^{-2}$
Impact strength			
notched (DIN 53453-			
Charpy)	No break	No break	No break
Hardness – Shore D	66	61	60

Note. lb(wt)in.$^{-2}$ × 0.0703 = kg(wt)cm^{-2}. Properties measured after conditioning at 20°C and 65% R.H., 14 days.

laundering is given in Table 14-9. The bonding was carried out on a press with 5 lb(wt) in.$^{-2}$ (350 g(wt)cm^{-2}) clamping pressure and a temperature of 293°F (145°C) measured at the bond. Bond strength was measured on strips 1 in. (25 mm) wide by delaminating in a tensile testing machine at a rate of 8 in. min^{-1} (200 mm min^{-1}).

Methods of Application

Because the use of adhesives in fusible interliners is in the early stages of acceptance, standard methods of application have not yet evolved. Most activity is being carried out with powdered or granular adhesives. These are applied by several methods.

In the simplest method, the powdered adhesive is charged to a hopper and "sprinkled" on the interliner fabric as it passes beneath. Vibrating screens and plates in combination with mechanical devices are used to maintain a relatively uniform distribution of adhesive. The fabric is then run through a convection oven or banks of infrared heaters to melt the adhesive on the fabric. With this method, it is difficult to get uniform distribution of the adhesive; further, the particle size of the resin adhesive is critical. If the particles are too fine, they sift through the fabric; if too coarse, excessive amounts must be applied and cause stiffness or loss of "hand" in the final laminate (43). A particle size of 300 to 500 microns (ca. 35 to 50 mesh) is generally used.

A method gaining wider acceptance is the so-called powder point process. In this method, the powdered adhesive is doctored onto a cool roll having depressions on its surface (gravure roll). This roll is in contact with a heated roll, and the interliner fabric travels between. At the point of contact, the hot roll causes the adhesive to melt and transfer to the fabric. Sometimes the fabric is preheated to reduce the quantity of heat which must be supplied by the hot roll.

Table 14-9. Bond strength^a of various face fabrics to cotton interliner (42)

	Bond strength, lb(wt)in.$^{-1}$ (g(wt) 25 mm^{-1})		
Face fabric	Original	After 20 dry cleanings	After 20 wash cycles at indicated temperature
Cotton	5.5 (2500)	4.0 (1800)	3.3 (1500) 60°C
Polyester/cotton	4.2 (1900)	2.9 (1300)	2.9 (1300) 40°C
Acetate fabrics	2.0 (900)	1.8 (800)	1.5 (700) 40°C
Wool	7.7 (3500)	5.7 (2600)	4.6 (2100) 40°C

^{a} Nylon adhesive applied at 0.46 oz yd^{-2} (15 g m^{-2}).

Using this technique, almost any pattern of adhesive or size of dot desired can be laid down on the interliner fabric. Of course, a gravure roll having the desired dot size and spacing must be prepared for each pattern. Adhesive powders used in this process are finer in particle size than those used for "sprinkling" and are usually less than 200 microns (ca. 80 mesh).

Another method used to apply the adhesive is rotary screen printing. This is similar to a silk screening process. Here, the adhesive is either in solution or suspended in a vehicle, usually water, to form a dispersion, and is forced through a metal screen having the desired pattern. After application, the adhesive dots must be dried to remove the water and heated to fuse them to the fabric. This method also allows an almost infinite variety of adhesive patterns to be used.

The third method used to apply the adhesive to the interliner fabric is to spray it from a solution. The solution is formulated with highly volatile solvents so that loss of solvent during spraying causes the resin to come out of solution and form filaments. These filaments form a random web of adhesive on the interliner fabric.

Summary

By whatever method applied, polyamide resins are gaining increased acceptance as the preferred adhesive in fusible interliner fabrics, and this new application should consume increasing amounts of specialty nylon resins.

PRESSED AND SINTERED PARTS

Using techniques similar to those in powder metallurgy, it is possible to prepare polyamide composites with properties significantly different than standard materials. Polyamide powders with an ultimate particle size of 4 to 10 microns are admixed with up to 80 vol % of fillers or other modifying ingredients, cold pressed in closed dies, and sintered to form the final part.

Powders are prepared from nylons-66 and -610 by solution and precipitation (44). The powder is compacted and ground to a granular product in order to increase the bulk density and allow easier mold filling. Inorganic fillers such as mica and molybdenum disulfide can be added to reduce creep or modify frictional characteristics. Powdered metals can be added to improve thermal conductivity, alter electrical properties, and so on (45).

The powdered nylon is introduced into closed dies and pressed at room temperature at pressures of about 2820 kg (wt) cm^{-2} [40,000 lb (wt) in.$^{-2}$]. If lower pressures are used, parts of controlled density and pore size can be prepared. Advantage is taken of this characteristic to produce porous parts saturated with lubricants, inks, and other materials for special applications. After pressing, the part has sufficient strength to withstand moderate handling without breaking.

It is important that the parts be sintered in a nonoxidizing atmosphere; otherwise, excessive oxidation occurs and physical properties are significantly impaired (46). A preferred medium for sintering is a hydrocarbon based oil. Typical sintering conditions are 30 min at 257°C (495°F) using a 2-hr heat-up and cool-down. If the moisture content of the powder is greater than about 1%, parts frequently will crack during this operation.

The physical properties of pressed and sintered nylon parts are generally inferior to those of a standard injection molded specimen (46). However, because of the control over porosity, the ability to hold up to 50% by volume of lubricants, and the improved creep and frictional characteristics provided by special fillers, sintered nylon parts find use in applications such as bearings, bushings, cams, and wear plates and give performance superior to standard nylons.

ROTATIONAL MOLDING

While rotational molding is a well-established process in the plastics industry, the use of powdered plastics did not start until about 1959 or 1960. Prior to that, the process was linked almost exclusively to vinyl plastisols.

Low-density polyethylene was the first powdered plastic material to be used in rotational molding on a commercial basis. In 1963, it is estimated that about 10 million pounds (ca. 4500 tons) of powdered plastic materials were used in rotational molding, and by 1967 the amount had grown to 75 million pounds (ca. 35,000 tons). Estimates as high as 250 million pounds (ca. 115,000 tons) have been made for 1972 (47). Along with increasing use of the process the other major trend in rotomolding is to larger parts. In 1968 parts weighing up to 175 lb (ca. 80 kg) were being molded on a routine basis, and in 1970 parts as long as 21 ft (ca. 7 m) and others weighing up to almost 900 lb (410 kg) were reported (48, 49).

Rotational molding of nylon has been accomplished on a commercial basis and shown to be feasible. However, it has not achieved very wide commercial acceptability. Probably less than 100,000 lb of nylon were used in rotocasting in 1970. Nylon-11 has been more widely used than any other type of nylon because of its low melting point. Rotomolding with nylon-12 has been reported (50), but there are at this writing few commercial applications probably because of its relative newness.

Rotational molding can be used to form almost any desired size or shape plastic article. It is especially useful in forming hollow articles such as containers, cylinders, ductwork, balls, and toys. The relative economics of rotational molding versus other methods have been compared by Nickerson (51). The process is basically simple in concept. A mold is filled with the thermoplastic powder; closed, and introduced into a high-temperature. The mold is rotated simultaneously about 2 axes causing the powder to be distributed over the inside of the mold. As the temperature of the mold increases, the powder begins to melt and fuse to the mold surface. When the material is completely fused the mold is transferred to a cooling chamber, chilled to below the melting point of the plastic, and the part removed. The inside surface of the mold thus becomes the exterior surface of the molded article. Details of the types of molds used, the various methods of heating the molds, and equipment available are described in the literature (47, 52).

A variety of thermoplastic powders are used in rotational molding. Polyethylene has been used the longest and accounts for a predominant share of the market. Ethylene/vinyl acetate and ethylene/ethyl acrylate copolymers are used where a soft, elastomer-like material is required. Polystyrene, acetals, cellulose acetate butyrate, polycarbonate and vinyl powders are also rotationally molded.

The properties of nylons-11 and -12 used in rotomolding are essentially the same as those used in coating processes (Table 14-3). Nylons are prone to thermal degradation and must be well stabilized when used for rotomolding. Special grades are usually offered for this application (53). In comparison with other thermoplastics, nylon offers superior strength, toughness, low temperature resistance, solvent resistance, and low permeability of hydrocarbons. Because of these properties, one of the major applications has been the rotomolding of fuel tanks for aircraft, autos, and motorbikes. To keep costs lower, composite fuel tanks have been developed having an inner surface of nylon and an outer surface of polyethylene. Because of its higher heat distortion and maximum operating temperature compared with other thermoplastics, nylon has also found application in heating and ventilation ducts for aircraft.

As in the case of coating powders, the relatively high price of nylon has restricted its use in all but the most demanding applications.

REFERENCES

1. Neumann, J. A., *Mod. Plast.* **27** (10), 85 (July, 1950).
2. Clements, Peter G. (to Schori Metallising Proc. Ltd.), Brit. Patent 643,691 (Sept. 27, 1950).
3. Gemmer, E., private communication.
4. Gemmer, E. (to Knapsack-Griesheim A.–G.), Ger. Patent 933,019 (Sept. 15, 1955).
5. Gemmer, E. (to Knapsack-Griesheim A.–G.) Brit. Patent 759,214 (Oct. 17, 1956; amended Aug. 21, 1957).
6. Gemmer, E. (to Knapsack-Griesheim A.–G.), Brit. Patent 862,494 (Mar. 8, 1961).

7. Gemmer, E., *Kunstoffe* 47, 510 (1957).

8. Gemmer, E. (to Knapsack-Griesheim A.–G.), U.S. Patent 2,844,489 (July 22, 1958).

9. Pascoe, W.R., *Mat. Des. Eng.* 51 (2), 91 (Feb., 1960).

10. Beike, H. (to Polymer Corp.), U.S. Patent 3,364,053 (Jan. 16, 1968).

11. Anon., *Ind. Eng. Chem.* 41, 1249 (1949).

12. Frantz, J. F., *Chem. Eng.* 69 (19), 161 (Sept. 17, 1962); *ibid.* 69 (20), 89 (Oct. 1, 1962).

13. Parent, J.D., N. Yagol, and C.S. Steiner, *Chem. Eng. Prog.* 43, 429 (1947).

14. Richart, D.S., *Plast. Des. Proc.* 2 (7), 37 (1962).

15. Schaaf, S., *Maschinenmarkt* 6 (21), (Jan., 1969).

16. Nagel, F. J. (to The Polymer Corp.), U. S. Patent 3,264,131 (Aug. 2, 1966).

17. Landrock, A., *Plastec Report* 13 (AD 431603), 96 (Jan., 1964).

18. Pettigrew, C.K., *Mod. Plast.* 43 (12), 120 (Aug. 1966).

19. Checkel, R. L., *Mod. Plast.* 36 (2), 125 (Oct., 1958).

20. Pettigrew, C. K., *Mod. Plast.* 44 (1), 152 (Sept., 1966).

21. Pascoe, W.R., *Met. Prods. Mfg.* 19 (1), 35 (Jan., 1962)).

22. Jacobson, M., J. Nagy, and A. R. Cooper, "Explosibility of Dusts Used in the Plastics Industry," Bureau of Mines Report of Investigations, 5971 (1962).

23. Miller, E.P., SPE Reg. Tech. Conf., "Plastic Powders II," 26 (Mar., 1967).

24. Deninson, E.E., SPE Reg. Tech. Conf., "Powder Processing of Plastic Materials," 20 (Oct. 25, 1965).

25. Oesterle, K.M., and I. Szasz, *Jour. Oil Color Chem. Assoc.* 48, 956 (Oct., 1965).

26. Lever, R.C., Paper presented at London Meeting, "Powder Coatings 1970."

27. Anon., *Metalloberflache* 23 (9), 277 (1969).

28. Kayser, J.A., and W.T. Groves, Paper presented at SAE Mtg., Detroit, Mich., Jan., 1968.

29. Stott, L.L., *Matls Methods* 41 (92), 121 (1955).

30. Stott, L.L., *Org. Finish.* 17 (6), 67 (1956).

31. Toder, I.A., N. I. Runyantseva *et al. Soviet Plast.* 1966 (3), 66.

32. Pascoe, W.R., Paper presented at SAE Meet., Chicago, Ill., June, 1960.

33. Crater, W. deC., SPE Reg. Tech. Conf., "Plastic Powders III," Tech. Papers, 42 (Mar. 12, 1969).

34. Dixon, J.B., *Mod. Metals* 25 (4), 89 (1969).

35. Gilbert, L.O., Rock Island Arsenal Lab, Tech. Rep. 62-3183 (AD288243) Sept. 20, 1962.

36. Anon., *Paint Varnish Prod.* 52 (6), 44 (1962).

37. Anon., *Plast. Des. Proc.* 8 (11), 17 (1968).

38. Anon., *Mod. Plast.* 39(7), 104 (Mar., 1962).

39. Gemmer, E., *Kunstoffe* 59 655 (1969).

40. Anon., *Rohm and Haas Reporter* 27 (1), 21-23 (Jan/Feb, 1969).

41. Schaaf, S., *Kunstoff-Berater* 11, 238 (1968).

42. Anon., Tech. Bulletin, GRIL-TEX® Powder (in Eng.), Emser Werke AG, Postfach CH-8039, Zurich (May, 1970).

43. Smith D. K., *Textile Chem. Colorist* 1 (27), 619 (Dec., 1969).

44. Stott, L. L., and L. R. B. Hervey (to The Polymer Corp.), U. S. Patent 2,639,278 (May 19, 1953).

45. Stott, L. L., and L. R. B. Hervey, *Mat. Meth.* **40,** 108-11 (Oct., 1952).

46. Stott, L. L., *Mod. Plast.* **35** (1), 157 (Sept., 1957).

47. Wittnam, Chas. A., *Plast. Des. Proc.* **9** (7), 32 (July, 1969).

48. Anon., *Plast. World* **28** (7), 42 (July, 1970).

49. Anon., *Mod. Plast.* **47**(10), 86 (Oct., 1970).

51. Nickerson, J. A., *Mod. Plast. Encyclo.* **45** (14A) 825 (1968).

52. Wright, V., *Mod. Plast.* **45** (7), 64 (March, 1968).

53. Crater, Willard deC., *SPE Tech. Papers* **15**, 547 (1969).

CHAPTER 15

Forming of Nylons

WALTER M. BRUNER

INTRODUCTION

In this chapter the term "forming" is used to mean reshaping a nylon preform under suitable pressure. Normally, the forming temperature is below the melting region but high enough to induce ductility. For example, nylon-66 may be reshaped in a simple forging process at a pressure of about 16,000 lb (wt)in.$^{-2}$ (1,100 kg (wt) cm^{-2}) and a temperature of about 347°F (175°C). Nonmelt forming results in well-shaped and durable articles from crystalline nylon homopolymers and copolymers.

Forming results in a degree of strength and toughness unattainable by melt processes. One example is package strapping of formed nylon-66 now marketed to replace steel strapping with advantages of package stability and safety. Other commercial examples include garment tags, tape rule housing, insulators, and harness ties. Short forming cycles and relatively simple equipment offer economies to offset preform costs. In 1970 about fourteen million pounds of formed nylon articles were manufactured and sold worldwide.

BASIC CONSIDERATIONS

Although shaping nylon by using metal forming processes is now commercial practice, the observed facts in relation to physical structure and properties are little understood. Also, within the nylon family of polymers, there are wide differences in performance. For example, some plasticized nylons act like rubbers in their tendency to spring back almost immediately after leaving the die. At the other extreme, high modulus nylons are difficult to form free of fracture.

Formability is that quality of a solid resin that enables it to accept a new shape under heat and pressure and to hold that shape under the expected end-use conditions. The property of "ductility" — the extent to which a material can be reshaped without fracture — is a necessary but not sufficient condition for formability which also requires form stability. However, of the basic properties that influence the formability of nylons, ductility in both tension and compression is particularly important. A convenient indication of ductility is that part of the stress-strain diagram in which further application of force produces a large change in elongation.

The Polymer

Because ductility is important in selecting the nylon, specifying preform shape, and setting operating conditions, improving ductility to a satisfactory degree is a prime consideration. Figure 15-1 shows that the ductile range of nylon-66 is lengthened to very nearly the entire stress-strain diagram simply by heating the preform. In effect, heating lowers the yield stress and increases elongation (Chapter 10, p. 335) so that shaping can occur without fracture. For most forming operations, nylon-66 preforms should be 50 to 100°C below the Fisher-Johns melting point. Glass-reinforced nylons form best at preform temperatures about 25°C higher than those normally needed for unreinforced nylons.

The effect of crystallinity on the stability of formed polymers has been reported by several investigators. Ito (1) observed that the strain recovery on annealing formed articles of noncrystalline resins was about 90%, but in crystalline resins it was only 20%. Normally, nylons-6, -66, -610, and -612 have adequate crystallinity for form stability.

High molecular weight also helps retention of shape and improves resistance to fracture. The relatively high molecular weight nylons respond to a variety of forming operations, but nylons of low molecular weight are limited to processes of relatively low displacement, such as shearing and coining.

Several investigators have proposed various tests to measure formability, and hence predict process suitability. For example, Warshavsky and Tokito (2) have used a cup test to measure "cold drawability." Broutman and Kalpakjian (3) also tried to define limiting draw ratio in terms of thinning resistance. A useful formability index for nylons is the ratio of the modulus of elasticity (Young's modulus) to the yield strength. Applying this to familiar materials like aluminum, the result is 450 to 670:

	$lb(wt)in.^{-2}$	$kg(wt)cm^{-2}$
Aluminum		
Modulus	$8\text{-}10 \times 10^6$	$5.6\text{-}7.0 \times 10^5$
Yield strength	$15\text{-}18 \times 10^3$	$1.0\text{-}1.3 \times 10^3$
Ratio	450-670	
Nylon-66, dry, $23°C$		
Modulus	4.1×10^5	2.8×10^4
Yield strength	11.2×10^3	0.77×10^3
Ratio	37	

As shown above, the corresponding value for dry nylon-66 at room temperature is much less. Thus, by this index aluminum is about 15 times as formable as this grade of nylon-66.

A practical formability index that correlates well with end-use performance is the change in length on heating ($100°C$, 1 hr) a nylon ribbon uniaxially rolled 2 X. By this test recovery averaged 0.99% for unmodified nylon-66 and up to 7.5% for a modified nylon-66.

Impact forging in open dies is limited by a tendency to fracture. However, under high hydrostatic pressure, the preform becomes ductile and can be forced through a die. This need for support is significantly different from metal forming. For example, an article may be formed from a brass workpiece in an open die, but a nylon preform of the same shape may fracture when impacted in such a die. However, the same article may be formed of nylon, provided the preform is supported in a closed die. Youngman (4) also points to the advantages of closed-die rather than open-die forming. An exception to this general rule is free drawing, a well-established method of pulling an unsupported preform to give elongated nylon articles of high tensile strength. In free drawing, draw ratio, and hence force, is controlled to elongate below the ultimate strength.

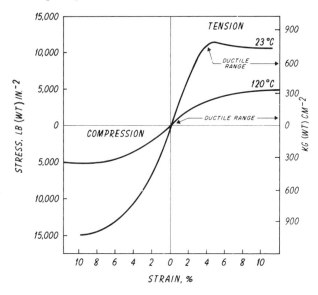

Fig. 15-1.   Stress-strain curves of a nylon-66 resin in tension and compression, dry, as-molded.

The Process

Punch and die design must take account of the very considerable stresses in forming nylon, especially the glass-reinforced grades. Since stresses are dynamic, fatigue occurs and may lead to failure of hardened tool steel at points of stress concentration. When slender punches are needed, stability problems may arise, such as punch wander because of inaccurate alignment or distorted preforms. The important ratio of punch length to diameter may be increased by selecting steel of high elastic modulus, such as a hardened tool steel or tungsten carbide. The punch nose is a critical area. Also, the area ratio between punch shank and head is a design factor. A large ratio spreads the load over a large pressure pad but requires more of the relatively expensive punch steel.

Die design must also consider the tensile forces on the inner die wall and provide shrink ring reinforcement if stresses exceed about 200,000 lb(wt)in.$^{-2}$ (14,140 kg(wt)cm^{-2}) for tool steel.

For most operations, solid dies are much preferred to split dies because the latter require large clamping forces. Also, split dies tend to leave axial lines or flash. Where notch effect may occur to weaken a die, effort should be made to eliminate it by steel selection and design. As a last resort, dies may be divided into two or more parts. Obviously, dies should be firmly seated and supported in die sets to control bending stresses. High friction during forming or ejection may be controlled by lubrication with mineral oil, silicone oil, or polyalkylene glycols.

Either mechanical or hydraulic presses may provide the cycling force needed to shape articles at production rates. The crank-driven mechanical press, generally cheaper than a hydraulic press of equal capacity, is well suited for small, fast cycle runs, sometimes combining operations, such as coining, with piercing, forming, or blanking. For small articles the sharp blow of the mechanical press is preferred to the squeezing action of the hydraulic press. The limited stroke length can be a disadvantage.

Hydraulic presses are more suitable for articles which require long working strokes or dwell time and where large forces are needed. They are readily equipped with force-limiting devices to control overloading and possible die damage.

The Product

Formed articles are generally tougher and stronger than those molded, extruded, or thermoformed. The degree of strengthening is directly related to the extent of resin movement in the die and to the direction of movement. The affected strength properties include flex durability and tensile, compressive, and impact strengths. Hence, in contrast to molded articles, the thinnest areas can be of the greatest strength and toughness because the greatest resin movement and orientation have occurred there. However, a high degree of displacement in the die can lead to reduced toughness in the transverse direction. In free rolling a slight increase in width may also occur to induce biaxial orientation and increase lateral toughness (See Chapter 8, p. 303). Nonmechanical properties are only slightly affected, such as electrical, thermal and surface properties, and moisture absorption.

The added strength arises from increased orientation (Chapter 8, pp. 300 and 304), which in turn results from molecular movements. Some evidence of these molecular shifts is clearly visible. For example, in a crystalline preform a photomicrograph of a section viewed with crossed light-polarizers shows the spherulitic structure typical of a product solidified from a melt. When the preform is displaced in a die, this spherulitic structure streaks out in shear. Where high displacement occurs in a fully oriented article, the spherulitic structure disappears completely. For this reason, oriented nylon appears translucent, almost transparent, when compared with unoriented nylon.

Another forming effect is springback, which may occur as the article is ejected from the die. Springback is elastic return in the direction of displacement to relieve stresses built in during forming. The amount of springback is the difference, after discounting thermal shrinkage, between a given dimension of the shaped article and of the die in which it was shaped. Most materials, even metals, show springback, but some plastics show it in high degree. The effect depends on physical structure and molecular weight. High displacement processes, high forming temperatures, and slow forming rates tend to limit springback. Springback is a factor to reckon with in design, so that adequate controls can be provided. Most springback problems are mitigated by one or more of the following measures.

1. Increase preform temperature so that it is close to but below the crystalline melting point.
2. Choose a resin of high molecular weight and crystallinity.
3. Use fillers and reinforcing aids to the extent the application will allow.
4. Overform (extend dimensions) where the application permits.
5. Use lubricants, especially in press forming and die drawing.
6. Provide dwell time as needed by the formed article.
7. Select the forming process most suitable for the article.

FORMING PROCESSES

Although the art of forming useful nylon articles by nonmelt methods is relatively new, a wide range of processes has been scouted. The list of nonmelt processes and examples described here is not complete but is intended to illustrate areas of utility. The terminology is borrowed mainly from metals technology because the machines and tools are those now used in metal fabrication.

Forming processes may be classified as producing continuous, elongated products such as the various drawing, rolling, and swaging operations. Or, they may be classified as producing discrete articles. The latter class includes coining, extrusion, forging, heading, press forming, roll threading, and shearing. The process of heading is discussed in Chapter 17.

Die Drawing

Die drawing consists of pulling a tube, rod, or strip through a die of smaller diameter, reducing the outside dimensions, and increasing the length. For drawing most unreinforced nylons, both feed and die are operated at about 95°C minimum and the draw product quenched in air or water. In continuous operation, with a wire-and-cable type caterpillar or other hauling device providing the drawing force, linear running speed can reach 150 to 200 ft min^{-1} (about 46 to 61 m min^{-1}). The operation results in improved toughness and dimensional control. Generally, these property changes are larger than result from manipulating the conventional melt extrusion processes. Die drawing differs from free drawing in both process details and results because the die exerts a transverse force on the workpiece not present in free drawing.

Table 15-1 shows the dimensions of a tube made of a typical high viscosity nylon-66 as extruded and after drawing through solid dies of diminishing orifice diameter. The extruded tube varied 0.046 in. (0.117 cm) OD, but after drawing it to only about 15% diameter reduction, this measurement varied only 0.003 in. (0.0076 cm). Throughout these draw steps, the wall thickness remained almost constant.

Rolling

Rolling offers a method for orienting nylon starting with extruded rod, ribbon, or sheet. The process may be applied to improve rods, sheets and ribbons over a wide range of properties, and dimensions. In one example, Miles (5) rolled a sheet of nylon-66 (number-average molecular weight about 8450) in water at room temperature. The sheet increased about 600% in area and about 100% in tensile strength in the rolling direction. The rolled sheet was more transparent than the unrolled. Orientation was indicated by x-ray examination (see Chapter 8, p. 304).

Swaging

Swaging is the process of applying hammerlike blows radially to a preform. The preform, normally a tube or rod, is introduced between a pair of shaped dies which reciprocate rapidly in a spindle slot while revolving around the work. The process is illustrated in Figure 15-2. Swaging machines include the rotary type, which rotates the preform, and the stationary spindle type, which does not rotate either the dies or the workpiece.

Rotary swaging reduces the tube or rod to a round of smaller diameter. It may also give a conical section or other symmetrical shape. To accomplish this, the dies move outward by centrifugal force and then are driven inward by the backers striking the rolls. The dies may clap against the preform 1800 times per minute in large machines or up to about 4000 times in small machines. The process may operate continuously as in the straight reduction of tubular diameter. Or, it may operate on the ends of relatively short pieces to provide tapers, contours, and shoulders, and to attach ferrules and other fixtures. In metals technology, swaging may be used to form handles, furniture legs, ball-point pens and automatic pencils, punches, and to make control cables. Swaging can accomplish many jobs that are difficult by other processes, such as internal threads in tubing and deep drawn, thin-walled cups.

The stationary spindle swaging machine differs from the rotary in both operation and capabilities. The dies have the same radial reciprocating motion but do not rotate about the work. Also, the work being swaged does not rotate. Feeding forces are less than those of rotary swaging.

Table 15-1. Dimensions of die drawn nylon-66 tubing

Draw no.	Die size (in.)	OD (in.)	ID (in.)	Wall (in.)	Reduction in OD, %
0 (extruded)	- -	0.486-0.532	0.314-0.353	0.080-0.093	- -
1	0.450	0.473-0.481	0.307-0.314	0.080-0.091	6.3
2	0.400	0.432-0.435	0.276-0.279	0.080-0.093	14.9
3	0.348	0.378-0.384	0.206-0.218	0.088-0.098	25.1
4	0.312	0.340-0.345	0.108-0.180	0.088-0.095	32.8
5	0.275	0.302-0.306	0.130-0.134	0.087-0.094	40.2
6	0.255	0.282-0.285	0.116-0.120	0.091-0.096	44.4
7	0.238	0.260-0.261	0.085-0.090	0.088-0.094	48.8

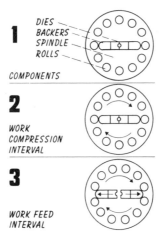

1

DIES
BACKERS
SPINDLE
ROLLS

COMPONENTS

2

WORK
COMPRESSION
INTERVAL

3

WORK FEED
INTERVAL

Fig. 15-2.   Schematic illustration of rotary swaging process.

The swaging action delivers tremendous energy from comparatively light-weight machines. For example, one commercial machine weighing 625 lb can deliver the equivalent of 8 tons of force. It does this by a great many light blows falling rapidly on the workpiece rather than a few massive blows. Swaging causes surface flow of the resin under pressure so that a burnishing action results. Thus, surface defects, tool marks, dents, and scratches may be smoothed away. Close tolerances are attainable in swaged articles, and there is no waste of material unless end trimming is needed.

Applied to nylons, swaging can shape elongated structures such as knitting needles and fasteners. It can thread articles and attach fixtures such as ferrules.

Operating temperature may be as low as room temperature, if the workpiece is reasonably moist, for good formability. To avoid crumbling, dry work should be swaged at temperatures of 65 to 120°C. As expected, internal friction raises the temperature of the workpiece so that at about 150°C the stabilizing "heat set" results.

In one example, a rod of nylon-66 when swaged from 0.5 to 0.25 in. (1.27 to 0.63 cm) in diameter, increased from 8000 to 20,000 lb (wt) in.$^{-2}$ (566 - 1415 kg (wt) cm^{-2}) in axial tensile strength.

Coining

Coining is a squeezing operation in which a well-defined imprint of the die is made on the workpiece. In many applications the workpiece surfaces may be completely confined or restrained by the die. Coining may include special operations, such as sharpening or changing a radius or profile, imparting an integral hinge, and toughening a projection. Normally, surface details of the dies are copied precisely, and, hence, the process may be well-suited to the manufacture of interlocking fasteners, printing characters, and similar articles which require dimensional accuracy.

In principle, any preform or workpiece that can be loaded above the compressive yield strength can be coined. Dry, unreinforced nylon-66 would require threshold pressures of about 10,000 lb (wt) in.$^{-2}$ (700 kg (wt) cm^{-2}) whereas glass-reinforced nylon would need over 26,000 lb (wt) in.$^{-2}$ (1840 kg (wt) cm^{-2}) based on published figures for compressive yield strength. As the yield strength increases, the area that can be coined using the same pressure is proportionately lower. In practice, up to three times the load calculated from the compressive yield strength may be needed to assure adequate resin movement to fill the die.

Coining should find extensive use where toughness and flex durability are needed, such as in hinged articles. Coined hinges in molded or extruded articles of nylon and acetal resin have been reviewed by Mengason (6). To coin hinges the workpiece is placed in a fixture and compressed until the hinge section reaches the desired thickness. The coining cycle is one second or less per tool mounted in a mechanical, pneumatic, or hydraulic press.

Good physical properties result when the workpiece is at ambient temperature (about 23°C). However, heated workpieces coin faster and at lower forces. Also, heat (up to about 150°C) is needed for nylon-66 when the workpiece is over 0.060 in. (0.15 cm) in thickness. Somewhat lower flex durability results at high workpiece temperatures.

When coined hinges are flexed, a very little drop in tear strength occurs until small fatigue cracks appear on the surfaces, and then tear strength may drop off rapidly. Coined hinges remain tough and flexible at temperatures as low as -40°C with only slight increase in stiffness.

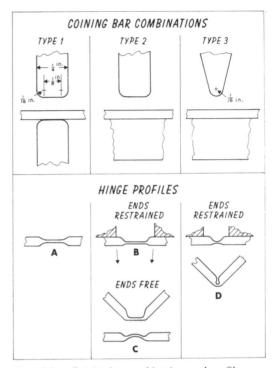

Fig. 15-3.   Coining bar combinations and profile.

For coining hinges, the die includes bars, a fixture to hold the bars, and a press to supply forming force. The bars, of hardness about R.C 45, should be polished and free of surface irregularities. Figure 15-3 shows three examples of coining bars and the hinge profiles produced by each. Type 1 bars produce hinges with high flex durability. The bars are generally of identical profile, each with flat coining surface and well-rounded corners. Increasing the width of the flat surface tends to raise flex durability, since the wider hinge has a larger bending radius and therefore lower outer fiber stress. Bars shown are generally wide enough to coin most hinges. For a bend of 180° in a coined article of over 0.60-in. (1.5-cm) original thickness, the flat should be increased to attain a larger bending radius.

In some applications an offset hinge may be needed. One method of offsetting is to coin the workpiece against a flat surface instead of another bar, as shown in Type 2 of Figure 15-3. Two different hinge profiles can result. Profile B results when the ends of the workpiece are restrained. The preferred flex direction is shown by the arrows. Profile C, shown in the formed and flattened positions, results if the workpiece ends are not restrained. Generally, offset hinges have less flex durability than centered hinges. Hinges from Type 3 bars are coined with less force than those from Types 1 and 2 but have the lowest flex durability because of smaller bending radius and resultant higher fiber stresses. Type 3 is useful when a relatively sharp 90° external corner is needed (profile D).

Typical of the commercial applications already developed in the United States is a tape measure marketed by H. K. Porter Co., Inc., Pittsburgh, Pa., under their Disston trademark. The housing is molded flat in one part, and two hinges are then coined to form the base and two sides of the case. The design reduces the number of parts and simplifies assembly. The tape measure, along with typical coined hinges in test panels, are illustrated in Figure 15-4. Another example of coining is shown in Figure 15-5, which shows sharply defined letters formed by squeezing nylon into a die. Such coined letters, numbers, and figures should find use in print band assemblies and character pads or wheels for business machines. The advantage of coining over injection molding is the convenience of switching from one pattern to another without expensive mold construction.

Extrusion

Here the term "extrusion" is used as in metals technology. A preform or work piece enters a die at a temperature below the crystalline melting point. Usually there is a significant temperature rise during deformation because of internal friction, but in no case is the temperature allowed to reach the melting point. Extrusion may be backward or forward or combined backward-and-forward. The term forward or backward refers to movement of resin relative to the punch travel. Guided by the die, the preform moves under steady, although not uniform, pressure.

In normal operation, a preform confined and supported by a die body is pressed by a punch into the die cavity. The required relative motion between punch and die usually results from the die being attached to the stationary bed and the punch to the reciprocating ram. The tool may deliver pressure rapidly as a sharp blow when mounted in a mechanical punch press; or, it may deliver pressure more slowly by a squeezing action when mounted in a hydraulic press.

Although the "cold" extrusion of plastic articles is relatively new, many principles of extruding metals apply reasonably well. The shape of the article usually determines the extrusion process. For example, cuplike articles may be produced by either backward or forward extrusion, and shaftlike articles and hollow shapes form well in forward extrusion. Resins of good ductility may form well in one-step extrusion, but others may require two or more steps with annealing and dwell time to reach a stable form.

Extrusion applies best when a large quantity of identical parts must be produced. Obviously, quantity requirements determine the extent of automation

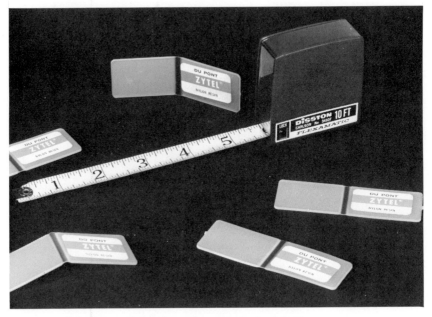

Fig. 15-4.   Coined tape rule housing and test panels of nylon-66.

that can be supported. At low production levels, some machining to supplement extrusion may prove economical. Also, extrusion may be combined with other operations, such as heading, to form articles of different diameters. Articles with large differences in cross-sectional areas and weight distribution are difficult to form economically from preforms of the smallest or largest diameter of the article. Instead, the most economical procedure is to select a preform of intermediate diameter, to reduce area in forward extrusion and to form larger sections by heading.

Figure 15-6 illustrates a die suitable for forward extruding a small cup using a melt extruded rod as a preform. When mounted in a mechanical punch press, the die has performed well for scouting preform shape and operating conditions. Well-formed cups were made from a nylon-66 with the preform at 210°C and the die at about 190°C. Die temperature was maintained by a resistance heater controlled by a thermocouple sender. The total forming force was 8000 lb (3,630 kg) corresponding to pressures of about 40,000 lb (wt) in.$^{-2}$ (2830 kg (wt) cm^{-2}) of preform cross section. The mechanical press rate was 90 cycles per minute. Nylon-66 extrudes well at preform temperatures of 175°C.

Figure 15-7 shows a golf tee, preform, and typical die for scouting operating conditions. The golf tee is an example of a stepped shaft. Preform temperature was 175°C for nylon-66. Both mechanical and hydraulic presses form well-

Fig. 15-5.   Raised letters formed by coining nylon-66.

shaped and dimensionally stable tees. The golf tee shows the expected increase in axial tensile strength and modulus (Table 15-2). Dimensional changes after boiling in water are very small, possibly because the effect of water absorption was balanced by recovery.

Forging

Forging is generally defined (7) as "essentially a process for enlarging and reshaping some of the cross-sectional area of a bar, tube, or other product form of uniform, usually round, section." Here, the term applies to reshaping a plastic preform under heat and pressure with moderate deformation, for example, gears from tube or rod sections. Forging methods are classified as open-die, closed-die,

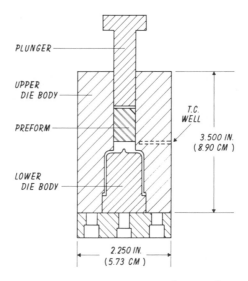

PLUNGER

UPPER
DIE BODY

PREFORM

T.C.
WELL

3.500 IN.
(8.90 CM)

LOWER
DIE BODY

2.250 IN.
(5.73 CM)

Fig. 15-6.   Forward impact extrusion, cup die.

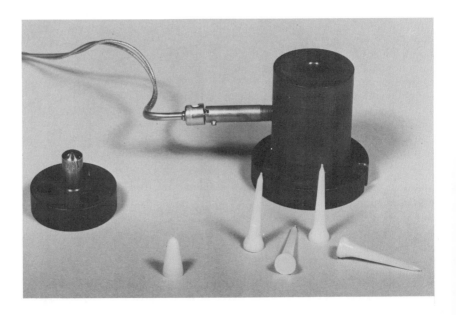

Fig. 15-7.   Golf tee, preforms, and scouting die.

Table 15-2. Properties and dimensions of forward extruded golf tee

	Nylon-66	
	Injection molded	Formed golf tee
Beam modulus, lb(wt)in.$^{-2}$	140,000	250,000
Stress at break, lb(wt)in.$^{-2}$	11,000	33,000
Elongation at break, %	300	60
Dimensional change after immersion in boiling water:		
Control length, in.[a]		1.869
Length after 5 min., in.[b]		1.868
Length after 30 min., in.[b]		1.871
Length after 60 min., in.[b]		1.870
Length after 60 min., in.[a]		1.869

[a] Measured at room temperature.

[b] Measured while still hot.

upset and roll forging, and ring rolling. Open die methods generally are limited to small deformations and preforms of low L/D ratios. The reason for limited use of open-die methods is the relatively low tensile strength of plastics, which may lead to cracks and splits if the preform is not supported.

Closed die forging has many possibilities at temperatures just below the crystalline melting point. In a closed die the preform moves and is reshaped under compressive forces without fracture. Usually only the forward surface may move unsupported to fill the die, and for some articles even those surfaces may be supported by rubber or other energy absorbing die components.

Figure 15-8 illustrates a typical scouting die to forge a gear of $2\frac{1}{2}$ in. (6.35 cm) diameter and $\frac{1}{2}$ in. (1.3 cm) face width. The preform is a cylinder of $2\frac{1}{4}$ in. (5.7 cm) OD, $\frac{3}{8}$ in. (0.95 cm) ID and $\frac{1}{2}$ in. (1.3 cm) height. With the preform of nylon-66 at 150°C about five seconds dwell time is needed in a hydraulic press. Such gears are 2 to 4 times as strong as molded gears when tested to failure in a static torque test (Fig. 15-9). In this test, the usual simple tooth shear does not occur; instead, massive destruction of the entire gear occurs at higher stress levels (Fig. 15-10). This failure pattern is evidence of radial orientation.

"Injecto Forming"

"Injecto-forming" is a dual-function forming technique by which a preform is molded in a cavity while a previously molded preform is simultaneously forged into a strengthened article in a separate die cavity of the same tool. The technique uses the normal clamping force available in every injection molding

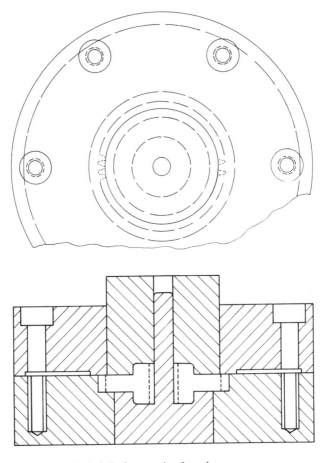

Fig. 15-8.   Typical die for scouting forged gears.

machine to provide stamping or forging pressure. No changes in the molding machine are needed, but the process does require a tool having both mold and die functions. Since the preforms need not be precisely shaped, the mold may be relatively simple. Figure 15-11 illustrates an "injecto-forming" tool designed and operated for demonstration.

Press Forming

Press forming is the term used for the process of shaping sheets by punch and die methods. Tensile, compressive, shearing, and bending forces, or combinations of

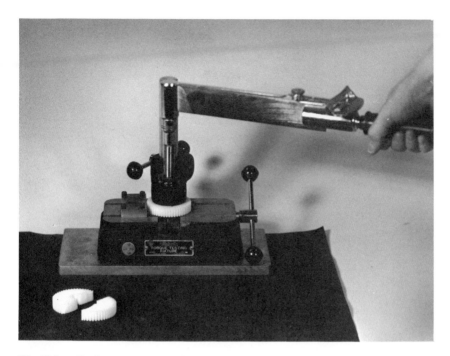

Fig. 15-9.   Static toruqe tester and test gear of nylon-66. Torqued to failure.

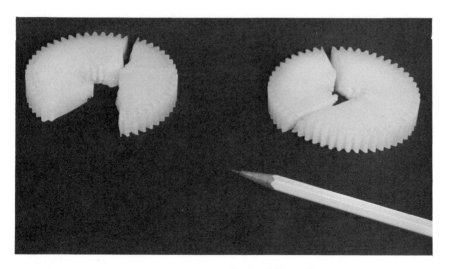

Fig. 15-10.   Failure pattern of torque gears of nylon-66 in static torque test.

527

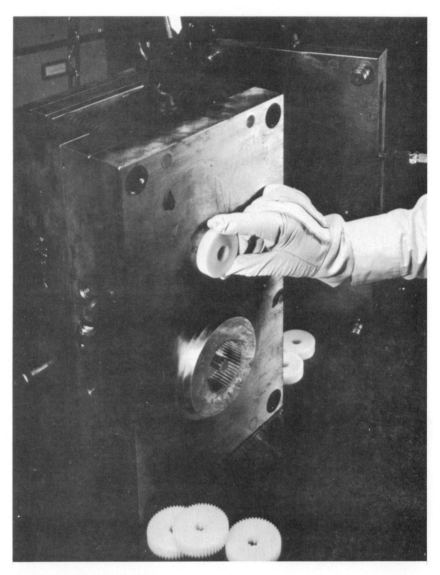

Fig. 15-11. "Injecto-Forming" tool.

these, may be required. Free stretching of the nylon may occur when the sheeting is subjected to high tensile forces to form some articles. Compound tools may perform cutting and trimming operations as a part of the shaping cycle. Rubber pad forming is a variation of press forming in which a block of rubber or polyurethane replaces one of the dies in a matched die set. The rubber block should be at least three times the depth of the article formed. Experience shows that rubber pad forming works well for articles of large area and for depths up to about 2 in. (5.1 cm).

Figure 15-12 illustrates springs of conical section (Belleville) and other shapes press formed of extruded sheeting of high molecular weight nylon-66. The Belleville springs serve as handle bearings typical of those commonly used as both bearing and washer in automobile door handles. The forming die illustrated in Figure 15-13 is of simple design and provides for clamping the sheet, forming the crown, and cutting the center and outer circles. The forming parts are of Vega steel. The sequence and force requirements are shown in Figure 15-14, as taken with the die operating in the Instron Tester. The maximum force recorded is 4300 lb (1950 kg). In operation the die was mounted in a 38-ton Walsh mechanical punch press operating at the rate of 90 cps. The die was at ambient temperature and sheeting at 60 to 80°C. There was no attempt to control moisture content of the sheet.

Roll Threading

Roll threading is a forging process applied to forming threads. As shown in Figure 15-15, the threaded faces of hardened steel dies are pressed against the

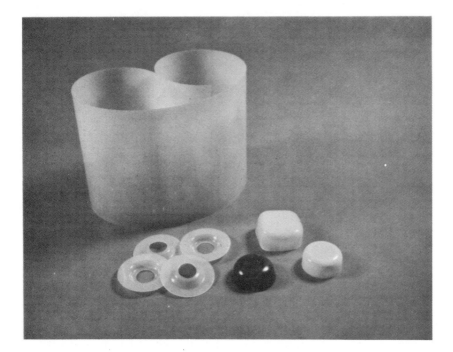

Fig. 15-12.   Extruded sheet of nylon-66 and press formed articles.

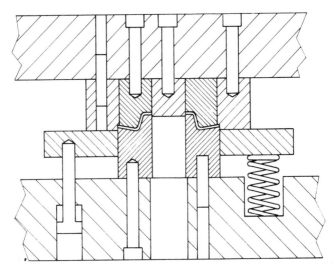

Fig. 15-13.   Die for press forming Belleville springs.

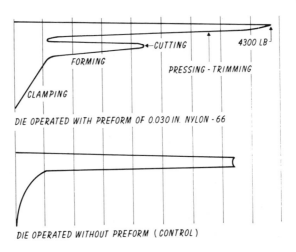

CHART SPEED = 2 IN./MIN   (0.847 MM/SEC)
X HD.   "   = 0.2 IN./MIN (0.0847 MM/SEC)
FULL SCALE = 5000 LB   (2270 KG)

Fig. 15-14.   Force versus traverse to form handle bearing.

workpiece, reforming the surface into threads as the workpiece rolls on the die faces. The die faces have a thread form which is the reverse of the thread to be produced. The die action is to penetrate the surface, displacing material to form the thread roots and forcing displaced material outward to form the thread crests.

In contrast with cut threads, rolled threads have improved strength, accuracy

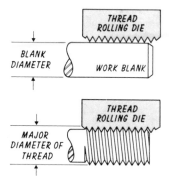

Fig. 15-15.   Schematic description of thread rolling.

and surface finish. They are produced at high rates with no material waste. Normally, the workpiece diameter is between the major and minor thread diameters.

For nominal 1/2 X 20 threads (20 threads per in.) a rod of nylon-66 resin 0.468 in. (1.2 cm) in diameter was threaded in a Reed three-cylinder die hand-fed machine (8). Oil temperature was 65°C, although this may be in the 40 to 120°C range with satisfactory results. Within 1 sec., the rod was threaded for 1 in. (2.54 cm) of end length, and reversed for threading the other end. The thread diameter increased or "built" to 0.492 in. (1.25 cm). Similar rods of glass-reinforced nylon (20% glass content) were threaded at these conditions and with equally good results.

When pulled to failure in an Instron tensile tester the above threaded nylon-66 bolt broke at 1840 lb (835 kg) tensile force without thread stripping (single nut chucked at each bolt end). The bolt of glass-reinforced nylon failed without thread stripping at 1880 pounds (852 kg) but with slightly lower elongation than unreinforced nylon. In other tests, machined threads frequently stripped when bolts of similar size and thread configuration were pulled to failure.

Shearing

Shearing is the cutting of plastic sheet between die components by which the sheet is stressed in shear between two cutting edges to the point of fracture. Shearing is commercially practiced to manufacture a number of nylon products, such as retainer rings. Figure 15-16 illustrates the process and defines terms for the die parts. Also, the Belleville springs (Figure 15-12) of press formed nylon-66 illustrate a shearing operation in both the center hole and trimming steps.

In shearing, the plastic sheet, subjected to tensile stresses on the die side and

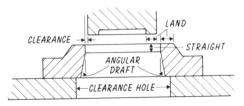

Fig. 15-16.   Parts of a typical shearing die.

compressive stresses on the punch side, stretches beyond the elastic limit. Plastic deformation and reduction in cross-sectional area follow. Punch penetration occurs before fracturing starts and reduces the cross-sectional area through which the cut is being made. Finally, fracturing starts through cleavage planes in the reduced area and becomes complete. Fractures should begin at both upper and lower cutting edges. If the clearance (distance between punch and die cutting edges) is correct for the resin being cut, these fractures spread toward each other and eventually meet, causing complete separation. Further punch travel carries the cut plug through the sheet stock and into the die opening.

Critical factors in the shearing operation include clearance, angular draft, land, straight, shear, and ductility. Proper clearance between cutting edges enables the fractures to meet, and hence the fractured portion has a clean surface. With too little clearance, more than one band of resin may be cut before separation occurs. With too much clearance, pinch-off may result in jagged edges. Generally, hard materials require more clearance and also may leave a burr on the die side of the sheet. Obviously nonhomogeneous sheeting may result in jagged edges, even with proper clearance.

Angular clearance enables the blank or slug to clear the die as shown in Fig. 15-16. It is usually ground 1 to $2°$ per side, depending on sheet thickness. Land is the flat surface adjacent to the die cutting edge. Its purpose is to reduce area to be ground or reground to keep a sharp cutting edge. Straight is that part of the die surface between the cutting edge and the angular clearance in a blanking or punching die. Experience suggests a minimum straight wall height of 1/8 in. (0.32 cm) on all materials less than 1/8 in. thick (0.32 cm). As a general rule, straight wall height for thicker sheeting may equal sheeting thickness.

Shear is the amount of relief ground off the punch face, mainly to reduce the shearing force, to limit stress on the tool, to enable thicker or more resistant sheeting to be punched on the same press or to allow use of lower rated presses. In other words, shear is a convenient means to control forces but not work needed in a cutting operation. Shear is normally defined as a fraction of the sheet thickness to be cut. For example, a shear of $t/3$ is one third the thickness, t, of the sheeting. The relationship of shear to punch load is shown in Figure 15-17.

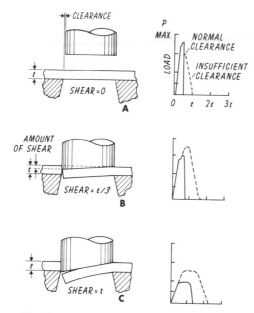

Fig. 15-17.   Loads for various degrees of shear.

Shear need not be limited to one-way punch shear. It may also be concave or convex and located in the die, or it may be concave and located on the punch. Or, it may be a balanced two-point shear located on the punch. Other shear patterns are possible and useful, according to the sheeting and configuration to be cut.

REFERENCES

1. Ito K., *Mod. Plast.* **43** (8), 154 (1966).

2. Warshavsky, M., and N. Tokita, *SPE J.* **26** (8), 55 (1970).

3. Broutman, L. J., and S. Kalpakjian, *SPE J.* **25** (10), 46 (1969).

4. Youngman, E. A., F. E. Condo, and P. M. Coffman, *SPE Tech. Pap.* **14**, 1–D (1968).

5. Miles J. (to E. I. du Pont de Nemours and Co.), U. S. Patent 2,244,208 (June 3, 1941).

6. Mengason, J., *SPE J.* **25,** 72 (1969).

7. Anon., Metals Handbook, *Forming,* Vol. 4, Ed. 8, American Society for Metals, Metals Park, Ohio, 1969.

8. Anon., "Thread and Form Rolling," Reed Rolled Thread Die Co., Holden, Massachusetts, Rolling Data 1-6A, 1964.

CHAPTER 16

Nylons as Binder Polymers

J. W. SPRAUER AND J. R. HARRISON

INTRODUCTION

"Binder polymers" are polymers used in adhesives, coatings, and miscellaneous applications which involve bonding materials together. They constitute a use area of high-molecular-weight polymers not encompassed by the major polymer areas of films, fibers, plastics, or elastomers.

"Nylon" is a coined word which has come into generic use to designate high-molecular-weight linear polyamides capable of being formed and drawn to high-tenacity fibers. From early work, the number-average molecular weight required to meet this definition is at least 10,000 (1,2). By this definition, the

535

polyamides prepared from dimerized vegetable oil acids and ethylenediamine and the like (3, 4, 5) are clearly not nylons, being highly branched and of relatively low molecular weight and being amorphous or of low degree of crystallinity. It has become common in the trade to refer to these materials as "polyamides." A less ambiguous term, "fatty polyamides," has been suggested by Peerman (6).

In recent years the distinction between "nylons" and "polyamides" has become blurred for three reasons:

1. Dimerized vegetable oil acids ("dimer acids") that are now available in high degrees of purity are capable of forming essentially linear, relatively high-molecular-weight polyamides (7, 8).

2. High purity "dimer acid" has been copolymerized with longer chain polymethylene diacids, diamines, and amino acids resulting in relatively high molecular weight interpolymers (9, 10).

3. There has been a proliferation of new nylon compositions having special properties (11, 12).

In general, our remarks will encompass properties and uses of nylons having binder polymer properties regardless of the exact composition. Properties and uses of "fatty polyamides" have been recently reviewed (6) and will not be considered further in this chapter.

PROPERTIES OF NYLON BINDER POLYMERS

Crystallinity

The common image of nylons arises from the properties of the homopolymers, including toughness, hardness, inertness, abrasion resistance, and solvent resistance, and depends to a large degree upon their crystallinity. Binder polymers usually require greater flexibility, lower melting point, and better solubility than the homopolymers. Modification in properties of nylons by copolymerization was recognized by early workers (13, 14, 15). Copolymerization results in disordering of chain regularity which leads to a lower degree of crystallinity. The crystal structure and morphology of nylons are discussed in Chapter 8. Crystallinity is reviewed briefly here to permit an understanding of the modifications leading to the special properties of binder nylons.

The crystal structures of nylon-66 and nylon-610 were established by Bunn and Garner (16, 17). The molecules are arrayed in a fully extended planar zigzag linked by hydrogen bonds alternately to two adjacent molecules to form sheets that are stacked in a triclinic cell structure. Such a structure is disordered by random introduction of polymethylene chains of different lengths, and by head to tail inversion of the amide group through copolymerization of amino acid and diamine-diacid combinations. Infrared evidence (18, 19) shows that the amide groups maintain essentially full association, by hydrogen bonding and dipole-dipole interaction, even into the melt. Hence, chain disorder results in contraction and twisting of the hydrocarbon chains away from the fully extended planar zigzag array. Other methods of disordering, and thereby of reducing crystallinity, are blocking of hydrogen bonding by N-substitution (20-23); introducing side groups, especially bulky side groups, along the chain; and introducing bulky linking groups between the amide groups.

Disordering a given polyamide structure reduces the melting point and broadens the melting range. Crystallization of the disordered structure becomes sluggish, nucleation is slow, and the structure frequently quenches to the amorphous state upon rapid cooling from the melt. However, crystallization will almost always occur on aging, especially above the α-relaxation temperature (cf. Chapter 9); on reheating slowly; on annealing at some optimum temperature intermediate to α-relaxation and crystalline melting; on very slow cooling from the melt; or on evaporating from solution well below the melting point. Extremely disordered polymers have such diffuse melting points and such low heats of fusion that the crystallinity becomes difficult to detect by x-ray methods. But such low crystallinity can still be evaluated by differential thermal analysis (DTA) or differential scanning calorimetry (DSC).

Figure 16-1 shows DTA behavior (24) of typical nylon binder polymers on melting after careful crystallization by very slow cooling from the melt (at less than 1°C per minute). Fast cooling from the melt results in the behavior of Figure 16-2(a) showing complete absence of crystallization. Slow reheating results in the behavior of Figure 16.2(b) showing a crystallization exotherm followed by endothermic melting. On the other hand, slower cooling from the melt results in normal exothermic freezing (Figure 16.2(c)) in which the freezing point depression is somewhat dependent upon cooling rate. One may cool rapidly from the melt to the vicinity of the freezing point to observe isothermal crystallization after an induction period (Figure 16.3). Generally, there is an optimum temperature for nucleation and crystallization. One would also expect the rates to be influenced by stress-induced orientation.

Related observations have been made on nylon homopolymers but the rates are generally much faster. Rapid crystallization well below equilibrium melting temperatures has been called "cold crystallization" by Dole (25). "Cold crystallization" is necessarily somewhat incomplete. Inoue (26) has observed isothermal nucleation and crystallization of nylon homopolymers, showing an influence of thermal history well above the melting point on nucleation and crystallization rates. This effect is presumably the result of retention of some order in the melt and diminishes at high melt temperature. Liberti and Wunderlich (27) have studied the effect of history and fast heating rates on nylon-6 melting points.

The nature of the crystallinity of nylon-66 and the effect of crystallinity

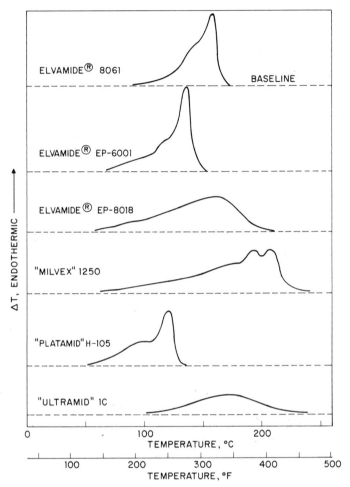

Fig. 16-1. Thermograms of typical nylon binder polymers (Du Pont 900 DSC; normalized 25-mg sample, 0.5°C/in., 10°C/min heating).

upon its properties has been studied by Starkweather and others (28-31; cf. Chapter 9). Density and infrared data indicate a typical degree of crystallinity of about 40 to 55%, depending upon thermal history. These data combined with calorimetric data (32) lead to an estimate of 45 cal/g for the heat of fusion of perfectly crystalline nylon-66 (33).

In comparison, the disordered nylons having desirable binder polymer properties have diffusely melting crystallinity, as shown in Figure 16-1. If the crystallinity is highly developed, binder nylons have heats of fusion in the range of 5 to 20 cal/g.

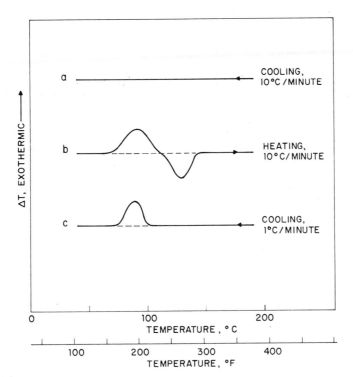

Fig. 16-2. Thermograms of sluggish crystallization of a nylon-6/66/610.

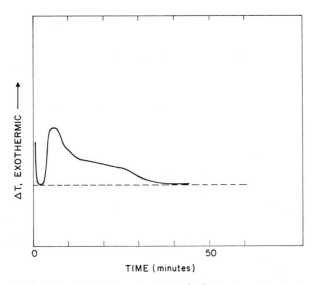

Fig. 16-3. Isothermal crystallization of a nylon-6/66/610 (Du Pont 900 DSC; 65°C).

539

The second-order transitions of nylons are reviewed in Chapter 9. The α-relaxation (34, 35) is especially important in determining the properties of nylon binder polymers at room temperature. Structural disordering or lengthening the links between amide groups lowers the α-relaxation temperature as compared to nylon-66. Absorbed water at ambient humidity substantially further depresses the α-relaxation temperature and plasticizes the composition. Most useful nylon binder polymers have their α-relaxation near or below room temperature, so that they are normally in a leathery or rubbery state. Nylons containing large proportions of aromatic or alicyclic ring structures between the amide groups have high glass transition temperatures and are consequently too hard, stiff, and inflexible for most binder polymer requirements.

The β- and γ-relaxations of binder nylons have not been extensively studied but appear to be little changed from those of the homopolymers. These relaxations, of course, importantly influence physical properties of all nylons at low temperature.

Solubility

Solvents with strong hydrogen-bonding properties, especially strong hydrogen donors such as formic acid, phenols, mineral acids, and fluoroalcohols, dissolve nylon homopolymers. Because of the crystallinity, higher temperature generally favors solution. For example, even nylon-66 is soluble in methanol under autogeneous pressure at 180°C, and benzyl alcohol and similar high boiling alcohols dissolve many nylons at the normal boiling point.

The disorder, that is, reduced crystallinity, of nylon binder polymers has a major influence in increasing solubility in common solvents at relatively low temperatures. For example, the more disordered nylon-6/66/610 terpolymers are soluble in warm methanol, ethanol, etc. (15). The solutions are metastable at room temperature, though time to gel varies widely with polyamide composition. The gels are presumably crystalline and hydrogen bonded; gelation is reversible by heating. Probably in all cases there is some temperature below the melting point at which the nylon and solvent are miscible in all proportions, though this has rarely been defined because the solutions become extremely viscous above 20 to 40% concentration. Molecular weight (at least above about M_n = 10,000) has little influence on equilibrium solubility, but can influence rate of solution.

Mixtures are frequently better than single solvents. That is, they give more gel resistant solutions at a given temperature, dissolve more nylon at equilibrium at a given temperature, and probably attain complete miscibility with nylon at lower temperature. Alcohol-chlorohydrocarbon mixtures, such as methanol and methylene chloride, ethanol and chloroform, or methanol and trichloroethylene, are especially effective. Mixtures of alcohols and aromatic hydrocarbons dissolve interpolymers containing large hydrocarbon groups, such as "dimer acid" copolymers. Addition of water retards gelling of alcohol solutions of nylon-6/66/610 terpolymers though in some cases it also retards rate of solution and reduces solubility.

Polymer solubility has been the subject of several reviews (36-39). Nylon solubility probably can be rationalized in terms of thermodynamics (that is, solubility parameter) and the empirical hydrogen-bonding and polarity parameters, but his has not yet been systematically done. Nylons, in general, have high solubility parameter (38, 40), but the solubility is clearly dominated by crystallinity and hydrogen-bonding factors.

Physical Properties

The sluggish crystallization of nylon binder polymers renders their properties sensitive to thermal history. The low crystallinity, in any case, has very important influence on physical properties. Some hardness is retained, more or less proportional to the crystallinity. Crystallization eliminates tack and raises the yield point. It also leads to effective orientation under stress resulting in "cold drawing" behavior and undirectional strengthening. Low crystallinity results in high strain at high stress, that is, in high "work to break," equivalent to high toughness.

In a number of instances we have found that quenched or partially quenched materials crystallize under stress within the normal time of physical testing, substantially contributing to the observed physical properties. Apparently, the low crystallizability permits stress-induced crystallization, or crystal reorganization, on a reasonable time scale without development of internal flaws.

While much has been published on structure-property relations of nylons, specific information on nylon binder polymer compositions is indeed sparse (41). Commercial compositions are frequently proprietary.

Table 16-1 shows comparative typical data on several nylon binder polymer compositions (24). Physical properties were observed by appropriate ASTM procedures, where available. Heats of fusion were observed on dried, encapsulated, annealed samples (that is, program-cooled at $0.5°C$ min^{-1} or less) in the Du Pont 900 DSC. Comparative penetrometer softening temperatures were observed on samples melted in a thin layer on aluminum foil on an open hot plate well above the melting point, rapidly quenched by transfer of the foil to a dry-ice cooled aluminum block, and transferred cold to a Du Pont Thermal Mechanical Analyzer (Du Pont 900 penetrometer accessory). This is an adaptation of a procedure originally described by Edgar (42). The penetrometer softening temperatures correspond approximately to the α-relaxation temperatures observed at low frequency.

Definition of melting points of polymers having low diffusely melting

Table 16-1. Comparative typical product data on nylon binder polymers

Polymer:			Elvamide® 8061
Company:			Du Pont
Composition:			6/66/610
	ASTM test procedure	Sample conditioning	
Initial modulus, lb(wt)in.$^{-2}$	D1708	a	33,000
Yield, lb(wt)in.$^{-2}$	D1708		2,200
Elongation at yield, %	D1708		16
Tensile, lb(wt)in.$^{-2}$	D1708		7,400
Elongation at break, %	D1708		300
Hardness, Shore D	D2240	a	64 to 70
Abrasion loss, mg (CS17, 1000 g 1000 cycle)	D1044	a	4
Brittleness temperature, °C	D764	a	−56
Water absorption, % gain (equilibrium, 98% R.H., 23°C)	- -	b	10.6
Solution stability, hr (time to gel at 20% concen.)	- -	b	
in methanol	- -		70
in $CH_3OH/CHCl = CCl_2$ (50/50, w/w)			> 200
Typical melt index, dg min.$^{-1}$ at °C (2160 g load)	D1238	b	5 at 190°C
DTA approximate melting range, °C	- -	b	90-165
Minimum flow temperature, °C	- -	b	158
Heat of fusion, cal g^{-1}	- -	b	9
Penetrometer softening temp., °C	- -	b	25

[a] Crystallized samples, 50% R.H., 23°C.

[b] Essentially anhydrous samples.

[c] At 5% conc; 8-10% dissolves at reflux.

crystallinity, such as shown in Figure 16-1, obviously requires more than one parameter. The area of the DTA melting endotherm is a measure of the heat of fusion. The temperature at the DTA maximum has no special significance, at least not with the less crystalline polymers. The temperature at onset of melting has significance in defining the lower limit of the melting range, though it is difficult to determine with precision. The upper limit of the melting range has

Table 16-1. Comparative typical product data on nylon binder polymers (continued)

Elvamide® EP-6001 Du Pont --	Elvamide® EP-8018 Du Pont --	"Milvex" 1250 General Mills "Dimer Acid Based"	"Platamid" H-105 Dr. Plate 6/12/+	"Ultramid" 10 BASF 6/66/PACM6
33,000	51,000	76,000	30,000	124,000
2,100	2,600	4,900	2,000	5,300
16	23	24	20	6
9,600	6,100	6,000	8,000	6,100
630	400	310	550	300
64 to 71	69	77	68	79
5	--	--	--	--
−69	--	--	--	--
8.9	2.3	0.8	7.0	14.4
0.5	Insol.	Insol.	0.5	25
3	30^c	Insol.	100	17
3 at 190°C	4 at 230°C	7 at 230°C	20 at 190°C	30 at 230°C
70-147	60-205	60-230	55-130	100-220
138	190	200	119	188
8	13	12	5	8
30	12	5	7	45

greater interest, but this is also difficult to determine with precision in DTA, requiring uncertain and laborious correction for the logarithmic decay of the differential temperature to the baseline (43). Nonetheless, it is useful to list an apparent DTA melting range, bearing in mind these limitations. Determination of the upper limit of the melting range on the microscope hot stage between crossed polarizers (ASTM D2117) also becomes very difficult and subjective at very low crystallinity. It has been found that the upper limit of the melting range can be objectively defined by placing a sample in the melt index apparatus (ASTM D1238) with the ASTM orifice and 2160 g load at a temperature about 20°C below the upper melting limit, program-heating at 1°C min^{-1}, and observing the temperature at which polymer begins to flow from the capillary orifice exit. This is called "minimum flow temperature." With substantially crystalline, sharp melting polymers, the hot stage microscope melting point and the minimum flow temperature coincide.

USES OF NYLON BINDER POLYMERS

As discussed in preceding paragraphs, to be useful as binder polymers, nylons require modification in the direction of lower melting point and better solubility, primarily to facilitate application, and generally also in the direction of better toughness and flexibility. Most uses fall into three major categories: thermosetting adhesives, thermoplastic adhesives, and solvent-applied binders.

Thermosetting Adhesives

One of the outstanding binder polymer uses of nylon is in "nylon-epoxy" adhesives, which are thermosetting structural adhesives widely used in the aerospace industry. Nylon-epoxy refers to various proprietary formulations of nylons and various epoxy resins (44-48).

Although nylon-epoxy adhesive formulations are proprietary, a composition having reasonably typical performance has been described by Gorton (49) comprising the following:

Zytel® 61 nylon resin*	86 parts
Epon® 828 epoxy resin†	12.7 parts
Dicyandiamide	1.3 parts

Dicyandiamide is a well-known epoxy resin curing agent which is relatively inactive at low temperature and is activated by heat. Gorton presents evidence that this composition cocures, that is, the epoxide reacts simultaneously with the nylon and the curing agent, at least under his cure conditions of 350°F for 1 hr.

Most nylon-epoxy adhesives are formulated in solution and cast and dried to produce an adhesive film. This is primarily a matter of convenience. Melt formulation at limited time-temperature exposure appears practicable. Solutions of the nylon-epoxy in proprietary formulations are used for primer applications.

Nylon-epoxy adhesives provide uniquely high adhesive performance (lap shear and peel strength) over a wide temperature range (44, 50) and especially down to cryogenic temperatures. Apparently they are the only adhesives to date

---

*Nylon-6/66/610 now sold by Du Pont as Elvamide® 8061.
† Shell Chemical commercial diglycidyl ether of bisphenol-A.

**Table 16-2. Federal specification MMM-A-132, type 1, class 1
required performance**

Tensile shear, 24°C (75°F), lb(wt)in.$^{-2}$	5000
Tensile shear, 82°C (180°F), lb(wt)in.$^{-2}$	2500
Tensile shear, −55°C (−67°F), lb(wt)in.$^{-2}$	5000
Fatigue strength, 24°C (75°F)	10^6 cycles at 750 lb(wt)in.$^{-2}$
Creep-rupture, 1600 lb(wt)in.$^{-2}$, 24°C (75°F), 192 hr	0.015 inch max.
Creep-rupture, 800 lb(wt)in.$^{-2}$ 82°C (180°F), 192 hr	0.015 inch max.
Tensile shear, 24°C (75°F) (after 30 day salt-water spray)	3600
Tensile shear, 24°C (75°F) (after 30 day at 49°C (120°F), 95-100% R.H.)	3600
Tensile shear, 24°C (75°F) (after 30 day immersion in water)	3600
T-Peel, 0.020 inch, Clad 2024-T3, lb(wt)in.$^{-1}$ average	50
Blister detection, 24°C (75°F), lb(wt)in.$^{-2}$	3600

capable of meeting Federal Specification MMM-A-132, Type 1, Class 1, "Adhesives, heat-resistant, air frame structural, metal-to-metal." For ready reference and orientation, these specifications are listed in Table 16-2.

For some honeycomb sandwich construction in some uses, adhesives must conform also to Military Specification Mil-A-25463, which covers strength, fatigue, and creep resistance of defined honeycomb structures. Nylon-epoxy adhesives, depending on formulation, have marginal difficulty in meeting some creep requirements.

The reasons for the uniquely high performance of nylon-epoxy adhesives have not been satisfactorily explained. Gorton (49) considered the effects of cross-linking variations, which are important but hardly sufficient per se to explain the performance. Specific adhesive forces are involved, especially in more complex proprietary formulations. Limited compatibility of nylon and epoxy suggests the probable importance of polyphase structure, varying with composition and formulation. From our own observations, cross-linking the nylon inhibits the already sluggish crystallization discussed in the preceding section. In fact, a cured unstressed structure is frequently not crystallized. Stress-induced crystallization leads to "cold drawing" to high elongation; namely, a tough failure giving high shear and peel performance. Thus, it appears the performance depends primarily upon adhesion and secondarily upon the interaction of crystallinity and cross-linking effects.

There have been reports on the sensitivity of the nylon-epoxy bond under long term exposure to moisture and to moisture under stress (51-57). Nonetheless, nylon-epoxy adhesives show a high degree of water resistance in meeting the MMM-A-132 specification, and they have a long record of successful service in the aerospace industry. Lehman and Trepel (58) found a nylon-epoxy to have outstanding performance for metal-to-metal lamination in hydrofoil construction, testing under water for five months and subjecting intermittently (one month total) to severe cyclic stress. In our laboratories, Deyrup (59) has observed a limiting applied stress below which the bond is indefinitely stable while simultaneously exposed to water. Most adhesive uses do not involve simultaneous exposure to high stress and high moisture except intermittently, suggesting that testing under water at constant high stress may be overly severe.

For some time, the aerospace industry has been seeking an adhesive with nylon-epoxy performance with milder curing conditions (60). This need arises from the adverse effects of high cure temperature on the aerospace aluminum alloys. New proprietary formulations curing at 250 to 280°F have been disclosed with indication of improved water resistance (61, 62).

The aerospace industry is unusual in the time and effort that can be justified in metal cleaning, assembly, and curing. One may assume that other applications of thermosetting nylon adhesives with less severe performance requirements and less meticulous application methods are under development.

Thermoplastic Adhesives

Can Side Seam Adhesive

Nylon binder polymers are capable of high performance as thermoplastic adhesives. An outstanding example is the complex continuous process for adhesively bonding lapped side seams of metal cans in a fraction of a second at a rate of about 500 cans per minute (63, 64, 65). Full details on the nature of the polyamide adhesive used in this process have not been disclosed. An available patent (66) shows early use of "superpolyamides" (that is, nylons) with nylon-11 being preferred. Various "dimer" acid nylon copolymers have been proposed (9, 67). Special benefit of a specific primer coating on the metal has been claimed (68). Some efforts to tailor polyester adhesives to this application have been described (69, 70).

The impetus of this development was the economic pressure to eliminate the soldered tinplate joint in order to use aluminum and untinned steel which could not be soldered (71). Early efforts were made with thermosetting compositions (72), but the economic necessity for fast application obviously favored a thermoplastic. Subsequent forming of the ends of the can and baking of the interior enamel impose severe demands on the strength of the adhesive bond and on the stability to environmental exposure. The cited references have emphasized the requirement of high peel strength.

The necessity for highest available peel strength for structural adhesive applications is not always recognized and appreciated. At the least, peel strength is a semiquantitative assessment of the resistance to stress concentration, implying also impact and fatigue resistance. The unusual capabilities of binder nylon adhesives have been demonstrated in the nylon-epoxy thermoset and the can seam applications, so that much wider use as structural adhesives can be predicted with some confidence.

A line of "dimer-acid-based" nylons has been offered with suggestion of broad thermoplastic adhesive applications on metals and plastics (73-75).

Textile "Fusibles"

A textile fusible is a fabric bearing a thermally bondable adhesive (76-78). A familiar and unsophisticated example is heat-sealable patching tape. To avoid stiffness, the adhesive layer is usually discontinuous. It can, for example, be an unsupported nonwoven web. More often, however, it is supported by one of the substrates to which it is applied as a paste from a rotogravure roll or by sprinkling a powder. The bonding operation is usually completed later, often by a different company.

This type of application is fundamentally different from what is known as fabric lamination, which is done continuously so that every piece cut is a uniform laminate. But in many garments, it is desirable to change the thickness from one part to another. Parts of a coat front, for example, are quite stiff, while others are limp. This effect is produced by bonding small pieces of fusible to the parts being stiffened, and not elsewhere. Thus, the garment manufacturer will die cut the fusible, the stiffening piece, and bond it by heat to the coat front. The adhesive is already on the diecut piece. It must, therefore, be a thermoplastic — laminating adhesives are thermosets — to permit its fusion to one piece of fabric and subsequent bonding to another.

No single adhesive will perform well on all fabrics, the better fusibles being those that work with the greater variety of fabrics. Most premium fusible resins are nylons, which are both versatile and durable. They give bonds that show little or no loss of strength after many cleaning or laundry cycles. They can be bonded on European electric presses and also, since they are plasticized by steam, on the steam presses common in the United States.

The resins are almost all interpolymers, although some very heavily plasticized homopolymers have been used. Most contain caprolactam as one of the ingredients. They are usually sold as powders (see Chapter 14).

Solution-Applied Nylons

The availability of nylons soluble in common industrial solvents opens a large number of potential uses. Industrial realization of this has been slow, no doubt because of economic considerations. Nylons with improved solubility are under development.

Thread Bonding (Thread Coating)

Nylon solutions are used to provide thin coatings of polymer on the surface of sewing thread. Both continuous filament and spun threads are bonded to keep the strands together and to provide a smoother exterior that will pass more easily through the sewing machine and the needle. In staple threads, the resin also contributes shear resistance between adjacent fibers, increasing the tensile strength. Cotton, nylon, and polyester threads can be bonded in this way for both light duty and heavy duty applications.

The resins include multipolymers such as Elvamide® 8061 (Du Pont) and postreacted resins such as alkoxymethylated nylon (Belding Chemical Industries, see Ref. 23), often plasticized or otherwise formulated. The resin in proportion of 5 to 10% of thread weight is added to the thread from an alcohol solution.

Test procedures are rather empirical, and the function of the bonding agent has not been defined numerically. The ultimate test is sewing: if the thread runs through the machine smoothly without flaking, the agent is satisfactory. Screening tests usually involve flexing or scraping the thread to check flaking and strand adhesion. There is no good test for smoothness, which is evaluated by sewing. Thus, although the functions of the bonding agent are difficult to quantify, some of the requirements are known. They include solubility, rapid drying (usually a few seconds at 200 to 250°F), lack of tack to permit subsequent removal from a bobbin without resistance, adhesion, and flexibility.

Closely related to thread bonding is the relatively small use of nylon solutions applied to cut edges of textile materials to prevent fraying and raveling.

Miscellaneous Uses

Nylons have found some use as lacquers on plastics, leather, and textiles to provide improved abrasion resistance. Abrasive resistant coatings on woods and metals are potential uses not significantly developed. Similarly, anticorrosive primers and coatings on metals are a significant potential.

Nylon solutions have been used to apply a relatively impermeable coating to the inside of rubber and plastic fuel tanks.

Composites are among our newer structural materials, owing their existence to the demand for higher strength-to-weight ratio, primarily in the aerospace industry. See Ref. 79 for a recent extensive review on epoxy resin composites. Soluble nylons have been used in various formulations with epoxies and phenolics to prepare composites. They have been used in wet lay-up, in filament winding, and in preparation of "prepregs." The preparation of "prepregs" is similar to that of adhesive films. Nylons in minor proportions flexibilize and toughen the more brittle epoxy and phenolic compositions. The systems are partially incompatible, no doubt polyphase blends, and incompletely cocured. "Prepregs" have been used, for example, in the manufacture of fishing rods and in the manufacture of aircraft interior structures.

Binder nylons have been used to prepare formulated plastic printing plates (80-82).

Low-melting nylons, melt-applied or solution-applied, form excellent color concentrates with a variety of colorants, useful especially in coloring higher melting nylon plastics. See Ref. 83 for a general discussion of colorants for plastics; see also Chapter 11.

REFERENCES

1. Carothers, W. H., and J. W. Hill, *J. Am. Chem. Soc.* **54**, 1579 (1932).

2. Bolton, E. K., *Ind. Eng. Chem.* **34**, 53 (1942).

3. Bradley, T. F., and W. P. Johnson, *Ind. Eng. Chem.* **32**, 802 (1940); *ibid.* **33**, 86 (1941).

4. Bradley, T. F. (to American Cyanamid), U. S. Patent 2,379,413 (July 3, 1934).

5. Cowan, J. C., *J. Am. Oil Chem. Soc.* **39**, 534 (1962).

6. Peerman, D. E., in *Encyclopedia of Polymer Science and Technology,* Interscience Publishers, New York, 1969, Vol. 10, p. 597.

7. Emery Industries, Inc., "Empol Dimer Acids," Cincinnati, Ohio (1967).

8. Fischer, F. M., and F. M. Linn (to General Mills), U. S. Patent 3,256,304 (June 14, 1966); U. S. Patent 3,297,730 (Jan. 10, 1967).

9. Peerman, D. E., and L. R. Vertnik (to General Mills), Can. Patent (May 27, 1969); Brit. Patent 1,189,846 (Apr. 29, 1970).

10. Anon. (to Schering A. G.), Brit. Patent 1,055,676 (June 18, 1967); 1,107,524 (May 27, 1968).

11. Sweeny, W., and J. F. Zimmerman, in *Encyclopedia of Polymer Science and Technology,* Interscience Publishers, New York, 1969, Vol. 10, p. 483.

12. Sweeny, W., in *Kirk-Othmer Encyclopedia of Chemical Technology* 2nd ed., John Wiley & Sons, Inc., New York, 1968, Vol. 16, p. 1.

13. Fuller, C. S., W. O. Baker, and N. R. Pape, *J. Am. Chem. Soc.* **62**, 3275 (1940).

14. Baker, W. O., and C. S. Fuller, *J. Am. Chem. Soc.* **64**, 2399 (1942).

15. Catlin, W. E., E. P. Czerwin, and R. H. Wiley, *J. Polym. Sci.* **2**, 412 (1947).

16. Bunn, C. W., and E. V. Garner, *Proc. Roy. Soc.* (London) A **189**, 39 (1947).

17. Bunn, C. W., in *Fibres from Synthetic Polymers,* R. Hill, Ed., Elsevier Publishing Co., New York, 1953, Chap. 11.

18. Trifan, D. S., and J. F. Terenzi, *J. Polym. Sci.* **28**, 443 (1958).

19. Cannon, C. G.., *Spectochim. Acta* **16**, 302 (1960).

20. Baker, W. O., and C. S. Fuller, *J. Am. Chem. Soc.* **65**, 1120 (1943).

21. Biggs, B. S., C. O. Frosch, and R. H. Erickson, *Ind. Eng. Chem.* **38**, 1016-19 (1946).

22. Wittbecker, E. L., R. C. Houtz, and W. W. Watkins, *Ind. Eng. Chem.* **40**, 875 (1948).

23. Cairns, T. L., S. D. Foster, A. W. Larchar, A. K. Schneider, and R. S. Schreiber, *J. Am. Chem. Soc.* **71**, 651 (1949).

24. Sprauer, J. W., C. C. Herrick, *et al.,* unpublished data, Electrochemicals Dept., E. I. du Pont de Nemours & Co.

25. Dole, M., *Kolloid Zeit.* **165**, 40 (1959).

26. Inoue, M., *J. Polym. Sci.* **55**, 753 (1961); *ibid.* A-1, 2697 (1963).

27. Liberti, F. N., and B. Wunderlich, *J. Polym. Sci.* **6**, Part A-2, 833 (1968).

28. Starkweather, H. W., Jr., and R. E. Moynihan, *J. Polym. Sci.* **22**, 363 (1956).

29. Starkweather, H. W., Jr., and R. E. Brooks, *J. Appl. Polym. Sci.* **1**, 236 (1959).

30. Starkweather, H. W., Jr., G. E. Moore, J. E. Hansen, T. M. Roder, and R. E. Brooks, *J. Polym. Sci.* **21**, 189 (1956).

31. Starkweather, H. W., Jr., J. F. Whitney, and D. R. Johnson, *J. Polym. Sci.* **A-1**, 715 (1963).

32. Wilhoit, P. C., and M. Dole, *J. Phys. Chem.* **57**, 14 (1953).

33. Dole, M. and B. Wunderlich, *Makromol. Chem.* **34**, 29 (1959).

34. McCrum, N. G., B. E. Read, and G. Williams, *Anelastic and Dielectric Effects in Polymeric Solids,* John Wiley and Sons, New York, 1967, Chap. 12.

35. Woodward, A. E., J. A. Sauer, C. W. Deeley, and D. E. Kline, *J. Colloid Sci.* **12**, 363 (1957).

36. Korshak, V. V., and T. M. Frunze, *Synthetic Hetero-Chain Polyamides,* translated by N. Kaner, Daniel Davey and Co., New York, 1964, p. 315.

37. Walker, E. E., in *Fibres from Synthetic Polymers,* R. Hill, Ed., Elsevier Publishing Co., New York, 1953, Chap. 13.

38. Gardon, J. L., in *Encyclopedia of Polymer Science and Technology,* Interscience Publishers, New York, 1965, Vol. 3, p. 833.

39. Wyart, J. W., and M. F. Dante, in *Kirk-Othmer Encyclopedia of Chemical Technology,* 2nd ed., John Wiley & Sons, Inc., New York, 1969, Vol. 18, p. 564.

40. Iyengar, Y., and D. E. Erickson, *J. Appl. Polym. Sci.* **11**, 2311 (1967).

41. Korshak, V. V., and T. M. Frunze, *Synthetic Hetero-Chain Polyamides,* translated by N. Kaner, Daniel Davey & Co., New York, 1964, p. 467.

42. Edgar, O. B., *J. Chem. Soc.* **1952**, 2638.

43. Vold, M. J., *Anal. Chem.* **21**, 683 (1949).

44. Riel, F. J., *SPE Tech. Papers* **7**, Part 27-2 (Washington, D. C.) Jan., 1961.

45. Lincoln, J. D., U. S. Patent 3,496,248 (Feb. 17, 1970).

46. Jaenicke, V. W. (to Whittaker Corp.), U. S. Patent 3,406,053, (Oct. 15, 1968).

47. Frigstad, R. A. (to Minnesota Mining and Manufacturing Co.), U. S. Patent 3,499,280 (June 10, 1969).

48. Lopez, E. F., U. S. Patent 3,371,008 (Feb. 27, 1968).

49. Gorton, B. S., *J. Appl. Polym. Sci.* **8**, 1287 (1964).

50. Rayner, C. A. A., in *Adhesion and Adhesives,* R. Houwink and G. Salomon, Eds., Elsevier Publishing Co., New York, 1965, p. 333.

51. Sharpe, L. H., Applied Polymer Symposia No. 3, 353 (1966).

52. Carter, G. R., ASTM, Special Tech. Pub. No. 401, 28, Philadelphia (1966).

53. Carter, G. F., *Adhesive Age* **10**, 32, Oct. 1967.

54. Hause, C. I., W. C. Pagel, and A. B. McKnown, ASTM, Special Tech. Pub. No. 401, 94, Philadephia (1966).

55. Wegman, R. F., W. M. Bodnar, E. S. Duda, and M. J. Bodnar, *Adhesives Age* **10**, 22,, Oct. 1967.

56. Bodnar, M. J., and R. F. Wegman, *SAMPE Journal* **5**, No. 5, 51 (1969).

57. Delollis, N. J., and O. Montoya, *J. Appl. Polym. Sci.* **11**, 983 (1969).

58. Lehman, A. F., and W. B. Trepel, Mater. Res. Std. **7**, No. 9, 383 (1967).

59. Deyrup, A. J., unpublished data, Electrochemicals Dept., E. I. du Pont de Nemours & Co.

60. Aker, S. C., Applied Polymer Symposia No. 3, 251 (1966), *Structural Adhesives Bonding*, M. J. Bodnar, Ed., Interscience Publishers, New York.

61. Hopper, L. C., *Natl. SAMPE Tech. Conf. Proc.* **2,** 63 (1970).

62. Seago, R., *Natl. SAMPE Tech. Conf. Proc.* **2,** 135 (1970).

63. Anon., *Prod. Eng.* **38,** 89 (1967).

64. Anon., *Iron Age* **199,** No. 10, 100 (1967).

65. Ellis, R. F., *Mod. Packaging* **43,** No. 3, 77 (1970).

66. Anon. (to American Can Co.), Brit. Patent 1,148,401, (Apr. 10, 1969).

67. Ess, R. J., and D. E. Floyd (to General Mills), U. S. Patent 3,397,816 (Aug. 20, 1968).

68. Anon. (to American Can Co.), Brit. Patent 1,148,402 (Apr. 10, 1969).

69. Jackson, W. J., Jr., T. F. Gray, and J. R. Caldwell, *J. Appl. Polym. Sci.* **14,** 685 (1970).

70. Battersby, W. R. (to United Shoe Machinery Corp.), U. S. Patent 3,329,740 (July 4, 1967); U. S. 3,437,063 (Apr. 8, 1969).

71. Bennett, K. W., *Iron Age,* Sept. 25, 1969, p. 90.

72. Groves, J. H. (to American Can Co.), U. S. Patent 2,934,236 (Apr. 26, 1960); U. S. 2,962,468 (Nov. 29, 1960).

73. Anon., *Chem. & Eng. News,* May 26, 1969, p. 16.

74. Anon., *Chem. Eng.,* June 16, 1969, p. 48.

75. Anon., *Package Eng.,* August 1969, p. 82.

76. Anon., *Chem. Week,* Apr. 6, 1968, p. 57.

77. Blennemann, D. in "International Symposium on Bonded and Laminated Fabrics," American Association of Textile Chemists and Colorists, Dec. 3-4 (1969).

78. Smith, D. K., *ibid.*

79. Lee, H. and K. Neville, *Handbook of Epoxy Resins*, McGraw-Hill Book Co., New York, 1968, Chap. 22.

80. Anon., *Chem & Eng. News,* Aug. 26, 1968, p. 62.

81. Anon., *Chem. Week,* Mar. 1, 1969, p. 21.

82. Anon., *Mod. Plast.,* Nov. 1969, p. 68.

83. Zabel, R. H., *Mod. Plast. Encycl.* **46,** No. 10A, 266 (1969).

CHAPTER 17

Treatment of Processed Nylons

E. M. LACEY, J. MENGASON, AND M. I. KOHAN

This chapter discusses procedures applied to nylon plastics subsequent to the initial melt fabrication. Forming is one such procedure already considered at length in Chapter 15.

ANNEALING AND ACCELERATED MOISTURE CONDITIONING: E. M. Lacey

Annealing and accelerated moisture conditioning are related because stress relief and crystallization occur in both operations. Absorption of water by nylons lowers the glass transition temperature (α-transition, Chapter 9) and allows relaxation processes to occur at a lower temperature. For this reason moisture conditioning reduces modulus and tensile strength but increases tensile elongation and resistance to impact. If the conditioned nylon is dried, however, it shows an increase in tensile strength and modulus similar to that observed after annealing. Any evaluation of the comparative properties of nylons must obviously be made on specimens that have comparable thermal histories and have been equilibrated to the same relative humidity.

The moisture conditioning treatments may accomplish as much stress relief as dry annealing at higher temperatures. However, the determination of the extent of stress relief is complicated by the dimensional expansion due to absorption of water.

Frozen-in Stress

Melt processing involves ultimately a solidification step with quenching of the surface. The low thermal conductivity of plastics necessarily means a slower rate of cooling for the core material. For any plastic this can generate internal strains because of the freezing of molecules near the surface into energetically unfavorable positions. For the crystalline nylons this also means that the degree of crystallinity and the density increase from the skin to the center of a melt fabricated part. Molecular orientation that occurs during extrusion and molding, particularly in cavities that involve long flow paths, induce frozen-in stresses. These internal stresses obviously vary in severity with such factors as the quench conditions, the shape and thickness of the part, and the fill rate. Relief of the orientation-induced stresses occurs over long periods of time in a manner analogous to creep at ordinary temperatures but takes place rapidly at the temperatures used for annealing or accelerated conditioning. However accomplished, stress relief is accompanied by anisotropic dimensional changes, the magnitude being related to the degree of stress relief. The change in crystallinity that also occurs causes isotropic changes in dimensions.

The dimensional instability of an internally stressed part can result in warpage during use especially when service temperatures are high. Where dimensional changes due to stress relief occur gradually over a long period of time, adaptation of the part to mating members may take place without problem. Avoidance of internal stress is important for thick, extruded shapes. Internal stress in such shapes may be partially relieved as a result of machining, cutting, or slicing operations during fabrication of component parts or preparation of prototypes. This can cause warpage which may occur immediately or over a period of time. Internal stress increases susceptibility to crazing in the presence of active agents such as aqueous zinc chloride solutions (2), glycolic acid, and others.

Laboratory procedures for determining the presence of frozen-in stress include annealing and chemical crazing tests. Molded parts are annealed by heat treating at 50°C below the melting point. Dimensional changes can then be measured and used, for example, to compare effects of different processing conditions on part quality. Annealing of both axial and circumferential sections of extruded rod stock can be used to define both magnitude and direction of the stress.

Stress cracking techniques are used for varied purposes such as characterizing the initial stress level, determining the thoroughness of the annealing treatment, and evaluating the improvement occurring from design changes, as in using a more generous radius. A variety of stress crazing agents have been used. Ensanian (1) found that aqueous hydrochloric acid, either alone or in the presence of alcohol and zinc dust, is an active stress crazing agent for nylons. Later Dunn and co-workers (2) found that 20 to 80% aqueous zinc chloride solutions were even more active stress cracking agents for internally stressed molded parts. Activity of the 80% zinc chloride solution is temperature dependent; 2 to 4 hr were required for cracking a nylon-66 at 23°C, 30 min at 30°C, and only 5 min at 50°C.

Different nylons vary in their resistance to stress cracking. Caprolactam homopolymers and copolymers rich in caprolactam are particularly sensitive. Nylon-66 is intermediate; and nylons-610, -612, and -11, less susceptible (8). Obviously, time, temperature, and concentration are important and must be considered in evaluating performance under test conditions.

Because frozen-in stresses are associated with orientation and anisotropy, physical properties may vary with the way the test specimen is aligned. The deflection temperature (ASTM D-648) can be particularly sensitive to internal stress, and specimens are frequently annealed before testing.

Annealing

Annealing is widely practiced as a commercial process in extrusion plants but less frequently in injection molding shops. Annealing of molded parts, that is, parts with thicknesses less than $\frac{1}{2}$ in. (1.27 cm), is ordinarily done in an oil bath at temperatures of 266 to 300°F (130 to 149°C) for nylons-6, -610, and -612 (melting point = 195–220°C) and 300–350°F (149–177°C) for nylon-66 (melting point = 260°C) (6, 10). It is carried out at higher temperatures (50°C below the melting point) for estimating frozen-in stress as noted above. The recommended annealing time is normally 15 min per 1/8 in. (0.32 cm) of thickness. Suitable heat transfer liquids include "Glyco Wax" S 932 (3), "Teresso" Oil 140 (4), paraffin (9), and other high-boiling hydrocarbons and waxes. The recommended annealing time is based upon experience in achieving adequate relief of stress in typical molded parts. Thicker sections may require more carefully controlled annealing conditions and may include controlled rates of cooling. The annealing of thick sections such as extruded stock shapes and profiles is common practice, but the techniques employed are considered essential elements in the manufacturing operation and are regarded as proprietary.

Inert gases such as nitrogen to avoid oxidation can be used for annealing although heat transfer is less efficient. They hold the advantage of eliminating the need to remove oil or wax from the annealed parts, but the cost of the gas, the increased time required, and the sealed equipment required have limited commercial use.

Although, as noted above, annealing is not commonly employed with injection molded parts, it is sometimes used because of the enhancement of a particular property. For example, gears and bearings are frequently annealed because a longer service life is attributed to this treatment. The reasons for improved performance have not been established but may relate to the higher flexural modulus and higher hardness resulting from increased crystallinity and to enhanced lubricity because of absorption of minute quantities of the annealing medium. Annealing also enhances service life in environments that contain stress corrosion agents and dimensional stability in applications that require close tolerances.

Injection molded nylon will show a density profile from the skin of the part to the core. The skin obviously has a lower density because of rapid cooling by the mold surface. Annealing increases percent crystallinity throughout the bar, but the skin-core density profile may be eliminated or inverted depending on the annealing conditions (5).

Because annealing involves heating at an elevated temperature in a dry atmosphere, it can lead to further polymerization of the nylon particularly at the surface. In one example (5) nylon-66 annealed in nitrogen for 30 min at $240°C$ showed an increase in $(\ln \eta_r)/c$ (0.5 g/100 ml m-cresol, $25.0°C$) from 1.20 to 1.36.

Moisture Conditioning

The effect of water on the properties of nylons is discussed in Chapters 9 and 10. Many of the characteristics of conditioned nylon such as high tensile impact strength, high Izod impact strength, and high ultimate elongation are considered desirable by end users. Specifications sometimes reflect a need for these properties and require some degree of moisture conditioning. This may be done to facilitate an assembly operation such as one involving a snap fit, or to bring about immediately the properties and dimensions for an "average" humidity.

In commercial conditioning, the most common practice is to immerse the parts in hot water until the required moisture content has been reached. This is normally well under the equilibrium moisture content for 100% R.H., for example, about 3% for nylons-6 and -66 which contain 8.5 to 10% water at equilibrium. The moisture is present nonuniformly with more of the moisture in the surface and less in the core. Obviously, dimensions and properties depend upon the distribution of moisture, but the differences are usually small and make use of this procedure reasonable.

It is often desirable to condition to equilibrium at a particular R.H. in order to determine the ultimate changes in a specific environment. These invoke use of a super heated steam or immersion in certain salt solutions.

The time required to equilibrate a nylon to a given relative humidity depends upon the nylon, the humidity, the part thickness, temperature, and the physical structure of the nylon. In general the higher the equilibrium level of moisture, the less time is required for equilibration. This follows from the fact that the penetration of water into the core of the part is less impeded by nonhydrated amide groups, and the rate of diffusion tends to increase with increasing amide group concentration. Thus, increasing relative humidity, increasing amide group concentration, and decreasing crystallinity tend to lower the conditioning time. It is obvious that increasing temperature and decreasing part thickness have the same effect; not obvious is the observation that smaller spherulite size for a fixed level of crystallinity also increases the rate of water absorption although it has no effect on the equilibrium absorption.

For nylons-6, -66, -610, and -612, there appears to be only a small effect of temperature on equilibrium absorption; for nylon-11, however, equilibrium values at 100% R.H. of 1.9 wt % at 20°C and 2.9% at 100°C have been reported (11). Presumably this is a function of the $CH_2/CONH$ ratio, and a similar difference would be expected for nylon-12.

The absorption isotherms for nylons-6, -66, and -610 at 20°C (68°F) and 100% R.H. are indicated in Figure 17-1 for two thicknesses (12). The relationship between temperature, time, and thickness for specific moisture levels are given for the same nylons in Figure 17-2 (12).

Nylon parts can be rapidly equilibrated to 50% R.H. with superheated steam in a pressure vessel. The problem of controlling the temperature and pressure of the superheated steam within narrow limits and the cost of the pressure equipment has, so far, largely restricted the steam technique to laboratory use. The procedure is uniquely advantageous for the study of test pieces when electrical properties are important. Steam temperatures ranging from 120 to 140°C are adequate to insure essentially complete stress relief. Specimens with

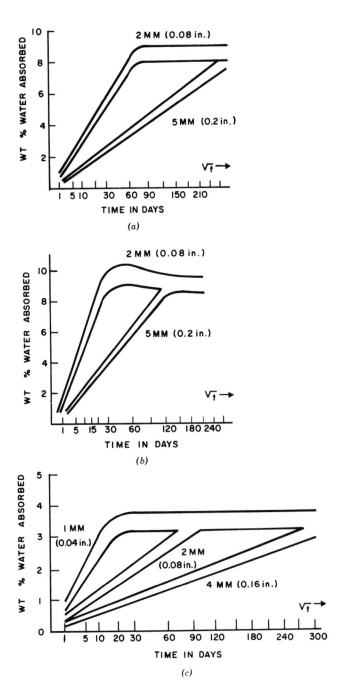

Fig. 17-1. Water absorption at 20°C, 100% R.H. of *(a)* nylon-66, *(b)* nylon-6, and *(c)* nylon-610 (12).

558

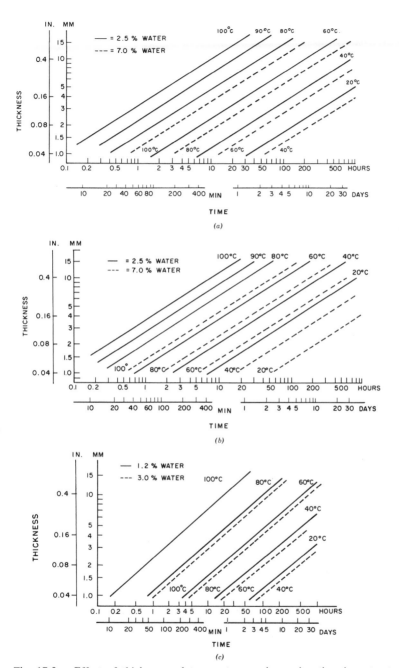

Fig. 17-2. Effect of thickness and temperature on immersion time in water to absorb specific levels of moisture for *(a)* nylon-66, *(b)* nylon-6, and *(c)* nylon-610 (12).

559

sections thicker than 3/8 in. (0.9 cm) are generally not treated because of the danger of degradation during longer equilibration times.

Another procedure (7) for conditioning nylons to 50% R.H. consists of immersing the nylon in a boiling aqueous potassium acetate solution to constant weight — normally one to two days. Compensation for extractables in nylon-6 or modified nylons may be necessary. The solution consists of 125 g of potassium acetate per 100 g of water (55.6 wt %) and boils at $121°C$. At this temperature the absolute pressure of saturated steam is 2 atm so that this solution corresponds to 50% R.H. All glass or glass-lined equipment is recommended to avoid contamination with iron or iron salts which color the solution and the nylon parts a dull red.

The density of the potassium acetate solution is 1.305 to 1.310 g/ml at $23°C$ $(73°F)$ and is maintained by incremental water additions when required. This obviously necessitates periodic cooling of at least part of the solution to $23°C$ $(73°F)$. It is also possible to check the concentration with a thermocouple if an adequately sensitive potentiometer is used.

The procedure has been widely used to bring test specimens to moisture equilibrium with 50% R.H. prior to physical property testing although absorption of a small amount of salt limits suitability when electrical properties are to be measured. The time required to bring parts of nylon-66 to equilibrium is shown in Figure 17-3 as a function of thickness. The high density of the solution necessitates the use of resistant stainless steel screening or other suitable materials to insure complete immersion of the nylon.

ASSEMBLY: J. Mengason

Assembly of the various components of a product can be of major importance. Designers must have a knowledge of the capabilities, limitations, and requirements of the various techniques used for joining parts of nylon to be able to apply them correctly and economically. Assembly should be considered early in the development of a design. In most cases a specific joint design or part geometry is required. Changes to a completed design to incorporate features for a specific technique are often expensive or impossible, particularly after completion of prototyping, testing, or tooling.

Assembly techniques are divided here into two general categories, permanent and recoverable. As the name implies, permanent techniques do not allow for separation and reassembly of parts once joined. These include, for example, spin and ultrasonic welding, wherein the nylon melts and fuses together producing a homogeneous joint. Recoverable techniques are those whereby parts can be assembled and disassembled a number of times. This number varies with the technique and can range from a few with self-tapping screws to thousands with threaded metal inserts.

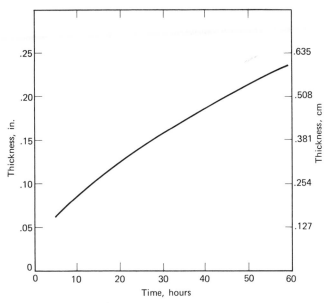

Fig. 17-3. Effect of thickness on time to absorb 2.5% water for nylon-66 in boiling potassium acetate solution (55.6 wt. %) (6).

The requirements of an application will normally dictate which category, permanent or recoverable, should be used. Beyond this, other factors such as part geometry, economics, aesthetics, and production requirements should be considered by the designer in the selection of an assembly technique. Discussion of these factors and of the assembly procedures most frequently used with nylons is included in this section. It is beyond the scope of this book to provide all of the technical details. They can be found in the numerous publications and manufacturers' literature available on the subject. References to a number of these are given in the text.

Permanent Assembly

Permanent assembly techniques can be divided into three categories: welding or heat fusion, heading, and cementing and solvent bonding.

Welding

Three welding techniques are described: spin welding, hot plate welding, and ultrasonic welding.

Spin Welding. Spin welding or friction welding, as it is also called, is a permanent, inexpensive, and highly reliable method for assembling circular nylon parts of unlimited diameters (1). The joints produced are leak free and can be made as strong as the basic unit. Double joints about the same axis of rotation can be formed simultaneously.

Spin welds are made by holding one section and rotating a mating section against it under pressure to generate frictional heat at the interface. After a film of melted nylon has formed, the spinning is stopped instantaneously and the melt is allowed to solidify under pressure. The actual welding takes less than one second. The overall cycle is determined primarily by the loading and unloading time. Production assembly rates for small parts may be as high as 60 pieces per min through the use of automatic part handling equipment.

The basic components required for a spin welder are a driving tool, a means of rotating the driving tool, a device to apply compressive force to the joint, a stationary chuck, and various control devices. Commonly used components are diagrammed in Figure 17-4. For prototype or low production volume, spin welding can be performed on a standard drill press with a driving tool and chuck. Machine lathes can be used for large diameter parts where low rotating speeds are required. High-volume applications require either a standard spin-welder or a drill press equipped with a timer and either a spin welding tool or an air cylinder.

There are three basic variables in the spin welding operation: rotational speed, joint pressure, and spin time. The minimum rotational speed and joint pressure can be approximated based on an average lineal velocity of 20 ft sec^{-1} (6.1 m sec^{-1}) and a joint pressure of approximately 700 lb (wt) in.$^{-2}$ (50 kg (wt) cm^{-2}) of projected joint area. Spin time is the duration of relative motion between the rotating surfaces. Once melting has occurred, relative motion must be stopped to allow the nylon to solidify under pressure because failure to arrest motion will cause tearing of the weld. Also, because of the low melt viscosity of most nylons, too long a spin cycle will cause the melted nylon to be thrown from the joint forming excessive flash. Spin times for parts of nylon are usually a fraction of a second.

There are two basic methods for spin welding which differ primarily in the type of driving tool used and in the control of spin time. These are the pivot and inertia methods. Although either can be used for any given application, the inertia method is generally used for parts with a joint diameter of 1 in. (25.4 mm) or greater while the pivot method is most often used when the diameter is less than one inch.

With the pivot method, the tool is constantly rotating at the desired speed. Spin time of the parts is controlled by a timer which is used to withdraw the tool when melt has been formed. A typical pivot tool consists of a driving element, with metal teeth and a spring-loaded pin through the center of the tool. The pin applies a load to the joint before the driving element engages the parts. At the end of the cycle, when the tool retracts disengaging the driving element, the pivot pin remains engaged for a sufficient time to allow the relative motion between the parts to stop and the weld to solidify under pressure.

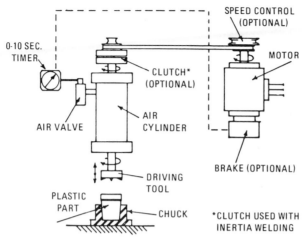

Fig. 17-4.   Basic equipment for spin welding (1).

In the inertia method, the spinning inertia tool is disengaged from the motor at the start of the weld cycle through a clutch in either the drive system or the inertia tool itself. When the tool engages the part, the kinetic energy of the freely spinning mass acting as a flywheel is converted into heat energy due to friction causing the nylon to melt. The relative motion ceases as a result of the total dissipation of the kinetic energy and the externally applied pressure. The pressure is maintained until the melt solidifies and then released when the tool retracts. At this time, the clutch engages bringing the mass to the proper speed for the next cycle. In this method, therefore, the spin time is controlled by the speed and the mass of the inertia tool. The size or weight of the tool can be approximated roughly by using 10 to 20 pounds of mass for each square inch $(0.7-1.4 \text{ kg (wt) cm}^{-2})$ of weld area. Since the energy of a rotating mass is equal to the moment of inertia of the mass times the velocity squared, a slight change of velocity can best be used to bring the system into balance.

There are several important requirements when setting up a spin-welding machine. The stationary chuck must be aligned so that the center axis coincides very closely to the spin axis of the driving tool. The length of travel of the driving tool to contact the workpiece should be as short as possible to prevent vibrations that can impair the weld. The rate of travel of the driving tool should be as rapid as possible with no lag in application of full pressure.

Joint design is one of the most important factors affecting weld quality. Many configurations will produce welds, but in order to obtain optimum weld quality, certain principles should be followed.

1. Use a maximum weld area with a minimum of material for economy.
2. Design self aligning parts to reduce molding and welding tolerances.

3. To accommodate any out-of-roundness, keep shape of at least one of the sections tubular with no brackets or ribs close to the joint.

4. Avoid butt joints because they tend to produce weak, brittle welds.

Figure 17-5 illustrates several commonly used joint configurations. The weld area is maximized by either tongue-in-groove or taper joints, which also improve alignment and reduce wobble during spinning. Prior to welding, parts should turn freely with good contact along all joint surfaces. Warpage or interference between the parts may result in poor joint quality.

Most joints will produce a degree of flash. If undesirable, flash may be automatically trimmed by a cutting tool incorporated into the spinning tool or trimmed after welding. When flash is light, it frequently can be trimmed by pushing the part through a sizing ring. Flash may be trapped or hidden by designing a reservoir within the joint or a shielding skirt outside the part.

Parts that are to be assembled must often be indexed relative to each other. This is possible with spin welding but it increases markedly the cost of equipment and the complexity of the operation. Use of an alternative technique such as ultrasonic welding should be considered.

Hot Plate Welding. Another fusion technique is hot plate welding (1). The surfaces of the parts to be welded are held against a heated tool until a thin layer of melt is formed, removed, and lightly pressed together allowing the melt to fuse and solidify creating a strong, permanent, leak-free joint.

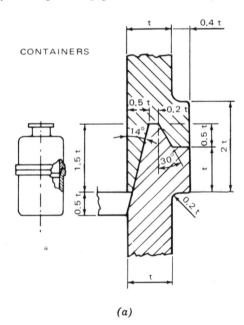

(a)

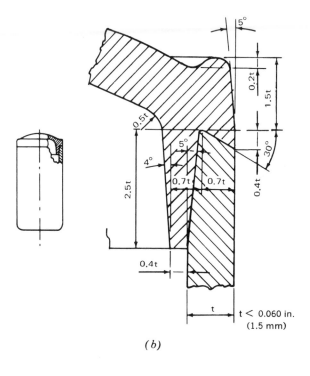

(b)

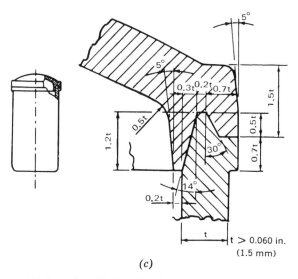

(c)

Fig. 17-5. Spin weld joint designs (1). (a) Mid-wall joint; (b) thin-wall joint; (c) end-plate joint.

A thermostatically controlled aluminum heating plate is suggested for hot plate welding since it maintains a uniform surface temperature. For best results, the surfaces of the tool may be coated with a thin layer of TFE fluorocarbon to prevent sticking of the molten nylon.

This technique has limited usage with nylons. Nylon-66 cannot normally be welded because brittle, low strength joints result due to oxidative degradation of the nylon at the joint interface. Other disadvantages of this technique are the comparatively long cycle and the necessity of removing flash. Since this technique is used primarily for irregularly shaped parts, flash trimming can be difficult and time-consuming. Hot plate welding is seldom used for high volume production, but is more often used for low volume applications and for fabricating large sections of material from which prototypes can be machined.

Ultrasonic Welding. Ultrasonic welding is one of the newest and fastest growing assembly techniques. The welding is very rapid, usually less than two seconds, and can be fully automated for high speed and high volume production. This technique, however, requires close attention to such details as part and joint design, welding variables, fixturing, and moisture content. For this reason, more detail is provided for ultrasonic welding of nylon than for other techniques.

In ultrasonic welding, high-frequency vibration and pressure are applied to the parts to be joined, and the heat generated at the interface melts the nylon (2). Pressure is maintained when the vibrations cease and the melt solidifies. This produces a strong, homogenous bond which approaches the strength of the original material.

The equipment consists of a power supply which converts low-frequency electricity of 60 Hz (cycles per second) to 20 kHz, a cycle timer, a transducer which converts the electrical energy into mechanical energy in the form of vibrations, and a "horn" which transmits the mechanical vibrations to the parts and applies pressure for proper solidification of the melted nylon. It is also necessary to have a base fixture to hold and to properly locate the parts under the horn.

The horn is a resonant metal section carefully tuned to vibrate at the frequency of the system, usually 20 kHz. It is generally constructed from high-strength titanium alloys, although aluminum is sometimes used. For nylons, they should be shaped to contact the part as close to the joint and as directly above the joint as possible with good contact between the horn and the part. A slightly raised area over the joint should be included if possible, in the design of the part to provide this contact (3). For parts with maximum dimension greater than approximately 3 in. (7.6 cm), special horn configurations such as slotted horns are usually needed. Large parts can also be welded with several clustered horns. With one technique, the horns, each with a transducer, are energized simultaneously from individual power supplies or sequentially energized from one power supply. The newest technique utilizes a cluster of horns attached to a single transducer which, when cycled, energizes the horns simultaneously (4).

Because of the crystallinity and high melting point of most nylons (365 to 518°F, 180 to 270°C), a high-energy input is necessary to obtain good welds. Proper joint design will minimize this requirement. Welders of sufficient power for most applications are commercially available. High amplitude of vibration of the welding horn is necessary to achieve adequate energy input in nylons. This is accomplished by use of high power, by horn design, or more conveniently by use of a second horn, called a booster, which is located between the transducer and the welding horn. Boosters that provide 2:1 or 2½:1 amplification are generally required except for very small parts. They frequently reduce substantially the time required for welding. Adequate pressure is also necessary, but if it is too high in relation to the amplitude, it can stall the horn and dampen the vibration. The proper amplitude to force ratio must be determined individually for each application.

A very important condition which must be considered when ultrasonically welding nylon is the moisture content of the parts. Nylon resins are hygroscopic and absorb moisture from the air after molding. Release of moisture during welding can lead to weak bonds. Nylon parts should therefore be welded dry-as-molded or dried prior to welding. Exposure for 24 hr to air at 73°F (23°C) and 50% R.H. before welding is sufficient to reduce the resultant weld strength by 50% or more (5).

Perhaps the most critical facet of ultrasonic welding is joint design. There are two basic types of joints – shear joints (3, 5) and butt joints. An important feature of all joints is a small initial contact area into which the energy is concentrated. The basic shear joint is shown in Figure 17-6 before and after welding. It is the preferred joint for ultrasonic welding of nylon. Initial contact is limited to a small area which is usually a recess or step in either of the parts for alignment. The two parts must not fit tightly. Welding is accomplished by first melting the contacting surfaces. The parts continue to melt along the vertical walls, telescoping together. The smearing action of the two melt surfaces eliminates leaks and voids and makes this the best joint for leak-free seals. Variations of the basic joint are shown in Figure 17-7.

The shear joint has the lowest energy requirements of all the types of joints and requires the shortest welding time. The initial requirement is low due to the small contact area. As the parts telescope together, the heat generated at the joint is retained until the vibrations cease because nylons are poor conductors of heat. Because of the smearing, telescoping action, the melted plastic is not exposed to air which would cool it too rapidly or cause oxidation.

The strength of the weld is determined by the depth of telescoping. This is controlled by the weld time as well as the part design. Joints can be stronger than the adjacent walls by designing the depth of telescoping 1.25 to 1.5 times the wall thickness. This is necessary because the tensile strength of the weld is

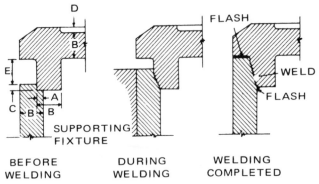

BEFORE   DURING   WELDING
WELDING   WELDING   COMPLETED

Fig. 17-6.   Ultrasonic welding "shear" joint.
Dimension A: .016 in. (.4 mm); this dimension is constant in all cases.
Dimension B: This is the general wall thickness.
Dimension C: .016-.024 in. (.4-.6 mm). This recess is to ensure precise location of the lid.
Dimension D: This recess is optional and is generally recommended for ensuring good
              contact with the welding horn.
Dimension E: Depth of weld = 1.25 to 1.5 B for maximum joint strength.

seldom 100% of the strength of the molded or extruded resin. The shear joint is therefore preferred for ultrasonic welding of nylon because of the low, nearly constant energy requirement, the short welding cycle, the high strength and leak-free seal obtainable, and the avoidance of oxidation.

There are several important aspects of this joint which must be considered. The top part should be as shallow as possible, in effect, it should be just a lid. The walls of the bottom half must be supported at the joint by the base holding fixture in order to avoid the expansion of this part during welding. Non-continuous or inferior welds result if the upper part slips off of the lower part, or if the stepped contact area is too small. Modifications to the joint such as those shown in Figure 17-8 should be considered for large parts because of dimensional variation or for parts where the top piece is deep and flexible.

Suitable allowance must be made in the design of the joint for the flow of molten nylon displaced during welding. When this flash cannot be tolerated for

Fig. 17-7.   Variations of basic "shear" joint (5).

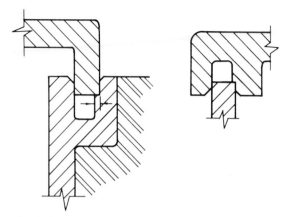

Fig. 17-8.   Variations of "shear" joint for large parts (5).

aesthetic or functional reasons, traps similar to the ones shown in Figure 17-9 can be designed into the joint.

The second basic type of joint is the butt joint which is shown in Figure 17-10. The tongue-in-groove provides the highest mechanical strength. Although the butt joint is quite simple to design, it is difficult to produce strong bonds or leak-free seals in nylon parts with this joint. A "V"-shaped bead or "energy director" on one of the two mating surfaces provides the small initial contact area necessary for energy concentration and rapid melting. The nylon in the energy director melts first and flows across the surfaces to be joined. However, because nylon crystallizes very rapidly, the melt can solidify before sufficient heat is generated to weld the full width of the joint. Consequently, with the exception of the area under the energy director, only spotty welding occurs. During the welding, the melt is exposed to air. This not only accelerates the solidification of the nylon but can also cause oxidative degradation, and the net result is a brittle weld.

For these reasons, the butt joint is best used for parts designed with interrupted joints requiring only nominal strength. It can be used on large parts where part to part dimensional variations or joint design limitations make use of the shear joint impractical.

Fig. 17-9.   Flash traps for "shear" joint (5).

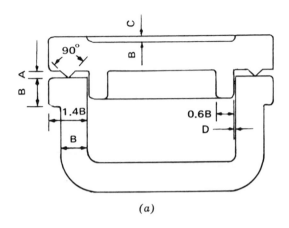

(a)

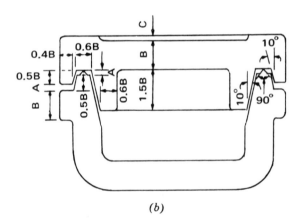

(b)

Fig. 17-10. Two types of butt joints with energy directors (5).

(a) Dimension A: 0.016 in. (0.4 mm) for B dimensions from 0.060 to 0.125 in. (1.5 to 3 mm) and proportionately larger or smaller for other wall thicknesses.

    Dimension B: General wall thickness.

    Dimension C: Optional recess to ensure good contact with welding horn.

    Dimension D: Clearance per side, 0.002 to 0.005 in. (.05 to .13 mm).

(b) Dimension A: 0.016 in. (0.4 mm) for B dimensions from 0.060 to 0.125 in. (1.5 to 3 mm) and proportionately larger or smaller for other wall thicknesses.

    Dimension B: General wall thickness.

    Dimension C: Optional recess to ensure good contact with welding horn.

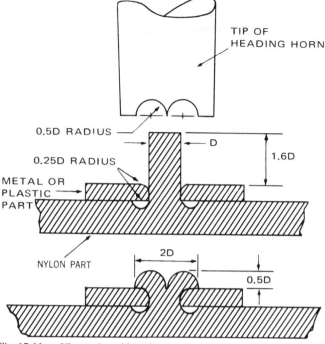

TIP OF
HEADING HORN

0.5D RADIUS

0.25D RADIUS

METAL OR
PLASTIC
PART

NYLON PART

D

1.6D

2D

0.5D

Fig. 17-11.   Ultrasonic staking (5).

Heading

Ultrasonic Heading. Ultrasonic equipment can also be used for heading or staking (6) to join nylon parts to parts of dissimilar materials, usually metal, or to other nylon parts (Figure 17-11). A stud on the nylon part protrudes through a hole in the second part. A specially contoured horn contacts and melts the top of the stud. The nylon then forms a rivet like head, the shape of which depends on the contour of the horn. This produces a tight joint because there is no elastic recovery as occurs when cold heading.

Stud welding (5), another technique that can be used in lieu of heading when both parts are nylon, is shown in Figure 17-12. This is, in fact, only another version of the shear joint. When designed with the depth of weld equal to the radius of the pin, the stud will break rather than shear out of the hole. A number of these can be welded at one time with one horn, or, if widely spaced, with a cluster of horns as described for regular welding.

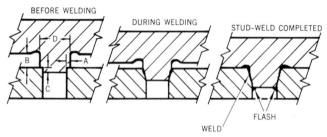

Fig. 17-12.   Ultrasonic stud welding (5).

Dimension A: .010 to .015 in. (0.25 to 0.4 mm) for D up to .5 in. (23 mm).

Dimension B: Dept of weld. B = 0.5 D for maximum strength (stud to break before joint
failure.

Dimension C: .016 in. (.4 mm) minimum lead-in.

Dimension D: Stud diameter.

In addition to simple stud welding, this technique can be used to lock a third
piece between two parts of nylon or, as separate molded rivets, to rivet a part to
one of nylon (Figure 17-13).

Cold Heading. A heading technique that requires less expensive equipment is
cold heading (1). As with ultrasonic heading, this technique can be used to form
strong mechanical joints between either similar or dissimilar materials, but it is
more difficult to produce and maintain tight joints. Cold heads are produced by
applying a compressive load to the end of a shaft. The load, which exceeds the
yield strength of the material, causes the shaft to permanently deform into a
rivet-like head.

The cold heading device (Figure 17-14) consists of two major elements, a
spring loaded sleeve which preloads the area around the shaft to assure a tight
fit, and the heading tool which, passing through the sleeve, cold heads the end of
the shaft. It can be used with a simple arbor press or, for more automatic

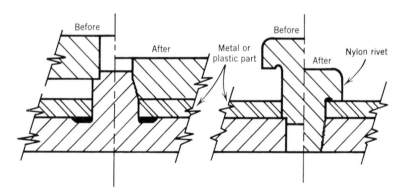

Fig. 17-13.   Stud welding used to hold or rivet a part to one of nylon (5).

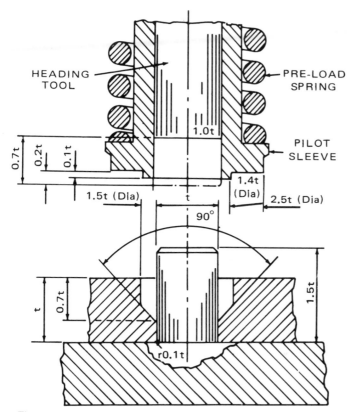

Fig. 17-14.   A typical cold heading tool (7).

operation, in a punch press or similar equipment. The rate of loading is not critical, but it should not be an impact load. Rates as high as 20 in. (508 mm) per minute have been used successfully. The force required for heading is directly related to the compressive yield strength of the particular nylon used. It increases slightly as heading progresses due to an increase in the cross-sectional area of the shaft. The force can be reduced by forming at elevated temperatures. At 200°F (93°C) the required force is about 50% of that needed at 73°F (23°C). Elevated forming temperatures can also be used as a means of obtaining a tighter assembly. Forming at temperatures above the use temperature is recommended because a formed head and shaft tend to recover their original shape when exposed to temperatures above that used in forming.

Another method of obtaining tighter fits is to reduce the cross-sectional area of that portion of the shaft which is going to be deformed. In this way, the load is required to exceed the yield stress of the nylon. The portion of the shaft not deformed is therefore subjected to a lower stress causing less elastic deformation and less recovery when the pressure is removed. The cross section can be reduced by tapering or coring the shaft. Keeping the shaft length before heading as short as possible in relation to the diameter is another way to improve tightness.

Cementing and Solvent Bonding

Cementing or solvent bonding is occasionally used to join nylon to nylon or to dissimilar materials such as wood, metal, or other plastics. This process is particularly applicable when joining large or complicated shapes. In such instances, these techniques are often the only solution to the joining problem. It is best suited to low-volume production or prototype purposes because of the long, labor-consuming bonding procedure which is not easily or economically automated.

The procedure generally consists of a number of steps; surface preparation such as cleaning or roughening of the surface, application of adhesive to one or both surfaces (with solvents allow time for the surfaces to soften), assembling, clamping in fixtures to exert uniform pressure on bonding areas, and curing, often for 24 hr. The curing time can be reduced by heating or by application of ultrasonic energy.

Five basic types of adhesive joints are shown in Figure 17-15. They should be designed with good contact and as large a surface area as possible to obtain maximum strength. For the latter reason, butt joints are generally the least desirable.

If the appearance of an assembled part is important, consideration must be given to the methods of application of the solvent and mating of the parts. Both of these factors can appreciably increase assembly costs. Poorly applied or excess solvent which runs or is squeezed from the joint will mar the surfaces. Too little solvent can produce poor bond strength.

Because of the excellent solvent resistance of most nylons, only a limited number of solvents are available. These tend to be toxic or otherwise hazardous and require special handling procedures and safety precautions. The recommended solvents for bonding nylon to nylon (8, 9) are aqueous phenol, solutions

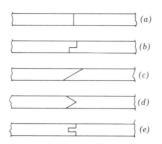

Fig. 17-15.   Basic types of adhesive joints. (a) Butt, (b) lap, (c) scarfed, (d) "V", (e) tongue-in-groove.

of resorcinol in alcohol, and solutions of calcium chloride in alcohol sometimes bodied by the inclusion of small percentages of nylon. Various commercial adhesives, especially those based on phenol-formaldehyde and epoxy resins, are used for bonding nylon to other materials. For information on specific bonding agents, plastic material suppliers or processors should be contacted. Care is required in handling phenols, which are toxic and are rapidly absorbed through the skin.

Recoverable Assembly

Integral Fasteners

There are three types of mechanical fastening techniques which have integral assembly features: snap fits, press fits, and molded-in threads (10). These are simple, economical, and rapid methods for assembling plastic to plastic or plastic to metal and are fully applicable to nylons. They require only simple, inexpensive assembly tools or, in some cases, no tools at all.

Snap Fits. Snap fit joints are not used as frequently as they could be due to lack of knowledge of the design possibilities or underestimation of the strength that can be obtained. Properly designed snap fit joints can lead to reduced assembly costs without sacrifice of functional requirements.

Two basic types of snap fits are shown in Figure 17-16, cylindrical and cantilevered lug snap fits. A recess in the form of a molded or machined undercut or hole is used on one part while the mating one has a protrusion which snaps into the recess. Snap joints may be designed as either temporary or permanent joints based on the return angle. A return angle of 45° in a cylindrical joint approaches permanency. True permanence is achieved with a return angle of 90° in which case the interfering area must be sheared or broken to separate the two parts.

During assembly snap fit joints pass through an interference or stressed condition. The allowable interference depends on the geometry or wall thickness of the parts, the temperature, the moisture content, and the strength or yield point of the nylon at the assembly conditions. It can be approximated using the formulas given in Refs. 1 and 12. For best performance of lug snap fits, the cantilevered segment should be tapered to reduce and distribute the bending stress.

Unless the nylon parts are to be moisture conditioned prior to assembly, it is best to base the assembly design on the properties of dry nylon because parts are often assembled soon after molding while still relatively dry. Consideration must then be given to the effects which normal moisture absorption and resultant dimensional growth will have on the interference and pullout strength of the joint in use.

Within the strength limitations of the nylon, the separation or breaking force

CYLINDRICAL SNAP-FIT JOINT

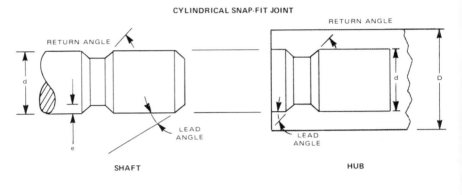

SHAFT   HUB

CANTILEVERED LUG SNAP-FIT JOINT

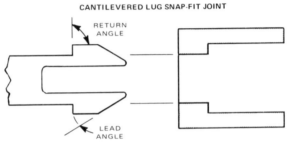

Fig. 17-16.   Typical snap fit joints (11).

of a snap fit is a function of design or part geometry. The return angles greatly influence pullout strength. Unlike press fits, the pullout strength does not decrease with time. A limitation to the strength of snap fits occurs when the undercut in the nylon part must be snapped from a mold without excessive distortion or breakage. Special ejection features are sometimes necessary to accomplish this.

Press Fits. Press fitting or as it is also called, force fitting, can be used to obtain strong and possibly pressure tight joints at a minimum of cost. This type of joint generally consists of two dissimilar materials as when a metal shaft is pressed into a gear of nylon or a bearing of nylon is pressed into a metal sleeve. One part passes into a slightly smaller hole. This interference causes a degree of expansion or contraction of one or both of the materials and is dependent on the respective elastic moduli. The interference limits for nylon-66 at room temperature and average moisture conditions can be approximated using the curves given in Figure 17-17. For other types of nylons, different conditions, or more exact limits, the designer should refer to specific formulas (1, 12) for press fitting for allowable interference, assembly force, joint strength, and dimension considerations.

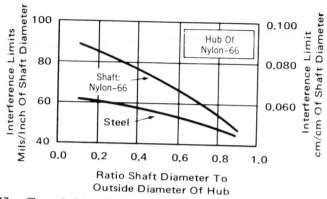

Fig. 17-17.   Theoretical interference limits for press fitting nylon-66 (1). Based on yield point and elastic modulus at room temperature and average moisture conditions.

Initial assembly force and joint strength are a function of the same factors affecting snap fits. The design and production of press fits are less complicated and therefore less expensive than snap fits because recesses or undercuts are not required. Unlike the snap fit, a press fit remains in the interference or stressed condition. On a smooth surfaced joint creep or stress relaxation of the plastic acts to reduce the effective interference. This, in turn, reduces the joint pressure and the holding power. With knurled or roughened joint surfaces, the plastic creeps into the grooves and valleys. The result is an effective increase in the coefficient of friction which compensates for the loss of joint pressure.

Due to moisture absorption and the high coefficient of thermal expansion of nylons relative to metals, changes in temperature and moisture content affect the dimensions and therefore, the effective interference of a press fit. In the case of a metal shaft pressed into a nylon part, the strength of the joint would be reduced by expansion due to heat, moisture, or a combination of the two. Data necessary to calculate dimensional changes due to these factors can be found in literature available from the appropriate nylon resin supplier.

Molded-in Threads. Another assembly technique with integral assembly features is the threaded joint. External and internal threads can be molded economically in parts of nylon. External threads can be molded by the use of thread unscrewing devices or more simply by parting the mold symmetrically about the center axis of the thread. Internal threads can also be molded through the use of an unscrewing device or a collapsing core. Although unscrewing mechanisms increase the cost of an injection mold and generally increase molding cycle time, the higher costs can frequently be offset by the ease of assembly. Threaded joints in nylon parts can be assembled by hand or by the use of common tools and can be readily disassembled with no damage to the threads.

Although almost any thread profile can be produced, the Unified Thread Standard is the most suitable for molded parts. Threads of Class 1 or 2 of this standard are adequate for most applications and are easily molded. Fine threads should be avoided as they are difficult to mold and can be easily damaged or cross-threaded. The root of the thread should be radiused as much as possible to reduce stress concentration.

Nonintegral Fasteners

There are many types and variations of nonintegral mechanical fasteners that can be used with nylon parts. These include self-tapping screws, threaded metal inserts, rivets, and clips. It is beyond the scope of this book to cover all in detail, therefore, only basic types are described. More complete listings and details can be found in manufacturers' literature.

Self-Tapping Screws. Perhaps the most widely used fasteners are self-tapping screws (1). Their use often results in low assembly costs because the screw cuts or forms a thread as it is inserted. This eliminates the need for molded or separately tapped internal threads. They can be divided into two basic categories: thread cutting, which taps or cuts a mating thread as the screw is driven, and thread forming, which mechanically displaces material as the screw is driven. Both types can be used with nylons although the thread-cutting screw is preferred because of lower boss stress and driving torque. The thread-forming screw is better suited for repeated assembly. It is also less expensive.

The optimum hole diameter for thread-cutting screws is equal to the pitch diameter of the screw. For thread-forming screws, the hole diameter should equal 0.8 times the screw diameter. Smaller hole diameters may be used to increase the area of thread engagement when the depth of penetration in the nylon part is limited. For optimum stripping torque, the engaged length of the screw and the outer diameter of the boss should be 2.5 times the pitch diameter. The presence of oil on the surface of screws, washers, or the nylon can cause a decrease in stripping torque by as much as 50%. Therefore, for high strip-to-drive ratio, which is desirable, all parts should be relatively free of oil or other lubricants.

Threaded Metal Inserts. When a joint must be assembled and disassembled often, threaded inserts should be used. They are generally made of brass, aluminum, or steel and contain internal threads which accept standard machine screws. There are several methods of inserting and holding the insert in the nylon part. They can be pressed in, in which case the outer surface is usually knurled. These are subject to the same requirements and limitations of basic press fits. Similar to these are expansion inserts, which expand against the sides of the hole in the nylon when the screw is inserted. Self-tapping inserts of the thread-cutting or thread-forming variety are similar to self-tapping screws with regard to installation and performance. Recently, special types of inserts have been developed which can be spun or ultrasonically inserted into nylon (6).

MACHINING: J. Mengason

Nylons can be easily machined using standard machining practices for metals. The decision to produce a part by machining or by other methods of fabrication such as injection molding is therefore based on factors other than machinability. Because machining is relatively expensive, due primarily to the use of skilled labor, it is used only when other fabricating techniques are more expensive or cannot produce the required shapes, tolerances, or surfaces.

Machining usually becomes the least expensive method of fabrication when the quantity of parts to be produced is small. In such cases, the cost may be less than that of building a mold and injection molding the parts. When high-speed automated machining equipment is used, the economical lot size increases. Although this type of equipment is usually more expensive, the requirement for skilled labor is greatly reduced. Occasionally, machining can be used economically as a secondary operation for such tasks as drilling holes when the complexity of the part would require an extremely complicated and, therefore, expensive mold.

Most nylons can be machined using techniques normally employed with soft brass. They can be turned, drilled, sawed, reamed, milled, punched, threaded, and tapered. High speeds and low feed rates give the best results. Because of low thermal conductivity, heat can build up rapidly and must therefore be kept to a minimum or removed by acceptable coolants. Water or soluble oils are generally used. Although they will permit higher cutting speeds, coolants are generally not necessary to produce work of good quality.

Sharp tools with generous clearance are necessary. Insufficient clearance causes compression of the nylon, and an increase in tool wear and heat due to friction. Because of the high coefficient of thermal expansion of nylons relative to metals, excessive heat generation can cause the stock to expand such that when cooled, the dimensions of the finished part will be incorrect. For this reason, it is advisable to measure dimensions at room temperature for maximum accuracy. Since nylons have a much lower modulus of elasticity than metals, the stock should be well supported during machining to prevent deflection and resultant inaccuracies. Vibration and chatter must also be avoided.

Carbide- and diamond-tipped tools can be used to advantage when machining nylons. Tools should be sharp with honed or polished surfaces where they come into contact with the stock to reduce frictional drag and temperature. The shape of the tool should be such that continuous-type chips are generated. Large, positive rake angles combined with the proper speeds, feeds and depth of cuts will usually accomplish this. Specific recommendations can be found in a number of sources including *Modern Plastics Encyclopedia* (13) and *Machining of Plastics* (14) by Dr. A. Kobayashi, and literature from stock shape and resin suppliers. Dr. Kobayashi concludes that nylons have excellent machinability because of the wide range of conditions that produce continuous chips, the smoothness of machined surfaces, and the small amount of tool wear. The conditions to produce continuous chips are less critical in nylon-6 and -610 than in nylon-66 (14).

The effects of moisture and post-mold shrinkage on dimensions are described in detail in other sections of this book and apply to machined parts as well as those produced by other techniques. Where maximum dimensional accuracy and stability is required, moisture conditioning and annealing should be considered.

Some machining operations tend to create burrs on the part. They are best avoided by maintaining sharp cutting edges, providing generous chip clearances and, in some cases, backing up the stock. Burrs can be removed by a number of techniques including singeing, melting, vapor blasting, honing, tumbling, and hand-carving or scraping.

SURFACE COLORING AND DECORATING: M. I. Kohan

Nylon surfaces are colored or decorated for reason of appearance, identification, information, or improvement in specific properties. Dyeing, painting, vacuum metallizing, electroplating, hot stamping, and inking techniques can be employed. Guidelines on the more common techniques follow, but the differences in solvent resistance and surface quality of the various nylon modifications can mean, for best results, a procedure specific to the individual nylon. It is often advisable to consult with the supplier of the decorating material as well as the supplier of the resin. Sources of decorating materials are tabulated in annual compendia such as *Modern Plastics Encyclopedia* and *Modern Packaging Encyclopedia*.

Adequate preparation of the surface may be the key to success. Cleaning and priming are the procedures normally involved, but some finishing techniques may require a preliminary surface roughening. This is best accomplished by chemical etching, and a 1 to 3-min dip in a strong acid solution at about 150°F (66°C) followed by a hot water rinse may suffice. More precise control of etching has been accomplished via the "satinizing" process (1), which involves immersion in an acid solution that contains a finely divided, inert solid and subsequent exposure to hot air. A microscopically rough surface is achieved with improved mechanical adhesion to the coating but without embrittlement of the substrate or loss of the aesthetic quality of the topcoat. Process details may vary with the nylon and the intended coating. For this reason and because hot acid solutions are involved, care is essential to achieve safely the desired enhancement of adhesion. Table 17-1 provides examples of "satinizing" and its effect on the nylon surface.

Table 17-1. "Satinizing" of nylons (1)

	Nylon-66						Nylon-6		Nylon-6 (Anionic Polym'n)	
	Compound	Wt %	Compound	Wt %	Compound	Wt %	Compound	Wt %	Compound	Wt %
Solution dip Components:										
Acid	p-Toluene sulfonic	0.3	Conc. phosphoric	10.0	Conc. sulfuric	1.0	p-Toluene sulfonic	0.5	p-Toluene sulfonic	0.5
Finely divided, inert solid	"Dicalite"	0.5	"Dicalite"	0.5	"Dicalite"	0.5	"Dicalite"	0.5	"Dicalite"	0.5
Wetting aid	Dioxane	3.0	"Triton" X-100	0.1	"Triton" X-100	0.1	Dioxane	3.0	Dioxane	3.0
Solvent	Perchloroethylene	96.2	Water	89.4	Water	98.4	Perchloroethylene	96.0	Perchloroethylene	96.0
Treatment:										
Temp., °C (°F)	100 (212)		95 (203)		95 (203)		121 (250)		121 (250)	
Time, sec	30 60		60		30		10 30		10 30	
Air exposure 60 sec, 121°C (250°F)	Yes No		Yes		Yes		Yes Yes		Yes Yes	
Rinse and dry	Yes Yes		Yes		Yes		Yes Yes		Yes Yes	
Ca. 10 sec in hot tap Water and air dry										
Surface roughness										
10^{-6} in.	28-34 23-27		14-20		7-9		14-24 17-30		18-46 18-34	
10^{-6} cm	71-86 58-69		36-51		18-23		36-61 43-76		46-117 46-86	
Glossmeter reading	9 7.5		46		47		20 10		27-51 13	

Note: Surface roughness of untreated polymer
10^{-6} in. Nylon-6: 2-16, 5-41; Nylon-6 (Anionic): 7-32, 18-81
10^{-6} cm
Glossmeter reading of untreated polymer Nylon-6: 70; Nylon-6 (Anionic): 69-92

Dyeing

Dyeing of nylon is, of course, commonplace in the fiber industry. Many different kinds of dyes, each with its own set of application and performance characteristics, can be used with nylon fibers (2). In the plastics industry dry coloring (Chapter 11) is more common, but dyeing is also used to achieve a marbleized effect, to provide a variety of colors in short runs, and to facilitate identification of size or shape.

Even low-cost, general-purpose household dyes often prove to be satisfactory when applied to nylon plastics according to the manufacturer's directions for the dyeing of cloth. Dyes can be selected, however, to decrease the soak time required and to improve penetration and uniformity of color. Disperse azo dyes and water soluble acid and acetate dyes have been cited in the plastics trade literature (3-5). Particularly rapid dyeing to full, bright shades with good light fastness has been achieved with a group of neutral-dyeing premetallized acid colors in an aqueous bath containing benzyl alcohol (6, 7).

It is usually advisable to clean the nylon article first by immersing in a one percent soap solution for 5 to 10 min at 140 to 160°F (60 to 70°C) and then rinsing in clear water at the same temperature. The specific dye, the bath temperature, dye concentration, residence time, agitation, additives to assist penetration, and the nylon article are all factors in the dyeing operation. The conditions given in Table 17-2 are reported to yield full, bright shades on nylon-66 in 1 min or less. Lowering temperature, decreasing the benzyl alcohol content, or lowering the dye concentration obviously leads to lighter shades. Mild agitation of the dye bath, avoiding contact of the immersed parts, and a post-dyeing rinse are all desirable to assure uniform exposure to the dye.

Table 17-2. Conditions for dyeing nylon-66 to full, bright shades (Ref. 6)

"Capracyl" dye[a]	Benzyl alcohol, (ml)	Bath temp. (°F)	Bath temp. (°C)	Time (sec)
Yellow NW	2	100	38	60
Orange R	4	150	66	60
Yellow GW	4	150	66	60
Yellow 3RD	4	150	66	10
Olive Green B	4	150	66	60
Red G	4	150	66	60
Red B	2	100	38	60

[a] 2.0 g in 99 ml water and 1 ml glacial acetic acid. These are premetallized acid dyes that can be applied from neutral solution, but the dilute acetic acid solution increases the rate of absorption of dye without harmful effect on the nylon.

Continuous dyeing (7) of nylon jacketed wire and cable has been described that involves immersion times under 3 sec with the same dyes listed in Table 17-2. Dye concentrations of 1 to 3%, a bath temperature of 175°F (80°C), about 5% benzyl alcohol, and up to 1% acetic acid in a circulating dye system (Figure 17-18) are suggested. Dye penetration of about four microns was observed with jackets of nylon-610. This is adequate for any diameter jacket and gives rise to a simple plot relating dye requirement to the diameter of the nylon jacket (Figure 17-19). A decrease of 0.5% in dye content, initially at the 2 to 3% level, is acceptable and permits control of color by periodic make-up. Loss of benzyl alcohol is similarly compensated because the effect of benzyl alcohol on rate of dye absorption becomes small above 4%. Dyes typically contain a large fraction of inert material which will increase in concentration in the bath as more dye is added. This is tolerable until the inerts amount to at least 10% of the bath. Assuming a 3% initial dye concentration of which two-thirds is inert material and assuming make-up dye is added each time the nominal concentration falls to 2.5%, lead to the estimate of 24 make-ups to reach a concentration of 10% inerts.

The amide group concentration in the nylon affects dyeability, and nylon-66 is dyed more readily than nylon-610 which in turn dyes more readily than nylon-11. The same factors that influence solvent and hydrolysis resistance (Chapter 2, p. 68-71), such as percent crystallinity and degree of perfection of the molecular configuration, also play a role in dyeability. Thus, a thin section from a cold mold will dye more rapidly to a greater depth than a thick section

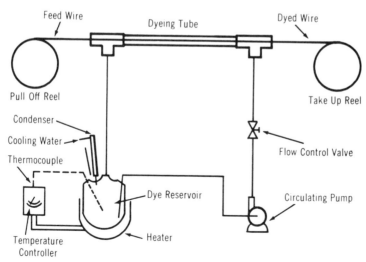

Fig. 17-18.   Schematic diagram of apparatus for continuous dyeing of nylon jacketed wire (7).

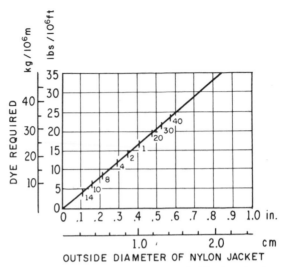

Fig. 17-19.   Dye required versus jacket diameter (7). Numbers in plot refer to American wire gage size for the conductor enclosed in building wire (THWN-THNN) constructions.

from a hot mold, and in general nylon-6 dyes somewhat more rapidly than nylon-66. Variations in surface characteristics clearly pose the problem of irregular dye pick-up.

Dyeing does not normally impair the mechanical properties of nylon plastics. Abrasion resistance, toughness, and insulating properties of nylon jackets are not adversely affected by dyeing (7). Because nylon jacketed constructions find varied uses, each with its own set of property requirements, the dyed jacket is best checked in each specific application (7).

Painting

Painting of nylon may be indicated to achieve a specific color effect, to integrate the part with other painted parts as in the case of automobile bodies, or to provide an aesthetic or weatherproofing topcoat otherwise difficult to obtain, as in the case of cast nylon. Because of their excellent solvent resistance and high heat distortion temperatures, nylons are relatively easy to paint.

The technique employed depends on the effect desired and the kind of topcoat selected, and working closely with the individual paint supplier is advisable. For example, if an acrylic lacquer finish is to be used on nylon-66, Du Pont (8) recommends that the part first be dipped for 1 to 3 min in water or a 25/75 by volume mixture of water/isopropyl alcohol to improve uniformity of adhesion. A 1 ± 0.2 mil (0.025 ± 0.005 mm) coat of Du Pont 828-1967 Grey (or colored analog) Sealerless Primer is then spray coated on using about 30% by volume of an aromatic diluent such as xylene or toluene. Baking for 45 min at 325°F (163°C) provides a satisfactory substrate, which can be sanded if desired, for application of the acrylic finish.

Where an alkyd enamel finish is to be used, the aqueous dip should be avoided. No pretreatment is necessary except perhaps wiping with a hydrocarbon solvent such as naphtha to remove dirt. A 1 ± 0.2 mil (0.025 ± 0.005 mm) coat of Du Pont 764-1751 Oxide Non-Sanding Enamel Primer is then spray coated on using about 20% by volume of an aromatic diluent. Baking for 45 min at 275°F (135°C) yields a surface ready for the enamel finish.

The choice of primer may differ and the baking time or temperature may have to be lower with nylons that soften at a lower temperature than nylon-66 in order to avoid shrinkage or warpage. "Satinizing" affords an alternative to priming.

A good, general discussion (9) of painting is available that points out that a hard, well-adhering paint can reduce extensibility and impair toughness. The effect of painting on mechanical properties may be negligible depending on the paint, part, and use, but it should not be overlooked.

Metallizing

Metallic coatings are best applied to thermoplastics by vacuum evaporation or by electroless deposition followed by electroplating (10). Not only is a decorative effect and a conductive surface obtained, but the radiation and thermal resistance, weatherability, and dimensional stability are also improved.

Vacuum metallizing is accomplished by heating a metal until its vapor pressure is at least 10^{-2} mm Hg in a chamber under high vacuum (0.5 μ). The surfaces to be coated must lie within the mean free path of the molecular emission. A wide variety of metals requiring vaporization temperatures ranging from 268°C for cadmium to 2059°C for platinum have been used, but aluminum (1188°C) is probably the metal most often evaporated. Substrates with smooth surfaces lacking sharp corners and free of contamination or mold release agent are required. Fillers, reinforcing agents, or regrind that contains dirt particles may cause surface irregularities which will become even more obvious after metallizing. To help provide a smooth surface and to avoid degassing of the substrate, a base lacquer coat is first applied. The composition of these lacquers is proprietary with metallizers. The lacquer coat is thin and will not cover up ripple effects due to weld lines, imbedded particles, or other surface defects. After deposition of a thin metal layer [as little as 5×10^{-6} in. or 0.13 μ will completely mask the surface underneath (9)], a protective coating is finally applied. The total thickness of all three layers is normally less than 7×10^{-3} in. (0.18 mm).

Small nylon parts can often be vacuum metallized without predrying, but large parts with thin walls are best dried to avoid warpage and iridescence, a loss of brilliance that occurs when the metallic film is distorted (9, 10).

Electroless deposition is a process by which a metal in aqueous solution is caused to "plate out" on a substrate without the aid of an externally applied voltage or the driving force of electrochemical displacement. Deposition takes place even on nonconducting surfaces because of a chemical reduction selectively catalyzed at the surface of the substrate. Details of the procedures employed have been provided (10); in contrast with the vacuum process, electroless deposition requires a surface that has been roughened by mechanical or chemical means. Little data on the efficacy of roughening techniques are available (Table 17-3).

Table 17-3. Effect of surface treatment and roughness on peel strength of electroplated nylons

Nylon	Treatment	Surface roughness		Peel strength		Ref.
		$(10^{-6}$ in.)	$(10^{-6}$ cm)	$(lb(wt)in.^{-1})$	$(kg(wt)cm^{-1})$	
6	Vapor blasting, 200 mesh abrasive	53	135	2.5	0.4	11
6	Vapor blasting, 400 mesh abrasive	29	74	1.5	0.3	11
- -	Chemical	- -	- -	8	1.4	12
66	Chemical[a]	- -	- -	5.5-12	1.0-2.1	13

[a] Etched in 80/10/10 (wt) ethylene glycol monoethyl ether/conc. sulfuric acid/water 30 sec at 78°F (26°C).

Other

Hot-stamping of roll leaf (14) is used principally to mark or decorate thermoplastics. Parts with raised or depressed surfaces are equally amenable to hot-stamping, and no special surface treatment is required. Temperatures of 300 to 375°F (150 to 190°C) and dwell times of 0.5 sec have been used with nylons melting above 400°F (204°C). Flat or curved surfaces are acceptable, and roll leaf especially suitable for reinforced nylons is available.

Silk screening, direct printing with a rubber stamp, and wipe-in painting are still other marking techniques that have been applied to nylons. The preferred procedure may vary with the specific nylon composition and marking fluid. If adhesion of the coating is a factor, surface roughening helps. The ink or paint supplier is the best source of information for these techniques which are not widespread in the nylon plastics industry.

REFERENCES FOR ANNEALING AND MOISTURE CONDITIONING

1. Ensanian, M. *Nature* 193, 161 (1962).
2. Dunn, P., A. J. C. Hall, and T. Norris, *Nature* 195, 1092 (1962).
3. Glyco Products Co., Brooklyn, New York.
4. Standard Oil Co. of New Jersey.
5. Starkweather, H. W., G. E. Moore, J. E. Hansen, T. M. Roder, and R. E. Brooks, *J. Polym. Sci.* 21, 189 (1956).
6. Anon., "Zytel® Nylon Resins, Design and Engineering Data," Plastics Department, E. I. du Pont de Nemours and Co., Technical Release 152, August, 1966.
7. Ecochard, F., Societe Rhodiaceta, Nylon Department, Tech. Report No. 53, February, 1950.
8. Dunn, P., and G. F. Sansom, *J. Appl. Polym. Sci.* 13, 1641 (1969).
9. Chapman, F., *Plast,* p. 406 (Oct., 1957).
10. Anon, Foster Grant Co., Tech. Bulletin No. 2-62.
11. Anon., Aquitaine – Organico, "Rilsan" Technical Data Sheet 2, No. 155, 1969.
12. Anon., BASF Technical Bulletins, "Ultramid A, B, and S Resins," Nos. 81121, 81114, and 81129, July, 1969.

REFERENCES FOR ASSEMBLY AND MACHINING

1. Anon., "Application Design," Plastics Department, E. I. du Pont de Nemours and Co., Dec., 1961.
2. Kolb, D. J., *Mod. Plast. Encycl.* 46 (10A), 718 (Oct., 1969).
3. Horvath, L. S., "Ultrasonic Welding of Delrin® Acetal Resin," Du Pont de Nemours International S. A., Technical Release D4011, September, 1967.
4. Anon., Sonobond Corporation, Bulletin SP2/2M, June, 1970.
5. Williams, J. K., and James Mengason, "Ultrasonic Welding," Plastics Department, E. I. du Pont de Nemours & Co., Technical Release 171, 1971.
6. Anon., "Ultrasonic Assembly," Branson Sonic Power Co., Bulletin S-888.
7. Anon., "Cold Heading," Du Pont de Nemours International, S. A., Technical Release 7, April, 1962.
8. Anon., "Zytel® Nylon Resins, Design and Engineering Data," Plastics Department, E. I. du Pont de Nemours & Co., 1962.
9. King, A. F., "Adhesives for Bonding du Pont Plastics," Plastics Department, E. I. du Pont de Nemours and Co., Technical Release 152, August, 1966.
10. Smoluk, G. R., in *Engineering Design for Plastics,* E. Baer, Ed., Reinhold Publishing Corp., New York, 1964, Chap. 17.
11. Anon., "Engineering Design," Plastics Department, E. I. du Pont de Nemours and Co., 1970.
12. Anon., "Plastics Reference Issue," *Machine Design* 40 (29), (Dec., 1968).
13. Maranchik, John, Jr., "Machining Plastics," *Mod. Plast. Encycl.* 46 (10A), 695 (Oct. 1969).
14. Kobayashi, Akira, *Machining of Plastics,* McGraw-Hill Book Company, New York, 1967.

REFERENCES FOR SURFACE COLORING AND DECORATING

1. Bruner, W. M. (to E. I. du Pont de Nemours and Co.), U. S. Patent 3,235,426 (Feb. 15, 1966); Bruner, W. M., and C. M. Baranano, *Mod. Plast.* **39** (4), 97 (1961).

2. Snider, O. E., and R. J. Richardson, in *Encyclopedia of Polymer Science and Technology,* Vol. 10, Interscience Publishers, a division of John Wiley and Sons, Inc., New York, 1969, pp. 424-427.

3. Anon., "Surface Colouring of 'Ultramid' Articles," BASF Tech. Infor. Bull., Feb., 1969.

4. Anon., "The Coloring of Gulf Nylon," Gulf Oil Co. Tech. Bull., Sept., 1968.

5. Anon., "Coloration of Plastics," Tech. Bull., Organic Chemicals Dept., E. I. du Pont de Nemours and Co., Oct., 1965.

6. Anon., "Rapid Surface Dyeing of Molded Nylon Parts," Infor. Bull., Organic Chemicals Dept., E. I. du Pont de Nemours and Co., June, 1966.

7. Anon., "Wire and Cable Coaters' Handbook," E. I. du Pont de Nemours and Co., 1968, pp. 98-106; Heyman, E., Paper presented at Annual Wire Association Meeting, Oct., 1965.

8. Moffett, W. K., Fabric and Finishes Department, E. I. du Pont de Nemours and Co., private communication.

9. Anon., "Painting Plastics Manual," Bee Chemical Co., Lansing, Illinois, 1962.

10. Goldie, W., *Metallic Coating of Plastics,* Electrochemical Publications Ltd., Hatch End, Middlesex, 1969, Vol. 1.

11. Goldie, W., *Metal Finish. J.* **11**, 265 (1965).

12. Anon., *Chem. Eng. News* **44** (24), 80 (June 13, 1966).

13. Anon., unpublished information, Plastics Department, E. I. du Pont de Nemours and Co.

14. Gladen, C. F., *Mod. Plast. Encycl.* **46** (10A), 708 (Oct., 1969).

CHAPTER 18

Designing with Nylons

J. H. CRATE

INTRODUCTION

A speedometer take-off gear made of nylon-66 such as the one in Figure 18-1 has been operating for five years transmitting the number of revolutions of the automobile drive shaft to a flexible cable which in turn feeds the rpms to the speedometer which then displays the velocity of the car to the driver. The take-off gear performs silently, effectively, and quietly and is one of the many component parts of the auto that we take for granted. But, like the million other parts in today's automobiles, there was a time when it did not exist. However, once the need arose, a process started that eventually resulted in the gear as we now know it; this process is called design, or, more specifically, product design.

Automobiles are only one example where product design is relevant; actually it touches and influences and, for that matter, shapes all products that are mass-produced regardless of the raw material — metal or plastic; the nature of

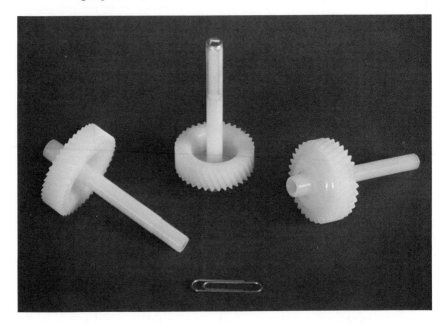

Fig. 18-1.   Speedometer take-off gear of nylon-66.

the device — toothbrush or computer; and the quantities made — one or ten million. Within the design process, different areas are encompassed including the functioning of the part (does the lever on the paint sprayer pivot freely for accurate control, is the force required and the shape suitable for hand operation?), the mechanical design (what size bearing should be used for minimum wear, and is the handle sufficiently rigid to withstand the loads involved?) and the appearance (is the part visually attractive, easy to clean?). Nylon, like other plastic and metallic materials, provides a physical means for the designer to satisfy the needs of the product.

Like all the creative arts and sciences, product design, its techniques and patterns, has evolved from the beginnings of man, starting with the simple hunting knife and developing to today's complex space craft. It is evolutionary too in another sense, that of materials. The first large-scale raw materials were the metals, and it was on these that the occurrence of the industrial revolution was based. Through hundreds of years these materials came to be considered the basis of the product designer's efforts — he played with metal toys, he was taught design using metals as a physical reference point, he worked professionally with metals because they were available, cheap, and the know-how was most nearly complete. With the introduction of plastics, no effect whatever occurred on the designer's thinking other than that this was a decorative material or a more readily available one in lieu of others, such as their use as a replacement for billiard balls. The designer, if he considered them at all, saw them only as a novelty. Though large business developed around plastics in the early twentieth century, particularly the nitrates and acetates, they were in fields apart from those of the hard goods product designer — clothing fitments, cosmetic appliances like brushes and combs, footwear, shirts and collars, and corset stays. However, in the early 1940s a number of events occurred that started to provide plastics with a completely new respect, particularly in the eyes of the designer: shortage of metals and labor, and the discovery of nylon. The designer at first found that he could substitute this new plastic for metals and that he could obtain equivalent if not better performance in applications such as gears and bearings.

Once metals were again available, he returned to the tried and true, and plastics were again consigned to the decorative aspects of our culture. Nylon continued to be used in gears and bearings and generally as a substitute material for another 15 years with little expansion into other applications. However, in the middle 1950s, the economies of thermoplastics were realized through high-speed molding and further incentive was given to their use in more diverse, performance-critical applications. An interesting note on nylon is that it was not generally considered a plastic during this time because it had proved satisfactory in gears for many years and had acquired a good reputation with manufacturers and designers, even in tradition-bound industries such as hardware and plumbing. Thus, it was readily accepted for many applications as a greaseless replacement for brass, and not only did it do the job, it inspired the development of the many varieties of engineering thermoplastics that are marketed today. But even then it was used as a replacement for other more expensive or short-supply materials. Today, however, nylons along with the other plastics have reached a new level of importance in a host of products manufactured worldwide. It is no longer solely a substitute for other materials in the sense that it is shaped to replace a metal part; it is a basis for the designer's efforts around which he determines the geometry of his part designs. This is certainly not true in all cases, but the approach is established, and its value has been proven by many applications which have been performing well for a number of years, such as this timing chain sprocket of nylon 66 shown in Figure 18-2.

The key to efficient product design is the proper use of material, that is, designing around the properties of the material to be certain that the design takes full advantage of these properties. The fact that many designers are doing this is seen in the plastic products now marketed. But there are many products where this is not the case because the designer, still educated on metals, unintentionally restricts his thinking and therefore does not make efficient use of plastics. This is the designer who hopefully will someday become alert to the new directions to be found with plastics, and it is to this designer that we must direct our attention.

Fig. 18-2.   Timing chain sprocket of nylon-66.

DESIGN FACTORS

Innovation and Combined Functions

To achieve a genuinely innovative design, the designer must first familiarize himself with the materials' properties, processing behavior, and adaptability to secondary processing methods or assembly techniques. Only with this knowledge will he be able to relate the materials to the forms required for the product under development. In this way too he will be able to see the likenesses and differences between metals and plastics. He will find that he can use the same basic engineering formula with plastics that he uses with metals (3). On the other hand, he will see that plastic parts are more sensitive to the environment in which they are used and their design will require more careful considerations of temperature, moisture levels, and the duration of loading. Finally, he will learn that to be practical, both functionally and costwise, plastic parts must be designed with a lower factor of safety than is used with metals, and he must pay more careful attention to the refinement and simplicity of the design. The same process leads to an understanding of the similarities and differences among nylons and to the best choice of a particular nylon for a particular design.

Innovation is essential to the success of part design even in metals, but with plastics in general and nylons in particular it takes on a new dimension by virtue of the flexibility and scope of the molding process. As an example, a single processing operation can provide us with a gear, bearing, cam, and shaft in one piece with each portion having the surface finish, strength, and resilience required by its function. Parts designed in this manner are said to have "combined functions." By comparison, a similar product of metal would normally be made up of four separate parts which are then assembled. The designer must think in terms of combining functions to obtain the true value of the nylon and to gain an advantage over competitive materials.

Doing something the same old way with a nylon is incompatible with improved product design - the designer must take another look; he must innovate to find the real advantages. The speedometer take-off gear still turns because the designer chose a nylon with the proper combination of properties — strength, resilience, wear resistance, and low coefficient of friction — and he shaped the part by combining functions that resulted in an innovative solution at the right price.

The Stress-Strain Curve

The core of the majority of structural design problems centers around the stress-strain curve of the material under consideration. The formulas that are used are based on a curve that is a straight line and has the same slope in both compression and tension. In other words, the modulus of elasticity of the material is a constant and equal whether under tension or compression. In the case of metals such as steel, this is a valid condition; however, in the case of most plastics, and this would include nylon, this assumption is not correct, particularly at high levels of stress or strain. Therefore, it is highly desirable to design from data from the stress-strain curve rather than from single-point property values which are useful primarily for comparison of materials. (Examples of such stress-strain curves are shown in Chapter 10 and are available from raw material suppliers for specific nylon compositions. An analysis of the dependence of tensile properties on temperature and moisture can be found in Chapter 9.) When selecting the curve for reference, the one which applies to the end-use conditions of the part in question must be considered. If the part were an electrical connector which was to be snap-fitted to a metal chassis, the snap-fit elements would be designed around the material properties existent during the assembly operation. If these conditions were 73°F (23°C) and 50% R.H., the stress-strain curve at these conditions would be used to calculate the wall thickness of the snap fit. Knowing the deflection, the strain can be calculated from the standard deflection formula and the stress picked off the stress-strain curve. If the conditions at assembly are not known and could vary between 60 and 110°F (15.6 and 43°C) and 0 and 50% R.H., the curve that gives the most severe conditions should be used. For example, for a given strain, the lower the temperature and humidity, the higher the level of stress; hence, the part should be designed on the basis of the stress-strain curve at 60°F (15.6°C) and 0% R.H. On the other hand, if a fixed load is to be applied to the part and the amount of maximum deflection is critical, data at the higher levels of temperature and humidity, where the material is more flexible, would be used.

Tolerances

The designer must check carefully the need for close tolerances and whether more precise tolerances will be worth the additional cost. Regardless of the economics, it may be unreasonable to specify close production tolerances on a part designed to operate through a range of environmental conditions. Dimensional changes due to temperature variations alone can be three to four times as great as the specified tolerances. Also, in many applications, close tolerances with nylon plastics are not as critical as with metals because of their resiliency.

Many factors control the production of precision parts, and some of these are noted below:

1. A design for a plastic part should indicate the conditions under which the dimensions shown must be held. For example, a drawing should state that dimensions shown are to be as specified after molding, after annealing, after moisture conditioning, and so on.

2. Over-all tolerances for a part should be shown in inches per inch or in millimeters per millimeter, not in fixed values. As an example, a title block might read, "All decimal dimensions ± 0.00 X in./in. or ± 0.00 X mm/mm (not ± 0.00 X in. or ± 0.00 X mm), unless otherwise specified."

3. Only those tight tolerances required for specific dimensions should be labeled as such. Less important dimensions can be controlled by over-all tolerances.

Generous molding tolerances should be allowed in any areas which will be machined after molding. Production variables such as the number and size of cavities in a mold also affect tolerances. Where compromises in tolerances may be acceptable from a performance standpoint, discussion of such tolerances with the molder may result in economies. For example, the use of a multicavity mold is usually an economical production method. But, as the number of cavities per mold increases, so must the tolerances on critical dimensions. An increase of 1 to 5% per cavity is about average. For example: dimensions of a part produced in a single cavity mold may be held to ± 0.002 in./in. (± 0.002 mm/mm). When the number of cavities is increased to 20, the closest tolerances obtainable may be ± 0.004 in./in. (± 0.004 mm/mm).

Appearance and Structure

The appearance of molded nylon has had an important effect on its acceptance as a marketable material. Its distinctive, natural, bonelike coloring served to identify it from other plastic materials when first introduced. A roller, gear, or bearing of nylon and its color became a mark of quality in the local hardware store and the commercial warehouse. Today, it can be molded or extruded in any color desired, though most manufacturers maintain an inventory on only the most popular ones. Not only single colors but blends of color can be used to achieve decorative effects. An excellent example is the gunstock in Figure 18-3, which has been commerical for a number of years. It uses a variety of colors which blend sufficiently in the molding machine to provide a smooth, mottled surface that closely resembles the grain in wood.

The shape of nylon parts is limited only by the need to remove the part from the mold; hence, undercuts, unless their need warrants the additional tooling expense, should be avoided. The designer is free to use the range of shapes from precise geometric to free form with the knowledge that it can be flawlessly reproduced through the molding process. However, the surface quality of a part can be affected by structural elements such as wall thickness, ribs, and fillets (radiused internal corners) because of their effect on shrinkage during molding. Such internal elements can cause surface depressions or other visible distortions. The wall thickness must be held constant whenever possible. In part (a) of Figure 18-4, the melt cools evenly throughout the part and the solid conforms to the shape of the mold. In part (b), the heavy section cools after the thinner has solidified, and when the core of the heavier section becomes solid, depressions occur on the outside surface. This problem could be overcome by relocating the

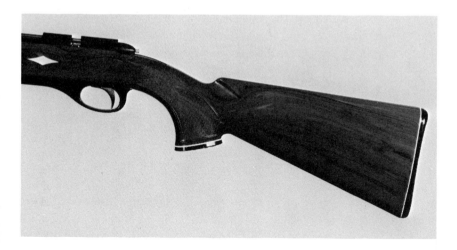

Fig. 18-3.   Gunstock of nylon-66.

gate in the heavier section as shown in part (c). However, the gate location is often restricted, and this solution can rarely be counted upon to solve sinks caused by variations in wall thickness. Even a simple corner on a part can provide problems due to the variations in wall thickness that exist (Figure 18-5). Problems could be: molded-in stresses, warpage, sinks, voids, and even variations in tolerance. The warpage that would occur could look like that in Figure 18-6 due to the slower cooling of the long edge along the corner of the part. One solution to this situation is shown in the center sketch of Figure 18-7 — simply use a rounded exterior corner to provide an even wall thickness. This is not always feasible from a functional standpoint when a sharp exterior corner is required, but the solution shown in the right-hand sketch retains the corner by placing a small extension on the exterior curve.

Another constantly recurring design feature of molded parts is the use of ribs or reinforcements. The increase in wall thickness that occurs at the juncture of a rib and a wall (Figure 18-8) could cause problems similar to those created with a corner design. The rib should be made thinner to allow even cooling of the outer wall and to prevent a sink mark over the rib area on the exterior surface of the part. Gear design is an excellent example of the need for constant wall thickness (Figure 18-9). In sketch (a), the ideal place for gating is shown at the hub of the gear. If there is a molded-in steel shaft, the next best place to gate the gear is in the web [Sketch (b)]. Two less desirable places to gate the gear are in the edge of the gear (c) and one of the tooth profiles (d). The first will cause possible warpage due to unequal pressure throughout the gear when it is being molded, and the second could cause distortion to some of the gear teeth and eventual wear problems. Figure 18-10 shows one way to provide even wall thicknesses in a gear. Sometimes, two gears are combined. The design on the left contains two

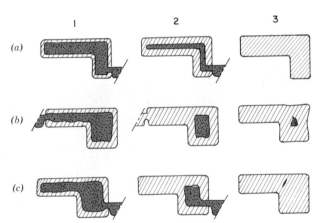

Fig. 18-4.   Effect of varying wall thickness on shrinkage: (a) constant wall thickness, (b) varying thickness, and (c) gate in thick section to compensate for variation in thickness.

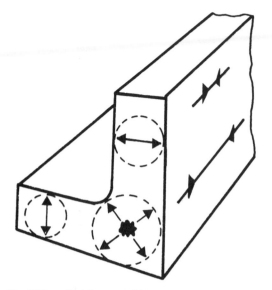

Fig. 18-5.   Shrinkage possibilities at corner of molding.

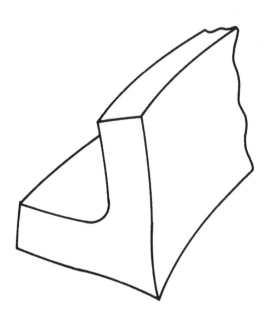

Fig. 18-6.   Example of shrinkage at corner.

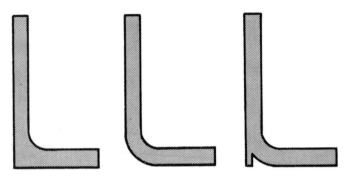

Fig. 18-7.   Alternative edge designs.

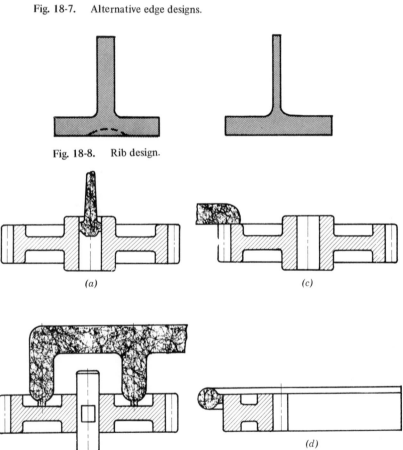

Fig. 18-8.   Rib design.

(a)

(c)

(b)

(d)

Fig. 18-9.   Gear designs: *(a)* hub gated, *(b)* web gated, *(c)* gated at edge of gear, *(d)* tooth gated.

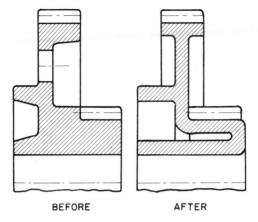

BEFORE   AFTER

Fig. 18-10.   Redesign of gear for constant wall thickness.

areas which are heavier in section and could cause warpage. These can be eliminated by moving the web of the smaller gear axially and centering the web of the larger one. This change should provide the even wall thickness and constant shrinkage that are desired to provide a part within tolerances.

The other basic factor essential to both good molding and also to good part design is the use of fillets on inside corners (Figure 18-11). The optimum size of the fillet is half the thickness of the wall of the part. Actually, a good rule of thumb is to make the inside fillet radius as large as possible up to a point that the heavy wall developed will not lead to the corner distortion problem that we have just discussed. Another general rule that can be applied to part design which is also helpful in molding a part is that the designer should, if possible, eliminate all abrupt discontinuities in the geometry of the part. This will not only provide a better flow pattern when the part is being molded, but it will also minimize stress concentration when the part is under load.

Assembly and Cost Considerations

The opportunities to minimize assembly and to combine functions are probably two of the most important considerations in the design of parts of nylon. As mentioned earlier, these are the factors which most often can make the design competitive with other products made of less expensive raw materials. The ability to mold the assembly function as an integral portion of the molded part immediately eliminates the need for accessory assembly devices. These methods of assembly include snap-fitting, spin and ultrasonic welding, and self-tapping screws, and are discussed in Chapter 17. Nylon can be assembled to other materials by means of integral fastening devices that also avoid the cost of clips, nuts, screws, or other special assembly devices.

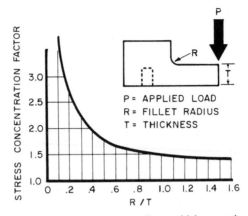

Fig. 18-11.   Stress concentration versus fillet radius to thickness ratio.

The costs of alternative materials to serve the same function can be compared as shown in Table 18-1. Stiffness depends on the first power of the elastic modulus if loaded in pure tension or compression as in a strut end loaded in tension, a sandwich type of construction where the skin is similarly loaded in tension or compression, and a shell type structure such as pipe. Table 18-1 ignores the fact that a combination of properties rather than a single property most often determines the choice of resin; it also disregards processing costs and service life. It is no substitute for a careful analysis of value-in-use (see Chapter 19).

SPECIFIC STRUCTURES

Some applications for nylon have been in use for many years, and the know-how required to design these parts has been well documented. The following examples of three of the most basic structures made of nylon will serve to illustrate the type of data available and the ease with which it can be used in basic engineering formulae.

Gears

Gears of nylon are designed with conventional formulas using the material properties existent at the temperature and humidity of the gear when it is in use. For example, a designer wishes to use a gear in an appliance that will be lubricated only once at the time of assembly. Nylon is a good choice here because of its excellent performance as a gear with only initial lubrication. The type of nylon to be evaluated is 66 because it has a long and successful history in gear applications.

Table 18-1. Cost comparison of alternative materials

Function	Comparison	Comment
Space filling	$$C_B = W_A \times \frac{d_B}{d_A} \times P_B$$	--
Strength Tensile or compressive loading	$$C_B = W_A \times \frac{d_B}{d_A} \times P_B \times \frac{S_A}{S_B}$$	Strength proportional to wall thickness
Flexural loading	$$C_B = W_A \times \frac{d_B}{d_A} \times P_B \times \left[\frac{S_A}{S_B}\right]^{1/2}$$	Strength proportional to (wall thickness)^2
Stiffness Tensile or compressive loading	$$C_B = W_A \times \frac{d_B}{d_A} \times P_B \times \frac{E_A}{E_B}$$	Stiffness proportional to modulus
Flexural loading	$$C_B = W_A \times \frac{d_B}{d_A} \times P_B \times \left[\frac{E_A}{E_B}\right]^{1/3}$$	Stiffness proportional to (modulus)^3
Impact	$$C_B = W_A \times \frac{d_B}{d_A} \times P_B \times \frac{I_A}{I_B}$$	

C = material cost per part
P = material cost per unit weight
W = part weight
d = specific gravity
S = tensile or compressive strength
E = modulus of elasticity
I = impact strength

Subscript A: material in existing part.
Subscript B: material proposed as a replacement.

The gear will be molded, 32 pitch (number of teeth per inch of pitch diameter), 20° full depth, 2.5 in. (6.35 cm) pitch diameter, 80 teeth, and will have a 0.50 in. (1.27 cm) face width. The gear will operate at 2000 rpm, average room temperature and moisture with initial lubrication, and an expected life of 1000 hr. Expected gear loadings are an operating torque of 30 in.-lb (wt) (0.346 m-kg (wt)) and a stall torque of 100 in.-lb (wt) (1.15 m-kg (wt)).

Total expected life is first calculated:

2000 revolutions/min × 1000 hr × 60 min/hr = 120 million cycles

The pitch line velocity is then determined from:

$$\text{P.L.V.} = \frac{\pi \times 2.5 \text{ in. pitch diam.} \times 2000 \text{ rev/min}}{12 \text{ in./ft}} = \frac{1310 \text{ ft/min}}{(400 \text{ m/min})}$$

The maximum recommended bending stress of 4400 lb (wt) in.$^{-2}$ (316 kg (wt) cm^{-2}) for nylon-66 is taken from test data of bending stress versus life cycles for 32 pitch gears.

The K factor of 0.70 for nylon-66 is obtained from material manufacturers' literature which reflects conditions of lubrication, whether or not teeth are hob cut, operating speeds, and pitch. The form factor Y is similarly chosen to reflect the number of teeth on the gear and the form of the teeth, a value of 0.74.

With this information, the maximum torque based on bending fatigue can be determined from

$$T = \frac{S_b D_p f\, YK}{2P_d} = \frac{4400 \times 2.5 \times 0.5 \times 0.74 \times 0.70}{2 \times 32} = \frac{44.5 \text{ in.-lb (wt)}}{(0.512 \text{ m-kg (wt)})}$$

Now the maximum torque due to stall loading can be determined by using the yield point stress for nylon-66 at room temperature (73°F or 23°C) and 50% R.H. and again using the torque equation without the design factor K:

$$T = \frac{S_b D_p f\, Y}{2P_d} = \frac{8500 \times 2.5 \times 0.5 \times 0.74}{2 \times 32} = \frac{123 \text{ in.-lb (wt)}}{(1.415 \text{ m-kg (wt)})}$$

Both of these calculations show that the gear should perform satisfactorily in the intended use.

Bearings

Nylon bearings have found use in a variety of commercial applications including hemispherical bearings for auto ball-joints, food mixer bearings, housing-bearing combinations, wear-surfaces on integral gear-spring-cam units for adding machines, and clock-bushings. These uses are attractive in nylon because they can take advantage of the low coefficient of friction, self-lubrication, excellent mechanical properties, and freedom from toxicity and taste that permits its use in food handling applications where FDA or National Sanitation Foundation approvals are required.

The suitability of a material for bearing applications is determined by the material's PV limit which is the product of limiting bearing pressure lb (wt) in.$^{-2}$ and peripheral velocity ft/min (m/min), or bearing pressure and limiting velocity, in a given dynamic system. When the PV limit is exceeded, the bearing may fail due to melting, cold flow, or binding. Because of the effect of temperature and humidity on the bearing properties of nylon, a generous safety factor must be used in design. Details on PV limits, wear, and factors of safety can be obtained from material manufacturers literature, but here a few general guidelines will suffice:

1. Design bearing sections as thin as consistent with application requirements to maximize heat conduction through plastic material adjacent to the bearing surface.

2. Metal/plastic bearing interfaces run cooler than plastic-plastic interfaces because heat is conducted from the interface more rapidly by metal than plastic. Metal/plastic bearings have higher PV limits than plastic-plastic bearings.

3. Provision for air circulation about the bearing can bring about cooler operation and increased bearing-life.

4. Lubrication can increase the PV limit of nylon multifold depending on type and quantity of lubrication. Nevertheless, nylon is used in many nonlubricated bearings. Where lubricants are used, these must be stable at the bearing temperature.

5. Water is not an effective boundary lubricant for nylon because it does not wet the plastic bearing surface. It should not be used to lubricate bearings of metal on nylon because rusting or corrosion of the metal can lead to rapid abrasive wear.

6. For sleeve bearings the length of the bearing should not exceed the shaft diameter.

7. Sliding unlubricated surfaces of nylons on zinc or aluminum surfaces is not recommended.

8. Bearing clearance is essential to allow for thermal expansion or contraction and other effects. For sleeve bearings of nylon operating against steel in a steel or cast-iron bearing, a rule of thumb for installed, room temperature clearance is C_d = 0.0005 in. + 0.025t + 0.001d or, in metric units, C_d = 0.0125 mm + 0.025t + 0.001d, where C_d = diametral clearance, in. (mm); t = wall thickness of plastic, in. (mm); and d = shaft diameter, in. (mm). For other bearings use 0.3 to 0.5% diametral clearance.

9. Surface grooves should be provided in the bearing so that wear debris may be cleared from the bearing area. For lubricated bearings, the groove can increase the supply of lubricant. Bearing pressure will increase with grooving.

10. For bearing applications in dirty environments, use of seals or felt rings can increase bearing life if they are effective in preventing penetration of dirt into the bearing.

As with other applications, bearings in nylon should be considered as one of a group of functions within the product under study, not as a part to be added to the assembly. By so doing, the designer may be able to incorporate the bearing function as part of a frame or housing, thereby reducing the number of parts and subsequently the manufacturing cost of the product.

Pressure Vessels Under Long-Term Loads

As previously noted, it is essential for the designer to take note of the end-use requirements and environment of a part before attempting to determine its geometry. This is particularly true of a pressure vessel where safety is such a critical factor. In this example, we will determine the side wall thickness of a gas container which must meet these requirements: It must retain pressure of 106 lb (wt) in.$^{-2}$ (7.45 kg (wt) cm^{-2}) for ten years, at 150°F (65°C).

The outside diameter of the cylinder is 0.70 in. (1.78 cm), and the length is 2 in. (5.08 cm). Because the part will be under pressure for a long period of time, we cannot safely use short-term stress-strain data but should refer to creep data or, preferably, long-term burst data from actual pressure cylinder tests. Figure 18-12 shows data typical of this sort for nylon-66 which plots hoop stress versus time to failure for various moisture contents at 150°F (65°C). Actually, nylon-66 would be a good candidate for this application because it has high impact strength in the 50% R.H. stabilized condition and one of the highest yield strengths of the unreinforced nylons. Referring to the curve, we find a hoop stress level of 2700 lb (wt) in.$^{-2}$ (190 kg (wt) cm^{-2}) at 10 years and this can be used as the design stress. The formula for the thickness of a pressure vessel is

$$t = \frac{P D_O}{2 S + P} \times \text{F. S.}$$

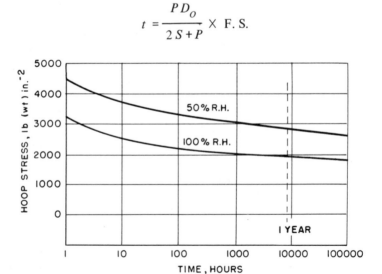

Fig. 18-12.   Hoop stress versus time for nylon-66 at 50 and 100% R.H.

where

t = wall thickness (in.) (cm)
P = internal pressure $(lb(wt)in.^{-2})\ (kg(wt)cm^{-2})$
D_O = outside diameter (in.) (cm)
S = design hoop stress $(lb(wt)in.^{-2})\ (kg(wt)cm^{-2})$
F.S. = factor of safety = 3

$$t = \frac{(106)\,(0.70)\,(3)}{(2)\,(2700)+106} \quad \text{or} \quad \frac{(7.45)\,(1.78)\,(3)}{(2)\,(190)+7.45}$$

$$= 0.04 \text{ in.} \qquad \text{or} \qquad 0.11 \text{ cm}$$

The ends of the cylinder present a more complex design problem, particularly if the cylinder is to stand upright because the best shape is a hemisphere. Were a flat surface used, it would buckle over a period of time and the cylinder would become unstable. The best solution, therefore, is to mold a hemispherical end with an extension of the cylinder or skirt to provide stability (Figure 18-13).

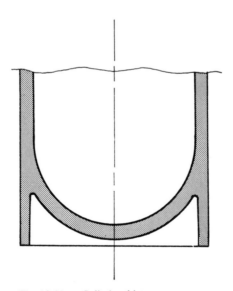

Fig. 18-13.   Cylinder skirt.

REFERENCES

1. Peckham, M., *Man's Rage for Chaos,* Chilton Books Div., Chilton Co., Philadelphia, Pa., 1970.
2. Ogorkiewicz, R. M., *Engineering Properties of Thermoplastics,* Wiley-Interscience, a division of John Wiley and Sons, New York, 1970.
3. The best and most reliable examples of application of standard engineering formulas to the design of plastics are to be found in the design literature of the raw material suppliers.

CHAPTER 19 _____

Economic
Considerations

G. L. GRAF, JR.

INTRODUCTION

The rate of growth in the use of plastics in the United States is shown for the years 1950-1970 in Figure 1-1 (Chapter 1, page 4). The rapid growth of plastics compared with other materials is emphasized by the usage trends for the same two decades shown in Figure 19-1. The acceptance of plastics as a material of construction, with a growth rate 2 to 6 times that of the other materials is due to their recognized value-in-use.

The designer of a plastic part which provides the optimum service at the

607

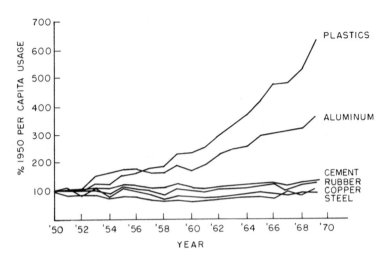

Fig. 19-1.   Basic materials usage (data from Ref. 1).

lowest cost to the user must consider material cost, processing cost, mechanical properties, and those special properties inherent in plastics (2). Some of these special properties include electrical and thermal insulation, favorable coefficients of friction, corrosion resistance, coloring without secondary operations, and lower maintenance cost. When the designer considers these items — material, processing, finishing, assembly, and maintenance costs in a design problem, the value-in-use of the material is brought into the picture. Nylons have proved themselves well suited to this design approach.

In the selection of materials, the designer should determine the cost required for equivalent performance in use (see Chapter 18, p. 601). Nylons are available in a wide range of compositions including glass fiber reinforcement. This selection provides the designer with considerable freedom to optimize the value-in-use.

Good part design is also important to economic production. The part should have the simplest possible configuration and minimum permissible section thickness to reduce material content, simplify production, and minimize tooling cost. Simplification of the design, however, should not be carried to the point that expensive assembly of multiple components is required. The variety of processing techniques applicable to nylons should be used to the fullest.

MATERIAL COST CONSIDERATIONS

A comparison of costs for competitive engineering materials (Table 19-1) shows a 25-fold spread from the least to the most costly. Nylons-66 and -6, which comprise the major fraction of the nylons used, are in the middle of the spectrum based on cost per unit volume. The number of nylons which are technically capable of manufacture is very large (see Chapters 2 and 11 and the Appendix), but cost and property requirements have so far limited the number of commercially available homopolymers to those shown in Table 1-4. The costs for five unmodified nylon homopolymers and for copolymers are compared in Table 19-2.

Material cost is a major element of the sales dollar for any plastic article. The Society of Plastics Industry in its biennial studies "Financial and Operating

Table 19-1. A partial comparison of engineering material costs

	Price, ¢/lb [a]	Cost, ¢/in.[3]
Polypropylene	21–26	0.68–0.98
ABS resins	30–45	1.14–1.72
Magnesium AZ 91-B	31.5	2.06
Modified polyphenylene oxide	59–75	2.34–2.87
Steel – Drawing Qual. Sheet	9.81	2.78
Aluminum SAE-309	28.75	2.74
Nylon-66 or -6	73–83	3.01–3.42
Polyacetal	65	3.34
Polycarbonate	75	3.25
Zinc SAE 903	18.50	4.40
Polysulfone	100	4.48
Polyphenylene oxide	150	5.75
Brass – yellow #403	47.25	14.49

[a] Representative pricing in the United States for first quality base material on July 1, 1971.

Table 19-2. Comparative cost of some nylon polymers

Nylon	Price, $/lb [a]	Cost, ¢/in.[3]
66 or 6	0.73–0.83	3.01–3.42
610	1.13–1.18	4.44–4.64
612	1.10–1.13	4.17–4.28
11	1.35–1.55	5.07–5.82
12	1.35–1.55	4.97–5.71
Copolymers	0.73–2.18	3.01–3.48

[a] Representative pricing in the United States for first quality unmodified base materials on July 1, 1971.

Ratios – Plastics Processing Companies" has consistently reported that the average material cost is 35 to 50% of the sales dollar, depending on the process used to manufacture the article. This applies specifically to nylon; hence, the designer and the end user should get the maximum level of utility from the material. Generally, when a nylon is chosen, the breadth of properties results in an article whose value is considerably in excess of the basic material cost.

Although many nylon articles are produced from unmodified resin, a large number of applications require the use of modified or colored compositions. It is not unusual for these special materials to be offered by the polymer supplier at a premium price. Where colored articles are required with only moderate color matching tolerances, consideration may be given to the surface coating of natural granular resins with pigments or blending with pigment-resin concentrates to obtain the desired colored feed stocks. Where color is desired only for coding or identification purposes, articles molded from natural resin may be surface dyed.

Modified compositions may be obtained economically in some situations by in-house compounding. As an example, glass-reinforced compositions with higher glass fiber content may be blended with unreinforced resin to achieve intermediate levels of reinforcement.

Since material is a considerable element of production cost, care must be used in handling both virgin and rework feed stocks to the various processes. The previous chapters on processing suggest procedures which conserve and protect the materials being handled. Where stock shapes are being used for production, consideration should be given to the recoverability and market value of all cuttings, trim, and misprocessed stock. It may be sold or arrangements may be made to have it reworked.

Costs are ultimately influenced by part design and performance in end use. It is essential that the designer know the end-use service conditions and the performance requirements of the part in order to design for the optimum use of the material (Chapter 18). This will result in minimum part sections and the maximum number of articles per pound of resin. The maximum yield of useful articles will then depend on good design as well as good process control and efficient use of rework material as discussed in the processing chapters.

PROCESSING COST CONSIDERATIONS

A number of processes are used for the production of articles of nylon. The processes may be divided into three general categories: melt processing, monomer casting, and nonmelt processing. The melt processes, in order of their volume of material consumption, are molding (Chapter 5), extrusion (Chapter 6), and powder processing (Chapter 14). Nonmelt processing, which is less important to nylon plastics than melt processing, includes forming (Chapter 15), stock shape machining, assembly, and other secondary operations (Chapter 17) applied to melt-produced basic shapes. The choice of nylon and fabrication technique depends on production volume, part design, service conditions, and cost.

Melt Processing of Nylons

Injection Molding

Injection molding is one of the most widely used processing techniques whether judged on the basis of pounds of material used, the numbers of articles produced, or the range of application of the products. In considering the economics of injection molding, one must evaluate the cost of materials, processing, and tooling. Material costs are discussed above. The economics of tooling and processing is treated mathematically in an article by Goettel (3).

Processing costs are determined by the molding cycle and the size and value of the equipment used in the molding process. Cycles are dependent on the heat transfer process in converting granular polymer to a melt and then cooling this melt in the mold cavity to the desired end-use article. The economic value and size of the machine used depend on the number of cavities in the tool and the required delivery rate of molten material. The matter of basic molding cycle is dealt with in Chapter 5.

One finds few references in the literature to the sizing of injection molding machines, specifically where polymers with the high crystallinity and melt temperatures typical of most nylons are concerned. The standard capacity rating for injection molding equipment based on polystyrene should be reduced when considering the unit for processing nylon. If a general purpose screw is used in a reciprocating screw machine, a projected polymer delivery rate 50 to 75% of that for polystyrene should be assumed. Some relaxation from this level of reduction can be obtained by the use of screws specially designed for the processing of nylon polymers. This type of screw, as described in Chapters 5 and 6, offers shorter cycles with lower, more uniform polymer temperatures. An analysis (4) of screw performance with high melting, crystalline polymers has indicated that with specially designed screws used in a 20:1 L/D barrel, it is of further advantage to utilize 75% or less of the potential travel of the screw during the injection stroke. The use of a shorter ram travel reduces the melt recovery time and the overall molding cycle. This implies that a larger machine will be required for the specific molding job, and it might be assumed that a higher cost would be encountered. However, the differential increase in investment versus the cycle reduction is such that lower costs per pound of material processed are achieved. Similar economic improvement should be available from equipment with higher L/D ratios for the barrel.

The shortest molding cycle is not always the preferred one; dimensional or property control requirements may be overriding.

Tool design as well as operating procedures affect the productivity of the molding process and the final cost of the molded article. In selecting tools, production and tooling costs must be compared. One has the opportunity to choose between molds with integral cavities, unit dies with insert cavities, or family type molds. The selection usually depends on product quality, sales volume, and part and tool costs. Consideration must also be given to the fact that if the article is to be custom molded, the customer bears the cost of the tool.

For large volume requirements, molds with integral cavities usually give low overall costs. Smaller production volumes are generally suited to insert cavity operations. However, the use of insert cavity tooling is generally not applicable where colors or special compositions are required. This results from the fact that a particular part is molded in a common mold base with a number of other items. The same resin, usually a general purpose one, must be suitable for all these parts. The cycle used may not necessarily be that which would be the optimum for the part you are buying. Quality may be somewhat less than is desired for critical applications.

In family type tooling a number of different parts, usually for the same application, are included in a single mold. This reduces tool costs; however, if the parts do not all have similar section thicknesses, the molding cycle may be longer than is necessary for certain parts produced within the molding shot. Family molds to some degree have the attributes and problems of the insert type mold and molding.

In the case of the integral cavity molds, one has a number of opportunities in tool design. The tool may be designed for automatic or normal semiautomatic operation. It may have a standard knockout runner system, or it may have a melt runner system which discharges only finished parts and no runner.

The selection of tooling for automated operation generally depends on the total production volume, part design, part complexity, molding equipment available, and the experience of the molding company. Complex parts requiring special core retracting mechanisms or unscrewing devices can generally be automated but only at a considerable investment. Production of parts without the necessity for having a runner system which must be recovered after each cycle has many advantages. However, the cost of this special design may counterbalance the economics of a standard runner system.

Extrusion

Extrusion processing yields shapes such as rod, sheet, and tubes as well as profiles, film, coated stock, and laminates. As in the case of injection molding, the economics of the extruded product is dependent on the cost of materials, processing, and tooling. In many instances, the extruded products are further processed mechanically into a variety of intricate parts.

From the processor's viewpoint, material costs are usually major elements of the ultimate product price. In the extrusion process for certain types of products it may be necessary to use nylon compositions which have special melt or solid state physical characteristics. These compositions usually bear a premium price. Hence, minimizing startup losses and prompt recovery of rework materials are of greater importance for an economical operation. All feed stock, whether virgin resin or rework, must be kept clean and dry to avoid generation of unusable scrap. Coating and laminating require especially careful control since the resin applied to a substrate is irrecoverable.

The continuous nature of the extrusion process generally results in low conversion costs. However, the cost of extrusion for a nylon can be expected to be higher than those for polyethylene or polystyrene for a number of reasons. These higher costs are due to the high melting point and crystallinity of the polymer and the need for dry feedstock.

The high melting point and crystallinity of nylons results in a large thermal energy demand from the extruder. As a result, it can be expected that the hourly output for any specific extruder will be 50 to 75% of its rating for lower melting or amorphous polymers. The energy requirements for the extruder per pound per hour output will be greater; the maximum output can be expected from extruders with high L/D ratio and high horsepower.

In extrusion, the pressure at the die and in the cooling section of the process is generally an order of magnitude lower than those used in injection molding. Therefore, the moisture content in the polymer as fed to the extruder and in the melt must be sufficiently low to preclude bubbling of the extrudate prior to freezing of the extruded shape. Virgin extrusion grade polymer, as delivered to the processor, will have the required low moisture content. However, rework and partially processed lots of material will usually require drying at a nominal cost with equipment specially designed for hygroscopic materials.

Powder Processing

Processing of powdered (Chapter 14) or granular nylons involves deposition and fusion onto a substrate, fusion and coating the interior of a hollow mold, or pressing and sintering into a basic shape. We shall consider only the first two techniques.

In the first case a nylon coating is applied to a substrate using either a fluidized bed or electrostatic deposition. Traditionally, a high-cost coating is applied to a low-cost substrate which is unsuitable for processing by other techniques. The coating process may be continuous or intermittent. The material and processing costs are usually higher than those for normal melt processing. The advantage for the process results from the fact that the articles being coated are generally difficult or impractical to make solely out of plastics. Hence, the value obtained cannot be truly measured against the other techniques.

The coating thickness is generally over 0.005 in. (0.13 mm) if obtained via fluidized bed, 0.001 to 0.005 in. (0.02 to 0.13 mm) via electrostatic deposition of powder and may be even thinner if one of the lower melting, alcohol-soluble nylons is applied from solution. The comparative economics of powder processing versus solution coating will depend on the relative costs for powder nylon and solvent plus granular nylon and high- and low-temperature curing processes. Typical costs for a powdered nylon-11 and soluble terpolymer are $2.50 and $1.85 per pound, respectively. A discussion of the comparative economics of powder coating will be found in Ref. 5.

The fusion coating technique known as rotational molding is used primarily for the production of large hollow articles. Either powder or granular resin is applicable to the process. Low tooling costs and long processing cycles are typical. To date, nylon has not found wide range usage in this process.

Other

Compression molding is a method of melt processing which is not used widely with nylons. The long cycles required in the conventional process to melt the granular polymer, mold, and cool have generally been uneconomical. More recent developments in which the melt is generated independently and delivered to the mold for pressing and cooling have reduced processing costs. In general, compression molding has been restricted to the production of flat stock of moderate thickness.

Blow molding is a combination of molding and extrusion processes. The processing technique has found wide usage, especially with low cost polymers. Although the conversion costs for blow molding are low, the value-in-use for blown nylon containers has not been sufficiently high to generate any sizeable markets.

Monomer Casting

Casting of monomers to produce nylon articles (Chapter 13) is well suited to the production of large items with thick cross sections or hollow objects at low conversion costs. Material costs are lower since the feed stock is a monomeric form of the nylon rather than a polymer which must be melted prior to molding. The processing cycles are relatively short, especially when heavy section articles are considered. The casting process offers the opportunity of lower equipment investment since lower operating pressures are used. Since lower pressures are used, tooling costs can be lower. As the section thickness of parts decreases, the economic advantage decreases. Melt processing, such as injection molding, becomes more competitive.

The number of monomers which may be cast is limited, but foamed and filled compositions are readily produced.

Nonmelt Processing of Nylons

Nonmelt processing includes both techniques for conversion of a stock shape into a new form by machining, pressure forming, or thermoforming, and secondary operations such as assembly, annealing, or decorating.

Shape Formation

Machining of stock shapes of nylon is of two types: stamping or blanking and screw machining. In both cases, the feed stock or raw material is a semiconverted form of the polymer such as sheet, rod, tube, and so on. The choice of these processes depends largely on the volume of the application.

The raw material costs are relatively high since the nylon has already been subjected to one processing step in the fabrication of the feed stock. In addition, processing losses due to trim or cuttings can be high. As an example, if 1 in. (25 mm) OD × 0.5 in. (13 mm) ID washers are blanked from a single 1.125-in. (29-mm) strip with 0.0625-in. (1.6-mm) spacing between blanks, the ratio of product to trim is approximately 1:1. The trim may be sold for reprocessing or may sometimes be reworkable in an integrated operation, but the original conversion cost to strip is lost.

Stamping or blanking operations are relatively high-speed, low-cost processes. Screw machining, especially when automated, is relatively low cost. Both types of processing also have lower tooling costs than the melt processing techniques. It is general experience that low volume production requirements will favor the machining of parts. For a further discussion of this area, see Refs. 6 and 7.

Nonmelt pressure forming (Chapter 15) represents a combination of some of the older as well as newer processes for transforming plastic materials into end-use articles. Press forming, cold heading, coining, die drawing, and roll forming are examples that in general employ techniques and tools developed in the metals industry. The material costs will be higher than those for melt forming processes because a partially transformed shape serves as the feed stock. Higher material costs are usually compensated by low processing and tooling costs.

Nonmelt processing via stamping, cold heading, and so on, is a high-speed method of conversion that utilizes equipment with relatively lower investment requirements than those for injection molding. This combination of factors can be expected to result in low converting costs. The tooling for nonmelt processing can be expected to be less than injection molding tooling. Although the unit cavity costs may be comparable for the two processes, the processing rate for the nonmelt forming process reduces markedly the number of cavities required for a specific production volume. Hence, the part cost including tool write-off will be potentially lower.

Thermoforming is somewhat unique to the plastics family. In this process, sheet stock is heated and shaped into the end use article. Pressures up to approximately 60 lb (wt) in.$^{-2}$ (4.2 kg (wt) cm^{-2}) may be used to deform the plastic. This is at least an order of magnitude lower than the pressures used in the forming processes discussed above. As with other nonmelt forming processes, its material costs are relatively high. Tooling costs are low and converting costs may be low, the magnitude depending on the article produced and the conversion process.

Secondary Processing

Although plastics converting processes are very flexible and adaptable to the production of a variety of complex shapes and articles, it is occasionally necessary or desirable to join two or more parts together. The components may be joined economically by riveting, welding, cementing, mechanical interference fitting, and other techniques. Of course, the conventional mechanical fastening with metal screws or rivets and plastic or metal bolts is applicable. With the exception of cementing and mechanical fastening, the joining processes have no added material costs, are capable of short cycles, and have relatively low over-all cost. When the production volume is large, the use of automated feed and processing systems can be a means of cost reduction.

The joining of two parts by riveting or snap fitting requires only an investment in a low cost arbor press, a jig for application of compressive force, and an operator. If one of the nylon components has a molded-in stud for riveting, a choice can be made between cold heading or the application of heat. In either case, production rates are usually high.

Hot plate, spin, or ultrasonic welding are flexible and widely used processes. A higher level of process sophistication results in increased equipment investment. They have a short production cycle and are readily tooled for manual or automatic operation. In either case, processing costs are low.

The bonding of nylon plastic articles can also be accomplished with solvents, adhesives, or cements. In all cases an added cost is involved for the materials. The use of adhesives offers the opportunity of short assembly cycles. The properties of the joints are dependent on the specific adhesive used. Cementing with either solvent or a chemically reactive bonding agent requires a curing period usually under clamping pressure. Although the processing cycle may be accelerated by the use of heat, the necessity of using fixtures, curing racks, and possibly heating tunnels means added expense. This technique requires added material and processing costs as well as increased operating investment and is, accordingly, less attractive.

Under certain more demanding engineering applications, it is necessary or advisable to stabilize the part with respect to its operating environment. This stabilization may require annealing at some temperature higher than anticipated service conditions. Or, because all nylons absorb water with a concomitant change in volume, moisture conditioning may be desirable. The processing cost and associated investment will depend on the degree of treatment.

Decorating of nylons (Chapter 17) encompasses a number of techniques for changing the appearance of a nylon article and adding to its value. The dyeing, hot stamping, and inking processes can be relatively low in cost and investment requirement. On the other hand, painting, vacuum metallizing, and electro-plating are more costly processes which require a considerably higher level of technical competence and investment. It is not unusual for these processes to be set up as independent departments within a company. Companies that do custom processing also exist.

Cost Estimating

With a preliminary design completed and a material of construction chosen, the designer is at a point where he can assess the probable cost of a nylon article utilizing the preliminary estimating technique outlined in Ref. 8.

This estimating technique is applicable to melt processes and nonmelt techniques that require heating. It is based on the historical observation that material cost represents 35 to 50% of the cost of a fabricated article. This relationship suggests the equation

material cost X factor = purchase price

Factors for some of the fabrication processes used for nylon are shown in Table 19-3.

It was indicated previously that material cost is 35 to 50% of the sales dollar for the average fabricated plastics part. This represents a material cost factor range of 2 to 3. The broad spread of the "overall range" in Table 19-3 reflects differences in material prices, part weights and specifications, production volume and costs, and finishing costs.

The probable average part factors and the basic equation should be applied only to the production of uncomplicated parts with average dimensional tolerances, little or no finishing, and in volumes of moderate to large size (10,000 units per year or greater). For small production volumes, other processes such as

Table 19-3. Purchase price/material cost

	Material cost factors	
	Overall range	Probably average part range
Injection molding	1.5-5	2-3
Extrusion	2-5	3-4
Compression molding	2-10	4-6
Thermoforming	2-10	3-5

machining may be more economical. For this case, see Refs. 5 and 6.

A more refined estimate of the cost of a nylon article would require consideration of material, conversion processing, post-processing, inspection, packing, shipping, and tooling costs. The development of such detailed costs requires a more intimate knowledge of the processes for plastics fabricating than the average designer has. This activity will usually be the responsibility of an experienced estimator, or assistance may be solicited from a fabricator.

For those interested in "order-of-magnitude estimating," the technique is described in Ref. 8. A more detailed discussion of estimating procedures and techniques for establishing machine-hour rates will be found in Refs. 9 to 12.

REFERENCES

1. Anon., "Survey of Current Business," Department of Commerce, U. S. Government Printing Office, Washington, D. C. (issued monthly).
2. Graf, G. L., Jr., *Metals Eng. Quart.* **2**, 36 (1962).
3. Goettel, J. C., *SPE J.* **14**, 32 (July, 1958).
4. Graf, G. L., Jr., unpublished data.
5. Anon., "Powdered Polymers — Rapidly Emerging Worldwide Markets," De Bell and Richardson, Inc., Enfield, Conn. 1971.
6. Carlyon, G. C., *Mod. Plast.* **38**, 90 (Oct. 1960).
7. Benkelman, W. D., *Mod Plast.* **44**, 118 (Sept. 1966).
8. Graf, G. L., Jr., in *Engineering Design for Plastics*, E. Baer, Ed., Reinhold Publishing Corp., New York, 1964, Chap. 19.
9. Graf, G. L., Jr., Technical Report No. 114, "Estimating Procedures and Estimates of Selling Price for a Molded Part in Zytel® Nylon Resin," Plastics Department, E. I. du Pont de Nemours & Co., Inc. (July, 1963).
10. Tucker, S.A., *Cost Estimating and Pricing with Machine-Hour Rates,* Prentice-Hall, Englewood Cliffs, New Jersey, 1962.
11. Daniel, R. L., and M. J. Cooper, *SPE J.* **23**, 75 (April 1967).
12. Steed, M., *Plast. Tech.* **16**, 49 (April 1970).

CHAPTER 20 _____

Nylon Applications

E. TURNER DARDEN

MARKETS FOR NYLON PLASTICS

Consumption

Nylon-66 was introduced as a molding powder in 1941, and a host of homopolymers, copolymers, and modifications have followed. The nylons have offered an ever-broadening set of property combinations that have spurred steady growth in spite of a constant increase in the number of competing materials available. Examples of applications made possible by the development of new nylon compositions include hammer handles using glass reinforcement, automotive air conditioner hoses made with tubing of plasticized nylon, and textile machinery gears made of nylon modified to provide greater toughness.

    Consumption of nylon plastics in the United States for the years 1962-1970 (Figure 20-1) has shown an annual growth rate of about 12%. A further doubling or trebling in the decade of the 1970s has been projected (1).

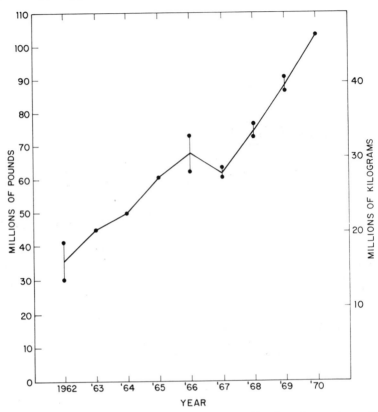

Fig. 20-1.   United States consumption of nylon plastics (data from January issues, *Modern Plastics*, 1963-1971; ranges reflect different estimates in successive years.)

Market Analysis

Historical United States consumption of nylon, by industry, is shown in Table 1-1 (Chapter 1, p. 3). A more detailed breakdown of the industries served by nylon plastics is provided in Table 20-1 and illustrates the broad range of applications open to the diversified family of nylon plastics. The biggest users of nylon are the automotive, appliance, electrical, and machinery industries.

Table 20-1. Areas of application of nylon

Appliances	Electrical
Laundry equipment	Industrial controls
Cooking equipment	Wiring and associated devices
Dishwashers and disposers	Industrial connectors
Cooling equipment	Batteries
Consumer electronics	Telephone parts
Housecleaning equipment	Switches
Small kitchen appliances	
Sewing machines	Hardware
Personal care and grooming	Furniture fittings
	Door and window fittings
Automotive	Tools
Chassis parts	Lawn and garden implements
Power plant	Boat fittings
Decorative body parts	
Functional body parts	Machinery
Electrical parts	Agricultural
Fuel system	Mining and oil drilling
Instrumentation	Food processing
Heating, ventilating, air-conditioning	Printing
Accessories	Textile processing
	Engine parts
Business equipment	Pumps, valves, meters, filters
Business machines	Air blowers
Vending machines	Material handling equipment
Office equipment	Standard components
	Gears
Consumer products	Cams
Kitchen utensils	Sprockets
Toys	Bearings
Sporting goods	Gaskets
Apparel fitments	Pulleys
Personal accessories	Brushes
Photographic equipment	
Musical instruments	Packaging
Brush bristles	Aerosol
Film for cooking	Film and coated substrates
Fishing line	

APPLICATIONS AND PROPERTIES

Relationship of Applications to Properties

Nylons are noted for their outstanding combination of properties, including a high degree of toughness, abrasion resistance, heat resistance, and solvent resistance. Seldom does a single property determine which plastic material is chosen for an application. Actual end-use property requirements are complex, and several of the properties listed in Table 20-2 may be essential to satisfactory performance in a given application.

Table 20-2. Key properties of nylons

Toughness
Fatigue resistance
Low friction
Abrasion resistance
Resistance to oils and solvents
Stability at high temperatures
Fire resistance
Creep resistance
Drawability
Good appearance
Good molding economics

Although the best known properties of nylon plastics are toughness, low friction, and resistance to abrasion, heat, and solvents, there are other properties of commercial importance. These include the high tensile strength of certain oriented nylons, which is equivalent to that of steel, and the low density and easy fabricability which permit nylon to compete with metals that are nominally less expensive. The properties of nylons are discussed in previous chapters, especially Chapters 10 and 11. How these properties are used in typical applications is our concern here.

Illustrative Applications Based on Specific Properties

The applications of nylon discussed in the following pages are arranged by the principal property being illustrated. As already noted, most applications require a combination of the properties afforded by the different nylons.

Toughness

The impact resistance of nylon varies from good to excellent, depending on the moisture content and the composition. Absorption of moisture enhances

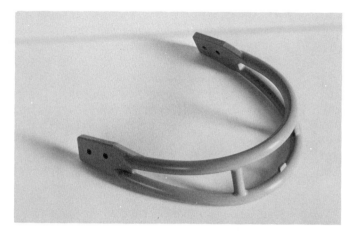

Fig. 20-2. Football face guards are made of nylon-6 because of its great toughness. (Photo courtesy Rawlings Sporting Goods Co.)

toughness. Copolymers, terpolymers, plasticized compositions, and modified compositions with high levels of toughness in the dry state are available. Glass fiber reinforcement may improve or reduce the toughness, depending on fiber orientation, the amount incorporated, adhesion between the nylon and the glass, and the nature of the application.

Some illustrative applications requiring great toughness are football face guards, hammer handles, and housings for small tools and appliances. Football face guards (Figure 20-2) are molded of nylon-6 and have the toughness and

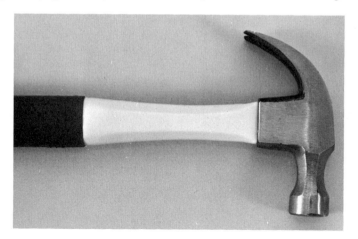

Fig. 20-3. Hammer handles of glass-reinforced nylon are superior to the wood they replaced. (Photo courtesy Ennis Manufacturing Co.)

stiffness required for a safety item of such critical importance to the athlete. Hammer handles (Figure 20-3) are made from a modified nylon-66 containing 33% glass fibers. They not only withstand the repeated impacts of use but also the force of the steel wedge driven into a slot in the end to make a tight fit with the head of the hammer. Housings for electric shavers, power tools, transistor radios, and many other small tools and appliances made of nylon successfully resist the abnormal but not unexpected shock of being dropped. In addition to toughness colorability, good appearance, electrical insulation, and other properties may also be essential to meet the complex needs of these consumer products.

Resistance to Fatigue and Repeated Impact

Nylon has outstanding resistance to fracture by repeated impacts at a level below that required to cause failure by a single shock. This property makes it suitable for applications such as hammer handles (above), automobile door striker plates, and other uses requiring thousands of impacts. It is also well suited for oscillating machinery parts, gears, sprockets, and other uses where millions, rather than thousands, of impact loadings are involved.

Nylon-66 and a polycarbonate were subjected to repeated tensile impacts at a level 60% of that required for fracture in a single blow (2). The polycarbonate, which has very high toughness in a standard impact test, failed on the

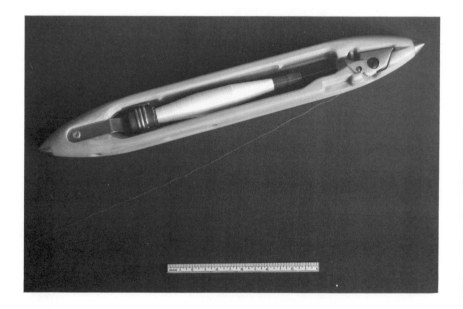

Fig. 20-4. Textile shuttles of glass-reinforced nylon-66 withstand frictional heat and millions of severe impacts. (Photo courtesy La Technica Electro-Textil, Barcelona)

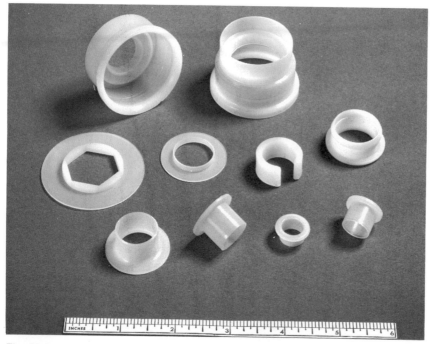

Fig. 20-5.   These nylon bushings and bearings are used on a single power lawnmower. (Photo courtesy McDonough Power Equipment Co.)

thirty-seventh impact but the nylon-66 was not broken after being impacted 250 times.

Additional applications of nylon requiring resistance to fatigue or repeated impact are a backing of a nylon-6/66/610 terpolymer for metal printing plates, and a textile shuttle of glass-reinforced nylon-66 (Figure 20-4).

Low Friction

Nylons display a low coefficient of friction in contact with many other materials, and are sometimes called "self-lubricating" for that reason. This property is responsible for the use of nylons in many sliding applications, including many where lubrication is not permissible. Journal bearings, bushings, gears, and cams are typical applications which make use of this property. The collection of friction-reducing parts shown in Figure 20-5 are all from a single tractor lawnmower.

Additives are sometimes employed to enhance the natural lubricity of nylon. Chief among these are molybdenum disulfide, graphite, and "Teflon" TFE resin. Sliding guide shoes contained within the rubber handrails of moving stairways (Figure 20-6) are molded of nylon-66 containing molybdenum disulfide and are reported to outlast the unmodified nylon by a factor of three.

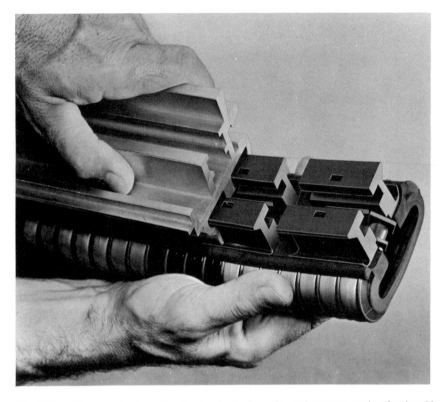

Fig. 20-6. These guide shoes for the handrail of moving stairway are made of nylon-66 with molybdenum disulfide added for lubricity. (Photo courtesy Polymer Corp.)

Abrasion Resistance

When used within their PV (pressure × velocity) limitations, nylons display remarkably good resistance to wear. Some highly specialized applications that utilize this property are sliding shoes or gibs for elevators, tubing for conveying abrasive yarns in textile mills, coatings for steel machinery parts, timing screws for packaging machinery, automotive timing sprockets, and wire insulating jackets. All are illustrated below.

The elevator gib (Figure 20-7) employs the abrasion resistance of unmodified nylon-66. This 8-in. shoe slides on a steel rail running the length of the elevator shaft, absorbing side thrusts and preventing the car from swaying, or tilting due to uneven load distribution. Nylon gibs eliminate the need for heavy lubrication, formerly required with metal or synthetic rubber gibs. A yearly application of a wax-based lubricant to the rails is still required to prevent rusting and avoid

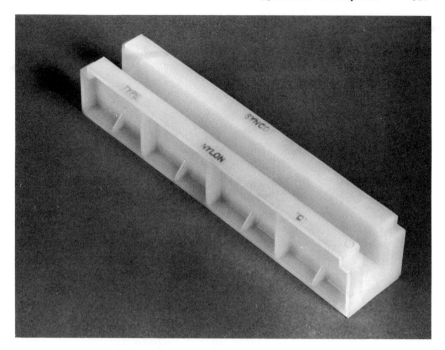

Fig. 20-7.   Nylon elevator gibs slide thousands of miles against steel rails, with minimum lubrication. (Photo courtesy Nylube Products Co.)

subsequent pick-up of solid particles by the nylon. These gibs slide thousands of miles before being replaced.

The automotive timing sprocket shown in Figure 20-8 has a nylon-66 rim and teeth molded onto an aluminum hub. The hub gives the necessary rigidity in thin section, while the nylon gives improved sprocket and chain life, plus quieter operation. Similar sprockets drive plate-top conveyor chains in food processing plants, where no lubrication is permissible. Since water is often present, nylon-610 is used for its low water absorption.

In the manufacture of carpets, yarn is conveyed from a creel or rack containing about 1000 large spools to a single tufting machine. The distance from each spool to the machine is about 100 feet. Nylon tubing about one-quarter inch in diameter guides these yarns to their destination (Figure 20-9). Despite the abrasive character of many carpet yarns, the tubing is not perceptibly worn in five years, even at the bends. Nylon-610 and nylon-612 are preferred because of their low moisture absorption and minimal expansion which, if greater, would produce "festooning" of the tubing with an intolerable increase in yarn tension.

Nylons-610, -612, -6, and -66/6 are used as thin extruded jackets over poly(vinyl chloride)-insulated building wire (Figure 20-10). The superior abrasion

Fig. 20-8.   Timing sprocket for automobile cam shaft has nylon-66 teeth for long wear and noise reduction. (Photo courtesy Chrysler Motors)

resistance of nylon permits a reduction in the thickness of the PVC. The resulting construction has a smaller overall diameter, permitting conduits to carry three times as many wires as formerly, and facilitating the process of pulling the wires through the conduit. This reduces the chance of damaged insulation and its attendant risks (4).

Many package filling machines use timing screws (Figure 20-11) to deliver empty containers to the filling location at precisely the right rate. The screw must slide against the side of the can, bottle, or carton without defacing it, and without itself being worn excessively over a long period of time. Because of the massive size of these screws (up to six inches in diameter), and the number of design variations required, they are machined from nylon-66 or nylon-6 rod rather than being molded. Their superior performance compared with polished steel or phenolic resin justifies the cost.

The frictional characteristics of nylon have been combined with the strength of steel through the fluidized bed coating technique (Chapter 14). Coating a steel driveline slip spline for a truck (Figure 20-12) with nylon-11 increases its

Fig. 20-9.   These 100-ft tubes made of nylon-610 convey abrasive yarn to a carpet tufting machine. (Photo courtesy M & Q Plastics, Inc.)

operating life at least 800%. Although lubrication is still required, the power losses and frictional heat are reduced because the high spots of the nylon do not break through the lubricant film as do those of steel (3).

Resistance to Oil and Solvents

Nylon plastics are generally impervious to attack by solvents and chemicals other than strong acids. They slowly absorb small amounts of water and traces of hydrocarbons, the former increasing and the latter decreasing as the amide

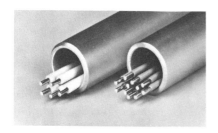 

Fig. 20-10.   Abrasion-resistant nylon jacket over PVC insulation reduces diameter of wires, permitting more wires in same conduit (left), or a smaller conduit (right). (Photo courtesy Du Pont Co.)

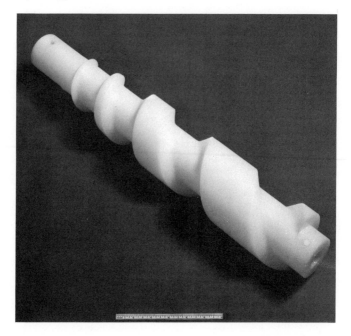

Fig. 20-11.   This timing screw machined from a rod of nylon handles empty containers on filling machine. (Photo courtesy Ernst Timing Screw Co.)

group concentration in the nylon is increased (see Chapter 3, p. 89). Permeability to most gases is generally low (see Chapters 10 and 12).

Extruded tubing, usually of plasticized nylon-11, nylon-6, or nylon-66/6, is overbraided with synthetic yarn or metal wire, and usually jacketed in a subsequent extrusion step to produce hose for hydraulic, pneumatic, or process use (Figure 20-13). Unlike the competing elastomers, a single type of nylon is suitable for either ester or hydrocarbon-based hydraulic fluids, eliminating the possibility of costly error. Hydraulic hoses must in some cases operate at 250°F (120°C) yet withstand flexing at arctic temperatures. The low permeability of nylon to halogenated refrigerants (for example, "Freon") makes it suitable for use in air conditioning systems. Millions of feet of such hose are used annually for automotive air conditioning.

Nylon-6 and nylon-11 gasoline tanks of irregular shape are made by rotational molding (Chapter 14) using finely powdered resin in a hot mold (Figure 20-14).

Nylon film is beginning to find its place in food packaging. Largely because of its low permeability to oxygen and edible fats, preservation of baked goods in fresh condition up to two months without refrigeration is possible (Figure 20-15). Greases do not penetrate nylon packages. Sterilization of the sealed package is made possible by the high melting points of nylons. Cooking of meats, poultry, and other foods in bags made of nylon-66 film, which withstands high oven temperatures, promises new convenience for the home chef.

Packaging of nonfood products, including greased machinery parts, in nylon-11, nylon-6, or nylon-66 film shows promise for the future because of the grease resistance and the puncture resistance of such films.

Stability at High Temperatures

Stability at high temperatures refers to both form retention and resistance to oxidative degradation. Cooking utensils utilize the former property, and under-the-hood automobile parts, the latter. These and other examples are discussed below.

Form stability in hot cooking oils and all other foods is required of the nylon-66 utensils shown in Figure 20-16. These familiar household items withstand thousands of cycles of alternate exposure to hot grease and hot detergent solutions. Freedom from toxic additives is of course a further requirement. Silverware baskets for commercial dishwashers are made of nylon-66 with improved hydrolysis resistance and are exposed to innumerable cycles of scalding water and detergents.

Fig. 20-12.   This heavy duty driveline slip spline for a truck is coated with nylon-11 for longer life. (Photo courtesy Aquitaine Chemicals, Inc.)

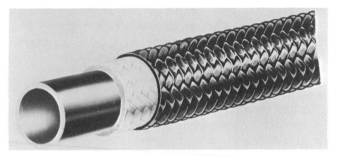

Fig. 20-13.   Core of plasticized nylon-11 or nylon-66/6 in flexible hydraulic hose permits smaller diameter than former elastomeric cores. (Photo courtesy Polymer Corp.)

Electrical coil forms for countless uses are made of nylon-66, usually protected with an antioxidant against prolonged heating (Figure 20-17). In cases requiring stiffness at unusually high temperatures, glass-reinforced nylon-66 is used. Automobile horn coil bobbins are an example of the latter.

Handles, housings, and other parts in contact with hot metal components are often made of glass-reinforced nylon because of its exceptionally good retention of stiffness at high temperatures, combined with toughness and grease resistance.

The automotive valve stem oil seal shown in Figure 20-18 must resist lubricating oil and its degradation products at extreme engine temperatures. The material is a heat-stabilized (that is, stabilized against oxidation) 33% glass-fiber-reinforced nylon-66.

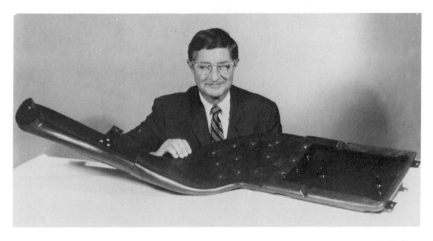

Fig. 20-14.   Large ducts for trucks and automobiles are rotationally molded from nylon-11. (Photo courtesy Aquitaine Chemicals, Inc.)

Fig. 20-15. Baked goods are preserved for two months without refrigeration in nylon-11 film using infrared sterilization. (Photo courtesy Aquitaine Chemicals, Inc.)

Fig. 20-16. Spatula blades and spoons of nylon-66 withstand highest cooking temperatures. (Photo courtesy Ekco Housewares Corp.)

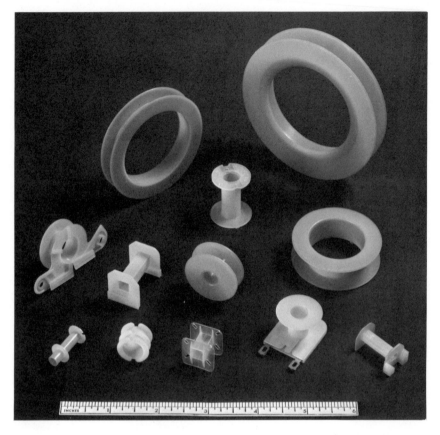

Fig. 20-17.   Countless electrical coils are wound on nylon-66 forms. (Photo courtesy Du Pont Co.)

Fire Resistance

All nylons are low in flammability and generate very little smoke or noxious fumes. Some are classified as self-extinguishing in certain standard tests. These characteristics are important in uses where extreme overheating can occur, and where fire would be a serious hazard. Additives have been employed to make fire retardant compositions for applications such as fittings for aircraft interiors.

Television tuner parts (Figure 20-19) must resist prolonged high temperatures in normal use and, in event of a rare malfunction, may be exposed to destructive heat. The use of self-extinguishing nylon-66 reduces the danger of fire in such circumstances. Although the nylon parts may be destroyed, they will not contribute significantly to a fire.

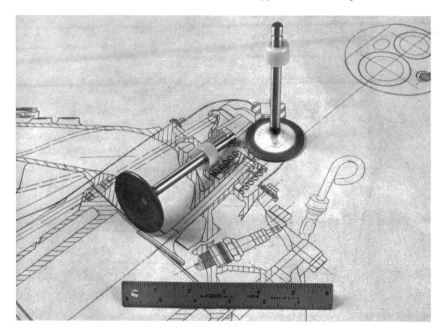

Fig. 20-18.   Automobile valve stem oil seal of glass-reinforced nylon-66 functions at extreme engine temperatures. (Photo courtesy TRW, Inc.)

Resistance to Creep or Relaxation

The stress necessary to cause permanent deformation depends on both temperature and time. Thus, a stress below that required to cause immediate nonrecoverable strain will cause permanent distortion if sustained long enough. This is commonly called creep. Similarly, the stress required to maintain a constant deformation decreases with time. Resistance to creep or stress relaxation in nylons improves with increasing crystallinity.

Many nylon applications require creep resistance in combination with other essential properties. Cylindrical pressure containers for liquid butane are sealed by crimping a sheet metal top over the lip of a nylon cylinder. The elongation and impact resistance of the nylon are utilized in preventing fracture at the moment of crimping. In other cases, nylon gaskets and seals maintain leakproof joints for months or years by their resistance to creep under compression. Furniture casters depend on this property plus a high modulus, as do gears which remain loaded when idle.

Fig. 20-19.   Self-extinguishing nylon-66 is used for television tuner parts. (Photo courtesy Du Pont Co.)

Drawability

When molecular orientation is achieved by cold-drawing or otherwise forming nylon at a temperature below its melting point, high tensile strengths are obtained. The tensile strength in the direction of orientation is up to ten times that of the undrawn nylon and higher than many grades of steel. This useful characteristic is employed to make strapping for industrial packaging. Nylon-66 strapping has much higher recoverable elongation than steel. Thus, the strapping shown in Figure 20-20 maintains its tension when the size of the package is reduced. Other advantages are easy removal by cutting and disposal by incineration or chopping.

Another useful oriented form is monofilament. This finds numerous applications such as brush bristles, tennis rackets, woven screens, and strings for musical instruments. Fishing lines with the strength of steel wire plus the flexibility to permit reeling without kinking are well known (Figure 20-21). A fluorescent dye is sometimes incorporated to make the line more highly visible to the angler against a dark background and less visible to the fish against the light background above.

Good Economics

Raw material cost is only one factor in molding economics (see Chapter 19). The other major factor is the cost of conversion to a useful shape. Nylon has two

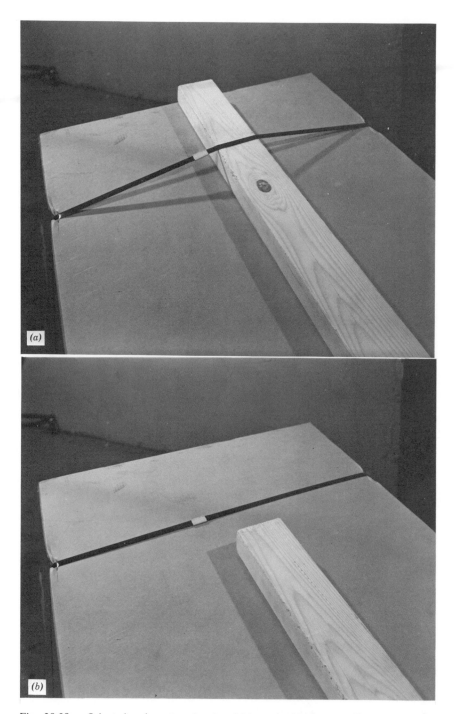

Fig. 20-20. Oriented nylon strapping has high strength plus a resilience that keeps packages tightly bound. In this demonstration, no slack is produced on removing the board. (Photo courtesy Signode Corp.)

637

economic advantages over metals — lower density and, normally, lower processing cost. At a price per unit weight slightly higher than brass, nylon costs only one-fifth as much on a volume basis. Even those die-casting alloys which are less expensive than nylon on a volume basis often cannot compete because of extensive secondary operations required to produce a finished part. At the very least, die-cast parts usually must be treated to remove flash. Often, drilling,

Fig. 20-21.   Fishing line of oriented nylon has the tensile strength of steel plus flexibility, low visibility, freedom from corrosion. (Photo courtesy Du Pont Company.)

tapping, and other machining operations are required, whereas the same part can be produced from nylon in a single operation by injection molding.

Even where low-volume production prohibits injection molding, nylon can be more economical than metals. Extruded strip in a complete range of widths and thicknesses is available for die-cutting and punch-press operations. Extruded rods and tubes are produced for automatic screw machine operations as well as for miscellaneous machining of special items.

Other Applications

Nylon-66 and nylon-6, often in special high-molecular-weight grades, are extruded to produce rods, bars, tubes and sheets for machining, die stamping and other fabricating techniques common to metals.

Nylon-6 can be made from its liquid monomer in low-pressure molds to produce large items that would be impossible by injection molding (Chapter 13). Large machinery parts and stock forms for machining are produced in this manner. An interesting application of the technique is the manufacture of nylon millwork such as window shutters and folding doors (Figure 20-22). The items, being rot-proof and of one-piece construction, are longer lasting than wood. Shutters require infrequent painting and are not subject to attack by moisture, fungi or insects.

Nylon-6/66/610 terpolymers are .alcohol-soluble and have been used as binders for glass fibers in such objects as fishing rods and vaulting poles. Fabrics coated or impregnated with these nylons can be permanently joined to other fabrics by pressing with a hot iron. "Iron-on" clothing patches and replacement pockets are familiar to the modern housewife. A nylon solution is sometimes sprayed onto the edges of stacks of cut fabrics in garment factories to prevent fraying or sliding (see Chapter 16).

Nylons foams are known but have been little used to date due to their comparative difficulty of preparation. Nylon-6 rigid foam saturated with lubricating oil has been used as a bearing material. The same material has been tested, apparently with success, as the center layer of a sandwich panel between sheets of glass-reinforced polyester. Plasticized foams of nylon-6 have been tested for boat fenders because of their great toughness and resilience. Again, manufacturing techniques have been a limitation.

SUMMARY

Nylons comprise a broad class of materials, each with an excellent combination of physical properties. This leads to a multiplicity of varied applications which continue to expand. Only a small proportion of the possible nylons have been

Fig. 20-22.　Millwork, such as these folding doors, is made by polymerizing nylon-6 in the mold. (Photo courtesy Du Pont Co.)

prepared and evaluated, and of these only a comparative few have been introduced to the market. At this writing, nylon-12 and nylon-612 have recently been introduced, and others are in advanced stages of development. Nylons made from cyclic or branched diamines and acids are known and may someday be commercial molding and extrusion resins.

Chemical modification of the amide linkage is possible and has been practiced for years on a small scale (see Chapter 11). Manipulation of molecular weight gives added control over processing and physical properties and contributes to the versatility of nylon. The properties of a given type of nylon may be modified in various ways with additives. It may be nucleated, plasticized, lubricated, protected against hydrolysis or oxidation, or rendered more fire resistant. Properties may also be modified with fillers or reinforcing agents, or by blending one nylon with another.

The commercial applications of nylons number in the thousands, and we have discussed only a few typical examples in this chapter. Other applications are extremely diverse; fasteners for automobile trim, mixing valves for washing machines, loom parts, license plate bolts, bar-stool swivels, hose couplings, lawn sprinkler parts, instrument feet, lighter fluid containers, football shoe cleats, ball bearing races, knobs, bobbins, cams, slides, guides, and hooks are a few random examples. Nylon moldings and extrusions are found throughout the home, the family car, the airplane, the office, the store, the farm, and the factory. The diversity and adaptability of nylons promise a continued steady growth.

REFERENCES

1. (a) Anon., *Mod. Plast.* 47 (1), 98 (Jan., 1970); (b) Fisher, J., *Mod. Plast.* 48 (2), 51 (Feb., 1971).
2. Heater, J. R., and E. M. Lacey, *Mod. Plast.* 41 (9), 123 (May, 1964).
3. Kayser, J. A., and W. T. Groves, unpublished paper No. 680118, delivered to SAE Automotive Engineering Congress, Jan., 1968.
4. Gibson, A. L., *The American Registered Architect,* July, 1967.

APPENDIX _____

Nylon Plastics Suppliers and Compositions

M. I. KOHAN

The table below lists the nylon plastic compositions of major suppliers in the United States as cited in trade and company literature. Colors and special compositions made upon request or still in the development stage are not included. Reprocessors and suppliers of unique compositions are identified. Sources of nylon articles, whether stock shapes, monofilament, strapping, film, cast objects, or other fabricated forms, are not included.

This tabulation illustrates the variety of compositions available to the nylon plastics industry. It can never be exact because both materials and sources appear and disappear in response to the needs of the marketplace. Annual compendia that list processors as well as suppliers and products include *Plastics World Directory of the Plastics Industry, Modern Plastics Encyclopedia,* and *Plastics Technology Processing Handbook,* but only the last named attempts to list specific proprietary compositions and their properties.

Coding systems are sometimes obvious but are explained only where they have been explicitly provided in company publications. The absence of an entry in the additive column means either that none is used or that none is specifically indicated by the manufacturer. Only additives unique to the given composition and not those present in the base resin are identified.

Company and location	Trade name	Code	Nylon	Modification		Comment
				Base resin	Additive	
Adell Plastics Inc. Baltimore, Md.	"Adell" Nylon	--	--	--	--	Reprocessor
Allied Chemical Corp., Plastics Division, Morristown, N.J.	"Plaskon" Nylon	8200	6	--	--	Med. visc., inj. molding
	"Plaskon" Alpha Nylon	8200C	6	8200	Incr. crystall.	Higher strength, faster molding
	"Plaskon" Nylon	8200HS	6	8200	Heat stabilizer	
	"Plaskon" Nylon	8200HS-1	6	8200	Heat stabilizer	Lighter color than 8200HS
	"Plaskon" Nylon	8200MS	6	8200	MoS_2	Improved lubricity
	"Plaskon" Nylon	8200P	6	8200	Plasticizer	Tougher, more flexible
	"Plaskon" Nylon	8202	6	--	--	Low visc., inj. molding
	"Plaskon" Alpha Nylon	8202C	6	8202	Incr. crystall.	Higher strength, faster molding
	"Plaskon" Nylon	8202HS	6	8202	Heat stabilizer	
	"Plaskon" Nylon	8202HS-1	6	8202	Heat stabilizer	Lighter color than 8202HS
	"Plaskon" Nylon	8202MS	6	8202	MoS_2	Improved lubricity
	"Plaskon" Nylon	8203	6	--	--	High visc., extrusion
	"Plaskon" Alpha Nylon	8203C	6	8203	Incr. crystall.	Stiffer, faster line speeds
	"Plaskon" Nylon	8204	6	8214	Heat stabilizer	Low visc., inj. molding, tougher, more flexible
	"Plaskon" Nylon	8205	6	--	--	Very high visc., extrusion
	"Plaskon" Nylon	8205MS	6	8205	MoS_2	Improved lubricity
	"Plaskon" Nylon	8206	6	--	Plasticizer	Med. visc. for flexible monofil
	"Plaskon" Nylon	8207	6	--	--	Med. visc., extrusion
	"Plaskon" Nylon	8211	6	8205	Heat stabilizer	Very high visc., extrusion
	"Plaskon" Nylon	8214	6	--	Plasticizer	Low visc., inj. molding
	"Plaskon" Nylon	8215	6	--	Heat stabilizer	High visc., extrusion
	"Plaskon" Nylon	8216	6	8206	For gloss	High gloss 8206

Trade name	Grade			Additive	Application
"Plaskon" Nylon	8220	6	8221	Heat stabilizer	Wire jacketing
"Plaskon" Nylon	8221	6	--	--	Extrusion
"Plaskon" Nylon	8230	6	--	6% glass	Low visc., inj. molding
"Plaskon" Nylon	8231	6	--	14% glass	
"Plaskon" Nylon	8233	6	--	30% glass	
"Plaskon" Nylon	8250	Copolymer	--	--	Extr. and inj. molding, flexible, impact resistant
"Plaskon" Nylon	8251	Copolymer	--	--	Extr. and inj. molding, flexible, impact resistant
"Plaskon" Nylon	8252	Copolymer	--	--	Extr. and inj. molding, flexible, impact resistant
"Plaskon" Nylon	8253	Copolymer	--	--	Extr. and inj. molding, flexible, impact resistant
"Rilsan"	BMN	11	--	--	Inj. molding
"Rilsan"	BMN P20	11	BMN	Plasticizer	Semi flexible
"Rilsan"	BMN P40	11	BMN	Plasticizer	Very flexible
"Rilsan"	BMN G8	11	BMN	Graphite	Improved lubricity
"Rilsan"	BMN G9	11	BMN	Graphite	Higher loading of graphite
"Rilsan"	BMN Y	11	BMN	MoS_2	Improved lubricity
"Rilsan"	BMNY BZ TL	11	BMN	Heat stabilizer, bronze bead	Improved thermal conductivity
"Rilsan"	BMN Black T	11	BMN	Heat stabilizer	
"Rilsan"	BMN TL	11	BMN	Heat stabilizer and light stabilizer	
"Rilsan"	BMN T5	11	BMN	Ht and lt stabilizers	Better mold release
"Rilsan"	BMN D	11	BMN	Release agent	Better mold release
"Rilsan"	BMN GF	11	BMN	Foaming agent	For very light items
"Rilsan"	BMF	11	--	--	Inj. molding, low visc.
"Rilsan"	BMV	11	--	--	Inj. molding, high visc.
"Rilsan"	BMV P10	11	BMV	Plasticizer	Slightly flexible

Aquitaine Chemicals Inc.
Glen Rock, N.J.

Explanation of code
B,K = manuf. reference
Z = glass reinf.
M = molding
EC = extrusion on cables
ES = blow molding, extrusion of profiles
F = low visc.
N = med. visc.
V = high visc.
HV = very high visc.
P10 = slightly flexible
P20 = moder. flexible
P40 = very flexible
G8,G9 = graphite
Y = MoS_2

(Continued)

Company and location

D = mold release
E = antistatic
L = light stabilized
T = temp. resistant
X = fungicidal
W = nonflam.
BZ = bronze bead
GF = blowing agent

Trade name	Code	Nylon	Modification		Comment
			Base resin	Additive	
"Rilsan"	BMHV P40	11	--	Plasticizer	Very high visc., flexible
"Rilsan"	KMF	11	--	--	Rigid, low visc.
"Rilsan"	KMF T6	11	KMF	Ht and lt stabilizers	
"Rilsan"	KMV TL	11	--	Ht and hydrolysis stabilizers	High visc.
"Rilsan"	GSMR	11	--	--	For rotational molding
"Rilsan"	ZM 30	11	--	30% glass	
"Rilsan"	ZM 30 W3	11	ZM 30	30% glass, self-extinguishing	
"Rilsan"	ZM 30 Black TL	11	ZM 30	30% glass; ht, lt, and hydrolysis stabilizers	
"Rilsan"	ZM 23 G9	11	--	23% glass, graphite	
"Rilsan"	ZM 43 G9	11	--	43% glass, graphite	
"Rilsan"	BESN	11	BESN	--	Extrusion
"Rilsan"	BESN P20	11	BESN	Plasticizer	Semi-flexible
"Rilsan"	BESN P40	11	BESN	Plasticizer	Very flexible
"Rilsan"	BESN Black T	11	BESN	Heat stabilizer	
"Rilsan"	BESNO TL	11	BESN	Ht and lt stabilizers	
"Rilsan"	BESN G9	11	BESN	Graphite	Improved lubricity
"Rilsan"	BESV	11	--	--	High visc.
"Rilsan"	BESV Black T	11	BESV	Heat stabilizer	

Trademark	Product	Nylon type	Code	Additives	Description
"Rilsan"	BESHV	11	--	--	Very high visc.
"Rilsan"	KESV Black TL	11	--	Hydrol. stabilizer	For cable coating
"Rilsan"	BECN	11	--	--	Semi-flexible
"Rilsan"	BECN P20	11	BECN	Plasticizer	
"Rilsan"	BECN Black T	11	BECN	Ht and lt stabilizers	
"Rilsan"	BECN TL	11	BECN	Ht and lt stabilizers	
"Rilsan"	BECV P40	11	--	Plasticizer	Very flexible
"Rilsan"	BESNO A	11	--	"A" grades modified for food applications	
"Rilsan"	BESVO A	11	--		High visc.
"Rilsan"	BESHVO A	11	--		Very high visc.
"Rilsan"	CIESNO A	Copolymer	--	--	
"Rilsan"	BEFN	11	--	--	For monofilament
"Rilsan"	BEFF	11	--	--	For monofilament, low visc.
"Ultramid"	B3	6	B3	--	Low visc., inj. molding
"Ultramid"	B3K	6	B3	Ht, UV, and hydrolysis stabilizers	
"Ultramid"	B3S	6	B3	Nucleat. agent	Fast molding
"Ultramid"	B35	6	--	--	Low visc. but higher than B3
"Ultramid"	BF	6	--	Stabilizers	Jacketing composition
"Ultramid"	B4	6	--	--	Med. visc., inj. molding
"Ultramid"	B4K	6	B4	Ht, UV, and hydrolysis stabilizers	
"Ultramid"	B4W	6	B4	stabilizers	Better ht resistance than B4K
"Ultramid"	B5	6	--	--	High visc., extrusion
"Ultramid"	B6	6	--	--	Very high visc., cross-linked
"Ultramid"	B4K Black 490	6	B4K	1% MoS$_2$	Better lubricity

BASF Corp.
Paramus, N.J.

Explanation of code

B = nylon-6
A = nylon-66
S = nylon-610
2 and 3 = low visc.
35 = low to med. visc.
4 = medium visc.
5 = high visc.
6 = very high visc. (cross-linked)
S = rapid crystall'n

(Continued)

Company and location	Trade name	Code	Nylon	Modification		Comment
				Base resin	Additive	
F,K,W,H = heat, UV light, and hydrolysis stabilized	"Ultramid"	B4K Black 590	6	B4K	2% carbon black	Weather resistant B4K
F,K = ht stability 3-5× better than unstab.	"Ultramid"	B4K Black 690	6	B4K	10% graphite	Improved lubricity
	"Ultramid"	KR1311	6	B3	Plasticizer	Low visc., inj. molding, flexible
W = ht stab. 10-20× better than unstab.	"Ultramid"	KR1149/2	6	B3S	Plasticizer	Higher impact strength than B3S
H = somewhat better ht. stab. than W	"Ultramid"	TR9010	6	B3	ca. 8% monomer	Flexible version of B3
	"Ultramid"	KR1309	6	B4	ca. 8% monomer	Flexible version of B4
	"Ultramid"	B5 Black 53010	6	B5	3% carbon black	Weather resistant B5
G5 = 25% glass fiber	"Ultramid"	KR1302/153	6	B4	15% glass and ht stabilizer	
G6 = 30% glass fiber	"Ultramid"	B3 WG5	6	B3	25% glass and ht stabilizer	
G7 = 35% glass fiber	"Ultramid"	B3 WG6	6	B3	30% glass and ht stabilizer	
G10 = 50% glass fiber	"Ultramid"	B3 WG7	6	B3	35% glass and ht stabilizer	
	"Ultramid"	B3 WG10	6	B3	50% glass and ht stabilizer	
	"Ultramid"	KR1346/203	6	--	20% chalk and ht stabilizer	
	"Ultramid"	KR1346/303	6	--	30% chalk and ht stabilizer	

"Ultramid"	A3	66	--	--	Low visc., extrusion or inj. molding
"Ultramid"	A4	66	--	--	Med. visc., extrusion
"Ultramid"	A6	66	--	--	Very high visc., extrusion
"Ultramid"	A3K	66	A3	Stabilizers	
"Ultramid"	A4K	66	A4	Stabilizers	
"Ultramid"	A4H	66	A4	Stabilizers	
"Ultramid"	KR1101	66	A3	Better natural color	
"Ultramid"	KR1225/1	66	A3	Better natural color and rapid crystallization for fast molding	
"Ultramid"	KR1342	66	A3K	For better friction properties	
"Ultramid"	KR1120/5	66	A4H	Stabilizers	Better color than A4H
"Ultramid"	KR1180	66	A4K, A4H	3% MoS_2	Improved lubricity
"Ultramid"	A4K Black 490	66	A4K	1% MoS_2	Improved lubricity
"Ultramid"	A4K Black 590	66	A4K	2% carbon black	Weather resistant A4K
"Ultramid"	A4K Black 690	66	A4K	10% graphite	Better lubricity
"Ultramid"	KR1307/153	66	A3	15% glass and stabilizers	
"Ultramid"	A3HG5	66	A3	25% glass and stabilizers	
"Ultramid"	A3 XG5	66	A3	25% glass and stabilizers and fire retardant	

(Continued)

Company and location	Trade name	Code	Nylon	Modification		Comment
				Base resin	Additive	
	"Ultramid"	A3 WG5	66	A3	25% glass and stabilizers	Low visc., monofil extrusion
	"Ultramid"	A3 WG7	66	A3	35% glass and stabilizers	Med visc., extrusion
	"Ultramid"	A3 WG10	66	A3	50% glass and stabilizers	High visc., extrusion
	"Ultramid"	S2	610	--	--	Inj. molding
	"Ultramid"	S3	610	--	--	
	"Ultramid"	S3K	610	S3	Stabilizers	
	"Ultramid"	S4	610	--	--	
	"Ultramid"	6A	66/6 co-polymer	--	--	Soluble in hot mixtures of alcohol and water
	"Ultramid"	1C	66/6/PACM6	--	--	Soluble in alcohol
Belding Heminway Co. New York, N.Y.	"Moleculoy"	66-01	66	--	--	Inj. molding
	"BCF" Nylon	808	66	66	(Alkoxy-alkyl substitution increases with increasing code no.)	
	"BCF" Nylon	809	Alkoxy-alkylated	66		Alcohol soluble, cross-linked by heat and acid catalyst
	"BCF" Nylon	818	Alkoxy-alkylated	66		
	"BCF" Nylon	819	66	66		
	"BCF" Nylon	829	66	66		
	"BCF" Nylon	1107	11	--	--	Low visc.
	"BCF" Nylon	1157	11	--	--	Medium visc.
	"BCF" Nylon	1177	11	--	--	High visc.
	"BCF" Nylon	1107-P40	11	1107	Plasticizer	
	"BCF" Nylon	1157-P40	11	1157	Plasticizer	
	"BCF" Nylon	1177-P40	11	1177	Plasticizer	

		Type			
Celanese Plastics Co. Newark, N.J.	"Celanese" Nylon	1000-1	66	--	--	Inj. molding, extrusion
	"Celanese" Nylon	1000-2	66	1000-1	Surface lubricant, release agent	
	"Celanese" Nylon	1003-1	66	1000-1	Heat stabilizer	
	"Celanese" Nylon	1003-2	66	1003-1	Surface lubricant, release agent	
	"Celanese" Nylon	1200-1	66	--	--	High viscosity
	"Celanese" Nylon	1300-1	66	--	--	Tough, fast molding
	"Celanese" Nylon	1310-1	66	--	Control crystallinity	Tough, fast molding
	"Celanese" Nylon	1310-2	66	1310-1	Control crystallinity	
	"Celanese" Nylon	1500-1	66	--	Lubricant, mold release agent	
	"Celanese" Nylon	1500-2	66	1500-1	33% short glass	
	"Celanese" Nylon	1503-1	66	--	Surf. lubricant, release agent	
	"Celanese" Nylon	1503-2	66	1503-1	33% glass, heat stabilizer	
	"Celanese" Nylon	1600-1	66	--	Surf. lubricant, release agent	
	"Celanese" Nylon	1600-2	66	1600-1	40% glass	
	"Celanese" Nylon	1603-1	66	1600-1	Lubricant, release agent	
	"Celanese" Nylon	1603-2	66	1603-1	Heat stabilizer	
					Lubricant, release agent	

(Continued)

Company and location	Trade name	Code	Nylon	Modification		Comment
				Base resin	Additive	
E.I. du Pont de Nemours and Co., Plastics Department Wilmington, Del.	"Zytel"	101	66	--	--	Inj. molding, extrusion
	"Zytel"	101 L1	66	101	Lubricant	Easier molding
	"Zytel"	101 L	66	101	Mold release agent	
	"Zytel"	102	66	--	Color stabilizer	
	"Zytel"	131	66	--	Color stabilizer and nucleating agent for fast molding	
	"Zytel"	131 L	66	131	Mold release agent	Very fast molding
	"Zytel"	103 HSI-L	66	101	Heat stabilizer, lubricant, mold release agent	
	"Zytel"	103 HS-L	66	101	Heat stabilizer, lubricant, mold release agent	
	"Zytel"	113	66	--	Heat stabilizer	Stability between 101 and 103. Better electricals than 103
	"Zytel"	105 BK-10	66	--	Carbon black	Weather resistant
	"Zytel"	106 BK-10	66	105 BK-10	Heat stabilizer	
	"Zytel"	121	66	--	Oxidation and heat stabilizers	
	"Zytel"	122	66	121	Lubricant	
	"Zytel"	42	66	--	--	High visc., extrusion

	Grade	Type		Additive	Description
"Zytel"	43	66	--	--	Extrusion coating grade
"Zytel"	31	610	--	--	Inj. molding, extrusion
"Zytel"	38	610	--	--	Higher visc., tougher than 31
"Zytel"	33	610	--	Heat stabilizer	Wire jacketing
"Zytel"	33H	610	--	Heat stabilizer	Higher visc., tougher than 33
"Zytel"	37W BK-10	610	--	Ht stabilizer and carbon black	Weather resistant
"Zytel"	37L BK-10	610	--	Ht stabilizer and carbon black	Lower visc. than 37W BK-10, weather resistant
"Zytel"	211	6	--	Monomer	Flexible for inj. molding, extrusion
"Zytel"	151	612	--	--	Inj. molding, extrusion
"Zytel"	151L	612	151	Mold release agent	
"Zytel"	158	612	--	--	Higher visc., tougher than 151
"Zytel"	158L	612	158	Mold release agent	
"Zytel"	153 HS-L	612	158	Heat stab., lubricant	Wire jacketing
"Zytel"	157HS-L BK10	612	--	Carbon black, ht. stab., lubricant, release agent	
"Zytel"	51 SE-1 BK-40	Copolymer	--	Flame retardant	Inj. molding, extrusion
"Zytel"	109L	Copolymer	--	Color stabilizer, lubricant, release agent	
"Zytel"	109 BK-10	Copolymer	109	Carbon black	Weather resistant

(Continued)

Company and location	Trade name	Code	Nylon	Modification		Comment
				Base resin	Additive	
	"Zytel"	141	Blend	--	--	For special combinations of strength and toughness
	"Zytel"	408	Modified 66	--	--	Better molding, toughness
	"Zytel"	409 BK-09	Modified 66	408	Heat stabilizer	Weather resistant
	"Zytel"	410 BK-10	Modified 66	408	Carbon black	Wire jacketing
	"Zytel"	58	Copolymer	--	Heat stabilizer	Flexible tubing, cable jacketing
	"Zytel"	91	Copolymer	--	Ht stabilizer and plasticizer	
	"Zytel"	3606	Copolymer	--	Ht stabilizer and carbon black	Weather resistant, for special wire constructions
	"Zytel"	63	Copolymer	--	--	Low melting, flexible, tough
	"Zytel"	69	Copolymer	--	Plasticizer	Lowest melting, most flexible
	"Zytel"	70 G-13	66	101	13% short glass	
	"Zytel"	70 G-13B	66	101	13% short glass (cube blend)	
	"Zytel"	70 G-33	66	101	33% short glass	
	"Zytel"	70 G-33HS1-L	66	70 G-33	Heat stabilizer	
	"Zytel"	70 G-33HR	66	70 G-33	Hydrolysis resistant	
	"Zytel"	70 G-43	66	101	43% short glass	High impact strength
	"Zytel"	71 G-13	Modified 66	408	13% short glass	High impact strength
	"Zytel"	71 G-33	Modified 66	408	33% short glass	
	"Zytel"	77 G-33	612	151	33% short glass	
	"Zytel"	77 G-43	612	151	43% short glass	
	"Elvamide"	8061	Copolymer	--	--	Alcohol soluble, binder polymer
	"Elvamide"	8062	Copolymer	--	Plasticizer	Alcohol soluble, binder polymer
	"Elvamide"	8063	Copolymer	--	--	Softer, more soluble than 8061

Company	Trade name	T	TMD-T (see Table 1-4)		Description	Notes
Dynamit Nobel Sales Corp., Norwood, N.J.	"Trogamid"			--	--	
Fiberfil Division, Dart Industries Evansville, Ind.	"Nylafil"	G-1/20	66	--	20% long glass	
	"Nylafil"	G-1/30	66	--	30% long glass	
	"Nylafil"	G-1/30/EM	66	G-1/30	Flow aid	
	"Nylafil"	G-1/30/HR	66	G-1/30	Hydrol. stabilizer	
	"Nylafil"	G-1/30/HS	66	G-1/30	Heat stabilizer	
	"Nylafil"	G-1/30/MS/5	66	G-1/30	MoS_2	Improved lubricity
	"Nylafil"	G-1/30/TF/12	66	G-1/30	"Teflon"	Improved lubricity
	"Nylafil"	G-1/30/TF/44	66	G-1/30	More "Teflon"	Improved lubricity
	"Nylafil"	G-5/30	66	G-1/30	UV and weather stabilizers	
	"Nylafil"	G-10/40	66	--	40% long glass	
	"Nylafil"	G-10/40/MS/5	66	G-10/40	MoS_2	
	"Nylafil"	G-2/30	610	--	30% long glass	
	"Nylafil"	G-2/30/MS/5	610	G-2/30	MoS_2	
	"Nylafil"	G-12/40	610	--	40% long glass	
	"Nylafil"	G-12/40/MS/5	610	G-12/40	MoS_2	
	"Nylafil"	G-3/30	6	--	30% long glass	
	"Nylafil"	G-3/30/MS/5	6	G-3/30	M_oS_2	
	"Nylafil"	G-13/40	6	--	40% long glass	
	"Nylafil"	G-13/40/MS/5	6	G-13/40	MoS_2	
	"Nylasar"	J-1/20	66	--	20% intermed. glass	
	"Nylasar"	J-1/30	66	--	30% intermed. glass	
	"Nylasar"	J-1/30/MS/5	66	J-1/30	MoS_2	
	"Nylasar"	J-2/20	610	--	20% intermed. glass	

Company and location	Trade name	Code	Nylon	Modification		Comment
				Base resin	Additive	
	"Nylasar"	J-2/30	610	--	30% intermed. glass	
	"Nylasar"	J-2/30/MS/5	610	J-2/30	MoS_2	
	"Nylasar"	J-3/20	6	--	20% intermed. glass	
	"Nylasar"	J-3/30	6	--	30% intermed. glass	
	"Nylasar"	J-3/30/MS/5	6	J-3/30	MoS_2	
	"Nylaglas"	S-1/30	66	--	30% short glass	
	"Nylaglas"	S-1/30/MS/5	66	S-1/30	MoS_2	
	"Nylaglas"	S-10/40	66	--	40% short glass	
	"Nylaglas"	S-2/30	610	--	30% short glass	
	"Nylaglas"	S-2/30/MS/5	610	S-2/30	MoS_2	
	"Nylaglas"	S-12/40	610	--	40% short glass	
	"Nylaglas"	S-3/30	6	--	30% short glass	
	"Nylaglas"	S-3/30/MS/5	6	S-3/30	MoS_2	
	"Nylaglas"	S-13/40	6	--	40% short glass	
	"Nylode"	NY-1/ASF/20	66	--	Asbestos	
	"Xylon"	6601	66	--	--	Inj. molding, extrusion
	"Xylon"	66N1	66	--	Nucleated	Faster molding
Firestone Synthetic Fibers Co., New York, N.Y.	"Firestone" Nylon	100-001	6	--	Very flexible	Inj. molding, extrusion
	"Firestone" Nylon	110-001	6	--	Flexible	Inj. molding, extrusion
	"Firestone" Nylon	200-001	6	--	--	Inj. molding, extrusion
	"Firestone" Nylon	336-001	6	--	--	Extrusion, med. visc.
	"Firestone" Nylon	340-001	6	--	--	Extrusion, med. high visc.
	"Firestone" Nylon	346-001	6	--	--	Extrusion, high visc.
	"Firestone" Nylon	348-001	6	--	--	Extrusion, very high visc.
	"Firestone" Nylon	415-001	6	--	15% glass	
	"Firestone" Nylon	430-001	6	--	30% glass	

Manufacturer	Trade name	Grade	Type		Additive	Description
Foster Grant Co. Leominster, Mass.	"Fosta" Nylon	435X	6	--	--	Low visc., fast setting, inj. molding
	"Fosta" Nylon	436	6	--	--	Low visc., inj. molding
	"Fosta" Nylon	438	6	--	--	Low visc., fast setting, inj. molding
	"Fosta" Nylon	446X	6	--	Nucleating agent	High visc., extrusion
	"Fosta" Nylon	457	6	--	Heat stabilizer	Low visc., inj. molding
	"Fosta" Nylon	512	6	--	--	Medium visc., inj. molding
	"Fosta" Nylon	523	6	--	Nucleating agent	Medium visc., inj. molding
	"Fosta" Nylon	534	6	--	Heat stabilizer	Medium visc., inj. molding
	"Fosta" Nylon	545	6	--	--	Relatively high visc., inj. molding, extrusion
	"Fosta" Nylon	556	6	--	Heat stabilizer	Relatively high visc., wire jacketing
	"Fosta" Nylon	567	6	--	Plasticizer	High visc., inj. molding, extrusion
	"Fosta" Nylon	578	6	--	--	High visc., extrusion
	"Fosta" Nylon	589	6	--	--	Very high visc., extrusion
	"Fosta" Nylon	641	6	--	Plasticizer, ht and lt stabilizers	High visc., wire jacketing
	"Fosta" Nylon	4103	6	--	30% glass	
General Mills, Inc. Chemical Division Minneapolis, Min.	"Milvex"	1000	Involves "dimer acid" (see Chap. 16, also Chap. 2, p. 00)	--	--	Low softening binder polymer, coatings
	"Milvex"	1225		--	--	Extrusion grade binder polymer
	"Milvex"	1235		--	--	Slightly higher softening than 1225
	"Milvex"	1250		--	--	Higher softening than 1235, film, filament

(Continued)

Company and location	Trade name	Code	Nylon	Modification		Comment
				Base resin	Additive	
	"Milvex"	1525	Involves "dimer acid" (see Chap. 16, also Chap. 2, p. 00)	--	--	Film grade
	"Milvex"	1535		--	--	Similar to 1525 but higher modulus
	"Milvex"	1550		--	--	Highest tensile strength
	"Milvex"	4000		--	--	Highest modulus, solution coating for abrasion resistance
Gulf Oil Corp., Chemicals Dept. Houston, Texas	"Gulf" Nylon	401	6	--	--	Inj. molding
	"Gulf" Nylon	401AX	6	401	Nucleating agent	Fast molding
	"Gulf" Nylon	401BX	6	401	Nucleating agent	Fast molding
	"Gulf" Nylon	402	6	--	Monomer	Inj. molding, flexible, tough
	"Gulf" Nylon	404	6	--	Heat stabilizer	Inj. molding
	"Gulf" Nylon	406 (NX1096)	6	--	Nucleat. and release agents	Fast molding
	"Gulf" Nylon	423A	6	--	Nucleat. and release agents	Fast molding
	"Gulf" Nylon	451	6	--	Lubricant	Inj. molding, tougher
	"Gulf" Nylon	451A	6	451	Monomer, lubricant	Rel. visc. of 50, inj. molding
	"Gulf" Nylon	452	6	--		Inj. molding, flexible
	"Gulf" Nylon	453	6	--	2% carbon black	Inj. molding, weather resistant
	"Gulf" Nylon	454	6	--	Heat stabilizer	Inj. molding
	"Gulf" Nylon	600	6	--	--	Rel. visc. of 50, extrusion
	"Gulf" Nylon	601	6	--	Heat stabilizer	Extrusion
	"Gulf" Nylon	603	6	--	Monomer	Extrusion, flexible

Manufacturer	Product	Nylon type	Related grade	Additives	Description
"Gulf" Nylon	606	6	--	--	Med. visc., extrusion coating, film
"Gulf" Nylon	607	6	--	--	High visc., extrusion
"Gulf" Nylon	609	6	--	--	Medium high visc., extrusion
"Hüls" Nylon 12	L1500	12	--	--	Inj. molding
"Hüls" Nylon 12	L1600	12	L1600	--	Inj. molding
"Hüls" Nylon 12	L1610	12	L1600	--	Powder
"Hüls" Nylon 12	L1640	12	L1640	Ht and weather stabilizers	
"Hüls" Nylon 12	L1640 Powder	12	L1640	--	Rotational molding
"Hüls" Nylon 12	L1700	12	L1700	--	Higher visc., inj. molding, extrusion
"Hüls" Nylon 12	L1700M	12	L1700	--	Monofilament grade
"Hüls" Nylon 12	L1801	12	--	--	Normal flow, higher visc. than L1700
"Hüls" Nylon 12	L1801F	12	L1801	--	Extrusion grade
"Hüls" Nylon 12	L1901	12	--	--	High visc., inj. molding
"Hüls" Nylon 12	L1901E	12	L1901	--	Extrusion
"Hüls" Nylon 12	L1901F	12	L1901	--	Film grade
"Hüls" Nylon 12	L1930	12	L1901	30% glass	
"Hüls" Nylon 12	L1931	12	L1901	30% glass	Better heat and weather resistance
"Hüls" Nylon 12	L1940	12	L1901	Ht and weather stabilizers	
"Hüls" Nylon 12	L1940 Powder	12	L1901	Ht and weather stabilizers	Rotational molding
"Hüls" Nylon 12	L1941	12	L1901	Ht and weather stabilizers	Better stability than L1940

Henley and Co.
New York, N.Y.

Explanation of Nylon-12 Code

L = homopolymer
N = copolymer
E = extrusion grade
F = film grade
M = monofilament grade
1st two nos. related to viscosity = $10 \times \eta_{rel}$ (0.5%, cresol)
3rd no:
1 = powder, stabilized
2 = plasticized
3 = contains glass
4 = stabilized
5 = contains MoS_2
6 = contains graphite
7 = contains antistat
8 = contains flame retardant
9 = contains other polymers
4th no: more than one additive

(Continued)

Company and location	Trade name	Code	Nylon	Base resin	Additive	Comment
	"Hüls" Nylon 12	L1950	12	L1901	MoS$_2$	Better lubricity
	"Hüls" Nylon 12	L1960	12	L1901	Graphite	Better lubricity
	"Hüls" Nylon 12	L1970	12	L1901	Antistatic agent	
	"Hüls" Nylon 12	L1980	12	L1901	Flame retardant	
	"Hüls" Nylon 12	L1990M	12	L1901	- -	Special monofilament grade
	"Hüls" Nylon 12	N1901	Copolymer	- -	- -	High transparency and flexibility
	"Hüls" Nylon 12	L2101	12	- -	- -	Very high visc., inj. molding
	"Hüls" Nylon 12	L2101E	12	L2101	- -	Very high visc., extrusion
	"Hüls" Nylon 12	L2101F	12	L2101	- -	Film grade
	"Hüls" Nylon 12	L2121	12	L2101	Plasticizer	Inj. molding, more flexible, higher impact strength
	"Hüls" Nylon 12	L2121E	12	L2101	Plasticizer	Extrusion
	"Hüls" Nylon 12	L2140	12	L2101	Stabilizers	Extrusion
Liquid Nitrogen Processing Corp. Malvern, Pa.	"Thermocomp"	PF-1002	6	- -	10% glass	
	"Thermocomp"	PF-1004	6	- -	20% glass	
	"Thermocomp"	PF-1006	6	- -	30% glass	
	"Thermocomp"	PF-1008	6	- -	40% glass	
	"Thermocomp"	PF-100-10	6	- -	50% glass	
	"Thermocomp"	PF-100-12	6	- -	60% glass	
	"Thermocomp"	PFL-4026	6	- -	30% glass, 10% TFE for better lubricity	

"Thermocomp"	PFL-4036	6	--	30% glass, 15% TFE for better lubricity
"Thermocomp"	PFL-4044	6	--	20% glass, 20% TFE for better lubricity
"Thermocomp"	PFL-4064	6	--	20% glass, 30% TFE for better lubricity
"Thermocomp"	PFL-4066	6	--	30% glass, 30% TFE for better lubricity
"Thermocomp"	PFL-4216	6	--	30% glass, 5% MoS_2 for better lubricity
"Thermocomp"	PFL-4218	6	--	40% glass, 5% MoS_2 for better lubricity
"Thermocomp"	RF-1002	66	--	10% glass
"Thermocomp"	RF-1004	66	--	20% glass
"Thermocomp"	RF-1006	66	--	30% glass
"Thermocomp"	RF-1008	66	--	40% glass
"Thermocomp"	RF-100-10	66	--	50% glass
"Thermocomp"	RF-100-12	66	--	60% glass

Company and location	Trade name	Code	Nylon	Modification		Comment
				Base resin	Additive	
	"Thermocomp"	RFL-4214	66	--	20% glass, 5% MoS$_2$ for better lubricity	
	"Thermocomp"	RFL-4216	66	--	30% glass, 5% MoS$_2$ for better lubricity	
	"Thermocomp"	RFL-4218	66	--	40% glass, 5% MoS$_2$ for better lubricity	
	"Thermocomp"	QF-1002	610	--	10% glass	
	"Thermocomp"	QF-1004	610	--	20% glass	
	"Thermocomp"	QF-1006	610	--	30% glass	
	"Thermocomp"	QF-1008	610	--	40% glass	
	"Thermocomp"	QF-100-10	610	--	50% glass	
	"Thermocomp"	QF-100-12	610	--	60% glass	
	"Nykon"	P	6	--	MoS$_2$	
	"Nykon"	R	66	--	MoS$_2$	
Monsanto Co. Hydrocarbons and Polymers Division St. Louis, Mo.	"Monsanto" Nylon	10V	66	10V	--	Inj. molding, extrusion
	"Monsanto" Nylon	20V	66	10V	External lubricant	
	"Monsanto" Nylon	10C	66	10V	Oxidation and hydrolysis stabilizers	
	"Monsanto" Nylon	10H	66	10V	Heat stabilizer	
	"Monsanto" Nylon	20H	66	10H	External lubricant	

Company	Trade name	Grade		20V	Additive	Application
	"Monsanto" Nylon	20M	66	20V	Mold release agent	Faster molding
	"Monsanto" Nylon	10L	66	10V	Color stabilizer	
	"Monsanto" Nylon	20N	66	20V	Nucleat. agent	
	"Monsanto" Nylon	50V	610	--	--	Inj. molding, monofilament
	"Monsanto" Nylon	50H	610	50V	Heat stabilizers	Wire and cable coatings
	"Monsanto" Nylon	55N	610	--	Mold release agent	Faster molding
Nypel, Inc. Conshohocken, Pa.	"Nypel" Nylon	--	--	--	--	Reprocessor
	"Nyreg"	--	--	--	Glass	
	"Nypelube"	--	--	--	TFE	
	"Micropel"	--	--	--	--	Powder
The Polymer Corp. Reading, Pa.	"Nylatron"	GS	66	GS	MoS$_2$	
	"Nylatron"	GS-21	66	GS	UV stabilizer	
	"Nylatron"	GS-HS	66	GS	Heat stabilizer	
	"Nylatron"	GS-51	66	GS	30% glass, ht and hydrolysis stabilizers	
	"Nylatron"	GS-51-13	66	GS	40% glass, ht and hydrolysis stabilizers	
	"Nylatron"	50-2	66	--	12.5% glass	
	"Nylatron"	50-9	66	--	30% glass	
	"Nylatron"	50-13	66	--	40% glass	
	"Nylatron"	52-5	--	--	--	
	"Nylatron"	GS-31	610	GS-31	MoS$_2$	**High PV performance**
	"Nylatron"	GS-33	610	GS-31	UV and ht stabilizers	
	"Nylatron"	GS-31-9	610	GS-31	30% glass	
	"Nylatron"	30-9	610	--	30% glass	
	"Nylatron"	30-13	610	--	40% glass	

(Continued)

Company and location	Trade name	Code	Nylon	Modification		Comment
				Base resin	Additive	
	"Nylatron"	GS-60	6	--	MoS$_2$	
	"Nylatron"	GS-60-9	6	GS-60	30% glass	
	"Nylatron"	GS-50/60	66/6 Blend	--	MoS$_2$	
	"Nylatron"	41	12	--	--	
	"Nylatron"	41-9	12	41	30% glass	
Wellman, Inc. Plastics Div. Boston, Mass.	"Wellamid"	--	--	--	--	Reprocessor
	"Well-Fibe"	--	--	--	Glass fiber	
	"Well-Sphere"	--	--	--	Glass spheres	
	"Well-A-Meld"	--	--	--	Both fiber and spheres	

INDEX